Lecture Notes on Data Engineering and Communications Technologies

Volume 272

The aim of the book series is to present cutting edge engineering approaches to data technologies and communications. It will publish latest advances on the engineering task of building and deploying distributed, scalable and reliable data infrastructures and communication systems.

The series will have a prominent applied focus on data technologies and communications with aim to promote the bridging from fundamental research on data science and networking to data engineering and communications that lead to industry products, business knowledge and standardisation.

Indexed by SCOPUS, INSPEC, EI Compendex.

All books published in the series are submitted for consideration in Web of Science.

Andrii Semenov · Iryna Yepifanova · Jana Kajanová
Editors

Data-Centric Business and Applications

Modern Trends in Financial and Innovation Data Processes 2025. Volume 2

Springer

Editors
Andrii Semenov
Faculty of Information Electronic Systems
Vinnytsia National Technical University
Vinnytsia, Ukraine

Iryna Yepifanova
Faculty of Management and Information Security
Vinnytsia National Technical University
Vinnytsia, Ukraine

Jana Kajanová
Department of Economics and Finance, Faculty of Management
Comenius University Bratislava
Bratislava, Slovakia

ISSN 2367-4512 ISSN 2367-4520 (electronic)
Lecture Notes on Data Engineering and Communications Technologies
ISBN 978-3-032-06607-7 ISBN 978-3-032-06608-4 (eBook)
https://doi.org/10.1007/978-3-032-06608-4

This Springer imprint is published by the registered company Springer Nature Switzerland AG
The registered company address is: Gewerbestrasse 11, 6330 Cham, Switzerland

Preface

This volume presents a multidisciplinary exploration of innovative strategies, technologies and policy tools that are intended to strengthen Ukraine's economic resilience, social development and technological advancement in the context of global and national challenges. The chapters cover a wide range of interconnected topics, including intelligent compliance systems, AI-integrated circular economy models, urban restoration initiatives, and sustainable leadership in higher education. Particular focus is given to Ukraine's post-war recovery, with discussions on smart city development, cybersecurity, energy efficiency, and regional competitiveness through tourism. The monograph also addresses key advances in data science, including neural network-based crisis prediction, intelligent audit technologies, and the part played by national scientific data ecosystems in promoting open science. Several chapters bridge the gap between digital innovation and industrial practice by showcasing new solutions in areas such as predictive maintenance, microprocessor design, and machine learning for audio processing. The monograph also examines strategic human resource management, accounting engineering and marketing perspectives, highlighting how socio-economic systems can adapt through transformative leadership and digital tools.

In the first chapter, "Intelligent Compliance as a Fundamental Tool for Ukraine's Economic Security", the authors examine the implementation of compliance processes in Ukraine's banking system. They identify the problematic aspects that banking institutions face when complying with financial monitoring legislation regarding customer due diligence (CDD) requirements. The authors provide descriptions of the processes of information search, data collection and due diligence within the CDD framework. Based on their concept of intelligent compliance, the authors propose an automated, high-tech approach to managing compliance risks. They developed a structural model of intelligent compliance processes based on Business Process Model and Notation (BPMN), constructed logical decision-making schemes, and integrated artificial intelligence methods to identify and predict risk-related situations. Deep learning algorithms, namely long short-term memory (LSTM) models, are proposed to enhance the detection of suspicious financial activities. Thus, the suggested intelligent compliance could be applied in ensuring Ukraine's financial stability, economic security and international reputation.

The subsequent chapter, "The Convergence of Artificial Intelligence with the Circular Economy", explores the current level of convergence between artificial intelligence and the circular economy in Europe. Eurostat's statistical information database was used for this analysis, with Poland and Sweden serving as case studies alongside other EU countries. The developed method is based on five key components: PIce (private investment and gross added value related to circular economy sectors); ETax (environmental tax revenues); WM (waste management operations); AIent (artificial intelligence by size class of enterprise); and AInace (artificial intelligence by NACE Rev. 2 activity). AIent takes into account the use of artificial intelligence in enterprises according to their size. This allows us to determine the effect of enterprise size on the implementation and use of AI, as well as its impact on production efficiency and sustainability. This chapter presents a method for evaluating the effect of convergence between the circular economy and artificial intelligence. Based on this, it is possible to carry out a ranking entitled "The effect of the circular economy, considering the effect of convergence with artificial intelligence". The results of this section are important for increasing the efficiency of artificial intelligence implementation in the economy and for calculating its impact.

The next chapter, "Urban Planning Principles and Models of Reconstructive Restoration of Urbanized Territories of Luhansk Region Destroyed by Hostilities, Based on Innovative Development", focuses on the Luhansk region of Ukraine and the cities and urban areas that have been severely damaged as a result of hostilities. This chapter employs comparative, graphical-analytical, experimental design and systematic analysis methods to examine the factors and conditions that influence the restoration of mass housing areas affected by hostilities. The generalisation of data obtained from statistical studies was also utilised. The authors developed methods for determining hierarchical levels and reconstruction targets, identified key stakeholders and types of reconstruction activities, and formulated the elements and components of the reconstruction information package. They also formalised design decisions on individual objects of reconstruction. The results obtained by the authors may be useful for developing programmes for the reconstruction of urban areas destroyed by military operations or natural disasters (floods, earthquakes, tsunamis, etc.).

In the next chapter, titled "Detecting Country of Origin Data Manipulation in Sanctioned Raw Material Supply Chains: Expert Survey Insights", the author presents the results of a comprehensive study into the issue of detecting and counteracting the manipulation of country-of-origin data in raw material supply chains under sanctions. The survey of Ukrainian experts identified the top 10 factors affecting supply chains in the context of circumventing sanctions through manipulation of country-of-origin data. The most significant factors identified were evasion and compliance technologies, layering of corporate structures, manipulation of trade volume indicators and unusual payment structures, as determined by the hierarchical clustering method. The author's findings allow for increased efficiency in business processes in the field of financial security, through the development of specialised analytical information technologies for identifying incidents of circumvention of sanctions and tariff restrictions.

In the subsequent chapter, "Transformational Leadership of a Higher School in the Implementation of Sustainable Development Goals (Marketing Aspects)", the authors present methodological approaches and formulate scientific recommendations to enhance Ukraine's higher education system's ability to demonstrate proactivity and intellectual leadership in the strategic planning and implementation of transformative processes in the economy and society for the country's post-war recovery in line with sustainable development priorities. The authors present their vision in the form of a model for the functioning of a higher education institution based on proactivity and transformational leadership, as well as the role of strategic marketing in this process. Marketing analysis and strategising value propositions in each component of the extended reproduction cycle will play an important role in determining priorities in the development of educational products of higher education institutions. The chapter presents a possible roadmap for reforming higher education institutions in Ukraine, which could facilitate communication with them in developed countries. These results will also be important for international companies financing sustainable development and innovation.

In the following study, "Accounting Engineering as a Tool for Innovative Support of the Transformation of Socio-economic Systems of Enterprises and Their Economic Activity", the authors provided a scientific rationale and developed methods and recommendations for forming an information support system to manage the economic activities of trading enterprises using accounting engineering technology. The key constraints whose presence could negatively affect the economic activity of trading enterprises were identified. A detailed reference model for managing the economic activity of trading enterprises in the context of the digital transformation of business processes is based on the relationship between the accounting and analytical components of the information system via accounting engineering technology. A model was developed for the structure of the accounting component of informational support for managing the economic activity of retail enterprises. The results demonstrate the potential impact of accounting on innovative processes in the context of economic transformations.

In the next study, "Innovative Tools for Motivating Strategic Human Resource Development in IT Business", the authors propose a methodology for creating an incentive system based on the motivational aspects of the work of IT enterprise employees. The study identifies and analyses the factors that influence the qualitative parameters and quantitative indicators of the provision of human resources to IT enterprises, examining the nature of their influence and the structural and logical relationships between them. Using the method of dynamic series, the authors examine trends in IT industry development and identify the structural and logical relationships between the trends. The results are complemented by M. Porter's Diamond model method. Using a methodology based on a typological model of labour motivation, the authors built a motivational profile of IT company employees and identified common motivational types among those surveyed. The authors developed infographics to illustrate how incentive methods align with the motivational types of IT company employees. This formalises the process of selecting incentives based on an employee's motivational type and serves as a foundation for building an effective incentive system in the IT sector.

The subsequent chapter, "Smart Cities and Urban Resilience in Ukraine: Towards Sustainable Recovery and Digital Transformation", explores the evolution and current state of smart city development in Ukraine, highlighting achievements and structural barriers. It pays particular attention to the systemic challenges—technical, legal, economic and sociocultural—that hinder the widespread implementation of smart technologies. The smart-city functional model is described as a cyclical interaction of technological infrastructure, service applications, and citizen engagement. The chapter describes the comparative characteristics of smart, sustainable, and smart-sustainable cities. Comparative smart city index rankings of selected cities with similar Human Development Index scores are constructed using data from 2021. The smart city development trajectories and digitalisation challenges in six selected Ukrainian cities are examined using data from 2022. The typology of barriers to smart city implementation in Ukraine is presented. The authors have developed strategic directions to ensure the implementation of the smart city model. Although Ukrainian cities demonstrate significant innovation potential, large-scale implementation is limited by financial, institutional, technological and social factors. The chapter's results may be of interest to countries developing national smart city strategies for crisis conditions.

The next chapter, titled "Processing of Audio Recordings and Feature Extraction for Machine Learning", provides an engaging exploration of how machine learning techniques can be used to analyse human speech and emotional tone, particularly in the context of academic video content. The authors present a robust methodological framework for processing audio recordings, extracting relevant acoustic features and training classification models to detect emotionally significant segments. Utilization of open-source tools and development of a custom annotation platform demonstrated technical innovation and practical utility. Applying their approach to real-world educational data highlights how emotional analytics can enhance teaching evaluation, learner engagement and content personalisation.

In the next chapter, titled "University Rankings and Students' Perception of Educational Quality as Drivers of Reputation and Enrollment Strategy in Service Marketing", the authors conducted an inquiry of students at the Faculty of Economics at Taras Shevchenko National University of Kyiv, using Google Forms. Quota sampling was employed to select the participants, who were asked to respond using a Likert scale. The factors that influence students' decision-making were examined, and the alignment of global ranking indicators with students' priorities was assessed. The study found that students are not interested in international rankings such as QS,

ARWU, and THE. Instead, they prioritise information on the practical nature of the university's education, employment opportunities after graduation, and personal development during their studies. Educational materials, faculty publications and the level of academic support are not usually considered primary factors in the university selection process. The authors conclude that a comprehensive strategy is needed to improve the quality of education, increase student satisfaction, and encourage active participation in global research. The results of this study will be valuable to international scholars when analysing the current state and future prospects of higher education in Ukraine.

In the following study, "Innovative Post-war Recovery Plan for Ukraine: Demographic Challenges, Growth Drivers, Cybersecurity", the authors aim to assess the current state of Ukraine's human and labour resources, determine the role of digital transformation in post-war economic recovery and develop recommendations for providing economic sectors with highly qualified personnel while taking labour market challenges into account. It is noted that the IT market in Ukraine continued to develop during the war and is expected to grow further in the post-war period. Demographic challenges during this period will primarily affect pensioners, the wounded and those returning from the war. Specialised retraining and training technologies will be crucial for these individuals in the post-war period. Attention is also drawn to the need to develop information security technologies, the importance of which will increase. The results of this section will be important for financing and innovation specialists, as they describe the recommended approaches, methods and technologies for Ukraine in the post-war period.

The following chapter, titled "Microprocessor Device for Predictive Maintenance Systems for Electrical Machines", makes a forward-looking contribution to smart manufacturing by focusing on the development of a microprocessor-based device for the predictive maintenance of electrical machines. Based on the principles of Industry 4.0, the proposed solution combines physical modelling, signal processing and embedded systems to diagnose coil degradation by accurately analysing the winding quality factor. Using phase shift measurements and real-time signal evaluation enables condition-based maintenance, which significantly reduces downtime and operational risks. The authors demonstrate a practical and scalable approach tailored to industrial environments by implementing compact hardware based on the ATtiny2313 microcontroller. This work reflects the broader trend of convergence between cyber-physical systems and intelligent diagnostics in modern industrial infrastructure.

In the subsequent chapter, "Development of an Early Warning System for Crisis Phenomena in Banking Entities Using Neural Network Technologies", the authors set out to develop an early warning system that can identify predictors and signals of potential problems in banks' financial activities before they occur. This system uses Kohonen maps to construct a dynamic coordinate system for the distribution of core financial reporting indicators. The developed method is based on a tool that groups observations into homogeneous categories according to predefined characteristics.

The study employs a hexagonal topology on a two-dimensional grid. The inputs to this network are (1 × 16) dimensional vectors, where each vector element corresponds to one of a bank's 16 financial indicators, and the output neurons are the neurons of the Kohonen grid. The number of banks assigned to each cluster is presented in the form of a SOM Sample Hits plot. For example, the clustering of Ukrainian banks in 2022 is presented. Comparing the clustering of Ukrainian banks in 2022 with their financial activities in 2024 showed that the forecast of bank behaviour was highly predictable.

The next chapter, titled "Impact of Energy-Saving Technological Changes on the Level of Competitiveness of Enterprises", addresses a critical aspect of modern enterprise strategy: the integration of energy-saving technologies as a means of achieving competitiveness in an era characterised by rising energy costs and sustainability imperatives. Through empirical analysis and strategic modelling, the authors evaluate the impact of technological modernisation on key performance indicators such as cost efficiency, market responsiveness and export potential. Using data from Ukrainian industrial sectors, the study provides valuable insights into the economic advantages of energy-efficient innovations. By combining quantitative impact metrics with sector-specific analysis, the research provides practical recommendations for developing enterprise-level energy policies. The study emphasises the importance of aligning green transformation with innovation strategy to achieve economic resilience and ecological responsibility.

The following chapter, "Conceptual Strategies for Managing Innovation Processes in the Life Cycle of the EU Tourism Industry", offers a timely and insightful exploration of how innovation can be managed strategically throughout the life cycle of the European tourism industry. In an era characterised by digital disruption, climate imperatives and changing consumer expectations, the authors present a conceptual framework that aligns innovation processes with the key stages of tourism development: emergence, growth, maturity and transformation. Drawing on EU policy instruments and empirical data, the study emphasises the pivotal role of digitalisation, sustainability, and human capital in tourism innovation. By emphasising adaptive strategy design and regional policy alignment, the chapter offers a holistic approach to navigating complex industry dynamics. A notable strength of the chapter is its systems perspective, which positions innovation as a collaborative effort involving public institutions, businesses, and academia. The proposed strategies address internal and external forces of change, enabling tourism stakeholders to respond proactively to demographic, environmental and competitive challenges.

The chapter "Ensuring the Competitiveness of Coastal Regions Through the Development of the Tourism Business" offers a timely, multifaceted analysis of the role of the tourism sector in enhancing the competitiveness of coastal regions during post-pandemic recovery and sustainable economic transformation. Through the integration of statistical analysis, policy frameworks and regional development strategies, the author offers a data-driven perspective on the strategic role of tourism in maritime and coastal regions. The study emphasises the importance of public–private partnerships, the investment climate and innovation ecosystems in fostering long-term regional growth. A key strength of the chapter is its focus on the intersection between tourism business development and broader socio-economic priorities such as job creation, environmental stewardship and digital infrastructure.

The next chapter, titled "Intelligent Audit Technology Knowledge Base", explores the evolving landscape of auditing in the digital era. Here, increasing data complexity and regulatory demands necessitate intelligent, adaptive solutions. The authors present a comprehensive framework for an intelligent audit platform that is built on knowledge-based systems, domain-specific ontologies and AI-driven inference engines. Addressing the entire audit life cycle—from data acquisition and regulatory interpretation to diagnostics and expert evaluation—the proposed system facilitates a transition from static reporting to dynamic, data-centric validation. Practical relevance and scalability are emphasised through key features such as layered audit documentation classification, anomaly detection through logical inference, and ERP platform integration. This approach is firmly grounded in international auditing standards, ensuring compliance with regulations while encouraging innovation.

The subsequent chapter, "Features of the Functioning of the National Repository of Scientific and Research Data as Part of a Multi-DB Scientific Ecosystem for the Development of Open Science in Ukraine", provides a timely and insightful exploration of Ukraine's efforts to establish a robust, data-centric infrastructure for open science by developing the National Repository of Scientific and Research Data (NRSRD). Positioned within a federated multi-database ecosystem, the NRSRD demonstrates a commitment to the FAIR principles, ensuring that research data is findable, accessible, interoperable and reusable. The authors provide a thorough analysis of the repository's architecture, integration mechanisms and legal basis, while also ensuring that its design aligns with European standards, such as the EOSC framework. Particular focus is given to data quality, citation, version control and provenance, highlighting the repository's ability to facilitate transparent and reproducible research. By connecting national infrastructure with global scientific networks, the chapter demonstrates how structured data governance can enhance the visibility and influence of Ukrainian research.

In the context of growing demands for high-speed image processing, secure data storage, and intelligent system design, the chapter titled "The Optical Dynamic Random Access Memory for Neurocomputers" offers a timely and innovative exploration of optical memory technologies. Focusing on dynamic optical random access memory (DRAM) deployed in fibre-optic lines, the authors address the limitations of conventional memory systems when applied to cutting-edge computing infrastructures. The chapter introduces a mathematical model of the memory base element, enabling precise analysis of light energy efficiency and system transfer functions. This approach lays the groundwork for designing high-performance optical memory systems suitable for neurocomputers, geoinformation energy control systems, and advanced optical electronics. The integration of optoelectronic solutions into multi-processor nanotechnology-based architectures reflects a forward-looking vision of scalable, energy-efficient computing.

Accurate measurement of gas concentrations is critical in many sectors, including environmental monitoring, industrial process control, healthcare, and public safety. The chapter entitled "Analysis and Systematization of Gas Concentration Measurement Methods" provides a comprehensive overview and structured classification of the primary methods used to detect and quantify gases. It focuses on five major groups: physicochemical, optical, electrochemical, acoustic and thermochemical techniques. Each method is examined in terms of its operating principles, metrological performance, application domains and practical constraints. Particular emphasis is placed on comparative analysis, with parameters such as sensitivity, selectivity, response time, equipment cost and ease of maintenance being evaluated. Notably, electrochemical methods are revealed to be the most balanced approach for practical deployment in a variety of operational contexts. The chapter also uses mathematical modelling to support sensor characterisation and performance evaluation, adding analytical rigour to the discussion. By offering a clear systematisation of current technologies, the study supports the informed design and implementation of gas sensing solutions.

The war in Ukraine has dramatically altered the country's business landscape, forcing companies to adapt quickly to unprecedented risks and structural changes. The chapter "Business Environment of Ukraine in War Conditions: Risks, Stabilization Factors, and Management Methods" offers a timely and in-depth analysis of the business environment in wartime, focusing on risk factors, stabilisation mechanisms and effective management strategies. Using a data-driven approach, the authors identify key sectors, such as retail trade, that have shown resilience and partial recovery. They also explore the role of relocation, economic expectations, and adaptive capacity in enterprise survival. The chapter introduces a multifactor econometric model to evaluate the influence of macroeconomic variables, such as employment, consumer spending, and interest rates, on business performance. The findings emphasise the urgent need for targeted interventions, including preferential lending programmes, export revitalisation efforts, and workforce development strategies. The chapter also outlines structural reforms aimed at strengthening industrial capacity and stimulating entrepreneurship through tax incentives and innovation support.

As financial transactions become increasingly fast, global, and digitised, the complexity and scale of financial crime continue to grow. The last chapter entitled "Transforming Financial Crime Prevention through Artificial Intelligence" addresses the urgent need for innovative solutions in combating illicit financial activity by exploring how Artificial Intelligence (AI) can support the prevention and detection of crimes such as money laundering, terrorism financing, and tax evasion. The authors provide a critical literature-based analysis that emphasises the integration of AI technologies with foundational compliance practices, particularly Know Your

Customer procedures. Drawing attention to the mounting challenges faced by financial institutions in processing vast data streams, the chapter highlights how advanced algorithms can enhance fraud detection, transaction monitoring, and risk assessment. The discussion also reflects on emerging threats posed by cryptocurrencies, whose anonymity facilitates new forms of exploitation in the digital financial landscape. While AI offers powerful tools, the study underscores the irreplaceable role of human judgment in navigating ethical and contextual dimensions of financial decision-making. Special attention is given to practical AI applications such as chatbots, which improve operational efficiency while supporting customer engagement and compliance.

Vinnytsia, Ukraine
Vinnytsia, Ukraine
Bratislava, Slovakia

Andrii Semenov
semenov.a.o@vntu.edu.ua

Iryna Yepifanova
yepifanova@vntu.edu.ua

Jana Kajanová
jana.kajanova@fm.uniba.sk

Contents

Intelligent Compliance as a Fundamental Tool for Ukraine's Economic Security

Ganna Panasenko and Viktoriia Hurochkina

Abstract The current state of compliance processes implementation in Ukraine's banking system is characterized through the lens of its impact on the country's economic security, fulfillment of international obligations, and adherence to recommendations in the field of anti-money laundering (AML) and counter-terrorist financing (CTF). The problematic aspects faced by banking institutions in complying with financial monitoring legislation regarding customer due diligence (CDD) requirements are identified. Fact-based data are provided to outline the resource expenditures (financial, time, and human) associated with compliance procedures, as well as the consequences of banks' violations of legal requirements in AML/CTF regulations. It is proposed to divide the customer due diligence process into separate stages, identifying "bottlenecks" in the fulfillment of key obligations by banking institutions. For each procedure, prospective solutions are formulated regarding the use of artificial intelligence (AI) technologies, with an emphasis on expected outcomes, such as reducing the time required for compliance procedures, improving the bank's operational efficiency, and enhancing customer service quality. To ensure the practical implementation of the proposed recommendations, an algorithmic approach for Intelligent Compliance is formalized. This approach includes the development of a structured process model, logical decision-making schemes, and the integration of AI methods for risk identification and prediction.

Keywords Compliance · Bank · Economic security · Artificial intelligence · Verification · Recommendations · Evaluation

G. Panasenko · V. Hurochkina (✉)
Kyiv Institute of Business and Technology, Kyiv, Ukraine
e-mail: v.hurochkina@wez.uz.zgora.pl; viktoriav2005@ukr.net

G. Panasenko
e-mail: panasenko@kibit.edu.ua

V. Hurochkina
University of Zielona Góra, Zielona Góra, Poland

A. Semenov et al. (eds.), *Data-Centric Business and Applications*, Lecture Notes on Data Engineering and Communications Technologies 272,
https://doi.org/10.1007/978-3-032-06608-4_1

1 Characteristics of the Current Situation

The modern financial environment is characterized by a high level of regulatory burden due to the increasing complexity of legal requirements, the strengthening of international financial oversight, and the growing risks associated with money laundering, terrorist financing, and cyber threats. According to research [1], the number of regulatory acts in the financial services sector continues to rise, requiring financial institutions to adapt rapidly and efficiently manage compliance risks. These regulatory changes have intensified, particularly after the 2008 financial crisis. Traditional methods of ensuring compliance no longer meet the demands of the modern regulatory landscape due to their low flexibility, high costs, and significant reliance on human factors. In response to these challenges, the financial sector is actively integrating Intelligent Compliance technologies, which leverage artificial intelligence (AI), big data, machine learning (ML), and automated analytical algorithms.

Intelligent Compliance is defined as an intelligent approach to regulatory compliance management that enables automated regulatory change monitoring, real-time risk analysis, and reduced non-compliance risks through adaptive algorithms. As noted in [2], AI-driven solutions can reduce compliance costs by up to 40% and decrease the time required for analyzing regulatory changes by 70%. Traditional compliance management methods often prove ineffective due to their dependence on manual analysis, human factors, and the complexity of harmonizing regulatory standards across jurisdictions. In response to these challenges, contemporary research proposes technology-driven approaches to compliance management.

Research in the field of Intelligent Compliance spans various domains, including the financial sector, risk management, and regulatory oversight. For instance, in financial technology, blockchain and Intelligent contracts offer potential solutions for real-time transparent compliance reporting and automated rule enforcement [3]. AI, big data, and blockchain technologies drive innovations in legal solutions, enhancing corporate compliance, contract management, and risk assessment [4]. Blockchain technology can provide a single source of truth for regulatory reporting, eliminating the need for data duplication and manual verifications.

Alongside blockchain implementation, another promising approach to optimizing compliance processes is the development of Legal Knowledge Graphs, which help structure and integrate large volumes of regulatory data. Technological advancements enable process optimization, improved data analysis, and enhanced compliance through automation, while also raising critical considerations regarding data privacy, AI bias, and transparency. These issues are particularly relevant in the highly sensitive domain of compliance related to Anti-Money Laundering (AML), Counter-Terrorist Financing (CTF), and the prevention of financing weapons of mass destruction (WMD). Progress in this area is assessed by the Council of Europe's MONEYVAL Committee, which serves as a permanent monitoring body evaluating measures to combat money laundering and terrorist financing. According to the MONEYVAL report from the fifth round of mutual evaluations (2017) and the second progress and reassessment report on Ukraine's technical compliance with

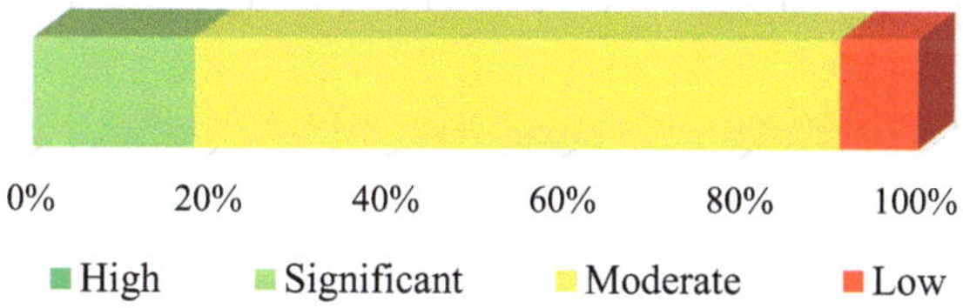

Fig. 1 Assessment of compliance with international AML/CFT standards (September 2020) [5]

MONEYVAL ratings (June 2020), Ukraine has received positive ratings. Specifically, national legislation was found to be fully compliant with 11 FATF Recommendations, largely compliant with 22, and partially compliant with 6 (Fig. 1) [5].

Despite this progress, Ukraine still needs to make significant improvements in more than 70% of the FATF Recommendations. The report also states that the country faces substantial money laundering risks, which directly impact its economic security. In this context, it is worth recalling the times when Ukraine was included in the FATF "blacklist" and the so-called "greylist".

The blacklist indicates that a country has serious strategic deficiencies in combating money laundering, terrorist financing, and the financing of weapons of mass destruction (WMD). As a result, the country is considered high-risk for financial and economic relations in the international environment. Furthermore, countries on the blacklist may be subject to economic sanctions and other restrictions imposed by FATF member states and international organizations.

The greylist includes countries that also have serious strategic deficiencies in addressing money laundering, terrorist financing, and WMD proliferation financing. However, they are under enhanced monitoring and actively cooperate with FATF to address these deficiencies.

In 2011, Ukraine was removed from the greylist, and since then, its AML/CTF legislation has been significantly improved and refined. However, risks remain, particularly in the banking sector, as the Ukrainian financial system is bank-oriented, with approximately 80% of total assets belonging to banking institutions. Banks are at the forefront of servicing both individuals and legal entities. Although the National Bank of Ukraine (NBU) has a clear understanding of risk and applies an appropriate risk-based approach to bank supervision, the effectiveness of compliance and AML risk management in banks remains unresolved, as evidenced by the scale of fines imposed for violations of AML/CTF requirements (Fig. 2).

The situation with enforcement measures by the National Bank of Ukraine, along with regulatory pressure, results from non-compliance and improper implementation of AML/CTF legislation, this is primarily due to:

- the prevalence of manual processes and procedures, which consume excessive time;
- difficulties in processing large amounts of information;
- use of outdated and inefficient IT systems.

Thus, it is evident that Intelligent Compliance technologies could become a key factor in improving the effectiveness of AML/CTF compliance in banks, reducing

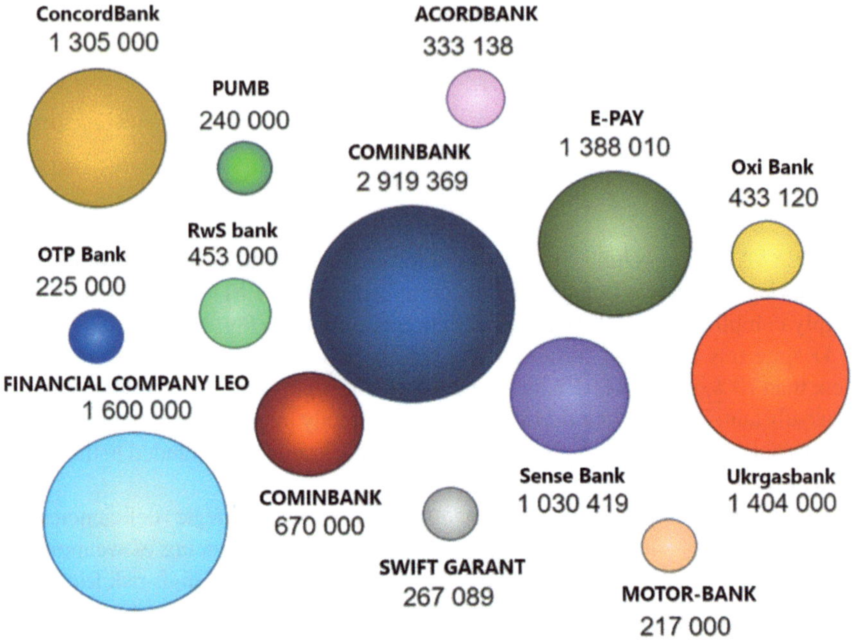

Fig. 2 Largest fines applied to banks and financial institutions in 2019–2024, EUR[1]

non-compliance risks and optimizing internal processes in the face of rapidly evolving regulatory standards. This will undoubtedly have a positive impact on the country's overall economic security.

2 Practical Aspects of Maintaining Compliance Procedures

Compliance costs in the AML/CTF sector are largely associated with the human factor rather than technology. According to expert estimates, employee-related components in banks (hiring and training) dominate, accounting for almost 70% of compliance costs, compared to just under 25% spent on technology (Fig. 3) [6].

This aligns with the findings of a study conducted by LexisNexis® Risk Solutions—"*The Future of Financial Crime in 2020*", which revealed that two-thirds of AML/CTF compliance budgets are allocated to personnel, while only one-third is spent on technology. On average, banks report that 38 employees are involved in AML/CTF compliance, whereas in large banks, this number can exceed 100 employees.

Data collected from various countries around the world clearly demonstrate that banks' compliance with AML/CTF legislation is a labor-intensive and high-cost

[1] According to the National Bank of Ukraine. URL: https://surli.cc/qlhieb.

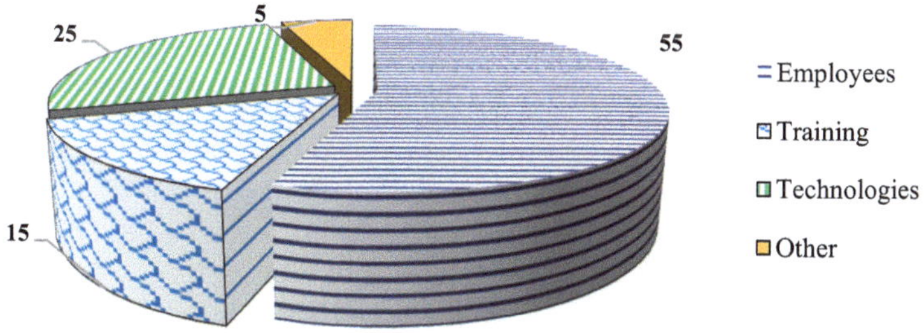

Fig. 3 Anti-money laundering compliance costs, % [6]

process. In recent years, these costs have increased significantly worldwide. For example, in 2023 alone, financial institutions in North America spent $65 billion, South America—$15 billion, the EMEA region—$85 billion, and the APAC region—$45 billion [7]. UK financial institutions spent £38.3 billion on combating financial crime in 2023, which is 12% more than the previous year and 32% more than in 2021, according to a report by LexisNexis Risk Solutions, a company specializing in data analytics [8]. Globally, AML/CTF compliance costs amount to $205 billion, significantly increasing the burden on institutions, especially due to tightening regulatory requirements and the growing complexity of financial transactions. At the same time, compliance professionals and responsible executives highlight a number of challenges that hinder the effectiveness and efficiency of AML compliance, including:

- outdated and inefficient IT infrastructure;
- poor data quality;
- ineffective internal tools;
- system failures.

Customer due diligence (CDD) and investigations account for two-thirds of the total AML/CTF compliance time and costs. According to a survey conducted by LexisNexis Risk Solutions, banks report that CDD is the most expensive and labor-intensive process, accounting for approximately 53% of total compliance costs (Fig. 4).

The increasing regulatory complexity and rising financial crime risks have led to a significant surge in AML/CTF compliance costs globally, with expenditures exceeding $205 billion. A major challenge for financial institutions remains the disproportionate reliance on human resources, where the majority of compliance budgets are allocated to personnel-related activities rather than technological solutions. Despite advancements in automation, compliance remains labor-intensive, particularly in CDD and investigative processes, which account for the largest share of costs and time consumption. The inefficiencies in outdated IT infrastructure, data quality issues, and fragmented internal systems further exacerbate compliance challenges, increasing operational burdens on financial institutions. The necessity for a more balanced approach integrating technology-driven solutions is evident, as

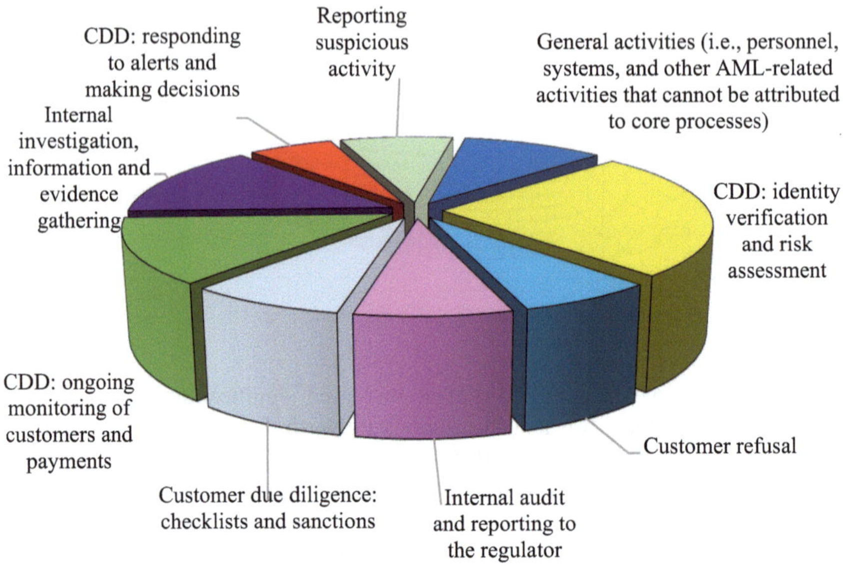

Fig. 4 Distribution of the cost of compliance processes in the field of AML/CFT, % [6]

automation and artificial intelligence have the potential to enhance efficiency, reduce compliance costs, and mitigate risks associated with financial crimes.

3 Compliance Procedures in Ukraine: Problematic Aspects and Promising AI Solutions

The Law of Ukraine "On Prevention and Counteraction to Legalization (Laundering) of Criminal Proceeds, Terrorist Financing, and Financing of the Proliferation of Weapons of Mass Destruction" [9] (hereinafter—the Law) establishes the obligations of banking institutions regarding CDD and the monitoring of their activities. However, as of today, there is no clear division of processes and procedures that would break down the fulfillment of these obligations into specific stages and actions. The recommendations of the National Bank of Ukraine provide only general guidelines and provisions, lacking a structured approach to the practical implementation of compliance requirements. Furthermore, there is insufficient understanding of the challenges and difficulties faced by practitioners in adhering to the Law, and there are no future-oriented solutions aimed at eliminating existing "bottlenecks" through the use of advanced technologies, AI and big data analytics methods. To address these methodological gaps and justify recommendations, we propose to conditionally divide the CDD compliance process into three parts, with a detailed description provided in the corresponding tables (see Tables 1, 2 and 3). These tables also identify the key problems and obstacles hindering the effective implementation of AML/CTF compliance requirements, propose enhancements to procedures using AI technologies, and outline the expected outcomes.

Thus, the data collection process is a fundamental stage of CDD, as its effectiveness directly impacts the quality of risk analysis and compliance with regulatory requirements. The use of AI technologies significantly optimizes data collection, enhancing its accuracy and speed. Among the key solutions are automated updates of customer information based on changes in sanctions lists and legal proceedings, risk categorization using AI, OCR document analysis for verification, automatic data collection from government registries and open sources (OSINT), AI-driven customer verification agents, and the implementation of blockchain technologies for secure data storage. By integrating automated technologies, the processing time for customer data can be reduced from 3–5 days to 30–40 min, significantly improving operational efficiency. Additionally, this reduces the risk of human errors, simplifies regulatory compliance, and enhances customer service quality. The implementation of such solutions enables the optimization of business processes and minimization of regulatory risks. A detailed breakdown of the process is presented in Table 2.

The comprehensive review is the final stage of CDD, involving a thorough analysis that takes into account updated information, an in-depth assessment of the customer profile, and the final decision on whether to proceed with cooperation. Currently, this process is conducted manually and includes a final screening round, an independent risk analysis following the "four-eyes principle", and the ultimate approval or rejection of the customer. This approach is time-consuming (6–7 days), labor-intensive, and prone to subjective errors.

The integration of AI technologies can significantly optimize this process [10, 11]. Machine learning enables the automatic detection of suspicious behavioral patterns and anomalous transactions, leading to more effective risk identification. Graph databases facilitate the analysis of relationships between companies and individuals, while deep learning allows for the evaluation of historical data to detect changes in customers' risk profiles. Additionally, AI algorithms streamline corporate structure analysis and help identify financial crimes and corruption risks at early stages.

Automating this process can reduce customer verification time from 6–7 days to 2–3 h, significantly improving operational efficiency. The use of digital technologies ensures data security and minimizes human error risks [10]. Moreover, the consistent application of automated procedures enhances regulatory compliance, reduces operational costs, and enables scalability in response to the growing volume of customer requests [11].

The optimization of the comprehensive review process, as presented in Table 3, will contribute to better AML risk management, improved regulatory compliance, and enhanced customer satisfaction through faster service delivery [12]. The implementation of AI at this stage will also help increase the bank's revenue by ensuring more efficient risk control and reducing false positives in transaction monitoring processes [12].

Table 1 Description of the process of "information search" within the framework of the CDD

	Search for information		Vulnerabilities, difficulties	Process execution time	AI solutions	Advantages
	Initial screening	Work with clients				
Current process	Detection of the presence of politically exposed persons (PEPs), persons associated with PEPs, sanctions, negative news about clients, watchlists and information on high-risk industries or jurisdictions, etc	Contacting customers with a request to provide the necessary KYC information	Not covering all databases, difficulty searching in multi-structured databases, different spelling of full names in different languages and in different databases, lack of language knowledge to search in databases of different countries. Multiple systems to view	2–3 days	1. Automated search and analysis of information. Using NLP (Natural Language Processing) to process textual sources, including news, social media, white papers. Automated Open Source Data Collection Systems (OSINT). Integration with global databases (World-Check, LexisNexis, Refinitiv) for customer verification 2. API integration for access to the database of sanctions lists	1. Detailed search for negative news, sanctions lists 2. Speed and simplification of search 3. The ability to limit the search to keywords, areas of activity, other details that optimize the search 4. Ability to compare different results 5. Generation of automatic reports 6. Improved data sharing and collaboration between compliance stakeholders
Implementation methods	Easy manual search for names in a database or on the web	Email, phone, personal contact, etc	Internet search does not cover all databases of government agencies and sanctions lists, the difficulty of comparing identification data that may be contained in different databases, the high probability of erroneous decision-making		Reduction of process execution time per client to 10 min	
Impact on work results банку	Reducing human error Increase customer satisfaction Expedited service Reduction of errors in compliance with legal requirements and reduction of fines Saving operating costs Reducing the risk of using a bank for money laundering					

Source Author's development

Table 2 Description of the "data collection" process within the CDD

	Data collection					Process execution time	AI solutions	Advantages
	Risk rating	Document verification	Collection of public information	Consolidation of customer data	Data Quality Checks			
Current process	Determining customer ratings based on different risk components	Verification of documents provided by clients and determination of the validity of documents	Collection of publicly available information to update existing information	Consolidation of customer information from different data sources before sending it for review	Data consistency, completeness, and more	3–5 days	1. Automated updating of customer data based on new changes in sanctions lists, PEP status, court cases 2. Using AI to categorize customer risks 3. Using computer viewing for automatic document processing (OCR) 4. Automated collection of information from state registers and open sources 5. Identification of beneficial owners through the analysis of corporate ties 6. Automatic data updates based on changes in regulatory policies 7. Using AI Agents for Customer Verification 8. Automated processes for reviewing documents and financial reports 9. Using blockchain technologies to maintain reliable records of customer verification	1. Automation can significantly streamline the KYC process, reducing the time and resources required to onboard new customers 2. Using centralized platforms for data management 3. Availability of structured data collection and processing procedures that ensure consistency and reduce the number of errors 4. Ability to get data from related areas and analyze them
Implementation methods	Simple and manual risk rating system, tools such as MS-Excel	Manual check on a computer	Secondary research, emails, phone calls, etc	Manually from multiple data sources	Manually using an overview		Reducing the process execution time per client to 30–40 min	

(continued)

Table 2 (continued)

	Data collection					Process execution time	AI solutions	Advantages
	Risk rating	Document verification	Collection of public information	Consolidation of customer data	Data Quality Checks			
Vulnerabilities	1. Significant expenditure of human time, labor intensity of analysis 2. High probability of human error 3. Difficulty integrating a variety of file formats 4. The available data is unreliable 5. The available data is unclear					Incorrect data entry Presence of a large amount of fragmented, unorganized data Unsecured communication channels Delaying the due diligence process		
Impact on the bank's performance	Reducing human error Improving operational efficiency Speeding up customer service Simplify regulatory compliance Attracting highly qualified personnel Simplification of business processes							

Source Author's development

Table 3 Description of the "due diligence" process within the framework of the CDD

	Due diligence			Process execution time	AI solutions	Переваги
	Final screening	Check	Approval			
Current Process	Final Screening Round Based on Updated Customer Information	Conducting a four-eye review or independent review based on the client's risk level	Verification of the final customer profile and granting permission	6–7 days	1. Preventive protection with the help of AI. Using machine learning to identify suspicious behavior patterns. Automatic analysis of transactions and identification of atypical financial transactions. Using graph databases to track relationships between companies and individuals 2. Intelligent solutions for the analysis of financial crimes and corruption risks 3. Using Deep Learning to Analyze Financial Transactions 4. Historical data analytics for risky changes in customer profiles 5. Analysis of corporate structures using machine learning algorithms 6. Identifying Abnormal Customer Behavior Using AI	1. Increased Accuracy: Automated decisions can help reduce human error by using advanced algorithms and data analysis techniques 2. Consistent application of customer data verification procedures, reducing the risk of non-compliance 3. Ease of scalability of automated solutions in the face of increased information flows or business expansion 4. Enhanced security: using digital technology to ensure a reliable and tamper-proof customer data storage and verification system
Nature of execution	Internal screening tool or manual screening	Manual verification	Manual verification		Reducing the process execution time per client to 2–3 h	
Vulnerabilities	1. Evolving complexity of meeting compliance and reporting obligations 2. Misinterpretation of documents 3. Increased adaptation time to new legal requirements and increased operating costs 4. Inconsistent application of verification procedures in different departments or in different employees 5. The complexity of mapping transactional networks and identifying counterparty risks 6. False positives 7. Significant time spent on correcting risk assessment errors 8. Delays in completing customer transactions					
Impact on results Bank Operations	Opportunities to increase income Better management of money laundering risks Simplify regulatory compliance Increase customer satisfaction Expedited service					

Source Author's development

The analysis of the CDD process reveals that its effectiveness remains limited by fragmented procedures, reliance on manual operations, and a lack of structured, technology-driven frameworks. Current practices in information search, data collection, and final due diligence are time-consuming, error-prone, and resource-intensive, thereby affecting compliance quality and operational efficiency. The integration of AI, including machine learning, natural language processing, and graph databases, offers a transformative potential by significantly reducing execution time, enhancing accuracy, and streamlining compliance workflows. Automated systems not only improve risk detection and decision-making but also ensure adaptability to regulatory changes, scalability in growing client environments, and reduced operational costs. A systematic application of AI-based solutions across all stages of CDD can thus strengthen banks' AML/CTF performance, improve customer satisfaction, and contribute to greater institutional resilience and economic security.

4 Automated and High-Tech Approach to Compliance Risk Management

Considering the presented concept of Intelligent Compliance and the high demands placed on compliance management systems—particularly in the context of preventing financial crimes, AML, and CFT—this section examines an automated and high-tech approach to managing compliance risks. The use of technological innovations based on machine learning and algorithmic methods not only enhances the efficiency of transaction monitoring but also significantly reduces the operational costs associated with regulatory compliance [13].

To increase the effectiveness, relevance, and accuracy of implementing the Intelligent Compliance concept, we propose algorithmization, built upon theoretical principles and practical examples, aimed at formalizing compliance processes in the form of concrete algorithms. The algorithmization process involves the development of a structured sequence of actions that enables efficient automation of data analysis, risk assessment, and the preventive protection of financial operations [14].

The next stage of the study focuses on detailing the key algorithmic approaches used to implement Intelligent Compliance. The methodology includes the development of a BPMN (Business Process Model and Notation)—based structural model of Intelligent Compliance processes, the construction of logical decision-making schemes, and the integration of AI methods to identify and predict risk-related situations [15, 16]. Thus, the subsequent analysis is focused on describing the algorithmization of the key components of the Intelligent Compliance system, which ensures effective regulatory compliance management and enhances the economic security of financial institutions.

In this context, the automation of financial transaction analysis, risk identification, and dynamic updating of compliance models becomes a necessity. MATLAB serves as a powerful tool for accomplishing these tasks due to its advanced suite of AI, ML, and big data analysis capabilities.

One of the key aspects of Intelligent Compliance is the collection and analysis of large volumes of financial data. MATLAB enables efficient interaction with various data sources, including:

- reading transactional data from databases and files (CSV, JSON, SQL);
- processing and normalizing numerical data for further analysis;
- detecting missing or anomalous values using AI-based models.

The process structure is as follows:

1. Data Collection—NLP processing, OSINT monitoring, database integration, AI-driven analysis.
2. Transaction Analysis—Three parallel branches: transaction evaluation, risk detection, anomalous behavior identification.
3. Risk Categorization and Graph Analysis—Classification of risks and mapping entity relationships.
4. OCR Verification → Beneficial Ownership Identification → Data Updates—Document validation, owner identification, and automated profile updates.
5. Report Generation → Intelligent Solutions → Corporate Link Verification—Compliance reporting, AI-driven insights, and company structure validation.
6. Risk Reevaluation → Decision (Suspicious Operation?)—Final assessment and regulatory compliance actions.

Detailing the sequence of process steps Intelligent Compliance:

1. Initialization and Data Collection. The process begins with the collection of input data from various sources, including the processing of textual information using Natural Language Processing (NLP), open-source monitoring (OSINT), and database integration for further analysis. This stage is foundational, as the quality of acquired data directly determines the accuracy of subsequent analytical operations.
2. Analytical Processing and Risk Detection. In the second stage, the collected data undergoes in-depth analysis. This includes transaction screening, risk identification, and graph analysis of relationships between companies and individuals. The implementation of AI allows for automation of suspicious activity detection processes and significantly reduces the likelihood of human error.
3. Diagnostics and Decision-Making. A decision gateway, "Suspicious Transaction?", serves as a logical branching point in the process. If a transaction is flagged as suspicious, it is routed to the preventive protection stage, where it undergoes a secondary review and a decision is made regarding further actions. If the transaction is deemed non-suspicious, it proceeds through the standard customer screening process.
4. Ongoing Processes: Preventive Protection and Staff Training. Unlike traditional models, this BPMN structure treats preventive protection and staff training as

continuous processes. Preventive protection operates in parallel with all analytical verification stages, ensuring the timely mitigation of potential threats. Simultaneously, staff training is conducted to keep employee competencies up to date with changes in legislation and methods of financial fraud.

5. Final Stages: Reporting and Regulatory List Updates. In the final stages of the model, OCR-based document verification, sanctions list updates, and corporate link analysis are performed. All collected data is integrated into an automated reporting system, enabling real-time responses to legislative changes and the provision of timely information to regulators.

The BPMN structural model of Intelligent Compliance processes is illustrated in MATLAB (Fig. 5) as a graph that visualizes both parallel and sequential business processes aimed at automating and improving the efficiency of risk detection, transaction monitoring, and staff training procedures. This model reflects the interaction between the key stages of the compliance system, ensuring its continuous operation and integration with modern big data analytics and artificial intelligence methods.

To identify suspicious transactions, the Naive Bayes classifier algorithm or more advanced models such as Random Forest and Neural Networks are used. MATLAB provides built-in functions for training and testing these models. For the detection of anomalous financial schemes, graph analysis is applied. MATLAB allows for the construction and analysis of graph structures representing interactions between transaction participants. It is an effective tool for building AI-driven solutions in the financial sector, significantly enhancing compliance with regulatory requirements and strengthening the bank's economic security.

The proposed BPMN-based Intelligent Compliance model ensures efficient risk management through the integration of modern AI technologies, automated data

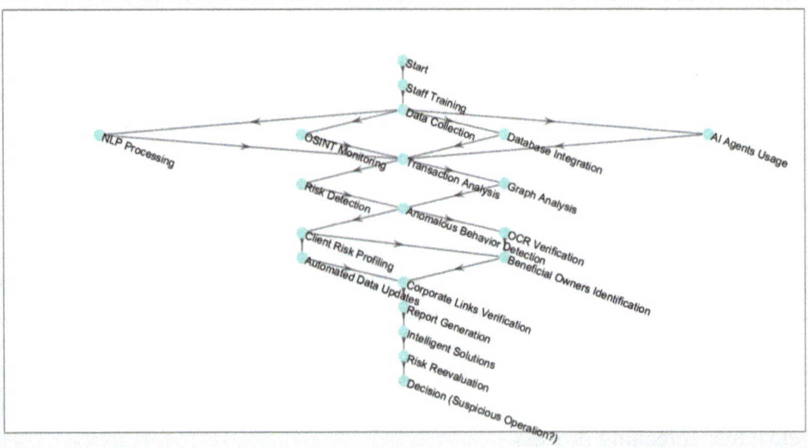

Fig. 5 BPMN-structural model of intelligent compliance processes in MATLAB. *Source* Author's development

analysis, and continuous oversight of compliance processes. Its key advantages include the parallel execution of critical processes, minimization of human error, and flexibility in decision-making. Implementing such a structure can significantly improve the effectiveness of the financial security system and ensure compliance with international regulatory standards.

Table 4 demonstrates the implementation of a full cycle of AI analytics for the Intelligent Compliance system within the MATLAB environment, enabling the automation of financial monitoring and risk management processes. At the initial stage, data loading and preprocessing are carried out using the readtable() function, which ensures the integration of financial transactions in tabular format and their subsequent normalization. The application of machine learning methods, such as Naive Bayes classification (fitcnb()), allows for effective risk prediction for each transaction, significantly enhancing the automation level of the compliance process.

The next stage involves graph-based relationship analysis using the graph() function, which enables the detection of hidden connections between transaction participants and the identification of potentially fraudulent schemes. This is achieved through the construction of networks between senders and recipients and their subsequent visualization. For improved anomaly detection, the anomalyDetection() function is used to analyze deviations from standard financial behavior and signal potential risks. The use of deep learning algorithms, particularly LSTM (Long Short-Term Memory) models, contributes to the increased accuracy in identifying suspicious financial activities. A critical component is the implementation of an automatic machine learning model update mechanism (retrainModel()), which enables the Intelligent Compliance system to autonomously adapt to changes in the financial environment and continuously improve risk classification accuracy. Additionally, a sanctions list database update mechanism via API (webread(), webwrite()) is introduced, ensuring up-to-date information on clients. The final stage is the construction of a reporting system and dashboards, implemented using the histogram() function, which allows for monitoring the distribution of transaction risks and evaluating the effectiveness of the compliance model.

Thus, the comprehensive application of AI methods in MATLAB enables the automation of financial control processes, reduces the risk of non-compliance with regulatory requirements, and optimizes the performance of Intelligent Compliance. The implementation of such solutions contributes to strengthening the country's economic security by minimizing financial risks associated with money laundering, fraud, and other economic crimes.

The development of AI and its integration into Intelligent Compliance processes significantly enhances the efficiency of financial transaction analysis, reduces risks, and automates monitoring procedures. MATLAB enables the implementation of a full AI analytics cycle covering all key stages of data processing: data loading, model training, risk prediction, graph-based relationship analysis, detection of anomalous transactions, model updating, result visualization, and personnel training.

At the first stage, input data preparation is carried out, including the loading of financial transaction data and the extraction of relevant features and risk labels. Next, the AI model is trained using ML methods, particularly the Naive Bayes algorithm,

Table 4 MATLAB codes for implementing the full cycle of AI analytics for intelligent compliance

Step	AI-method	MATLAB—command	Implementing AI in Intelligent Compliance
Downloading data	Preparing the entrance	readtable()	data = readtable('financial_transactions.csv'); % Downloading data features = table2array(data(:,2:end)); % Selection of features (financial parameters) labels = table2array(data(:,1)); % Tags (risky/non-risky operation)
Model training	Machine learning (ML)	fitcnb()	mdl = fitcnb(features, labels); % Bayes Model Training (AI Analytics)
Risk Prediction	AI-Analytics	predict()	predictions = predict(mdl, features); % Risk prediction for each transaction
Graph Relationship Analysis	Network analysis	graph()	G = graph(data.Sender, data.Receiver); figure; plot(G, 'Layout', 'force'); % Visualization of relationships between participants in transactions title('Graph-based Relationship Analysis for Risk Detection');

Definition of anomalies	Anomaly Detection	anomalyDetection()	risk_scores = anomalyDetection(features); % AI-Risk calculation function
			high_risk_indices = find(risk_scores > threshold); % Identify risky transactions
			disp("High-risk transactions detected:");
			disp(data(high_risk_indices,:));

Model updates	Self-learning AI	retrainModel()	new_data = updateComplianceData(); % Receiving updated financial transactions mdl = retrainModel(mdl, new_data); % Re-training the AI model

(continued)

Table 4 (continued)

Step	AI-method	MATLAB—command	Implementing AI in Intelligent Compliance
AI Analytics Visualization	Graphical analysis	histogram()	figure; histogram(risk_scores); % Building a histogram of risky transactions title('Risk Distribution Analysis');
Staff training	AI-driven Training	trainComplianceModel()	disp("Running AI-driven compliance training module..."); trainComplianceModel(); % Performing AI training for staff

Source Author's development

which enables the classification of operations based on their risk level. The following stage involves risk prediction, during which the model evaluates financial transactions and identifies potentially high-risk operations.

For deeper analysis, a graph-based approach is used to model relationships between payment senders and recipients, which is critical for detecting money laundering schemes. In addition, an anomaly detection method is applied to evaluate deviations in financial behavior from established regulatory parameters and to identify suspicious transactions.

To ensure the continuous relevance of the AI system, automated model updates based on new financial data are implemented, allowing algorithms to adapt to changes in the financial environment. Another important stage is the visualization of AI analytics, which includes building histograms of risky transactions and analyzing the risk distribution, thereby supporting informed decision-making in the area of financial monitoring. The final stage involves staff training through AI simulations, enabling compliance officers to more effectively identify risky transactions and improve their competencies in financial monitoring. Thus, the integration of AI analytics in MATLAB ensures a comprehensive approach to risk and compliance management, enhancing the efficiency of banking processes, reducing manual analysis costs, and ensuring adherence to regulatory requirements.

5 Conclusions

The implementation of Intelligent Compliance technologies is a strategically important direction for ensuring Ukraine's economic security in the context of tightening regulatory requirements, increasing financial risks, and the active adoption of digital technologies. Traditional compliance methods based on manual data processing are increasingly outpaced by automated approaches, which minimize the human factor, reduce operational costs, and enhance the accuracy of risk transaction analysis.

This study has assessed the current state of compliance procedures performed by banking institutions and has presented the evaluation results of Ukraine by the Council of Europe's MONEYVAL Committee regarding progress in the field of anti-money laundering and counter-terrorist financing. Particular attention is paid to the key challenges faced by banks in meeting regulatory requirements, especially the high costs associated with compliance processes. It is noted that the lack of unified and effective methodologies for implementing legislative requirements, combined with manual workload pressures, creates critical obstacles for financial institutions.

In light of the above, the study formulates recommendations for the application of artificial intelligence technologies, machine learning, graph databases, and blockchain in Intelligent Compliance processes. The recommendations are structured around dividing the compliance process into distinct stages, for each of which the key obstacles and regulatory compliance challenges are identified, along with justified proposals for improving procedures through the use of AI technologies and the expected results.

To facilitate the practical implementation of the proposed recommendations within Ukrainian banking institutions, an algorithmic approach for Intelligent Compliance is formalized. This approach includes the development of a structural model of Intelligent Compliance processes, the construction of logical decision-making schemes, and the integration of AI methods for the identification and prediction of risk-related situations.

Applying these recommendations in practice will enable the timely identification of fraudulent financial operations, enhance the effectiveness of measures to combat money laundering, terrorist financing, and the financing of weapons of mass destruction, and improve oversight of banking activities. Notably, intelligent algorithms allow for the real-time automated analysis of transactions, significantly reducing the risks of financial crime and reinforcing trust in the banking system.

The development of Intelligent Compliance will also contribute to the harmonization of national legislation with international standards, such as the FATF Recommendations, which is a critical factor in Ukraine's integration into the global financial system. The adoption of advanced technologies in financial monitoring ensures not only regulatory compliance but also the formation of a flexible risk management system capable of adapting to changes in the global financial environment.

Beyond the technological dimension, a key task remains the training of personnel and the implementation of AI-focused training programs to enhance the qualifications of compliance officers. This will improve the analytical capacity of professionals and ensure more effective control over high-risk financial operations.

Thus, Intelligent Compliance serves as a key tool for safeguarding financial stability, economic security, and Ukraine's international reputation. In today's conditions, the integration of advanced AI solutions into the financial monitoring system is not merely a technological innovation but a necessity for strengthening Ukraine's economic sovereignty and boosting its competitiveness in the international financial market.

References

1. OECD guidelines for multinational enterprises on responsible business conduct (2023) (in Ukrainian). https://odaold.od.gov.ua/wp-content/uploads/2023/11/oecd_guidelines_.pdf. Accessed 18 Mar 2025
2. Baker McKenzie announces spin-off of AI-powered legal risk solutions business SCOREalytics (2022). https://www.bakermckenzie.com/en/newsroom/2024/11/scorealytics. Accessed 07 Mar 2025
3. Pathak M (2024) Intelligent compliance: intelligent contracts for financial regulatory compliance. https://ssrn.com/abstract=4715505 or https://doi.org/10.2139/ssrn.4715505. Accessed 06 Mar 2025
4. Zhai Q (2023) Technology-driven legal innovation: the role of intelligent legal solutions in corporate compliance and governance. Int J Law Ethics Soc Sci
5. Report on the results of the fifth round of mutual evaluation, Ukraine (in Ukrainian) https://fiu.gov.ua/pages/dijalnist/mizhnarodne-spivrobitnictvo/zviti-pro-oczinku.html. Accessed 18 Mar 2025
6. Cutting the costs of AML compliance. https://risk.lexisnexis.co.uk/-/media/files/business/white-paper/cutting-the-costs-of-aml-compliance%20pdf.pdf. Accessed 18 Mar 2025
7. Cost-effective compliance: how AML case management systems cut costs. https://lucinity.com/blog/cost-effective-compliance-how-aml-case-management-systems-cut-costs. Accessed 18 Mar 2025
8. Report: the true cost of financial crime compliance. LexisNexis Risk Solutions (2024), 209 p
9. Law of Ukraine on prevention and counteraction to legalisation (laundering) of criminal proceeds, terrorist financing and financing of proliferation of weapons of mass destruction. https://zakon.rada.gov.ua/laws/show/361-20?lang=en#Text. Accessed 18 Mar 2025
10. Engelhardt-Nowitzki C, Kryvinska N, Strauss C (2011) Strategic demands on information services in uncertain businesses: a layer-based framework from a value network perspective. In: 2011 International conference on emerging intelligent data and web technologies. IEEE, USA, pp 131–136. https://doi.org/10.1109/eidwt.2011.28
11. Kryvinska N, Strauss C (2013) Conceptual model of business services availability vs. interoperability on collaborative IoT-enabled eBusiness platforms. In: Studies in computational intelligence. Springer, Berlin, Heidelberg, pp 167–187. https://doi.org/10.1007/978-3-642-34952-2_7
12. Voynarenko M, Hurochkina V, Sushkova OJ, Yepifanova I (2020) Beneficial ownership chain and its evaluation arrangement. Estudios de Economia Aplicada 38–4(1). https://doi.org/10.25115/eea.v38i4.3991. Accessed 18 Mar 2025
13. Ushenko N, Likhonosova G, Zahariev A, Shaulska L, Casey M, Hurochkina V (2023) Strategies for strengthening the economic security of business: taking into account the conditions of global financial challenges. Financ Credit Act Probl Theory Pract 6(53):300–317. https://doi.org/10.55643/fcaptp.6.53.2023.4178. Accessed 18 Mar 2025
14. Hurochkina V, Nahaieva O, Rymar O, Chornyi O, Makhanets L, Shyshpanova N (2021) Foreign state economic policy in the context of economic security. J Manag Inf Decis Sci 24(3):1–14. https://www.abacademies.org/journals/month-march-year-2021-vol-24-issue-3-journal-jmids-past-issue.html. Accessed 18 Mar 2025
15. Dzhedzhula V, Yepifanova I (2021) Optimization of energy saving potential of industrial enterprises. In: 2021 11th International conference on advanced computer information technologies (ACIT). IEEE, USA, pp 433–436. https://doi.org/10.1109/acit52158.2021.9548428
16. Dzhedzhula V, Yepifanova I, Kravchyk Y (2022) Use of the theory of fuzzy sets in determining the level of enterprise security. In: 2022 12th International conference on advanced computer information technologies (ACIT). IEEE, USA, pp 311–315. https://doi.org/10.1109/acit54803.2022.9913150

The Convergence of Artificial Intelligence with the Circular Economy

Viktoriia Hurochkina, Zakharii Varnalii, Piotr Kulyk, Oksana Cheberyako, and Dmytro Nikytenko

Abstract The purpose of the study is the systematization and scientific substantiation of theoretical aspects and practical recommendations for the development of a scientific and methodological approach to evaluating the effectiveness of the circular economy, taking into account the effect of convergence with artificial intelligence, which has been tested on the example of EU countries (2021). The study is devoted to the analytical evaluation and construction of the Ranking Effect of the circular economy with AI. The study was carried out by the method of monitoring the financial support of the implementation of artificial intelligence for the circular economy at EU enterprises, the directions of the use of AI, and at the same time the relationship with the effectiveness, namely the operations of the circular economy, was determined. In a study of the leaders of innovators in the circular economy with AI. In order to outline an effective model of AI implementation in the circular economy, a scientific and methodological approach to evaluation and ranking is proposed. It is emphasized that the convergence of artificial intelligence and the circular economy is primarily aimed at increasing the level of efficiency of resource use and reducing the ecological impact of enterprises on the environment as a whole.

V. Hurochkina (✉) · P. Kulyk
University of Zielona Góra, Zielona Góra, Poland
e-mail: v.hurochkina@wez.uz.zgora.pl

P. Kulyk
e-mail: p.kulyk@wez.uz.zgora.pl

Z. Varnalii · O. Cheberyako
Taras Shevchenko National University of Kyiv, Kyiv, Ukraine
e-mail: vzs1955@gmail.com

O. Cheberyako
e-mail: cheberyako@ukr.net

D. Nikytenko
National University of Water and Environmental Engineering, Rivne, Ukraine
e-mail: d.v.nikytenko@nuwm.edu.ua

A. Semenov et al. (eds.), *Data-Centric Business and Applications*, Lecture Notes on Data Engineering and Communications Technologies 272,
https://doi.org/10.1007/978-3-032-06608-4_2

Keywords Artificial intelligence (AI) · Circular economy · Efficiency evaluation · Resource optimization · Environmental impact reduction

1 Introduction

The convergence of artificial intelligence and the circular economy as an innovative approach to accelerating the implementation of the Sustainable Development Goals will contribute to the achievement of a climate-neutral economy by 2030. It is directly related to business processes aimed at reducing the impact of human activity on the environment and preserving resources [1, 2]. The circular economy, using the capabilities of artificial intelligence, involves the transformation of the "linear" model into a balanced one based on the "best" practices of the world in the process of waste recovery and recycling, stimulating the production of products with a long shelf life and reducing resource consumption [3, 4]. Convergent processes contribute to the emergence of new qualities and characteristics, that is, emergentity, the countdown of which is observed along the life cycle at the point of bifurcation [5].

The convergence of artificial intelligence and the circular economy is primarily aimed at increasing the level of efficiency in the use of resources and reducing the environmental impact of enterprises on the environment as a whole [6]. Structural modeling and systematization of artificial intelligence algorithms in circular economy models make it possible to highlight the innovativeness of cyclical business models at different stages of creation, and the implementation of dynamic pricing based on artificial intelligence will stimulate the behavioral aspects of top management to make effective management decisions and deepen the extended responsibility of manufacturers [7–9]. Artificial intelligence in circular economy models will contribute to the development of the waste management system, optimization in circular logistics, accurate demand forecasting, implementation of resource planning, improvement of product design to enhance the ergonomics of products and their packaging, implementation of exchange platforms and much more.

The processes of implementing artificial intelligence solutions in the field of circular economy require validation and approval of prototyping. The mechanism for implementing artificial intelligence solutions requires cooperation between government agencies, industrial companies, and research institutions to implement innovative solutions and create an appropriate legal and regulatory environment.

2 Related Works

In today's world, it is difficult to overestimate the importance of sustainable development and caring for the environment, society is increasingly interested in the concept of the convergence of artificial intelligence (AI) and the circular economy (CE). This convergence represents an exciting way to create more sustainable, innovative and convenient ways to use resources and optimize production. And the processes of their assessment and performance monitoring need further scientific and practical improvement.

Chauhan et al. [1] The economic potential of the implementation of the circular economy is based on the use of digital technological processes (big data, artificial intelligence (AI), blockchain and IoT). Thanks to the convergence of artificial intelligence in the circular economy, enterprises are transforming the circular economy.

Wilson et al. [2] the authors considered the technologies that today are drivers in the business ecosystem, where artificial intelligence for reverse logistics is highlighted as the main driver. Roberts et al. [5] pointed out that it is AI that is the decisive factor in the development of CE, and it will contribute to sustainable development in general through the creation of closed-loop models, business models and support a sustainable infrastructure that will develop circular scaling processes.

Another significant figure is Jane Ellenson, a researcher in the field of circular economy. She studies the possibilities of using AI to create more efficient waste and resource management systems, which will reduce pollution and increase recycling of waste in a circular cycle. Ghoreishi and Happonen [10] developed artificial intelligence mechanisms for effective integration into the circular economy at the product design stage, and Ali et al. [11] highlight the value of artificial intelligence for the transition to CE. Another important contribution to this field is provided by a team of researchers at the Massachusetts Institute of Technology (MIT), which is studying the application of machine learning and data analytics to improve recycling processes and resource utilization in CE. The authors directed their research to consider interesting aspects and note how foreign scientists have made a significant contribution to the development of this synergy between AI and CE. They help shape a new economic paradigm focused on sustainability and resource optimization for the benefit of our environment and society as a whole [12].

The authors [6] of this study have already presented a scientific and methodological approach to the assessment of the circular economy as a factor in strengthening the economic security of Ukraine in the post-war period, and the next step is devoted to the effectiveness of financial support and the processes of implementing artificial intelligence in the circular economy.

More over the issue of implementing AI for the CE is being actively discussed by consultancy [13].

3 Methodology

The methodological basis of the research is general scientific and special research methods, concepts, approaches, methods: circular economy theory, economic theory, modernization theory, theory of waste-free technological processes, systems theory, theory of fuzzy sets, science of artificial intelligence and its foundation, mathematical theory of calculations, theory algorithms and the creation of computers.

Information basis of the research: data of statistical information of Ukraine and Eurostat (Poland, Sweden and other EU countries for comparison).

4 Results

The scientific-methodical approach to identifying the effect of the circular economy, taking into account the effect of convergence with artificial intelligence, is a complex tool designed to analyze the interaction between these two key areas of development in the economy [14, 15]. This technique is based on several important components that help determine the effectiveness and contribution of both of these factors to the sustainability and sustainability of production. The author's methodology is based on 5 important components: PI_{ce} (Private investment and gross added value related to circular economy sectors), *ETax* (Environmental tax revenues), W_m Waste management operations, AI_{ent} (Artificial intelligence by size class of enterprise), AI_{nace} (Artificial intelligence by NACE Rev.2 activity).

The general methodology allows taking into account the interaction between these components to determine the effect of the circular economy and convergence with artificial intelligence [16–18]. This analysis can be an important tool for governments, businesses and researchers in developing policies and strategies aimed at achieving sustainable development and increasing economic productivity.

The component AI_{ent} takes into account the use of artificial intelligence in enterprises, taking into account their size. It allows you to determine how the size of the enterprise affects the implementation and use of AI and how it affects the efficiency and sustainability of production (Fig. 1) [7].

The component AI_{nace} is the Artificial intelligence by NACE Rev.2 activity. This component takes into account the use of artificial intelligence in economic sectors according to the NACE Rev.2 classification. It allows to establish how AI is integrated into individual sectors of the economy and affects their productivity (Fig. 2) [8].

In Fig. 3 the volumes of AI use in all types of economic activity, excluding the financial sector of the economy, are displayed. The main areas of application of AI technologies at enterprises today are [9]:

- performing analysis of written language;
- converting spoken language into machine-readable format (speech recognition);
- generating written or spoken language (natural language generation);
- identifying objects or persons based on images (image recognition, image processing);
- machine learning (e.g. deep learning) for data analysis;
- automating different workflows or assisting in decision making;
- enabling physical movement of machines via autonomous decisions based on observation of surroundings.

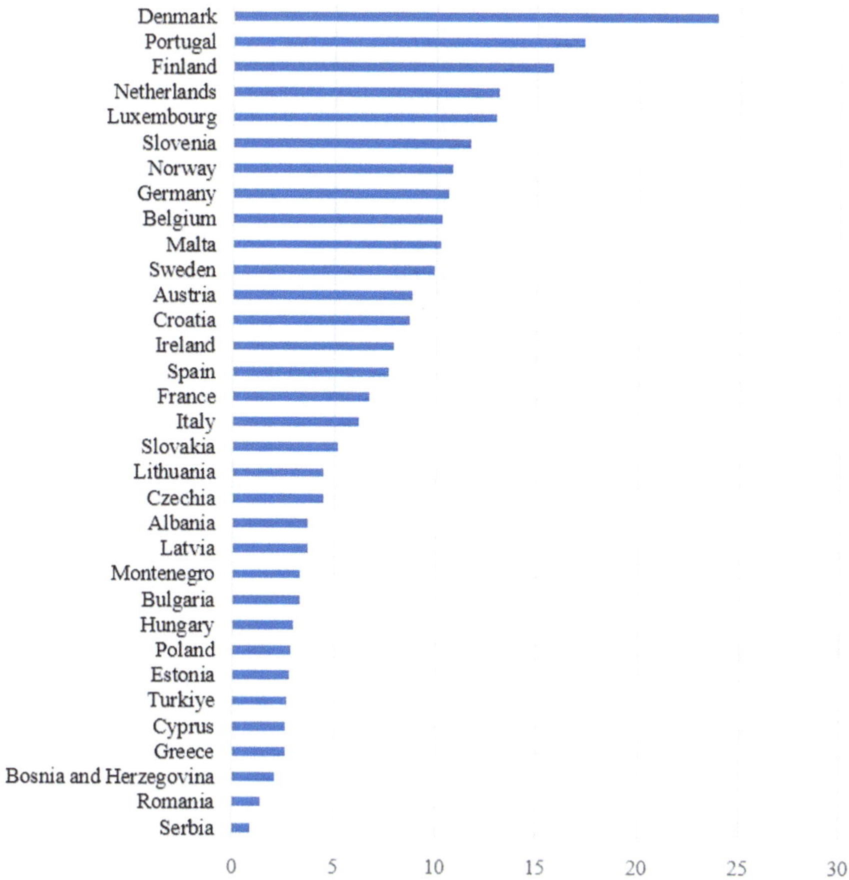

Fig. 1 Artificial intelligence by size class of enterprise [7]

The component PI_{ce} measures the amount of private investment and gross value added related to circular economy sectors in millions of euros. This makes it possible to estimate the economic contribution of circular sectors to the overall economy (Table 1).

The component *ETax* takes into account the amount of environmental taxation in millions of euros. Environmental taxation indicates how effectively the government collects funds to protect the environment and how this affects the support of circular and sustainable practices (Table 2).

The component W_m includes waste management operations. It assesses how waste management operations contribute to sustainable development and help avoid negative environmental impacts (Table 3).

Indicators of the development of the circular economy, taking into account the effect of convergence with AI, including financial flows from private investment in eco-innovation and environmental taxation of companies in the state budget, as a

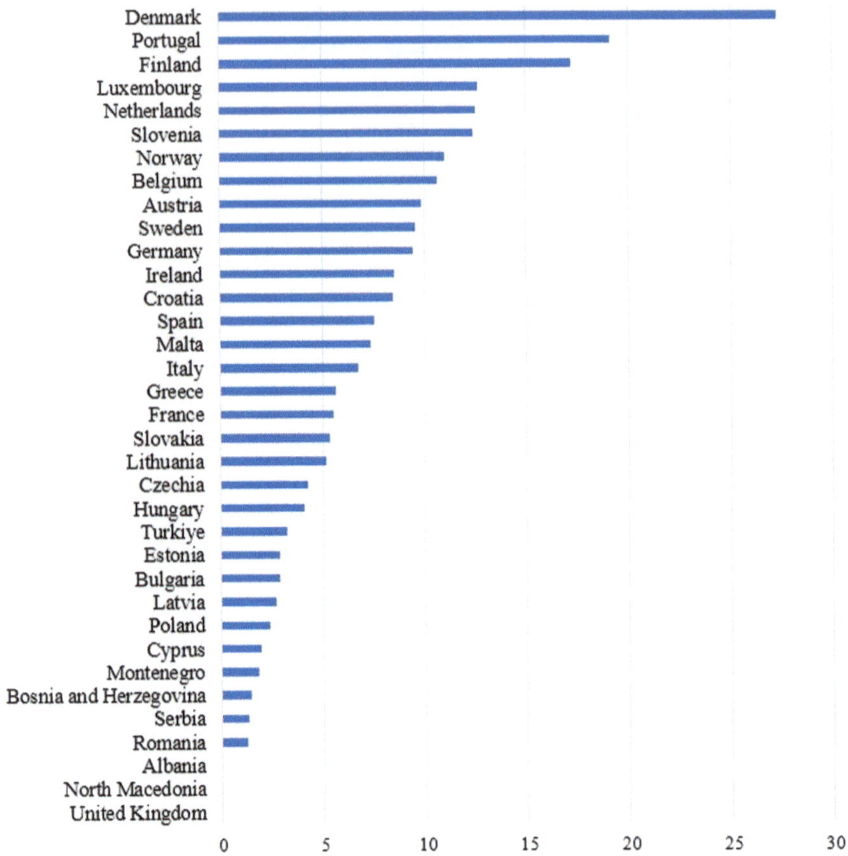

Fig. 2 Artificial intelligence by NACE Rev.2 activity [8]

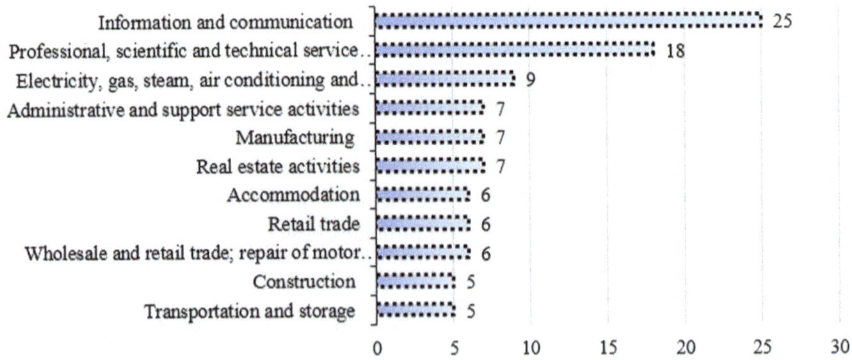

Fig. 3 Enterprises using AI technologies by economic activity, EU, 2021 (% of enterprises)

Table 1 Private investment and gross added value related to circular economy sectors

	2012	2014	2016	2018	2019	2020	2021	Absolute deviation	Relative deviation
Belgium	5612	4764	7154	6533	7511	7029	7251	1639	29.21
Bulgaria	211	250	304	394	419	380	395	184	87.20
Czechia	460	487	598	811	871	858	905	445	96.74
Denmark	2413	2607	2486	2735	2453	2960	3063	650	26.94
Germany	13299	14323	16386	34489	28970	29751	31507	18208	136.91
Estonia	196	112	132	195	187	192	200	4	2.04
Ireland	478	487	402	575	1704	2109	2699	2221	464.64
Greece	241	372	448	247	386	163	171	– 70	– 29.05
Spain	3423	3998	5008	5308	5889	5792	6108	2685	78.44
France	19945	18123	20257	18397	20050	20108	20405	460	2.31
Croatia	183	252	289	409	348	379	404	221	120.77
Italy	6 283	6855	8310	12,803	15,421	12,070	12,423	6140	97.72
Cyprus	121	33	58	92	73	50	51	– 70	– 57.85
Latvia	180	182	124	241	342	224	233	53	29.44
Lithuania	138	228	347	339	348	416	447	309	223.91
Luxembourg	312	354	515	432	800	681	722	410	131.41
Hungary	480	700	880	1130	1106	1022	1081	601	125.21
Malta	64	87	110	133	144	155	166	102	159.38
Netherlands	4774	4510	6257	7506	9313	8267	8700	3926	82.24
Austria	5100	5264	5517	5483	5667	5448	5474	374	7.33
Poland	1833	2499	2440	4325	4342	3727	3890	2057	112.22
Portugal	961	1202	1578	2004	2112	1731	1771	810	84.29
Romania	676	1331	899	918	1103	1060	1091	415	61.39
Slovenia	109	81	117	188	260	109	108	– 1	– 0.92
Slovakia	319	381	465	514	430	484	502	183	57.37
Finland	750	732	803	741	741	728	733	– 17	– 2.27
Sweden	1842	1643	2451	1816	1976	2200	2265	423	22.96
United Kingdom	18074	22746	27458	24723	30194	31936	33722	15648	86.58

result, it stimulates management at enterprises to implement the principles of circular economy, that is, waste management operations at enterprises using artificial intelligence. The author's methodology takes into account the weighted average volume of AI data processing operations by NACE Rev.2 activity and by enterprise size class.

The stages of the scientific and methodical approach to evaluating the convergence effect of the circular economy with the introduction of artificial intelligence can be determined by the following formulas:

Table 2 Environmental tax revenues

	2012	2014	2016	2018	2019	2020	2021	Absolute deviation	Relative deviation
Belgium	9735	10255	11469	12423	12628	11588	12537	2802	28.78
Bulgaria	1119	1221	1451	1470	1839	1859	1979	860	76.90
Czechia	3606	3347	3728	4129	4595	4148	4373	767	21.27
Denmark	10099	10622	11065	10933	10221	9892	9704	– 395	– 3.91
Germany	58198	58293	58442	59731	61119	57476	64714	6516	11.20
Estonia	489	533	645	709	890	657	717	228	46.64
Ireland	4173	4684	5117	5119	5020	4487	4953	780	18.69
Greece	6265	6628	6656	6822	7086	6226	7136	871	13.90
Spain	16325	19344	20793	22075	22018	19600	21265	4940	30.26
France	40946	43716	50125	56039	56327	50226	54421	13475	32.91
Croatia	1125	1380	1620	1844	1915	1648	1816	690	61.34
Italy	56251	58070	59481	58575	58299	50444	53383	– 2868	– 5.10
Cyprus	503	534	554	613	583	520	562	60	11.92
Latvia	660	790	908	983	900	916	918	258	39.08
Lithuania	548	634	748	900	921	954	1045	497	90.70
Luxembourg	1039	980	933	1028	1094	893	1048	9	0.87
Hungary	2533	2558	2942	3094	3308	2983	3096	563	22.21
Malta	205	240	277	322	346	297	289	84	40.74
Netherlands	21178	22216	23754	25877	27570	25366	26419	5241	24.75
Austria	7665	7975	8384	8784	9059	7975	8775	1110	14.48
Poland	10055	10562	11557	13474	13545	13330	16590	6534	64.98
Portugal	3682	3932	4819	5272	5418	4729	5025	1343	36.48
Romania	2667	3587	4085	4034	4732	4196	4730	2062	77.31
Slovenia	1389	1453	1569	1560	1615	1383	1453	64	4.61
Slovakia	1744	1932	2019	2226	2362	2294	2383	639	36.63
Finland	5949	5957	6709	6848	6730	6487	6289	340	5.72
Sweden	10311	9544	10341	9814	9779	9618	10218	– 93	– 0.90
Iceland	264	288	387	470	443	382	368	104	39.36
Liechtenstein	39	42	46	44	48	0	0	– 39	– 100.00
Norway	9262	8733	8044	8110	7630	6448	6984	– 2279	– 24.60
Switzerland	7204	7265	8558	8601	9039	9032	9179	1975	27.42

$$W_m = \frac{E_{ce}}{PI_{ce} + ETax} * 100\%, \tag{1}$$

$$E_{ce} = \frac{W_m}{100} * (PI_{ce} + ETax), \tag{2}$$

Table 3 Waste management operations EU

	2012	2014	2016	2018	2019	2020	2021	Absolute deviation	Relative deviation
Belgium	53.40	53.80	53.50	54.40	54.70	51.40	55.50	2.10	3.93
Bulgaria	25.00	23.10	31.80	31.50	34.60	35.20	28.20	3.20	12.80
Czechia	23.20	25.40	33.60	32.20	33.30	40.50	43.30	20.10	86.64
Denmark	42.50	45.40	48.30	49.90	51.50	45.00	57.60	15.10	35.53
Germany	65.20	65.60	67.10	67.10	66.70	70.30	67.80	2.60	3.99
Estonia	19.10	31.10	27.90	28.00	30.80	28.90	30.30	11.20	58.64
Ireland	36.60	39.80	40.70	37.70	37.40	40.80	0.00	– 36.60	– 100.00
Greece	17.00	15.40	17.20	20.10	21.00	0.00	0.00	– 17.00	– 100.00
Spain	29.80	30.80	33.90	34.80	38.00	40.50	36.70	6.90	23.15
France	37.70	39.70	39.70	40.70	41.00	41.70	43.80	6.10	16.18
Croatia	14.70	16.40	21.00	25.30	30.20	29.50	31.40	16.70	113.61
Italy	38.40	41.60	45.90	49.80	51.40	51.40	51.90	13.50	35.16
Cyprus	12.50	14.80	16.10	16.70	16.60	16.60	15.30	2.80	22.40
Latvia	14.60	27.00	25.20	25.20	41.00	39.70	44.10	29.50	202.05
Lithuania	23.50	30.50	48.00	52.60	49.70	45.30	44.30	20.80	88.51
Luxembourg	47.40	47.70	49.20	49.00	48.90	52.80	55.30	7.90	16.67
Hungary	25.50	30.50	34.70	37.40	35.90	32.00	34.90	9.40	36.86
Malta	14.80	11.70	12.70	10.40	9.10	10.90	13.60	– 1.20	– 8.11
Netherlands	49.40	50.90	53.50	55.90	56.90	56.90	57.80	8.40	17.00
Austria	57.70	56.30	57.60	57.70	58.20	62.30	62.50	4.80	8.32
Poland	12.00	26.50	34.80	34.30	34.10	38.70	40.30	28.30	235.83
Portugal	26.10	30.40	30.90	29.10	28.90	26.80	30.40	4.30	16.48
Romania	14.80	13.10	13.40	11.10	11.50	11.90	11.30	– 3.50	– 23.65
Slovenia	42.10	36.00	55.50	58.90	59.20	59.30	60.80	18.70	44.42
Slovakia	13.40	10.40	23.00	36.30	38.50	45.30	48.90	35.50	264.93
Finland	33.30	32.50	42.10	42.30	43.50	42.10	39.00	5.70	17.12
Sweden	46.90	49.30	48.40	45.80	46.60	38.30	39.50	– 7.40	– 15.78
Iceland	28.00	29.70	33.20	27.50	0.00	26.20	26.40	– 1.60	– 5.71
Norway	39.80	42.20	38.20	40.70	40.90	41.00	38.20	– 1.60	– 4.02
Switzerland	50.00	53.50	52.50	52.50	53.00	52.80	53.30	3.30	6.60
United Kingdom	42.60	43.40	44.00	44.10	0.00	0.00	0.00	– 42.60	– 100.00
Serbia	0.00	0.70	0.30	0.30	0.00	15.40	16.80	16.80	
Türkiye	0.00	0.00	9.20	11.50	11.50	12.30	12.30	12.30	
Albania	0.00	0.00	0.00	0.00	0.00	18.10	18.70	18.70	

where W_m is a waste management operations; E_{ce} is an effect of the circular economy; PI_{ce} is a private investment and gross added value related to circular economy sectors; $ETax$ is an environmental tax revenues, million euros;

$$E_{ce}^{AI} = \frac{W_m}{100} * (PI_{ce} + ETax) * \left(1 + \frac{AI_{nace} + AI_{ent}}{2}\right), \quad (3)$$

where E_{ce}^{AI} is an effect of the circular economy, taking into account the effect of convergence with artificial intelligence; PI_{ce} is a private investment and gross added value related to circular economy sectors; $ETax$ is an environmental tax revenues, million euros; W_m is a waste management operations; AI_{nace} is an artificial intelligence by NACE Rev.2 activity; AI_{ent} is an artificial intelligence by size class of enterprise.

In our opinion, to identify the impact of artificial intelligence on the development of the circular economy, we offer a formula for calculating the economic effect, which is based on determining the impact of the implementation of artificial intelligence with available funding (private investments and environmental taxation) on achieving a synergistic effect for the development of the circular economy.

According to the author's methodology Ranking Effect of the circular economy, taking into account the effect of convergence with artificial intelligence, EU, 2021 (Fig. 4).

Germany is the leader among the EU countries that implement waste management operations (recycling) at enterprises, and the top countries include: the Netherlands, Italy, France, Denmark, Belgium, Austria, Spain, Sweden. As evidenced by the results of the scientific-methodical approach to assessing the material effect of the circular economy, in the case of convergence with artificial intelligence, the volume

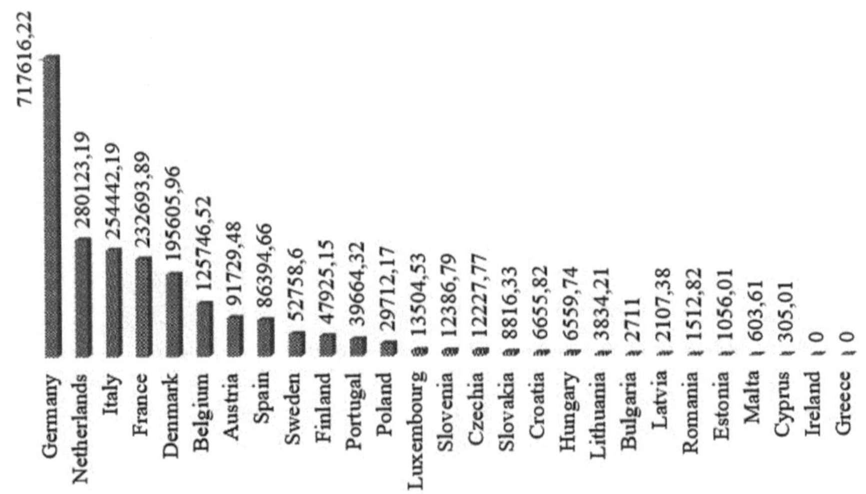

Fig. 4 Ranking Effect of the circular economy, taking into account the effect of convergence with artificial intelligence, EU, 2021 (million euros of enterprises)

of financing and the pool of opportunities of the circular economy become important, since it is Germany that has one of the largest indicators of waste management operations, respectively, which are required for the implementation of various types of investments.

5 Conclusions

The convergence of artificial intelligence and the circular economy is primarily aimed at increasing the level of efficiency in the use of resources and reducing the environmental impact of enterprises on the environment as a whole. Structural modeling and systematization of AI algorithms in circular economy models make it possible to highlight the innovativeness of cyclical business models at different stages of creation, and the implementation of dynamic pricing based on AI will stimulate the behavioral aspects of top management to make effective management decisions and deepen the extended responsibility of manufacturers. AI in circular economy models will contribute to the development of the waste management system, optimization in circular logistics, accurate demand forecasting, implementation of resource planning, improvement of product design to enhance the ergonomics of products and their packaging, implementation of exchange platforms and much more.

Acknowledgements The article was prepared within the framework of the National Research Foundation of Ukraine project "New geostrategic threats to human social security in a hybrid war and ways to prevent them" № 2021.01/0239 and University of Zielona Góra.

References

1. Chauhan C, Parida V, Dhir A (2022) Linking circular economy and digitalisation technologies: a systematic literature review of past achievements and future promises. Technol Forecast Soc Change 177. https://doi.org/10.1016/j.techfore.2022.121508
2. Wilson M, Paschen J, Pitt L (2022) The circular economy meets artificial intelligence (AI): understanding the opportunities of AI for reverse logistics. Manag Environ Qual Int J 33(1):9–25. https://doi.org/10.1108/MEQ-10-2020-0222
3. The Circularity Gap Report Austria 2019 Circle Economy & ARA
4. The Circularity Gap Report: An analysis of the circular state of the global economy 2018 Circle Economy. www.circularity-gap.world/2018
5. Roberts H, Zhang J, Bariach B, Cowls J, Gilburt B, Juneja P, Tsamados A, Ziosi M, Taddeo M, Floridi L (2022) Artificial intelligence in support of the circular economy: ethical considerations and a path forward. AI Soc. https://doi.org/10.1007/s00146-022-01596-8
6. Varnalii Z, Kulyk P, Nikytenko D, Cheberyako O, Hurochkina V (2022) Circular economy implementation as a strengthening factor in the economic security of Ukraine during the post-war period. In: 2nd International scientific conference on environmental sustainability in natural resources management (ISCES 2022), 01 Nov 2022, Riga, Latvia. https://iopscience.iop.org/issue/1755-1315/1126/1

7. Artificial intelligence by size class of enterprise (online data code: ISOC_EB_AI) Source of data: Eurostat. URL: https://ec.europa.eu/eurostat/databrowser/view/isoc_eb_ai/default/table?lang=en
8. Artificial intelligence by NACE Rev.2 activity. (online data code: ISOC_EB_AIN2) Source of data: Eurostat. URL: https://ec.europa.eu/eurostat/databrowser/view/ISOC_EB_AIN2/default/table?lang=en&category=isoc.isoc_e.isoc_eb
9. Correspondence table between the indicators of the OECD database on ICT Access and Usage by Businesses and the variables used in the Eurostat database. oecd.org. URL: https://stats.oecd.org/fileview2.aspx?idfile=988d6e51-8b82-4109-8b0a-2a74011bbb09
10. Ghoreishi M, Happonen A (2020) Key enablers for deploying artificial intelligence for circular economy embracing sustainable product design: three case studies. In: AIP conference proceedings, vol 2233. American Institute of Physics Inc. https://doi.org/10.1063/5.0001339
11. Ali ZA, Zain M, Pathan MS, Mooney P (2023) Contributions of artificial intelligence for circular economy transition leading toward sustainability: an explorative study in agriculture and food industries of Pakistan. Environ Dev Sustain. Springer Science and Business Media B.V. https://doi.org/10.1007/s10668-023-03458-9
12. Sysoieva I, Pozniakovska N, Mikluha O, Pukas A, Roleders V (2023) Social audit as a tool of civil society aimed at ensuring the sustainability. IOP Conf Ser Earth Environ Sci 1126(1):012031. https://doi.org/10.1088/1755-1315/1126/1/012031
13. Digital Debate (2023) The impact of artificial intelligence (AI) in urban environments. Mott Macdonald, 28 July 2023. URL: https://www.youtube.com/watch?v=NDCvJs-bwE0
14. Ellen MacArthur Foundation (2013) Towards the Circular Economy: Economic and Business Rationale for an Accelerated Transition, Cowes, UK
15. Watson R, Wilson HN, Smart P, Macdonald EK (2018) Harnessing difference: a capability-based framework for stakeholder engagement in environmental innovation: harnessing difference. J Prod Innov Manag 35:254–279
16. Kryvinska N, Bickel L (2020) Scenario-based analysis of IT enterprises servitization as a part of digital transformation of modern economy. Appl Sci 10(3):1076. https://doi.org/10.3390/app10031076
17. Stoshikj M, Kryvinska N, Strauss C (2013) Efficient managing of complex programs with project management services. Glob J Flex Syst Manag 15(1):25–38. https://doi.org/10.1007/s40171-013-0051-8
18. Voynarenko M, Hurochkina V, Dzhedzhula V, Yepifanova I, Menchynska O (2021) Applying fuzzy logic to modeling economic emergence. WSEAS Trans Bus Econ 18:424–433. https://doi.org/10.37394/23207.2021.18.43

Urban Planning Principles and Models of Reconstructive Restoration of Urbanized Territories of Luhansk Region Destroyed by Hostilities, Based on Innovative Development

Dmytro Shvydkyi, Vitaliy Shvets, Kostiantyn Sokolenko, Yevhen Rudniev, and Valeriy Sokolenko

Abstract The reconstructive restoration of urbanized areas of the Luhansk region destroyed by hostilities is a complex urban planning problem, since the scale of destruction is comparable to the size of a median European country. The principles and models of reconstructive restoration of urbanized territories of the Luhansk region, damaged by hostilities in the conditions of progressive technical degradation of buildings and structures, deepening discrepancy between the consumer and operational qualities of buildings with modern requirements and standards, taking into account the realities of the socio-economic situation in the region, on the basis of innovative development, are investigated. The basic urban planning principles and conceptual foundations for the restoration of urbanized areas and cities that have been destroyed are determined for the tasks of updating the system of urban settlements and the planning structure of the eastern region of Ukraine: safety, environmental friendliness, autonomy, energy efficiency, convenience, comfort, transport accessibility, and the creation of places of employment. It is proved that in the current conditions of post-war reconstruction there is a threat of primitivization of the territory and the formation of the so-called budget trap in cities. It is specified that the standard algorithm of urban planning design is complicated. The development of models and principles for the restoration of territories and settlements with different parameters

D. Shvydkyi · V. Shvets
Vinnytsia National Technical University, Vinnytsya, Ukraine
e-mail: v.shvets@vntu.edu.ua

K. Sokolenko · Y. Rudniev · V. Sokolenko (✉)
Volodymyr Dahl East Ukrainian National University, Kyiv, Ukraine
e-mail: sokolenko_1@snu.edu.ua

K. Sokolenko
e-mail: sokolenko@snu.edu.ua

Y. Rudniev
e-mail: rudnev_es@snu.edu.ua

A. Semenov et al. (eds.), *Data-Centric Business and Applications*, Lecture Notes on Data Engineering and Communications Technologies 272,
https://doi.org/10.1007/978-3-032-06608-4_3

and typological signs of destruction in order to formalize practical issues is an important task. This is a prerequisite for the creation of a bottom-up regulatory framework for recovery processes. The construction of matrices for analyzing planning units of the destroyed urban areas with a view to their restoration constitutes an updated basis for the restoration model. It is proposed to identify hierarchical levels and objects of reconstruction, generalize management entities and types of reconstruction activities, formulate elements and components of the reconstruction information package, and formalize design decisions for individual objects and elements of reconstruction. Methods for determining the effectiveness of residential area reconstruction in the context of post-war recovery should be differentiated in accordance with the defined goal and ensure innovative urban development.

Keywords City · Urbanized areas · Master plan · Urban development · Building typology · Destruction · Secondary construction waste · Recycling · Urban reconstruction · Reconstruction · Restoration

1 Introduction

The Donbas region of Ukraine is defined by two eastern oblasts—Donetsk and Luhansk. Its emergence and formation dates back to the 70 s of the nineteenth century, at the peak of the industrial revolution on the continent. It can be noted that the entire period of extensive development of the region had signs of semi-colonialism [1–3].

The resource was mineral deposits, primarily coal, ore, limestone, and refractory materials. The inland nature of the region and its convenient transport location were also important. The Industrial Revolution led to an explosive population growth and the formation of a powerful industrial complex, mainly mining and metallurgy. This resulted in a high level of urbanization of the territory, the formation of a dense network of urban agglomerations and communication routes. Two world wars that swept across the continent had a significant negative impact on the region. Imbalances in development and policy decisions led to the accumulation of a complex of urban planning, environmental, demographic, social and political problems that resulted in a deep crisis in the region. For a long time, Donbas has been teetering on the brink of crisis and catastrophe. The Russian aggression of 2014 plunged the region into chaos, and the situation can be assessed as catastrophic in terms of basic parameters. Statistical analysis reveals clear downward trends in the basic indicators of urban planning, demographic and economic conditions [4].

However, the end of the war promises the start of reconstruction and development of the affected areas. A methodological feature is the aspect that the greater the level of decline in basic indicators, the higher the indices and growth levels. This time interval has its limitations, and if the development momentum is not given at the initial stage, the situation may form a permanent crisis trend, which is unacceptable for the country. Crisis is a self-sustaining process that cannot produce development.

The priority task is to ensure innovative development of the region on the basis of comprehensive urban planning analysis.

Therefore, the issue and tasks of innovative development of the region are relevant, but taking into account the real situation, the available resource for the development of adequate urban development goals.

An analytical study of the state of the eastern regions allows us to identify a number of systemic problems, the most significant of which are [4–8]:

- reduction of the urban planning base, decline in industrial production;
- the established demographic trend of population ageing and shrinking of all age groups. Excess of the population mortality rate over the birth rate;
- unfavourable ecological condition of the territories, environmental pollution, growth of areas occupied by dumps, waste landfills, etc.;
- change in the function and shape of the region as a result of hostilities;
- large-scale destruction has led to the formation of landscapes of continuous destruction, depopulated and deprived of normal functioning;
- the formation of large-scale areas of destruction that require the implementation of a policy for managing construction waste from destruction.

Further, the study (chapter of the monograph) considers the concept of territorial restoration on the basis of advanced innovative development for the purpose of urban reconstruction of the region.

1.1 Problem Statement

The war in Ukraine has been going on for four years, with intense hostilities causing large-scale destruction every day. The prerequisites for global shifts are being formed, which initially start with the physical destruction of buildings, but subsequently lead to changes in the urban planning situation with catastrophic results. From a historical point of view, human civilization is an endless series of wars and armed conflicts, which were followed by processes of reconstruction and renewal, depending on the overall results of the next war. In the all-Ukrainian context, we should also expect the end of the phase of active hostilities, when the issue of rebuilding what has been lost will arise. Political statements such as 'we will rebuild everything that has been destroyed' are adjusted by life at the stage of considering what exactly is meant by this terminology. The fundamental approach is that the processes of destruction that have lasted for more than four years cannot be compensated for or rebuilt in a shorter period. Politicians and scholars are already formulating opinions that reconstruction will be a long process that will take years. The overall consequences may be felt for decades to come. There are obviously different hierarchical levels of objects and tasks of reconstruction, with a division by scale, duration, nature of impact, and actors involved. The complex problem of rebuilding the country requires an up-to-date scientific analysis of specific tasks and object solutions [9–11].

Restoring residential development in destroyed cities and towns in urbanised areas will pose a technical and organisational challenge. It lies in the scale of the destruction—the total destruction of neighbourhoods or blocks. The task is to determine the order of priority for renovation. The technical side of the problem is what rational scheme to choose for clearing the territory. Urban planning documentation, a project for detailed planning of a neighbourhood or quarter, operates within a timeframe of 2–3 years. If the task of clearing the territory of destroyed objects (buildings) is included, this will postpone the construction phase for the period of cleaning and dismantling. The country has no experience in solving such problems. The reconstruction of European countries after the Second World War is also not a relevant experience, as the typology of constructive solutions for building at the beginning of the last century had significant differences [12].

1.2 Analysis of Well-Known Studies and Publications

The experience of post-war reconstruction of countries is always specific and has an object-typological link. The overall socio-economic situation, financial and political conditions determine policy decisions (concepts). The analysis of the general principles and concepts of recovery and the results of regional development programmes is of undoubted scientific interest. Warning against wrong decisions of a fundamental nature is an important aspect of defining recovery tasks, models and approaches.

The Marshall Plan and the experience of rebuilding the European continent have been studied in detail in historical retrospect. The European Recovery Programme (ERP), more commonly known as the Marshall Plan, was implemented between 1948 and 1952. The USSR and the Eastern bloc countries refused to participate in the post-war reconstruction under the Marshall Plan for political reasons. Soon, political confrontation led to the emergence of the Cold War and the construction of the Iron Curtain. The USSR used an administrative-command model of development and governance, and the post-war reconstruction was based on the principles of so-called Soviet socialism.

As a result, the Marshall Plan led to the formation of the European Union. The United States gained a powerful ally during the Cold War, strengthening and recreating it on its own model. Thus, the main significance and result of the Marshall Plan assistance was not economic or administrative, but political. As is well known, in the competition between two political blocs and two political systems, the Atlantic Coalition emerged victorious. It is worth noting that Finland became a NATO member in 2024, thus closing the page on its longstanding neutrality.

A systematic analysis leads to the following conclusions. The Marshall Plan was the second of the known options for post-war reconstruction. It was put into effect 2 years after the end of the war and lasted for 5 years. The plan covered 18 participating countries (excluding the United States). The results of the plan's implementation were successful, but the reconstruction of the countries affected by the Second World War was not yet complete at the end of its validity.

The USSR used an administrative-command model of development and governance, and the post-war reconstruction was based on the principles of so-called Soviet socialism [13]. Ukraine and Belarus were the republics that suffered the most from the Second World War. Their territory was completely occupied. The front line moved across Ukraine twice. The consequences of the war were catastrophic. The building and design features at that time were characterized by manor, low-rise and partially medium-rise buildings in cities. The buildings were mostly made of stone and wooden structures. The rural population made up the majority—up to 70%—and thus served as a human and urbanization resource. The village was also self-sufficient in food and was harshly exploited in collective farms. In the context of the command economy, the primary tasks of reconstruction were to restore heavy industry, support the course of militarization, and the arms race. Five-year plans were developed in the paradigm of industrial growth. The social sphere was of a derivative nature, as was housing construction. The first steps in solving the housing problem were the so-called 'Khrushchevkas', the houses of the first mass production series of industrial construction. Mass housing construction of 4–5-storey buildings, commonly known as 'Khrushchevkas', began in 1957 - the housing problem was addressed 12 years after the end of the devastating war! As a result, from 1957 to 1963, the housing stock increased from 640 to 1184 million square meters of living space (i.e. almost twice as much), and more than 50 million people improved their living conditions. However, we should not forget about the functional and planning limitations of this development, which became apparent 30–40 years later.

In this way, the USSR restored its industry in a relatively short time at the cost of low living standards for its population. Structural systemic distortions eventually led to the collapse of the socialist system. The following problems can be identified from the considerable list of distinct problems caused by erroneous decisions in the context of the urban planning situation: the unresolved problem of houses of the first mass series of industrial development; chaotic development of cities, agglomerations, and industrial areas; formation of depressed areas of mono-industrial use; problems with water supply in urban areas; extreme technogenic load and environmental pollution of large areas of the territory; economic inequality in the development of territories [14–16].

The experience of rebuilding the Balkan countries after the collapse of socialist Yugoslavia is less relevant, as the scale and conditions of the conflict are not comparable.

With a total area of 603,628 square kilometers, Ukraine is the largest country in Europe. According to various estimates, 21% of its territory is occupied, which is approximately 126 thousand square kilometers. Out of 46 European countries, 14 are larger than this; 32 countries are smaller than the territory of Ukraine that is currently occupied. The scale of the destruction is also comparable to the volume of construction in some countries on the continent.

Thus, the factors and prerequisites that determine the directions, models and approaches to recovery are:

- the scale of destruction;

- the level of urbanization of the territory,
- the level of industrialization of the territory and the density of engineering systems and networks,
- trends in the demographic development of the recovery region,
- construction and design solutions of settlements,
- the presence of significant architectural and cultural heritage,
- the existence of a formed state policy towards recovery,
- the presence of a responsible civil society.

Financial and technical resources are also important.

1.3 Research Methods

The analysis and study of the place of reconstruction activity in the context of different phases of the life cycle of cities was carried out by the methods of comparative analysis, graph-analytical analysis, experimental design, systematic analysis of factors and conditions that determine the nature of the restoration of areas of mass housing development affected by hostilities; generalization of data obtained as a result of statistical studies. The information base of the study is the data obtained as a result of a statistical assessment of the scale and nature of the destruction in the eastern region of Ukraine. The object of the study is the regions of eastern Ukraine, in particular Luhansk region, cities and urbanised areas that have suffered large-scale destruction as a result of hostilities. The identification of aggregated territories that require the development of post-war recovery tasks was carried out using a typological analysis of the cities of the region and a graphical analysis of the scale and consequences of the destruction based on available information sources.

2 Characteristics of the Urban Planning Situation in Luhansk Region at the Present Stage

Luhansk region is 275 km long from north to south and 170 km from west to east, covering an area of 26.7 thousand square kilometers, which is 4.4% of the total territory of Ukraine. As of the beginning of 2022, the total population of the region was approximately 2,102,921 thousand people, which is 5.7% of the total population of Ukraine. Of these, the majority (87.2%) live in urban areas, while the rural population accounts for 12.8%. The population density in the region is expressed as 83.9 persons/km^2 as of 2014. Figure 1 shows the topography and zoning of the region.

Ukraine, in particular the Luhansk region, has long been experiencing a process of urbanization, characterized by the growth of cities and their increasing importance in public life. A high level of urbanization is typical for the Donbas region, where more than 80% of the population lives in urban areas. The settlement system

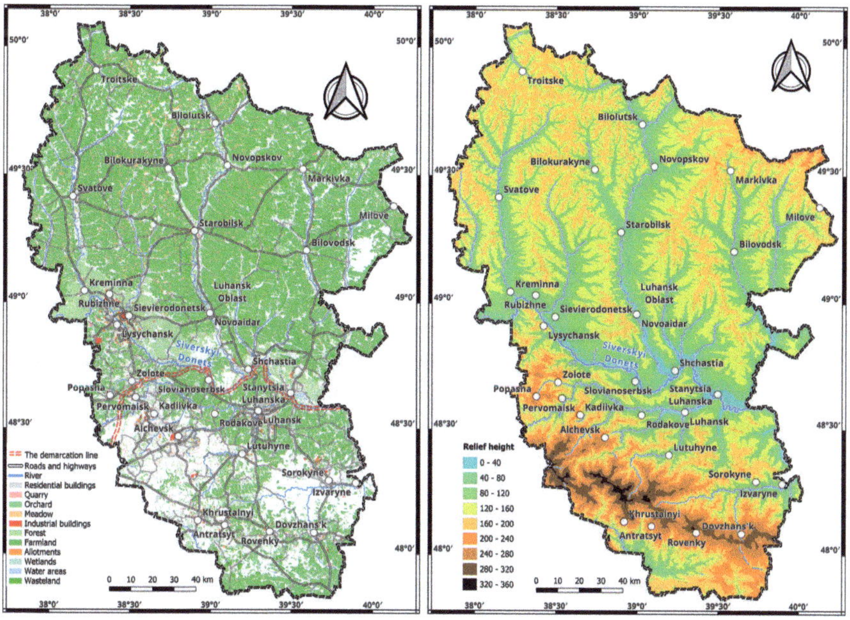

Fig. 1 Zoning scheme **a** and relief map **b** of the Luhansk region

in Luhansk Oblast was formed as part of a larger region of Eastern Ukraine, in particular its central urbanised zone, which also includes Donetsk Oblast. As a result of the historical process, the region is significantly distinguished by a dense network of settlements, industrial and utility zones, as well as transport and engineering communications, which form a complex macro-structure of the region. The availability of hard coal, non-metallic minerals, cement, glass raw materials and other construction materials contributed to the development of coal mining, chemical industry and ferrous metallurgy in the region. The main factors that influenced the development of the settlement system in the region include:

- linear-nodal structure of settlement, which was formed in accordance with the location of mineral deposits—mainly coal;
- sectoral structure and territorial concentration of the region's industry are determined by natural, geographical and historical conditions;
- the presence of several urban agglomerations;
- high population density in large cities;
- slow development of smaller urban settlements;
- a significant amount of pendulum migration;
- uneven distribution of cities across the region, with concentrations in the western, central and southern parts;
- disproportions in the size of settlements, with a significant part of the urban population concentrated.

The formation and development of the region's cities depended on the development of key industrial sectors, with large cities having a diversified structure, while smaller and medium-sized cities were specialized in one or more activities [17, 18].

The demographic situation in the country is generally characterized by sharply unfavorable and negative trends in population structures. Trends in the labour market in 2014–2021 include:

- shortage of skilled workers and young people due to labour migration, ageing of the available skilled workforce;
- lack of competitive wages;
- the duration of hostilities in the region, which is unpredictable and unstable;
- deterioration of the economic situation of city-forming enterprises—reduction of the urban planning base, reorientation from an industrial economy to a predominantly agricultural one;
- strengthening of the role of small businesses (mainly trade);
- shady employment and wages;
- low quality of educational services provided by vocational schools,
- inability to provide professional training in demanded professions;
- rising energy costs;
- lack of quality roads;
- remoteness of the region from markets.

Cities and agglomerations concentrate large industrial areas with convenient geographical location and sufficient engineering and transport infrastructure (Fig. 1). The above serves as a prerequisite for the creation of industrial clusters, which will help restore and rebuild the region's economy, create jobs, reduce migration, and create competitive products.

In the region, the largest amount of waste is generated by the coal mining industry and is concentrated in the form of waste heaps—spoil heaps that are sources of dust and environmental pollution. Covering a large area, waste heaps take land out of economic use. Most of the dumps are concentrated within the region's coal mining agglomerations (Fig. 2).

The absence of solid waste processing facilities, the concentration of waste in landfills with an increase in their number, the growing number of unauthorised landfills and the overloading of existing ones in the complex leads to an excessive man-made burden on the environment.

Figure 3 shows a map of closed mines in Luhansk Oblast. It forms the industrial framework of depressed industrial areas [19]. The few mines that are still operating are approaching critical age (100 years), and their economic and production prospects are questionable.

Figures 4 and 5 show how economic activity in the territories has degraded. The area affected by the hostilities is characterized by landscape degradation and reduced nighttime lighting levels.

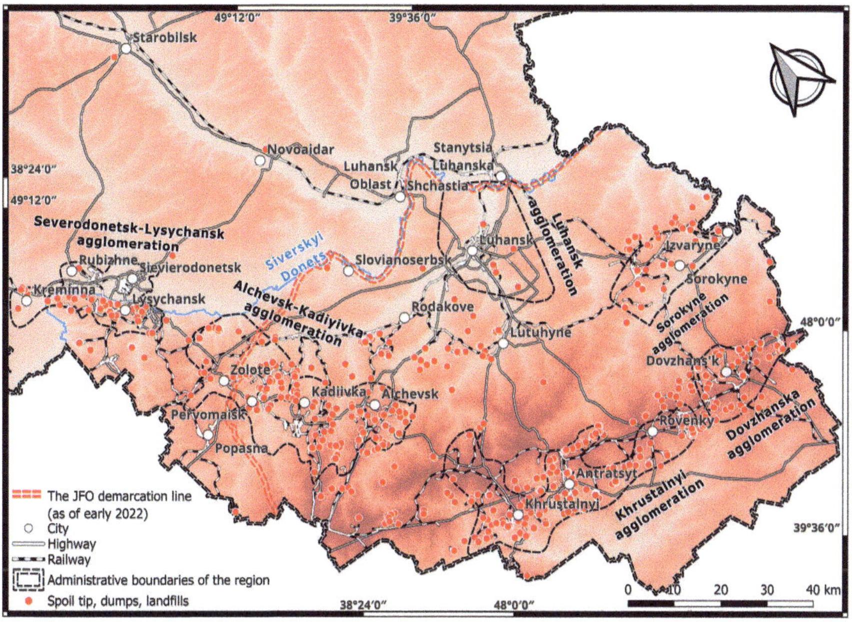

Fig. 2 The map of the concentration of spoil tips and landfills within the Luhansk region

2.1 Interim Conclusions

The Luhansk region is generally relevant to the industrial regions of Ukraine in terms of its characteristics and parameters. The level of urbanization in the region exceeds 80%. Reurbanisation is a threat of an archaic crisis spreading to depressed areas. The region is characterized by a dense network of settlements, industrial and utility zones, transport and engineering communications. The main factors that influenced the development of the settlement system in the region were: high population density in large cities; slow development of smaller urban settlements; significant pendulum migration; uneven distribution of cities across the region, with concentration in the western, central and southern parts; disproportions in the size of settlements, with a significant part of the urban population concentrated. Even before the conflict in 2014, the region had been experiencing steady downward trends in demographics, income levels, and production volumes. The situation in general has undergone a significant phase shift since the beginning of the conflict. In fact, Luhansk region is turning into an agricultural region. The level of environmental pressure is extremely high. Given the consequences of the war, a zone of continuous destruction is being formed in the region, which runs along the frontline. In some districts and communities, the amount of destruction exceeds the buildings that have survived.

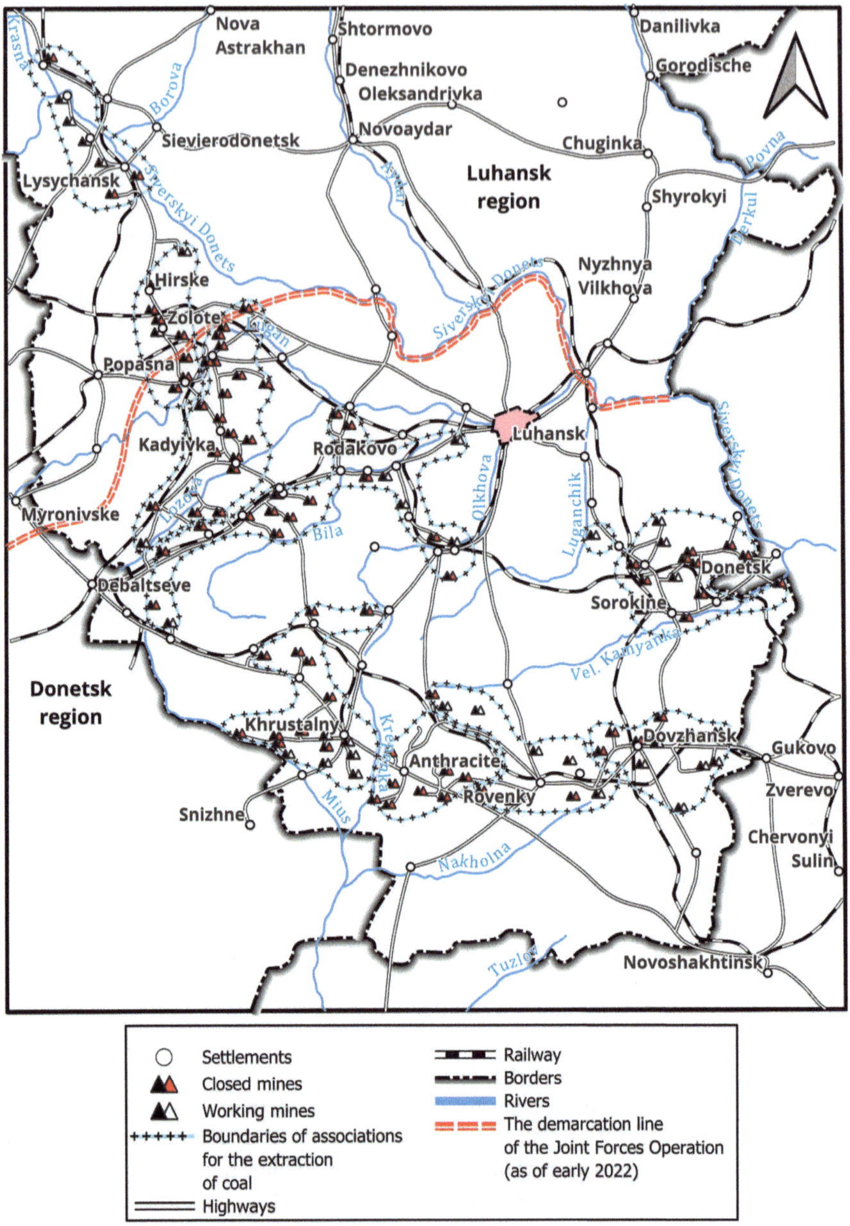

Fig. 3 Map of closed and operating mines in the Luhansk region

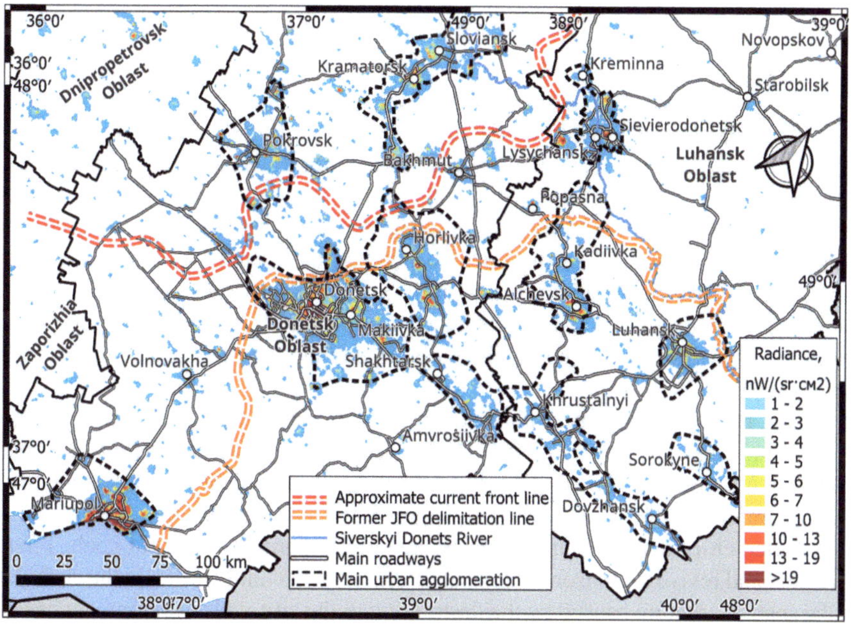

Fig. 4 Maps of night lighting on the territories of Donbas region, before the war in 2021

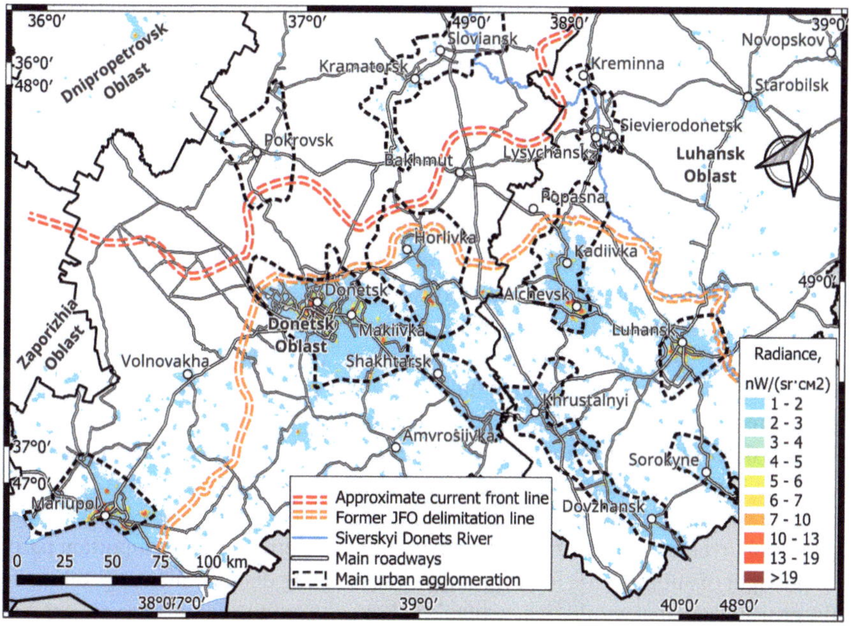

Fig. 5 Maps of night lighting on the territories of Donbas region, after the war in 2023

3 General Approaches and Concepts of Reconstruction

The professional urban planning community has developed an understanding that only targeted design aimed at meeting the socio-cultural needs of collective and individual residents can create a sustainable, comfortable living environment [20, 21].

Domestic experience is based on a prescriptive approach. Urban planning conditions were determined by management structures that planned and controlled the financing, development and construction of cities. Social issues were relegated to the background. And as has been repeatedly emphasised, the quality of the urban environment was deteriorating significantly. Problems were accumulating that still have no solution.

Today, the degree of uncertainty has increased even more. The consequences of the war. Unstable economy. The challenges of the social dimension of urban planning. A particularly acute problem is attracting investment in the implementation of programmes for the restructuring and restoration of the urban environment.

Affected cities that are being reconstructed must address two sets of goals. First, to restore and revive cities from crisis and destruction. Secondly, to adapt the changes in the spatial structure caused by the destruction to the city-wide development needs and social interests of the population. These needs and interests are diverse and often contradictory. Planning tasks should implement a social approach to solving them. In this case, the level of involvement of citizens in the development of solutions and support for successful projects increases. The votes of the voters in the local elections become really important. In cases where the reconstruction is focused on off-budget financing, the alignment of private investors' interests with the social attitudes of the population is of particular importance. Conflicts of interest between the city and the investor are resolved through planning regulations that govern the activities of private investors and define the basic rules and restrictions of development.

In market conditions, urban development projects are mostly implemented in the format of political and financial decisions rather than an architectural and spatial programme for the development and reconstruction of urban areas. Attractive urban objects—land, buildings and structures—become the subject of bidding and transactions, and investors are attracted on their basis. In modern market relations, an investor does not invest in a project unless its legal purity is ensured, unless urban planning regulations and existing restrictions are agreed upon.

The task is to find a compromise between the participants in urban planning processes. The parties are the city executive committees that defend the city's interests and the residents of the district, as their interests are often not the same as the city's. A potential investor in reconstruction and development projects is an important party with the right of subjectivity. In the process of developing detailed planning solutions at the project level that would be acceptable to both city authorities and private capital, the participation of a potential investor is a prerequisite for the success of the project.

In urban planning practice, two schools of compromise have emerged: the so-called European and the American. The European approach is aimed at social planning and is designed for centralised implementation, mainly with the help of state and municipal capital. The American approach involves regulating reconstruction through strict restrictions on the terms of sale and resale of urban land plots, but always with the participation of local governments and the population of the area. Ukraine has traditionally implemented the first type of decision-making. However, in the context of reliance on international assistance, the thoroughness and depth of urban planning decisions becomes especially important in terms of defining urban planning restrictions.

The process of implementing urban planning solutions in urban regeneration can be divided into three stages. The first stage is the development of an urban development policy strategy and a reconstruction concept. A programme is being prepared that regulates the mechanism for managing urban development and interaction with private investors. The second stage, the so-called 'design management', is aimed at organising the functional and planning space and developing measures aimed at restoring or transforming the building. At this stage, maximum involvement of all participants in the process is required to agree on the conditions, restrictions, and parameters of the restoration. The third stage of managing the implementation of urban development plans requires an individual approach to the development and implementation of individual local urban development plans.

For cities and settlements that have suffered large-scale destruction and severely restrict reconstruction measures, detailed planning projects can be developed when the customer is identified, the terms of reference are formed, and the technical conditions are determined. In this scenario, there is a pattern: the more restrictions are included in the project, the more difficult it is to implement a separate local plan. And this is where flexible plans are needed that can adapt to the individual needs of many customers.

The three objectives of residential redevelopment at the planning level of limited areas include:

- the rehabilitation of buildings and the living environment;
- social restructuring of planning and cultural services;
- and the renovation and modernisation of buildings and inner-quarter land plots.

Over time, and as buildings age, landscaped neighbourhoods become a haven for newcomers. This is the experience of a number of cities in Germany, France and other countries. In 40–50 years, the renovated urban areas were inhabited by immigrants. This confirms the principle that nature does not tolerate emptiness. But such areas become disadvantaged. A recursion of degradation is formed because investors leave disadvantaged areas.

Wave reconstruction is a method of preserving the indigenous population and strengthening the community core. For cities affected by hostilities, this principle seems reasonable and allows for maximum involvement and control of the recovery process by the community and residents.

The demographic situation in the cities was immediately distorted and will continue to be affected for a long time. With the overall decline in the population, the number of non-active, non-employed residents, such as pensioners and people with disabilities, is increasing disproportionately. At the same time, the number of children is decreasing. Design developments are aimed at ensuring individual characteristics of the living environment.

In the current conditions of developing plans for the restoration, development or reconstruction of the territories of destroyed cities, a new situation arises related to the tasks of land redistribution and changing the functional purpose of plots. Both for public needs and for private investors, there is a need to redistribute land use boundaries and implement urban development plans for a particular city territory. In this case, plots previously leased or owned on a long-term basis are withdrawn or bought out in accordance with civil and land laws. Accordingly, updated territory delineation projects are required. They are developed for built-up areas and other planning elements of the city in order to establish the boundaries of land use areas.

4 A Typological Analysis of the Destruction and Changes in the Territories of Ukraine Affected by the Hostilities

The main zones of cities are rural and industrial. Their location and contours determine the structure of the city's planning framework. The logic of war leads to the fact that industrial zones are subject to large-scale destruction due to their use as defence strongholds. At the same time, the transport infrastructure is being destroyed. Residential buildings and rural areas are being destroyed depending on their orientation in the offensive and defensive directions [22]. The ownership and affiliation of the destroyed buildings and facilities determines the conditions and parties to reconstruction. Industrial enterprises mostly have a private, corporate or mixed owner. Thus, reconstruction issues form a circle of corporate interest and decisions. Transport infrastructure is mostly publicly managed, with separate budgets and financing schemes. Residential development is the most diverse in terms of consumers and owners. In addition, housing is a socially sensitive part of real estate in settlements. The typological analysis of residential development [12] should be supplemented by the formation of land plots and zoning in detailed plans for the development of neighbourhoods and city blocks, to form a generalised criterion for the feasibility/possibility of rebuilding or reconstruction.

4.1 *Technical, Territorial, Environmental, Political, and Social Consequences*

The scale factor can have a decisive influence on the choice of rational restoration options.

For example, the destroyed private housing in Rubizhne, shown in Fig. 6, indicates the need for a comprehensive urban planning solution. The complete destruction of a block of private housing, including utilities. The residents were apparently displaced completely at random, and the possibilities and conditions for their return are questionable. Possible options include clearing and rebuilding the plots, changing the type of construction (multi-storey), changing the purpose of the plot with a change of functional use. Any urban renewal is possible only if there is a political settlement. There is a threat of a ruinous area. A socially important question is whether it is acceptable to have a new type of building zoning in the urban environment—ruins? According to the authors, it is necessary to determine the legal status of such sites and carry out reclamation of the territories.

An unnatural situation of the catastrophe of settlements is being formed, associated with an instantaneous (by historical standards) change in function and form [23].

Reconstruction should be based on the principles of advanced innovative development of growth clusters. The basis of development can be defined as the following resources.

Fig. 6 Destroyed estate buildings in Rubizhne

4.2 Urban Planning Principles and General Approaches to the Principles of Restoration of Urbanized Areas and Cities that Have Suffered Destruction

Three years of war is a long time, including for drawing interim conclusions. It can be assumed that military-civilian administrations—city, regional (MCA)—are not development tools. A military-civilian administration is created as a tool for crisis management in a difficult period of martial law or a state of emergency. For obvious reasons, the main functions of the MCA are to respond quickly to the current problems of the cities and regions. Military, security, social, and technical issues are given high priority. The administrations are working on programmes to restore the occupied territories. The main areas of these programmes are the establishment of administrative management, security measures, ensuring critical functions of the municipal economy—water, electricity, communications, gas and heat supply, and the restoration of social services. In other words, the focus is primarily on the urgent tasks of normalising the functioning of the territories. Regional and urban development is a much more complex task, operating in timeframes of decades, and is based on development programmes that define general principles, foundations and concepts of recovery. Unlike plans, a programme is not focused on specific indicators or targets. Programmes are aimed at implementing goals, achieving certain structural changes, and achieving qualitative parameters of the urban environment.

The international experience of post-war reconstruction has both positive and failed examples of reconstruction, due to a variety of reasons. Scholars and politicians agree on the basic postulate that reconstruction cannot be successful without community involvement, as evidenced by the modern European experience. There is a certain conflict between the administratively oriented management inherent in military-civilian administrations and the broad participation of local communities in the reconstruction of the destroyed territories. However, crisis management can provide an impetus—if a scientific and methodological approach is laid down, the proper conditions are created, and tasks and management structure are formed—in the form of a working group for urban development. The working group, a structural form (we are talking about an administrative-territorial unit—a city or an urban community), should definitely be headed by a chief architect, an urbanist with the appropriate professional education. The formalised goal at the first stage of crisis management is to update the city's master plan. The master plan as a document formalises the rules and objectives of urban planning activities.

The development, updating and approval of master plans is carried out in accordance with the procedure established by law and other regulatory legal acts [24–30]. It should be emphasised that the master plan is adopted by the local self-government body [28] and is designed for a period of up to 20–25 years. In this way, the local self-government body has an impact on the perspective directions of the city's development, based on the principles of market competition.

As a regulatory act, the master plan must comply with [31] (Article 4 of the Law 'On the Principles of State Regulatory Policy in the Field of Economic Activity'), i.e. be expedient, effective, balanced and predictable. It is the master plan that should systematise relevant information on demographic, social and economic parameters that characterise the actual state of development (decline) of the local community. Prospects for future community development programmes and plans, territorial areas of priority urban renewal and/or expansion should be based on realities. The demand and proposal for an investor at the stage of making a decision to invest is based on information from the master plan. Information on the housing market, real estate, non-residential facilities and plots is very sensitive.

The master plan allows us to correct the shortcomings caused by the so-called market failure. These include the fair price of land plots; the unwillingness of private investors to invest in the social and environmental components, which causes a decline in interest and ultimately reduces the attractiveness and value of land plots; and the problem of information asymmetry, when internal businesses are more knowledgeable than external investors. Having a comprehensive, strategically balanced and scientifically sound master plan will minimise these shortcomings.

For the areas of recovery affected by the hostilities, the main problems in urban planning are the reduction of the urban planning base: the number of people, industry, and degradation of transport infrastructure. Deterioration of basic indicators—quantitative and qualitative. That is, the distinguishing feature is recovery in the face of contraction. Nevertheless, urban planning principles such as improving the convenience of the population, the quality, safety and comfort of the environment, transport accessibility, and the availability of a range of services in accordance with regulatory indicators must be observed in the restoration tasks. The experience of restoring degraded (depressed, destroyed) areas shows that it is advisable to invest in places of employment, roads and engineering infrastructure, and to ensure the preservation of the natural environment.

The master plan addresses the issues of social and social development and improvement of the quality of the urban environment by planning the optimal structure of roads, pre-school and school education institutions, a system of green spaces, parks, and public recreation areas. The master plan sets out restrictions and environmental conditions for environmental protection.

The current regulatory and legal acts contain provisions on revitalization [32–34]. The problem is the emergency nature of the situation—a short time to develop plans and objectives. It is not enough to calculate the damage. Development and justification of reconstruction—it is advisable to start with master plans of settlements and communities as a basis.

The principles that should be considered when developing the main components of the master plan are: safety, environmental friendliness, autonomy, energy efficiency, convenience, comfort, transport accessibility, places of employment [35, 36].

In the current conditions of post-war recovery, there is also a threat of primitivisation of the territory and the formation of a so-called budget trap in cities. There are few (if any) examples of the renewal and transformation of single-industry towns. Factor #1—an active share of the population is leaving the territory. These are young people—students, high-level services (including medicine), and business. Even in the wealthy United States, there is a so-called 'rust belt' of cities that have fallen into decline since the end of the industrial period.

Studies [37, 38] raise the issue of common scientific and practical approaches to recovery tasks. In particular, it is noted that there is a certain terminological discrepancy between the categories of 'recovery', 'recovery plan', 'comprehensive recovery programme'. The basic challenges and reservations are the conditions and circumstances created by the war:

1. The population has decreased, and this negative trend continues.
2. A part of the industrial potential has been destroyed, for single-industry towns and outdated production cycles, and the economic feasibility of their reconstruction is questionable.
3. Infrastructure has been partially destroyed, but this problem is compounded by a decline in the population density of the territorial community and settlements. The task is to optimise and balance capacities in accordance with the population (demand), including for the future.
4. There is a need to define general approaches to preparing a decision on the reconstruction—full or partial—of the relevant settlement, the development of such a settlement in another location, or its liquidation, clearing and reclamation of the territory.
5. The task is to identify the bodies that prepare and make such decisions.
6. In order to determine the current priorities for reconstruction, it is necessary to coordinate planning (urban development) documents at the local, regional and national levels.
7. It is necessary to develop models, methods and principles of recovery for territories and settlements with different parameters and typological signs of destruction in order to formalise practical issues that arise at the recovery stage. This is a prerequisite for the creation of a bottom-up regulatory framework for recovery processes.

The content of the Programmes and, accordingly, the initial information for their development and processing contains a set of interrelated issues. At the initial organisational stages of reconstruction, the lack of qualified specialists, limited time to develop urban planning solutions, difficulty in correctly determining the scale, nature and consequences of the destruction, uncertainty of legal issues of property rights, succession, etc. will be burdensome. A general model for the restoration of settlements affected by large-scale destruction is shown in Fig. 7.

It is proposed to define hierarchical levels and objects of reconstruction, summarise the subjects of management and types of reconstruction activities, formulate elements and components of the reconstruction information package, and formalise design decisions on individual objects and elements of reconstruction.

Figure 8 shows a fragment of the matrix for analysing planning units (based on the example of Rubizhne). The city's general plan makes it possible to identify planning units that have common features, territorial delimitation and are combined by the nature of destruction by typological features. Generalised, deductive solutions to the

Definition of the settlement/administrative territorial community reconstruction object	**Definition of the governing entity - Military-Civilian Administration, Regional Military-Civilian Administration, City Council.**	**Development of a list of recovery packages - General Plan, Regional Development Strategies**	**Planning, design, coordination, approval of the restoration package - General Plan, Regional Development Strategies**	**Development/ reconstruction programmes and plans Of the settlement/ administrative territorial community reconstruction object**
Settlements, administrative and territorial communities or parts thereof	Working body, working group within the Military-Civilian Administration	Preparation of the Terms of Reference, technical specifications, and pre-project data.	Architects, City planners and urbanists, Utilities, Energy, Transport, Education, Social sphere, Healthcare, Taxation, State Emergency Service of Ukraine, National Police, City-forming enterprises, Business, Public associations	Streamlining the legal component of property rights, identifying owners
Characteristics that determine the tasks and priorities of reconstruction: Population; functions of vital support of the administrative-territorial community, district; industrial and economic significance of the regional or district level	Entities of all types of activity involved in the formation of urban planning decisions.	Defining the concepts, principles and objectives of reconstruction, taking into account regional realities.	Starting the business - signing contracts, agreeing on a schedule of activities, deadlines	Creating recovery programmes - list of objects
Conditional category by indicators (factors) of qualitative assessment % of the surviving housing stock; % of the population % of industry		Qualitative parameters: availability of free housing; lack of housing stock, analysis.	Search for investors and sources of funding. Updating upstream information, monitoring activities	The design part of the reconstruction projects. Estimated documentation. Contracts. Feedback - data updating, cadastres, databases

Fig. 7 Model of recovery of settlements affected by large-scale destruction

recovery strategy are defined in urban planning documentation at the city and community level. The separation of planning units and their independent study—with the development of reasonable options and directions for reconstruction—allows for the rapid accumulation of primary information. The work and analysis can be organised independently, with elements of complementarity, using various technologies and methods of study and analysis, with a full or limited level of analysis. An inductive approach—from simple to complex analysis and management decision-making—minimises the likelihood of unsuccessful decisions and reduces the time required to develop and update urban planning documentation.

Planning unit	Options	Condition. Typological description	Recovery concept
	Rubizhne Industrial and Pedagogical College. Founded in 1944. 730 students. 109 employees, including 50 teachers. Moved to the city of Kolomyia, Ivano-Frankivsk region.	The college is located on 6 hectares. Educational building (1939), workshops, dormitory, sports grounds, driving range. Completely destroyed. The main building is not subject to restoration	Change of functional purpose together with the site. *Restoration is possible only in the case of the formation of an educational institution with a complete reconstruction of the complex.
	Khimikov-Gorky residential area. Rubizhne Professional Chemical Technology Lyceum	Residential area Pivdennyi, Zabirkino. Zorya Research and Production Enterprise. Forms a separate planning district. Almost destroyed.	Rational structure of the site, convenient connections. Preservation of the functional purpose. Restoration is possible in case of reconstruction of the city-forming enterprise – Research and Production Enterprise 'Zorya'.

Fig. 8 A fragment of the matrix for analysing planning units of destroyed urban areas for the purpose of restoration (based on the example of Rubizhne)

4.3 *Innovative Approach. Redistribution of Territories and Objects. The Task of Creating Attractiveness in the Regions.*

Methods for determining the effectiveness of residential area reconstruction in the context of post-war recovery should be differentiated according to the goal. This may include creating comfortable living conditions; creating the preconditions for the development of existing general-purpose facilities; and ensuring the prerequisites for the expansion and renewal of the street and road network.

In Ukrainian practice, the method of comparative efficiency of reconstruction is used.

In foreign practice, methods are used to calculate the absolute efficiency of the reconstruction of residential buildings based on the difference between the rent before and after the modernisation of the housing stock and the costs of its implementation. The difference between the two practices is determined in the payback period [39, 40].

The demolition of housing stock and new construction in the areas to be reconstructed is advisable if the total cost of reconstructive transformation of the territories (taking into account the service life) does not exceed the cost of new construction in similar volumes:

$$\frac{C_n \cdot T_f}{T_s} - C_{c.o.} - (E_r - E_n) \cdot T_p \leq 0 \quad (1)$$

where:

- C_n—cost of new construction in the areas of recovery;
- $C_{c.o.}$—cost (expenses) for comprehensive overhaul of damaged buildings;
- E_r—expenses for the operation of buildings after their comprehensive overhaul;
- E_n—expenses for the operation of the new fund in the recovery areas;
- T_f—the final (residual) service life of the fund to be demolished;
- T_s—service life of the new facility;
- T_p—payback period is the new payback period for capital expenditures;

Thus, C_n—the cost of new construction in the restoration areas—corresponds to the indicative prices for new construction. The basic indicative prices can be minimised by the low initial cost of land plots; free (preferential) transfer of land plots to the developer; progressive planning, design and technical solutions; construction focused on target consumer focus groups; and the use of recycled materials from the demolition of destroyed buildings.

$C_{c.o.}$—The cost(s) of comprehensive overhaul of damaged buildings are significant and tend to increase due to: redevelopment of buildings; compensation for obsolescence; and physical damage of a burdensome nature.

E_r—the cost of operating buildings after their comprehensive overhaul. They will have the following characteristics: uncertainty; significant size compared to the new fund; upward trends due to general wear and tear; may be higher than normal due to a reduced number of users.

E_n—the cost of operating the new fund in the restoration areas. The advantages are predictability of costs, greater efficiency, and a standard service life (new is new).

T_f—final (residual) service life of the asset to be demolished - is probabilistically and statistically uncertain. It is determined on the basis of an expert assessment that cannot take into account all possible defects and damage.

The analysis of costs should be carried out for individual planning units (Fig. 8), with the development of recommendations for the settlement (territory) restoration programme.

5 Conclusions

The basic urban planning principles and conceptual foundations for the restoration of urbanised areas and cities that have been destroyed are defined for the tasks of updating the system of urban settlements and the planning structure of the eastern region of Ukraine: safety, environmental friendliness, autonomy, energy efficiency, convenience, comfort, transport accessibility, and the creation of places of employment. It is proved that in the current conditions of post-war recovery, there is a threat of primitivisation of the territory and the formation of a so-called budget trap in cities. There are few examples of positive renewal and restructuring of single-industry towns, and the scale of destruction creates factors of socio-demographic deterioration of reproduction conditions. An active part of the population, primarily young people, is leaving the territory. Non-active population groups are unable to provide a resource for recovery and form a recursion of crisis degradation. The task of optimising and balancing the functions and needs of cities in accordance with the population (demand), including for the future, should be based on the real parameters of the recovery areas. The author considers general approaches to preparing a decision on the reconstruction—full or partial reconstruction of the relevant settlement, development of such a settlement in another location, or its liquidation, clearing and reclamation of the territory. It is clarified that the standard algorithm of urban planning design is complicated. In order to determine the current priorities for restoration, it is necessary to coordinate planning (urban planning) documents at the local, regional and national levels. The development of models and principles for the restoration of territories and settlements with different parameters and typological signs of destruction in order to formalise practical issues is an important task. This is a prerequisite for the creation of a bottom-up regulatory framework for recovery processes. The construction of matrices for analysing planning units of the destroyed urban areas with a view to their restoration is an updated basis for the restoration model. It is proposed to identify hierarchical levels and objects of restoration, generalise management entities and types of restoration activities, formulate elements and components of the information package for restoration, and formalise design decisions on individual objects and elements of restoration.

References

1. Molchanov VB (2018) Ukrainian Donbas—the economic wonder of the second half of the 19th–early 20th century. Facets Hist 1(9):103–114 (in Ukrainian). https://doi.org/10.61655/2708-1249.1(9).2018.103-114
2. Kotygorenko V et al (2014) Donbas in the ethno-political dimension. Kuras Institute of Political and Ethno-National Studies, National Academy of Sciences of Ukrainem, Kyiv, 584 p (in Ukrainian). URL: https://ipiend.gov.ua/wp-content/uploads/2018/07/donbas_181.pdf
3. Alforov MA (2012) Urbanization processes in Ukraine in 1945–1991: Monograph. Donetsk Branch of the Shevchenko Scientific Society, Eastern Publishing House LLC, Donetsk, 549 p (in Ukrainian)

4. Center for Economic Strategy (2021) Analysis of the problems of economic development of Donetsk and Luhansk regions, 120 p (in Ukrainian). URL: https://ces.org.ua/vision-of-the-economic-strategy-of-donetsk-and-luhansk-regions/
5. KSE Institute (2024) Report on damages to infrastructure from the destruction caused by Russia's military aggression against Ukraine as of January 2024. KSE Institute, Kyiv, 39 p. https://kse.ua/wp-content/uploads/2024/05/Eng_01.01.24_Damages_Report.pdf
6. Sokolenko KV, Sokolenko VM, Filatiev MV, Chernih OA (2022) The influence of socio-demographic and external factors on the change of urban parameters of the Luhansk region. IOP Conf Ser Earth Environ Sci 1049(1):012079. https://doi.org/10.1088/1755-1315/1049/1/012079
7. Sokolenko KV, Sokolenko VM, Holodnov OI, Chernih OA (2023) Town-planning tasks and principles of restoration of urbanized territories of the Luhansk region, destroyed as a result of hostilities. IOP Conf Ser Earth Environ Sci 1254(1):012078. https://doi.org/10.1088/1755-1315/1254/1/012078
8. Denisov N et al (2017) Assessment of environmental damage and priorities for environmental restoration in Eastern Ukraine. In: Startsev O (ed) VAITE, 88 p (in Ukrainian). https://www.osce.org/files/f/documents/6/3/362581_0.pdf
9. Zavalnyi O, Nyzhnyk O, Vishnevsky D (2022) Regional model of territorial planning: influence of factors of military aggression and post-war perspectives. Spat Dev 2:84–93 (in Ukrainian). https://doi.org/10.32347/2786-7269.2022.2.84-93
10. Dreval I (2022) On the question of the development of concepts for the revival of the cities of Ukraine in the post-war period. Urb Dev Spat Plann 81:133–142 (in Ukrainian). https://doi.org/10.32347/2076-815x.2022.81.133-142
11. Pleshkanovska A et al (2024) Reconstructive activity in the context of urban life cycle phases: the case of Ukrainian cities. ACE Archit City Environ 18(54). https://doi.org/10.5821/ace.18.54.12127
12. Sokolenko KV, Sokolenko VM, Rudniev YS, Chernykh OA (2023) The influence of typological features of civil buildings on the feasibility of reconstruction in the cities of Eastern Ukraine. Bull Nat Univ Water Environ Eng 102(2):245–255 (in Ukrainian). https://doi.org/10.31713/vt2202320
13. Yatsenko VA (2005) Theoretical and practical ideas of the new urban planning of Donbas: its past, present and future. Urb Dev Spat Plann Sci Tech Collec 21:359–366. Kyiv National University of Construction and Architecture, Kyiv (in Ukrainian)
14. International Bank for Reconstruction and Development (2015) Ukraine: Urbanization overview. World Bank, Washington, DC. Retrieved from https://documents1.worldbank.org/curated/en/213551473856022449/pdf/ACS15060-REVISED-PUBLIC-eng-web-text-cover.pdf
15. Pleshkanovska A, Biriuk S (2022) Program for the reconstruction of the aging housing stock: from historic buildings to "Khrushchyovkas." InterConf 832–842 (in Ukrainian). https://doi.org/10.51582/interconf.19-20.01.2022.092
16. Ageieva HM, Kafiiev KP, Krivielov LI (2021) Reconstruction of the first mass-series buildings as a basis for sustainable development of city neighborhoods and quarters. Sci Const 27(1):32–40 (in Ukrainian). https://doi.org/10.33644/scienceandconstruction.v27i1.5
17. Zastavetska LB, Zastavetskyi TB, Zablotskyi BV, Dudarchuk KD (2019) Social and geographical aspects of development of urbanizational process in Ukraine. J Geol Geogr Geoecol 27(3):529–536. https://doi.org/10.15421/111877
18. Petryshyn H (2016) Changes in population in the context of urban development of Ukraine throughout the 20th and 21st centuries. Archit Stud 2(1):41–47. Retrieved from https://arch-studies.com.ua/web/uploads/pdf/Architectural%20Studies%202016_1-41-47.pdf
19. Layout of Mines in Luhansk Region (1980s) (n.d.). Mines and Quarries of Donbas (in Ukrainian). Retrieved from https://www.donmining.info/2019/05/shema-raspolozheniya-shaht-voroshilovgradskoy-oblasti.html?m=0

20. Dulko I (2023) Rebuilding narratives: the role of urban planning in post-war Ukraine. Masters thesis, University of Illinois at Chicago. Retrieved from https://www.researchgate.net/publication/371952476_Rebuilding_Narratives_The_Role_of_Urban_Planning_in_Post-war_Ukraine
21. Diomin M, Yatsenko V, Korotkova T (2023) Search of answers, why regional planning is basis of town-planning activity in construction of strategy of development of Ukraine. Urb Dev Spat Plann 82:3–16 (in Ukrainian). https://doi.org/10.32347/2076-815x.2023.82.3-16
22. Golodnov O, Shvets V, Sokolenko K (2023) Factors and prerequisites for the formation of new typological characteristics of the region of eastern Ukraine. Mod Technol Mater Des Const 35(2):120–129 (in Ukrainian). https://doi.org/10.31649/2311-1429-2023-2-120-129
23. Sokolenko V, Sokolenko K, Chernih O, Piddubnyi S (2021) The transformation of the engineering and planning organization of the territory of the Luhansk region in the conditions of the functions changing of the regional typology. E3S Web Conf 280:04003. https://doi.org/10.1051/e3sconf/202128004003
24. State Building Norms B.2.2–12:2019 (2019) Planning and development of territories (effective from 01 Sep 2019). Ministry of Regional Development of Ukraine, Kyiv, 208 p (in Ukrainian)
25. State Building Norms B.1.1-13:2021 (2022) Composition and content of urban planning documentation at the state and regional levels (effective from 01 Oct 2021). Ministry of Development of Communities and Territories of Ukraine, Kyiv, 81 p (in Ukrainian)
26. State Building Norms B.1.1-14:2021 (2022) Composition and content of urban planning documentation at the local level (effective from 01 Oct 2021). Ministry of Regional Development of Ukraine, Kyiv, 45 p (in Ukrainian)
27. State Building Norms B.1.1-15:2012 (2012) Composition and content of the master plan of a settlement [effective]. Ministry of Regional Development of Ukraine, Kyiv, 37 p (in Ukrainian)
28. Resolution of the Cabinet of Ministers of Ukraine No. 926 (2021) On approval of the procedure for the development, updating, amendment, and approval of urban planning documentation (as of May 3, 2023) (in Ukrainian). Retrieved June 21, 2023, from https://zakon.rada.gov.ua/laws/show/926-2021-n#Text
29. Law of Ukraine "On the Fundamentals of Urban Planning" (1992, with amendments) No. 52, Art. 683. (Version of 16 Oct 2020) (in Ukrainian). Retrieved from https://zakon.rada.gov.ua/laws/show/2780-12#Text
30. Law of Ukraine "On the Master Plan of Ukraine's Territory" (2002) (in Ukrainian)
31. Law of Ukraine "On the Principles of State Regulatory Policy in the Field of Economic Activity" (2003) No. 1160-IV (in Ukrainian)
32. Law of Ukraine "On the Peculiarities of State Policy to Ensure Ukraine's Sovereignty in Temporarily Occupied Territories in Donetsk and Luhansk Regions" (2018) No. 2268-VIII (as of May 7, 2022) (in Ukrainian). Retrieved from https://zakon.rada.gov.ua/laws/show/2268-19#Text
33. Resolution of the Cabinet of Ministers of Ukraine No. 1073 (2022) Procedure for managing waste generated due to damage (destruction) of buildings and structures as a result of hostilities, terrorist acts, sabotage, or works to eliminate their consequences (in Ukrainian)
34. Resolution of the Cabinet of Ministers of Ukraine No. 474 (2022) Procedure for carrying out dismantling works of objects damaged or destroyed as a result of emergencies, hostilities, or terrorist acts (in Ukrainian). Retrieved from https://zakon.rada.gov.ua/laws/show/474-2022-%D0%BF#Text
35. Ustinova I, Pleshkanovska A (2023) Urbicide and post-war reconstruction of residential buildings in Ukrainian cities: experience and prospects. Grail Sci 23:463–471 (in Ukrainian). https://doi.org/10.36074/grail-of-science.23.12.2022.82
36. Knayer T, Kryvinska N (2025) The difference between energy management systems and environmental management systems on the implementation of cross-sectional technologies in enterprises. Energy Rep 13:1691–1704. https://doi.org/10.1016/j.egyr.2025.01.030
37. Telnov A, Hurochkina V, Reshmidilova S, Yepifanova I, Yakivchenko A (2022) Ergodesign as a method for the design of the quality of products on the basis of benchmarking. J Hygienic Eng Des 40:27–35

38. Bertin M, Galli J, Rossi F (2022) Retracing reconstruction: Establishing an analytical method for the comprehension and systematisation of urban metamorphosis following extreme events. arXiv. https://doi.org/10.48550/arXiv.2201.04067
39. Apatenko TM, Beliubchenko OS, Haiko YI, Shyshkin E A et al (2016) Practice of innovative developments in the field of territorial-spatial development of cities and regions. In: Semenov VT, Lynnyk IE (eds) O. M. Beketov National University of Urban Economy, Kharkiv, 300 p (in Ukrainian)
40. Haiko YI, Zhydkova TV (eds) (2019) Problems and prospects of residential development under conditions of comprehensive urban reconstruction: monograph. O. M. Beketov National University of Urban Economy, Kharkiv, 248 p (in Ukrainian)

Detecting Country of Origin Data Manipulation in Sanctioned Raw Material Supply Chains: Expert Survey Insights

Ilona Dumanska

Abstract This paper aims to identify key factors that indicate practices of data manipulation, providing insights into the methods and motivations behind such activities. By analyzing the perspectives of various stakeholders involved in the supply chain, the study seeks to enhance understanding of the challenges and implications associated with data manipulation in the context of international trade and sanctions. The findings contribute to the development of strategies to detect and prevent such practices, ensuring the integrity and transparency of global supply chains. The purpose of this study is to develop a comprehensive approach to detecting and countering the manipulation of country of origin data in supply chains of raw materials under sanctions. This involves analyzing the key mechanisms of manipulation, identifying the main manifestations and influencing factors, and developing recommendations for effective monitoring and countering such manipulations. The main hypotheses of this manuscript are as follows: (1) manipulation of country of origin data in supply chains of raw materials under sanctions is a widespread phenomenon that significantly affects the effectiveness of international sanctions; (2) the main mechanisms of manipulation of country of origin data are the involvement of third parties, the organization of intermediary schemes, and changes in the corporate structure of enterprises; (3) effective cooperation and information exchange between customs authorities, logistics operators, customs brokers, banks, and private business entities are key to countering the manipulation of country of origin data. Objectives of the article include exploring the concept of country of origin and the importance of accurate identification in countering the circumvention of sanctions on raw material logistics, identifying key mechanisms of manipulation of country of origin data in supply chains based on scientific research and case analysis, pinpointing key manifestations of data manipulation in supply chains for circumventing sanctions, and identifying the top 10 key factors influencing supply chains that manifest in the context of manipulation of country of origin data to circumvent sanctions. The study also aims to conduct an expert assessment to identify the most important

I. Dumanska (✉)
Khmelnytskyi National University, Khmelnytskyi, Ukraine
e-mail: dumanskai@khmnu.edu.ua

A. Semenov et al. (eds.), *Data-Centric Business and Applications*, Lecture Notes on Data Engineering and Communications Technologies 272,
https://doi.org/10.1007/978-3-032-06608-4_4

determinants, such as access to quality education and technologies, which are essential for forming competitive human potential, and establish imperatives for priority monitoring of the identified sanctions.

Keywords Country of origin · Data manipulation · Sanctions · Supply chains · Raw materials · Expert survey

1 Introduction

In today's globalized economy, the issue of detecting manipulation of country of origin data in supply chains of raw materials under sanctions is becoming increasingly urgent. Russia, as one of the countries under international sanctions, is actively seeking ways to circumvent these restrictions by employing various schemes to manipulate country of origin data. The lack of official sources of information and access to reliable data on such manipulations creates significant challenges for international organizations and Ukrainian companies striving to ensure transparency and compliance in their supply chains.

One of the most pressing problems is the lack of access to official sources of information, which complicates the detection and counteraction of manipulations. In many instances, data on the country of origin of raw materials can be altered or concealed, allowing sanctioned countries to continue their activities under false pretenses. This not only undermines the effectiveness of sanctions but also poses risks to consumers and businesses, as dangerous or substandard materials can infiltrate supply chains.

In such circumstances, expert interviews become the sole reliable source of information on the manipulation of country of origin data. Experts from various fields, including supply chain managers, customs inspectors, customs brokers, specialists in the export-import operations of banks, and management of international business entities, provide valuable insights and practical experience that help to understand the current state of affairs and develop effective strategies to counteract manipulation. Their knowledge and experience are indispensable for developing a comprehensive approach to ensuring transparency and compliance in supply chains.

This research may be of interest to a wide range of individuals and organizations involved in logistics, customs control, compliance and international business. In particular, it will be useful for customs authorities, logistics operators, customs brokers, banks and private business entities seeking to ensure transparency and compliance of their supply chains. In addition, the study may be of interest to scholars and analysts studying the issues of international sanctions and their impact on the global economy. They can use the data to develop new methods to detect and counter manipulation of country of origin data. Finally, the study will be useful for government and international organizations involved in sanctions and their enforcement. They can interpret the research findings to improve their policies and procedures aimed at ensuring the effectiveness of sanctions.

2 Research Purpose and Main Hypotheses

The purpose of this study is to develop a comprehensive approach to detecting and countering the manipulation of country of origin data in supply chains of raw materials under sanctions. This involves analyzing the key mechanisms of manipulation, identifying the main manifestations and influencing factors, and developing recommendations for effective monitoring and countering such manipulations. The main hypotheses of this manuscript are as follows:

Hypothesis 1: Manipulation of country of origin data in supply chains of raw materials under sanctions is a widespread phenomenon that significantly affects the effectiveness of international sanctions (Sect. 3).
Hypothesis 2: The main mechanisms of manipulation of country of origin data are the involvement of third parties, the organization of intermediary schemes and changes in the corporate structure of enterprises (Sect. 4).
Hypothesis 3: Effective cooperation and information exchange between customs authorities, logistics operators, customs brokers, banks and private business entities are key to countering the manipulation of country of origin data (Sect. 5).

3 Examining the Evasion of Sanctions: Manipulation of Country of Origin Data in Raw Material Supply Chains

The ongoing conflict between Ukraine and Russia has precipitated a range of international sanctions aimed at curbing Russia's military activity and influencing its decision-making. However, the effectiveness of these sanctions is often undermined by the manipulation of country of origin data in supply chains, particularly concerning raw materials. This phenomenon creates loopholes that allow sanctioned entities to continue their operations, thereby weakening the intended impact of the sanctions. The manipulation of trade data has emerged as a prevalent practice among countries seeking to evade sanctions, undermining the integrity of international regulatory frameworks.

The manipulation of country of origin (COO) data refers to the falsification or misrepresentation of the geographical source of goods in international trade [1]. This practice is particularly insidious in supply chains involving raw materials, which are often subject to stringent regulations and oversight. According to a study [2], countries under sanctions frequently engage in renaming or rebranding products to obscure their true origins. This exploitation of regulatory gaps allows sanctioned nations to continue accessing global markets, thereby circumventing established sanctions.

Such maneuvers are notably prominent in the context of raw materials, where the COO plays a crucial role in compliance with international sanctions. The literature indicates that discrepancies between reported and actual origins of goods contribute to substantial economic impacts, with significant ramifications for sanctioned states and the global market as a whole [3]. Moreover, the impact of this manipulation extends beyond economic dimensions, as it also undermines the legitimacy of international law and compliance mechanisms designed to promote accountability.

The Russian annexation of Crimea in 2014 and its subsequent military actions in Eastern Ukraine have been met with a series of coordinated international sanctions by Western nations. These sanctions target various sectors, including defense, finance, and energy, with an emphasis on restricting access to critical raw materials [4]. Despite these initiatives, Russia has employed various tactics to obfuscate the trade of sanctioned goods, thereby diluting the expected effectiveness of these measures.

One notable example is the use of third-party countries to reroute shipments of raw materials [5]. By transiting through nations with looser import-export regulations, Russian entities are able to alter the COO documentation to reflect compliance with sanction regulations, effectively disguising the origins of the materials. Old reports indicate a marked increase in trade with countries such as Belarus and Turkey, where Russian exports are repackaged and subsequently exported to other global markets [6]. The manipulation of COO data in this context has allowed Russia to sidestep sanctions while simultaneously maintaining its status as a major player in the global energy market.

The manipulation of COO data under sanctions has significant economic implications not only for the sanctioned states but also for the enforcing countries. The efficacy of sanctions is often gauged by their capacity to undermine the economic stability of the target nation. However, studies suggest that the ability of sanctioned nations to adapt and exploit loopholes in trade regulations severely hampers the intended economic impact of sanctions [7]. In the case of Russia, projections indicated that the initial sanctions would lead to a substantial reduction in GDP; however, subsequent data revealed resilience in particular sectors due to trade circumvention tactics [8].

The loss of international revenue due to sanctions can result in unintended consequences for imposing nations, including reduced economic competitiveness within their markets [9]. Moreover, the manipulation of COO data affects global supply chains, leading to price distortions and inefficiencies as traders navigate increasingly complex networks to access raw materials. The ease with which countries manipulate their import-export documentation raises serious concerns regarding trade compliance and economic governance. According to recent research, advances in technology have facilitated the manipulation of COO data in unprecedented ways [10]. Digital platforms and blockchain technology, once viewed as tools for enhancing transparency and accountability in supply chains, are now increasingly exploited for deceptive practices. The manipulation of supply chain data can occur at multiple levels, from the documentation of origin to the reporting of prices and quantities [11].

For instance, the adoption of automated data management systems has made it easier for entities to fabricate origin documents, further obfuscating the true source of materials [12]. With increasing cyber capabilities, sanctioned nations have become adept at exploiting digital tools to maintain their presence in global markets, thus complicating enforcement measures. The implications of technology in sanctions evasion echo the sentiments of various scholars who emphasize the need for a comprehensive understanding of digital impacts on global trade regulation [13].

The analysis of COO manipulation during the Ukrainian–Russian conflict underscores the need for a coordinated response among Western nations to combat trade loopholes. Legislative frameworks governing international trade must evolve to address the complexities introduced by modern supply chains. Strengthening cooperation among customs authorities and enhancing information sharing can play a critical role in identifying and mitigating the risks associated with COO manipulation [14].

Moreover, international stakeholders should invest in the development of technological solutions that enhance transparency in the supply chain process. Innovative approaches such as blockchain technology could offer solutions for maintaining the integrity of trade data while reducing the potential for manipulation [15]. Collaborative efforts to establish best practices in supply chain transparency, coupled with a commitment to robust enforcement mechanisms, are crucial to improving the effectiveness of international sanctions [16].

In conclusion, the COO data in supply chains under sanctions is a pervasive issue with significant impacts on the international effectiveness of sanctions, particularly within the context of the Ukrainian–Russian conflict. As evident from recent literature, such manipulation not only undermines the economic pressures intended by sanctions but also poses challenges to international legal frameworks governing trade and compliance. Stakeholders must recognize the implications of these tactics and coordinate efforts to enhance the efficacy of sanctions in the global arena.

4 Mechanisms of Data Manipulation in Supply Chains: A Case Study Analysis and Factor Identification

4.1 A Case Study Analysis

Data manipulation in supply chains, especially with respect to sanctioned commodities, is a multifaceted problem for both regulators and global trading systems. This section describes the complex mechanisms through which data manipulation occurs, focusing on the roles of third parties, intermediary schemes, and manipulation of corporate structure.

Involvement of third parties. Data manipulation in supply chains, particularly for sanctioned commodities, is a multifaceted problem for both regulators and global trading systems. This section describes the complex mechanisms through which data manipulation occurs, focusing on the roles of third parties, intermediary schemes, and corporate structure manipulation.

Third countries are often instrumental in the manipulation of origin data within the supply chains of sanctioned commodities. These jurisdictions often offer regulatory leniency and favorable economic conditions that are attractive to companies seeking to circumvent international sanctions. For example, countries with weak enforcement mechanisms or countries without a strong legal framework provide a haven for coalition companies specializing in the rebranding of sanctioned materials. This coalition formation can involve multiple actors, including manufacturers, exporters, and logistics providers, who collectively develop strategies to conceal the true geographical origin of goods [9].

One illustrative case is the collaboration between firms in Eastern Europe and some Southeast Asian countries, where companies are known to set up front operations that facilitate the transshipment of sanctioned goods while creating a facade of legitimacy. This often involves combining multiple transactions to obscure the final path of the materials, effectively complicating the chain of custody and misleading regulators/Several documented cases demonstrate the effectiveness of these diversion strategies. For example, in a high-profile 2020 case, a network of companies operating out of Ukraine managed to redirect oil from sanctioned countries through a series of shell corporations established in neighboring countries. By manipulating shipping documents and relying on third-party logistics providers, these companies have successfully disguised the origin of the oil, allowing it to enter otherwise restricted European and American markets [5].

The implications of such successful maneuvers highlight a critical problem in the global regulatory framework, whereby the existence of third-party jurisdictions can seriously hinder compliance efforts. Thus, countries with minimal trade restrictions become major hubs for the illicit trade in materials that international law seeks to control [6].

Intermediary schemes. The use of intermediary schemes is another common tactic in attempts to manipulate data in supply chains. These schemes involve a range of actors, including freight forwarders, customs brokers, and even compliance consultants. Often, intermediaries act as buffers that insulate the main actors from direct administrative control, thereby allowing them to disguise the origin of goods.

The dynamic landscape of global trade is increasingly complicated by the proliferation of data manipulation practices. These practices can undermine regulatory frameworks, distort market dynamics, and facilitate the evasion of sanctions. Various stakeholders in the supply chain, including exporters, importers, regulators, and intermediaries, exhibit unique motivations for engaging in or combating data manipulation. This study seeks to elucidate ten key factors that suggest the presence of data manipulation, thereby improving understanding of the pertinent challenges and implications.

In various cases, intermediaries have engaged in data obfuscation either knowingly or through a lack of due diligence. For example, customs brokers may knowingly misclassify goods on import documents or provide false declarations of origin to facilitate transactions. The legal implications of this intermediary role can be serious, as intermediaries can be held liable for facilitating violations of international sanctions, but the line of liability remains blurred, often allowing the main companies to avoid detection [3]. The ethical implications of such manipulative practices are profound. Relying on intermediaries raises questions about corporate responsibility and governance. Many companies may claim that they have limited oversight of their logistics and supply chain operations; however, the rise of social responsibility standards requires a reassessment of these claims. Deceptive intermediaries not only contribute to the erosion of ethical business practices, but also undermine legitimate trade and economic stability [14]. In response to these challenges, regulators have begun to tighten the screws on intermediary transactions, pushing for greater transparency and accountability in supply chains. Initiatives such as the European Union's proposed rules on supply chain due diligence aim to impose stricter compliance requirements on intermediaries, potentially changing the dynamics of global trade networks [3].

Corporate structure changes. Corporate structure manipulation offers another layer of complexity to supply chain data manipulation using sanctions materials. Companies that use this tactic often establish multiple subsidiaries with varying degrees of interdependence, intentionally concealing the parent organization's involvement in direct transactions with restricted parties. This approach can involve the creation of shell companies that exist solely to facilitate transactions without a clear operational trail, complicating regulators' enforcement efforts.

A well-documented case involves a group of logistics service providers that changed their corporate identities during the imposition of sanctions, effectively dissolving the previous entities associated with sanctioned activities and re-designating new names and tax numbers. This tactical reshuffle can facilitate the continued operation of businesses that would otherwise be prohibited under international sanctions [7].

The case of a prominent shipping company illustrates this manipulation. After being sanctioned for its involvement in the embargo, the company restructured itself by creating several subsidiaries in several jurisdictions. In this way, it was able to hide its core business by facilitating supplies from sanctioned regions under the guise of legitimate trade. This not only allowed the company to maintain its operational capabilities, but also facilitated the illicit flow of materials into markets that strictly prohibited such imports [2].

Ultimately, the consequences of corporate structure manipulation go beyond the immediate scope of regulatory evasion; they contribute to a larger systemic problem, where legitimate businesses are increasingly disadvantaged by unfair competitive practices. As enforcement measures adapt and evolve, innovators in compliance monitoring and technology remain urgently needed to combat the ever-changing tactics employed by organizations intent on circumventing established norms [4].

4.2 Factor Identification

Analysis of the various cases reveals several critical manifestations of this manipulation, including the use of third-party intermediaries, the creation of complex intermediary schemes, and strategic changes in corporate structures. These mechanisms serve to conceal the origin of goods and evade regulatory scrutiny. Table 1 summarizes the key factors of the cases studied, highlighting the multifaceted nature of data manipulation tactics in the form of aggregate factors across the manipulation mechanisms outlined above.

Overall, the mechanisms for data manipulation in supply chains involving sanctioned raw materials are diverse and deeply rooted in the strategies used by different actors [19]. As discussed in this section, the roles of third parties, intermediary schemes, and manipulation of corporate structures not only facilitate the illicit trade in sanctioned materials, but also pose significant legal and ethical challenges. Future regulatory efforts must address these complexities, moving towards a comprehensive framework that takes into account the changing landscape of international trade and manipulation.

5 Insights from Expert Interviews as a Method for Detecting CCO Data in Supply Chains of Raw Materials Under Sanctions

In order to identify the 10 main factors influencing supply chains that manifest themselves in the context of COO data for the purpose of circumventing sanctions, we initiated a study of expert opinion. Data collection was carried out in the following timeframe: from September 4, 2024 to December 4, 2024. For these purposes, 5 expert groups of 5 people were formed by profile: customs, intermediaries, logistics support, banks, and business environment. At the same time, employees of various institutions and business entities were selected as experts. The expertise profile of the respondents is presented in Table 2.

The identification of the top 10 main factors influencing supply chains that manifest themselves in the context of COO data for the purpose of circumventing sanctions, based on an expert survey, was carried out in five stages. The sequence and description of the stages are shown in Fig. 1.

Based on Table 2, the initiators of the study developed a checklist of key factors of potential manifestations in the COO data to circumvent sanctions, corresponding to 10 factors per parameter. The following parameters were identified as: Involvement of third countries, Intermediary Schemes, and Corporate Structure Manipulation, the corresponding factors were selected based on the case study and literature sources in Sect. 3 of this publication.

The experts were asked to set the "+" mark as the appropriateness of the factor of potential manifestations in the manipulation of data on the country of origin to

Table 1 Key factors of potential manifestations of COO data to circumvent sanctions

№	Key factors	Sources
Involvement of third countries		
1	Discrepancies in shipping documentation	[5, 6, 8, 9, 13, 17]
2	Non-compliance with standardized reporting	
3	Anomalies in trade patterns	
4	Use of Shell companies[a]	
5	Over- and under-invoicing practices	
6	Abnormal consignment routes	
7	Complex multi-layered transactions	
8	Lack of transparency in beneficial ownership	
9	Use of unverified or inconsistent data sources	
10	Emergence of new trade relationships following sanction impositions	
Intermediary schemes		
1	Inconsistent reporting patterns	[1, 3, 10–14, 18]
2	Use of front companies[b]	
3	Altered documentation	
4	Discrepancies in financial transactions	
5	Geographic discrepancies	
6	Unusual payment structures	
7	Frequent changes in supplier and customer profiles	
8	Lack of transparency in supply chain relationships	
9	Regulatory evasion strategies	
10	Patterns of anomalous trade activity	
Corporate structure manipulation		
1	Inconsistent reporting metrics	[2, 4, 7, 12, 14]
2	Altered transaction histories	
3	Layering of corporate structures	
4	Duplicate billing practices	
5	Unexplained variances in data	
6	Transaction timing manipulation	
7	Non-transparent financial relationships	
8	Manipulation of trade volume indicators	

(continued)

Table 1 (continued)

№	Key factors	Sources
9	Use of misleading language in documentation	
10	Evasion and compliance technologies	

[a] in terms of: Covering up illegal activities (attempts to hide ownership and the true nature of transactions in order to circumvent sanctions)
[b] in terms of: Disguising the true origin of goods (attempts to create the illusion of legitimacy and make it difficult to track goods)

Table 2 Expert's profile

№	Expert group	Age	Professional experience	Positions
1	Customs services	Between 35 and 45 years old	Over 10 years	Customs inspectors
2	Logistics operators	Between 25 and 35 years old	Over 10 years	Logistics manager, import manager
3	Customs brokers	Between 25 and 35 years old	Over 15 years	Customs clearance agents
4	Banks	Between 35 and 45 years old	Over 10 years	Specialists of the department of service of export-import operations
5	Business practitioners	Between 35 and 45 years old	Over 15 years	Top managers of international business

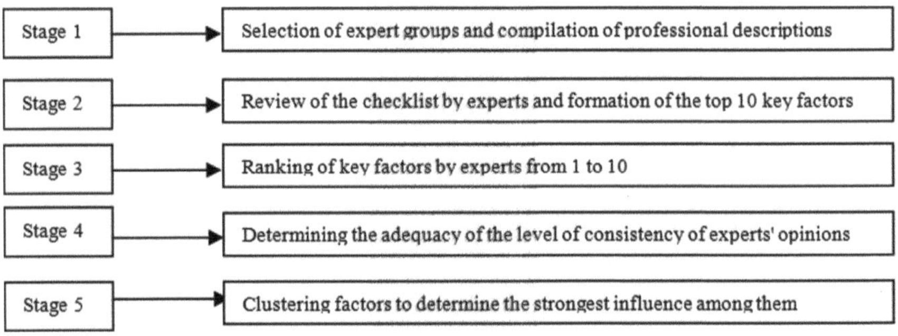

Fig. 1 Sequence and description of the stages of studying expert opinion

circumvent the sanctions that took place based on their experience and practice. The results of the experts' processing of the checklist are presented in Table 3.

The summary of the experts' responses based on the checklist identified the top 10 key factors of potential manifestations in the manipulation of data about the country of origin to circumvent sanctions. The selection of these factors was carried out on the basis of 5 marks, i.e. from all experts as confirmation of the support of such forms

Table 3 Results of the experts' processing of the checklist "Factors of potential manifestations in the manipulation of data about the country of origin to circumvent sanctions"

Key factors	Experts					Total
	Customs services	Logistics operators	Customs brokers	Banks	Business practitioners	
Involvement of third countries						
Discrepancies in shipping documentation	+	+	+			3
Non-compliance with standardized reporting	+			+		2
Anomalies in trade patterns	+				+	2
Use of shell companies	+	+	+	+	+	**5**
Over- and under-invoicing practices	+		+	+		3
Abnormal consignment routes	+	+	+		+	4
Complex multi-layered transactions	+			+	+	3
Lack of transparency in beneficial ownership				+	+	2
Use of unverified or inconsistent data sources				+	+	2
Emergence of new trade relationships following sanction impositions	+	+	+	+	+	**5**
Intermediary schemes						
Inconsistent reporting patterns		+	+		+	3
Use of front companies		+			+	2
Altered documentation	+	+	+	+	+	5

(continued)

Table 3 (continued)

Key factors	Experts					Total
	Customs services	Logistics operators	Customs brokers	Banks	Business practitioners	
Discrepancies in financial transactions			+	+	+	3
Geographic discrepancies	+	+	+	+	+	**5**
Unusual payment structures	+	+	+	+	+	**5**
Frequent changes in supplier and customer profiles	+		+		+	3
Lack of transparency in supply chain relationships	+	+	+	+	+	**5**
Regulatory evasion strategies		+			+	2
Patterns of anomalous trade activity	+	+	+			3
Corporate structure manipulation						
Inconsistent reporting metrics				+	+	2
Altered transaction histories		+	+	+		3
Layering of corporate structures	+	+	+	+	+	**5**
Duplicate billing practices		+	+		+	3
Unexplained variances in data	+		+	+		3
Transaction timing manipulation		+	+		+	3
Non-transparent financial relationships				+	+	2
Manipulation of trade volume indicators	+	+	+	+	+	**5**

(continued)

Table 3 (continued)

Key factors	Experts					Total
	Customs services	Logistics operators	Customs brokers	Banks	Business practitioners	
Use of misleading language in documentation	+	+	+			3
Evasion and compliance technologies	+	+	+	+	+	5

of manipulation in relation to all areas: customs, intermediaries, logistics support, banks, and business environment.

The following factors were identified: Use of Shell Companies, Abnormal Consignment Routes, Emergence of New Trade Relationships, Altered Documentation, Geographic Discrepancies, Unusual Payment Structures, Lack of Transparency in Supply Chain Relationships, Layering of Corporate Structures, Manipulation of Trade Volume Indicators, Evasion and Compliance Technologies, which are presented in Fig. 2.

To substantiate the assessment of the factors influencing the COO data to circumvent sanctions, we propose to use the Kendall mathematical modeling method, which is based on calculating the correlation coefficient as a measure of the linear relationship between random variables [20]. Kendall correlation is rank-based, that is, to assess the strength of the correlation, not numerical values, but the corresponding ranks are used. These can be, for example, rank estimates on a certain scale or ordinal positions of objects in a list. Kendall correlation allows us to establish whether there is a statistically significant relationship between these ordinal variables.

To do this, first of all, it is necessary to identify the factors influencing the COO data to circumvent sanctions. The factors identified by experts in Fig. 2 as significant were assessed for their significance in the range from 1 to 10. Accordingly, to determine the adequacy of the level of consistency of experts' opinions, the Kendall concordance coefficient was applied, and the results of the expert study are presented in Table 4, where two main indicators are calculated: D is the average value, which is calculated as follows: we obtained the indicator $d = 34.1$ (the sum of all ranks divided by $10 = 341/10 = 34.1$); d^2 is the quadratic value, which is calculated as the above-mentioned indicator D squared.

Checking the correctness of the matrix construction based on the calculation of the checksum (1)

$$\sum x_{ij} = ((1 + n) \times n)/2 = ((1 + 10) \times 10)/2 = 55, \tag{1}$$

where n is the number of factors.

The distribution of factors by significance is presented in Table 5.

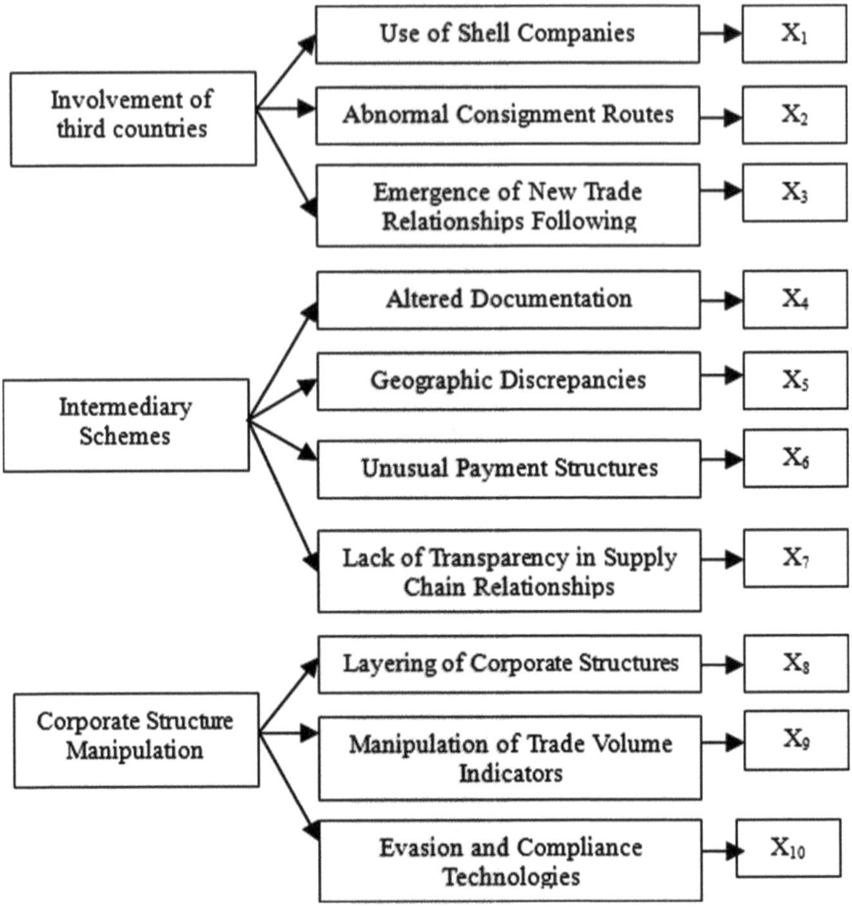

Fig. 2 Top 10 key factors of potential manifestations in the manipulation of country of origin data to circumvent sanctions: experts' choice

So, as can be seen from Table 5, the most significant factors were: Evasion and Compliance Technologies, Layering of Corporate Structures, Manipulation of Trade Volume Indicators, Unusual Payment Structures.

In turn, the least influential factors are such as Use of Shell Companies, Emergence of New Trade Relationships Following, Geographic Discrepancies.

It is taking into account the most influential factors that recommendations should be developed to counteract the COO data in order to circumvent sanctions.

Next, the adequacy of the level of consistency of expert opinions was determined based on the application of the concordance coefficient (W), which is an indicator of multiple rank correlation. W can take values in the range *from* 0 *to* + 1, *where* $W = 0$ means a complete lack of consistency in the assessments or judgments of experts, and $W = 1$—complete consistency. As the maximum permissible value of

Table 4 Expert's profile

Factors	Expert groups					Sum of ranks	D	d^2
	1	2	3	4	5			
X_1	4	5	5	8	4	26	– 8.1	65.61
X_2	9	9	7	6	8	39	4.9	24.01
X_3	4	4	5	3	6	22	– 12.1	146.41
X_4	5	5	6	6	5	27	– 7.1	50.41
X_5	3	5	5	5	4	22	– 12.1	146.1
X_6	8	8	10	9	7	42	7.9	62.41
X_7	6	7	7	5	4	29	– 5.1	26.01
X_8	9	9	8	10	8	44	9.9	98.01
X_9	9	9	10	8	8	44	9.9	98.01
X_{10}	9	10	9	8	10	46	11.9	141.61
$\sum$						341	X	858.59

Table 5 Ranking of factors influencing manipulation of country of origin data to circumvent sanctions by importance

Factors		Sum of ranks
X_{10}	Evasion and compliance technologies	46
X_8	Layering of corporate structures	44
X_9	Manipulation of trade volume indicators	44
X_6	Unusual payment structures	42
X_2	Abnormal consignment routes	39
X_7	Lack of transparency in supply chain relationships	29
X_4	Altered documentation	27
X_1	Use of Shell companies	26
X_3	Emergence of new trade relationships following	22
X_5	Geographic discrepancies	22

the concordance coefficient, we propose to take $W = 0.4$–0.6: (average consistency of expert opinions), $W = 0.1$–0.3: (low consistency of expert opinions), $W = 0.7$–1.0: (high consistency of expert opinions).

Calculations for assessing the average degree of consistency of opinions of all experts using the concordance coefficient are given below (2):

$$W = 12 \times S / \left(m^2 \times \left(n^3 - n\right)\right), \quad (2)$$

where $S = 858.59$ (sum of squared values), $n = 10$ (number of factors), $m = 5$ (number of experts).

$$W = (12 \times 858.59)/(25 \times (1000 - 10)) = 10{,}303.08/24{,}750 = 0.416.$$

Therefore, $W = 0.416$ indicates the presence of an average degree of agreement between experts' opinions.

To assess the significance of the concordance coefficient, we will use the Pearson agreement criterion (3):

$$\chi 2 = (12 \times S)/(m \times n \times (n + 1)), \tag{3}$$

$$\chi 2 = (12 \times 858.59)/(5 \times 10 \times (10 + 1)) = 10{,}303.08/550 = 18.73.$$

The calculated $\chi 2$ data are compared with the tabular value for the number of degrees of freedom $K = n - 1 = 10 - 1 = 9$ and at a given significance level $\alpha = 0.04$. Since the calculated $\chi 2$ is 18.73 > the tabular value of 16.9 (unified table of the Pearson criterion), then the $W = 0.416$ value is not random, and therefore the obtained results make sense and can be used in further research.

So, in the context of this study, using the Kendall correlation-regression analysis method, separate influencing factors in the context of manipulating data about the country of origin in order to circumvent sanctions are located in order of significance.

Kendall's correlation analysis allowed us to determine the factors by their significance, however, for a more consistent result, it is desirable to fulfill the following conditions:

- assess hidden (latent) factors, which can be a combination of several influential indicators;
- reveal possible relationships between factors and identify their groups by common characteristics;
- determine the general structure of factor influence.

To further detail the results and isolate hidden factors, the factor clustering method was used, which allows us to group factors into several main categories and determine which of them have the strongest impact on the development of practices of manipulating data about the country of origin in order to circumvent sanctions.

Calculation process for Hierarchical influence on human capital development.

1. Formation of input data. The results of assessing the significance of factors influencing human capital development were used for the analysis (Table 5): $X = [46, 44, 44, 42, 39, 29, 27, 26, 22, 22]$.
2. Normalization and data preparation. To conduct cluster analysis, a parameter is used—the factor significance assessment. The data are presented in the form of a vector (4):

$$d(A, B) = \sqrt{(X_A - X_B)^2}. \tag{4}$$

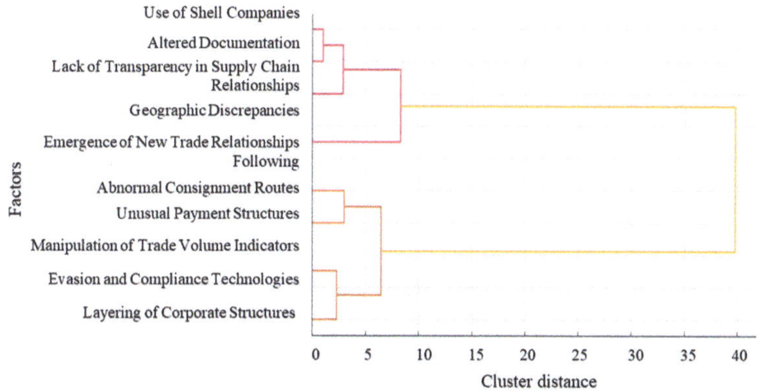

Fig. 3 Top 10 key factors of potential manifestations in the manipulation of country of origin data to circumvent sanctions: experts' choice

3. Using the hierarchical clustering method [21]. Hierarchical clustering is based on an algorithm for combining close objects into groups (clusters) [22]. We use the Ward method, which minimizes the variance within each cluster.

Stages of the Ward method:

- each factor is considered as a separate cluster;
- the distance measure between clusters is determined—the Euclidean distance method is used.
- the two closest clusters are combined to reduce the total variance within clusters;
- the process is repeated until all factors are grouped into a single cluster.

To analyze the resulting groups, a dendrogram is used—a clustering tree, which shows the levels of cluster association and their similarity (Fig. 3).

The dendrogram shows three main groups of factors:

1. Layering of Corporate Structures, Evasion and Compliance Technologies—these are factors that are directly related to the COO data in order to circumvent sanctions.
2. Unusual Payment Structures, Lack of Transparency in Supply Chain Relationships—these factors characterize the economic basis of the COO data in order to circumvent sanctions.
3. Altered Documentation, Use of Shell Companies, Emergence of New Trade Relationships Following, Geographic Discrepancies—these are systemic factors that depend on the sanctions policy regarding the country of origin of the raw materials.

Based on the expert interview insights, it can be argued that the identified factors suggest the potential for effective cooperation and information exchange between customs authorities, logistics operators, customs brokers, banks, and private businesses. This collaboration aims to counter the COO data through interaction and joint risk-based monitoring.

6 Conclusions

The COO data in supply chains under sanctions is a widespread problem that significantly affects the international effectiveness of sanctions, especially in the context of the Ukraine–Russia conflict. This manipulation undermines the economic pressure provided by sanctions and creates problems for the international legal framework governing trade and compliance. Stakeholders should be aware of the consequences of this tactic and coordinate efforts to improve the effectiveness of sanctions globally.

An expert survey identified the top 10 factors affecting supply chains in the context of COO data to circumvent sanctions. These factors include the use of shell companies, abnormal consignment routes, emergence of new trade relationships, altered documentation, geographic discrepancies, unusual payment structures, lack of transparency in supply chain relationships, layering of corporate structures, manipulation of trade volume indicators, and evasion and compliance technologies. Among the most significant factors were evasion and compliance technologies, layering of corporate structures, manipulation of trade volume indicators, and unusual payment structures.

Recommendations should focus on counteracting the most influential factors, such as layering of corporate structures and evasion and compliance technologies, which are directly related to the COO data. Unusual payment structures and lack of transparency in supply chain relationships characterize the economic basis of this manipulation. Altered documentation, use of shell companies, emergence of new trade relationships, and geographic discrepancies are systemic factors dependent on the sanctions policy regarding the country of origin of raw materials.

Expert interview insights suggest the potential for effective cooperation and information exchange between customs authorities, logistics operators, customs brokers, banks, and private businesses. This collaboration aims to counter the COO data through interaction and joint risk-based monitoring.

References

1. Koerber T, Schiele H (2024) Profound changes in global sourcing? The country of origin theory and its effects on sourcing decisions. J Bus Ind Mark 39(13):68–81
2. Lindsay-Poland J (2016) The politics of trade compliance: implications of sanctions for international law. J Int Law 14(2):225–249. https://doi.org/10.1017/ile.2015.5
3. Silva PM, Selden Z (2020) Economic interdependence and economic sanctions: a case study of European Union sanctions on Russia. Camb Rev Int Aff 33(2):229–251. https://doi.org/10.1080/09557571.2019.1660857

4. Giumelli F (2024) A comprehensive approach to sanctions effectiveness: lessons learned from sanctions on Russia. Eur J Crim Policy Res 30(2):211–228. https://doi.org/10.1007/s10610-024-09585-x
5. Coldal L (2023) The role of third countries in the effectiveness of economic sanctions with a case study of sanctions against Russia. Master's thesis, The University of Bergen
6. Khalid U, Okafor LE, Burzynska K (2022) Do regional trade agreements enhance international tourism flows? Evidence from a cross-country analysis. J Travel Res 61(6):1391–1408
7. Nguyen LQT, Ahmed R (2023) The impact of economic sanctions on foreign direct investment: empirical evidence from global data. J Econ Dev 25(1):79–99
8. Andermo E, Kragh M (2021) Sanctions and dollar dependency in Russia: resilience, vulnerability, and financial integration. Post-Sov Aff 37(3):276–301
9. Felbermayr G, Morgan TC, Syropoulos C, Yotov YV (2021) Understanding economic sanctions: interdisciplinary perspectives on theory and evidence. Eur Econ Rev 135:103720. https://doi.org/10.1016/j.euroecorev.2021.103720
10. Chang Y, Iakovou E, Shi W (2020) Blockchain in global supply chains and cross border trade: a critical synthesis of the state-of-the-art, challenges and opportunities. Int J Prod Res 58(7):2082–2099
11. Cui L, Yue S, Nghiem XH, Duan M (2023) Exploring the risk and economic vulnerability of global energy supply chain interruption in the context of Russo-Ukrainian war. Resour Policy 81:103373. https://doi.org/10.1016/j.resourpol.2023.103373
12. Park J, Shin B (2025) Conceptualizing digital sanctions as a new type of economic sanctions in the digital era: digital-related sanctions measures against Russia and their consequences. J Eurasian Stud 16(1):44–59
13. Wronka C (2022) Digital currencies and economic sanctions: the increasing risk of sanction evasion. J Financ Crime 29(4):1269–1282. https://doi.org/10.1108/JFC-07-2021-0158
14. Early BR (2021) Making sanctions work: promoting compliance, punishing violations, and discouraging sanctions busting. In: Research handbook on economic sanctions. Edward Elgar Publishing, pp 167–186
15. Dumanska I, Vasylkivskyi D, Hrytsyna L, Khmelevskyi O, Kharun O (2022) The impact of blockchain technology on the scenario development of a logistics enterprise. IJCSNS Int J Comput Sci Netw Secur 22(11):692–700
16. Al-Ababneh HA, Dumanska I, Derkach E, Sokhetska A, Kemarska L (2021) Integration of logistics systems of developing countries into international logistics channels. Acta Logist 8(4):329–340. https://doi.org/10.22306/al.v8i4.234
17. Dumanska I (2023) Export of agricultural products from Ukraine against the background of challenges to global food security: illicit trade, logistics and soil ecology during the war. In: Rural development: proceedings of the international scientific conference, pp 287–297. https://doi.org/10.15544/RD.2023.028
18. Newman AL, Zhang Q (2024) Secondary effects of financial sanctions: bank compliance and economic isolation of non-target states. Rev Int Polit Econ 31(3):995–1021
19. Sokolova A, Ratoshniuk T, Yepifanova I, Kravchyk Y, Ratoshniuk V, Dzhedzhula V (2024) Assessment of the efficiency of decentralization transformations in the rural areas of Ukrainian Western Polissia: current trends and challenges under the conditions of martial law. In: Lecture notes on data engineering and communications technologies. Springer Nature, Switzerland, pp 365–385. https://doi.org/10.1007/978-3-031-53984-8_16
20. McLeod AI (2005) Kendall rank correlation and Mann-Kendall trend test. R package Kendall, vol 602, pp 1–10
21. Dalmaijer ES, Nord CL, Astle DE (2022) Statistical power for cluster analysis. BMC Bioinform 23(1):205. https://doi.org/10.1186/s12859-022-04675-1
22. Shakhovska N, Yakovyna V, Kryvinska N (2020) An improved software defect prediction algorithm using self-organizing maps combined with hierarchical clustering and data preprocessing. In: Lecture notes in computer science. Springer International Publishing, pp 414–424. https://doi.org/10.1007/978-3-030-59003-1_27

Transformational Leadership of a Higher School in the Implementation of Sustainable Development Goals (Marketing Aspects)

Oksana Zamazii, Pavlo Izhevskyi, Valentyna Stadnyk, Vitaliy Yokhna, Vitalii Karpenko, and Eduard Brovarnyi

Abstract The aim of the work is to substantiate methodological approaches and develop scientific and methodological recommendations to increase the ability of the higher education system of Ukraine to demonstrate proactivity and intellectual leadership in strategizing and implementing transformation processes in Ukraine in line with the priorities of sustainable development. The initial hypothesis for forming recommendations on the priorities of sustainable development of Ukraine was the statement that in the modern world, social progress and environmental sustainability depend on the quality of education, which ensures the country's ability to form significant value propositions in the international exchange of goods and services. Such propositions include high-tech products that meet the requirements of environmental neutrality and resource efficiency. To confirm the hypothesis, a linear regression and variance analysis of the interdependence of the quality of higher education with the ability of countries to export high-tech products was carried out and Ukraine's rating was positioned in the evaluated set. The insufficient level of use of the potential of new knowledge in Ukraine for implementing technological transition in accordance with the requirements of the concept of sustainable development was indicated. We put forward the idea of transformational leadership of higher education in ensuring

O. Zamazii (✉)
Khmelnytskyi Cooperative Trade and Economic Institute, Khmelnytskyi, Ukraine
e-mail: O.V.Zamazii@xktei.km.ua

P. Izhevskyi
Leonid Yuzkov Khmelnytskyi University of Management and Law, Khmelnytskyi, Ukraine
e-mail: p_izhevskyi@univer.km.ua

V. Stadnyk · V. Yokhna · V. Karpenko · E. Brovarnyi
Khmelnytskyi National University, Khmelnytskyi, Ukraine
e-mail: stadnyk_v_v@ukr.net

V. Yokhna
e-mail: vyokhna@khmnu.edu.ua

V. Karpenko
e-mail: karpenkov@khmnu.edu.ua

A. Semenov et al. (eds.), *Data-Centric Business and Applications*, Lecture Notes on Data Engineering and Communications Technologies 272,
https://doi.org/10.1007/978-3-032-06608-4_5

the implementation of such a transition. We substantiated the essence, tasks and main directions of proactivity and leadership of higher education institutions in the processes of transformation of socio-economic systems and emphasized the priority of strategic marketing in the formation and positioning of the value of new educational products.

Keywords Transformation · Technological transition · Quality of education · Rating assessment · High-tech export · Proactivity of higher education institutions · Value propositions · Strategic marketing · Anticipatory management

1 Introduction

The development of societies and economies is currently confronted with global threats that could significantly undermine their resilience and security. This includes both climate threats, which annually result in multibillion-dollar losses to the global economy, and the technological transition driven by the urgent need to reduce anthropogenic impact on the environment. The primary goal of this transition is to abandon environmentally harmful, high-carbon footprint production that contributes to the intensification of the greenhouse effect, thereby worsening environmental conditions beyond reversible thresholds. According to the decisions of the Climate Summit in Paris, urgent measures are required to combat climate change and its consequences. In the context of these decisions, "all countries agreed to work to limit global temperature rise to well below 2 degrees Celsius, and given the grave risk, to strive for 1.5 degrees Celsius" [1]. This necessitates a gradual shift away from such technologies, emphasizing the use of renewable energy as the primary source of power. Simultaneously, the so-called "green" technological transition requires new engineering and technical solutions, implying intellectual effort and cutting-edge knowledge for generating and implementing innovations in technical designs, technological processes, and their instrumental-technological and organizational-logistic servicing, with information technologies playing a leading role.

In economically developed countries, scientific schools have long been established, working on fundamentally new technological platforms, developing comprehensive technological complexes for environmentally neutral production. Their participation in the global production of public goods involves producing new knowledge that materializes and brings high-value propositions to the market. This enables them to implement a sustainable development policy, evaluating progressive advancements through the lens of society's ability to achieve better economic outcomes while preserving the environment and improving the quality of life. The evident connection between the system of generating and implementing the economic potential of new knowledge, formed within the higher education system, and the country's ability to realize the concept of sustainable development underscores the importance of restoring such a connection for Ukraine. The country's economy, in its pre-war format and structure, is currently undergoing catastrophic destruction, and the demographic situation, exacerbated by forced migration, calls into question the country's ability to retain the necessary human capital for post-war revival, especially in its intellectual dimension. This makes research aimed at identifying the potential of higher education to ensure intellectual leadership in the strategic planning and implementation of transformational processes in the economy and societal life particularly relevant.

2 Literature Review

The issue of ensuring effective management of transformational processes in macroeconomic systems remains one of the most critical concerns for the scientific community in terms of identifying factors that determine the achievement of set goals in specific socio-historical circumstances. Scholars continually revisit this issue with each new phase of societal progress, comparing its dynamics in different countries and deducing cause-and-effect relationships to explain various outcomes. Increasingly, their research highlights the human factor, both in terms of its contribution to expanding sources of economic growth and the motivations that guide individuals in their actions [2–6]. Such research is being conducted in Ukraine, whose economy has degraded in the part that previously ensured its ability to create high-value-added products [7, 8]. The main reason for this is corrupt access to natural and material resources accumulated in previous periods, the use of which did not require significant investment costs and ensured high profitability without innovative renewal of production systems.

At the same time, the neglect of innovations as a source of development caused the deformation of the educational and scientific environment, the loss of a significant part of the country's scientific potential (due to the outflow of intellectually active members of society abroad) and the loss of interest of young people in acquiring engineering, technical and natural science professions. Ultimately, Ukraine has fallen into a "poverty trap" [9], where the absence of reliable property rights and corruption leads to the short-term nature of investments, businesses prefer shadow schemes, capital leaves the country, technological backwardness is preserved, the population's incomes decline, the domestic market contracts, and the economy stagnates [10]. Scientists had already pointed to these negative processes in Ukraine's economic structure even before the full-scale war [11, 12].

The war has further accelerated the processes of economic and technological degradation, significantly impairing the country's ability to achieve revival and sustainable development in the post-war period. To break out of this vicious circle of unsuccessful transformations in economic and societal life, it is crucial to clearly understand the ultimate goal of structural changes and identify their driving forces. In the current context, the main goal of transforming Ukraine's economy is not only to improve its ability to adapt to constant security challenges (including mechanisms for operational recovery/resource substitution at critical points of stability loss [13,

14]), but also to strengthen its resilience in [15] (and therefore competitiveness) in the strategic perspective, as emphasized in [15].

Therefore, defining the directions of structural changes in the economy and forming a mechanism for their implementation should be viewed through the prism of the economy acquiring new properties to ensure its stability in new conditions of global economic interactions, where geopolitics takes center stage, and major economic players declare goals to preserve ecosystem stability [16]. Higher education, by taking a proactive role in shaping development trends and preparing professionals with the necessary set of competencies to implement defined strategic tasks, is expected to play a significant role in addressing these challenges. Education was and remains the main competitive advantage of Ukraine.

It forms the quality of human capital, determining the ability of people to create new knowledge and find their practical application. That is, to form new points of economic growth, transforming the usual processes into more productive and resource-saving ones.

This is indicated, for example, by the components of the World Digital Competitiveness Ranking (World Digital Competitiveness Ranking). Three factors are subject to evaluation:

- knowledge embodied in know-how becomes the basis of future discoveries and new technologies;
- the level of development of digital technologies;
- readiness for future changes, which assesses the level of competence in the use of digital technologies.

In 2021, in this rating, Ukraine took 54th place among 64 countries, and the best position was for the "Knowledge" component—37. For the other two, the rating was 58. And this indicates a lag in the processes of practical application of the knowledge that is created in the field of education. At the same time, the dynamics of all rating indicators are positive—in 2019, the overall rating was 60 among 63 countries, knowledge—40, technology—61, and readiness for change—62 (in 2022–23, such an assessment was not carried out in Ukraine due to the war) [17]. That is the connection between the processes of creating new knowledge and their use for the development of digital technologies, which can find various practical applications, is gradually improving. But in today's world, it is no less important to develop other knowledge-rich areas In its current state, Ukraine must not lose the ability to produce high-tech products, as domestic production of such goods forms the basis of the country's defense capability.

Research indicates a close connection between a country's Education Index and its Environmental Performance Index in a set of countries included in the ranking. However, there is a somewhat weaker correlation between the Environmental Performance Index and the Index of High-technology Export. Authors [18] acknowledge this but do not analyze the reasons for such inconsistency. For Ukraine in its current state, it is crucial not to lose the ability to produce high-tech products, as domestic production of such goods forms the basis of the country's defense capability.

Given the tasks facing Ukraine today, it is important for higher education to ensure a tighter correlation between the outcomes of education, expenditures on the implementation of educational-professional programs, and the strategic goals of the country's development. This is emphasized by References [19–21]. Unfortunately, as research has shown [22], such correlation has not been achieved in Ukraine yet, with an unjustified excess of spending on training specialists in humanities compared to engineering-technical and natural sciences. This diminishes the productivity of Ukraine's collective intellectual capital in addressing the challenges of economic resilience and sustainable development. Certain measures to address this deformation through state regulation were declared by the Cabinet of Ministers of Ukraine in Resolution No. 1146 dated 24.12.2019 [23], but their full implementation was hindered by the full-scale war. Currently, among the numerous problems considered most obvious for urgent resolution, scientists highlight the mismatch of volumes and quality of higher education training to the modern and post-war needs of Ukraine, its regions, considering the consequences of the war, changes in population size, relocation of enterprises, and business in various sectors of the economy [24].

Concurring with this conclusion, it is important to note that the task of higher education should be considered much broader—not only to align with the urgent needs of ensuring the stability of the national economy but also to form an intellectual platform for the post-war revival of Ukraine's economy in line with sustainable development goals. The author's position is that higher education institutions should function in the mode of transformational leadership.

This term belongs to the semantic field of leadership theory [25] and pertains to characteristics of individual personalities [26, 27]. Therefore, its use to describe organizations that unite a certain number of individuals with different behavioral patterns requires a separate interpretation of its essence. Additionally, positioning higher education as part of institutions contributing to the achievement of sustainable development goals also requires additional argumentation, considering the priority of these development goals in Ukraine. The interconnectedness of all these issues has defined the goals and tasks of this research.

3 Research Objective

To outline methodological approaches and formulate scientific methodical recommendations to enhance the ability of Ukraine's higher education system to demonstrate proactivity and intellectual leadership in the strategic planning and implementation of transformative processes in the economy and societal life for the post-war recovery of the country in line with sustainable development priorities.

4 Research Methodology

The theoretical and methodological basis of this study is grounded in the theory of economic development, focusing on its provisions related to the endogenous nature of economic growth and the contemporary interpretation of the sources and constraints of this growth. The research methodology encompasses several theoretical concepts and methods to address various aspects of the defined objectives.

Specifically:

- To determine the target function of transformative processes in Ukraine's economy and identify the specific demands of the Ukrainian economy for professional training, considering its current state, analytical approaches of statistical analysis were employed, guided by the priorities of the sustainable development concept.
- To position the role and tasks of higher education in the transformative process, linear regression and variance analysis methods were utilized.
- To give concrete content to the definition of "transformative leadership of higher education" within the frameworks of leadership theory, knowledge economics, and the concept of endogenous innovative development, gnoseological analysis and abstract-logical methods were applied, relying on the conclusions of the dispersed knowledge concept and information asymmetry.
- To structure tasks and visualize the proactivity and transformative leadership of higher education institutions, structural-functional analysis methods and graphical modelling were employed.

5 Results and Discussion

The concept of sustainable development is now recognized as best meeting the requirements of maintaining a balanced interaction between humans and nature. It incorporates the views of progressive representatives of the scientific community and policymakers on how to limit the detrimental impact of humans on nature and preserve an acceptable quality of life for future generations. The main thesis of the concept is the increasing role of the social component in development tasks. Not only in the context of solving social problems (such as hunger, unemployment, health preservation, etc.) but primarily in changing worldview priorities. Among these priorities, quality education (ranking fourth in the list of sustainable development goals) stands out, enabling individuals to realize their intellectual potential in various professional activities requiring specialized knowledge and skills.

If preschool and school education instill basic knowledge and foster an interest in self-realization in students, the mission of higher education is much broader. Higher education ensures the expanded reproduction of the country's intellectual capital, serving as a producer of new knowledge and a driver of new global trends in the economy and societal life. Higher education must meet societal needs for quality professional training for the new generation of economic participants by substantiating the content of educational-professional programs with knowledge that forms the basis of graduates' future professions and successful self-realization in society In the modern information-intensive economic space, the higher education system significantly determines the sustainable development of the country.

In the context of sustainable development goals, knowledge specific to each profession should be complemented by an understanding of the close connection between the consequences of human activity and the state of the environment. This includes an awareness of the urgent limitations that society must impose on those who exploit non-renewable resources recklessly for profit. The complexity in presenting such knowledge in the process of professional training yields results.

As noted above, during the analysis of the impact of the education system on compliance with environmental sustainability standards in over a hundred countries in the ranking, a close connection between the country's Education Index and its Environmental Performance Index was identified [18]. In countries where the correlation is high, environmental priorities are not only declared at the policy level but are also filled with specific actions when making management decisions in the business environment. These countries predominantly adhere to the principles of environmental neutrality. This is evident in the relocation of industrial production to other countries, with a rise in the share of high-tech services rather than resource-intensive production in the export structure. At the same time, these countries remain producers of high-tech products, actively promoting them in the markets of other countries. Obviously, this ability is ensured by the high intellectual potential of engineering and technical workers in relevant enterprises who implement the knowledge acquired in higher education into original technological solutions.

To confirm this conclusion, data from the ranking assessment of countries based on the level of education and the level of high-tech product exports were used, published annually in the Global Indices list [28, 29]. The reflection of the interdependence of the rankings of both indices is presented in Fig. 1 (the highest ranking of both indices, reflecting the country's best position in the total sample, has a value of 1; countries with the highest ranking values are located in the bottom-left part of the figure). The results of the conducted linear regression analysis for the arrays of Global Indices with a sample of 60 countries are described by the equation $y = 0.719x + 8.238$. The information reliability level, determined by the coefficient of determination, is $R^2 = 0.559$, indicating a moderate correlation between the factors. The correlation coefficient is positive, $r = 0.748196$, indicating a direct relationship between the phenomena (a higher education ranking in a country is mostly associated with a higher rating for high-tech exports). To verify the adequacy of the results of the linear regression analysis, we applied analysis of variance using the Fisher criterion (degrees of freedom are 1 and 55) with a value of 69.94 (compared with the tabular critical value of Fisher). Assuming a significance level of $A = 0.05$ (confidence probability of 0.95), we obtained a result of $2.24e{-11} < 0.05$, allowing us to assert the adequacy of the model by this criterion. Checking the model for statistical significance using the Student's t-test, we obtained a probability value of $A/2 = 0.025$, identical to the previous check, $2.24e{-11} < 0.05$, indicating the statistical significance of this coefficient. In conclusion, our calculations confirmed

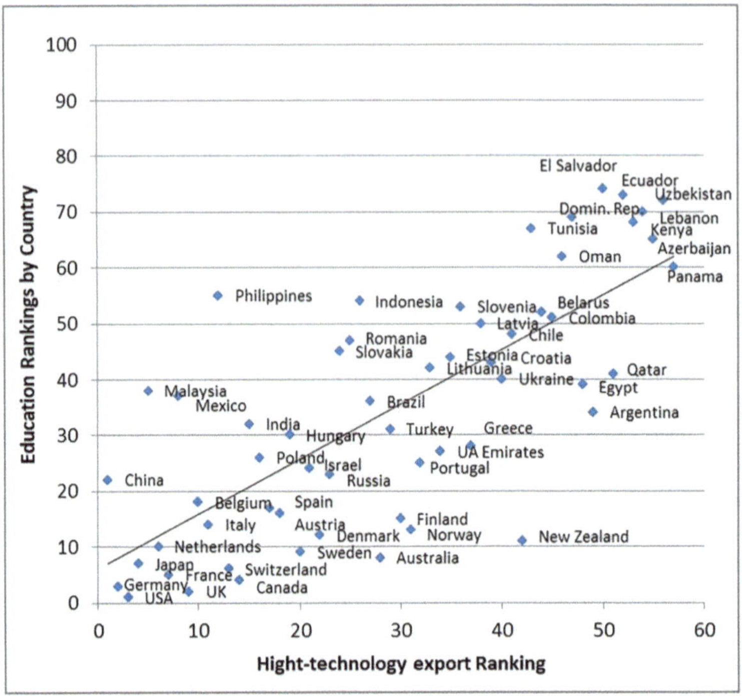

Fig. 1 Ranking dependence between education index and high-technology export index of countries. *Source* Built by the authors according to data [28, 29]

the assumption of a correlation between the quality of education in countries and their ability to participate in the international market with proposals for high-tech products. Ukraine also demonstrates this ability.

The results of the conducted linear regression analysis for the arrays of Global Indices with a sample of 60 countries are described by the equation $y = 0.719x + 8.238$. The information reliability level, determined by the coefficient of determination, is $R^2 = 0.559$, indicating a moderate correlation between the factors. The correlation coefficient is positive, $r = 0.748196$, indicating a direct relationship between the phenomena (a higher education ranking in a country is mostly associated with a higher rating for high-tech exports). To verify the adequacy of the results of the linear regression analysis, we applied analysis of variance using the Fisher criterion (degrees of freedom are 1 and 55) with a value of 69.94 (compared with the tabular critical value of Fisher). Assuming a significance level of $A = 0.05$ (confidence probability of 0.95), we obtained a result of $2.24e{-}11 < 0.05$, allowing us to assert the adequacy of the model by this criterion. Checking the model for statistical significance using the Student's t-test, we obtained a probability value of $A/2 = 0.025$, identical to the previous check, $2.24e{-}11 < 0.05$, indicating the statistical significance of this coefficient. In conclusion, our calculations confirmed the assumption of a correlation between the quality of education in countries and their ability to participate in the international market with proposals for high-tech products. Ukraine also demonstrates this ability.

As seen in Fig. 1, Ukraine ranks 42nd in the overall rating of high-tech product exports among 200 countries. The education index is 0.52 (maximum 1.0), suggesting that Ukraine's intellectual potential is quite high for generating new knowledge and implementing it in engineering, technology, and innovation. However, this potential is underutilized, as the main export commodities currently consist of raw materials and their primary processing products. Sustainable development requires the ability of economic entities to present themselves in international markets with more exclusive and significant propositions of consumer value, incorporating advanced technologies [30].

For Ukraine in its current state, it is important not to lose the ability to produce high-tech products—because the own production of such products is the basis of the country's defense capability. From the perspective of realizing the potential of new knowledge, it is also important to observe the connection between a country's spending on education (as a percentage of GDP spent on education) and its ability to be a participant in the international market for high-tech products. To determine the correlation between these factors through their rating evaluation, we utilized the existing array of Global Indices data [28, 29]. The results of comparing rating evaluations for a sample of 54 countries are presented in Fig. 2.

The result is described by the equation $y = -0.004x + 4.844$ with a coefficient of determination $R^2 = 0.04$, indicating a practical absence of a connection between the researched phenomena. This seemingly irrational result has a logical explanation when analyzing the structure of education spending presented in [31]. As a rule, the worse a country's ranking in the selected sample, the larger the share of its spending on primary and secondary education. Expenditures on professional, higher, and postgraduate education in developing countries are relatively low, especially in combination with their low GDP levels. It is evident that the quantity and quality of vocational training for their scientific, research, and engineering personnel are insufficient for the development of competitive products in the markets of other countries, thereby limiting their export potential in technological transfer.

However, this explanation cannot be applied to the ranking positions in Fig. 2 for countries such as China, Hong Kong, Malaysia, and some others. Despite their low ranking in education spending (75–100 out of 200 countries), they have a high rating for the High-Technology Export Index. To understand the reasons for this deviation from the previously described trend, it was deemed appropriate to analyze a sample of countries from the TOP-10 exporters of high-tech products (excluding the United States and Mexico, for which education spending data are not available). These countries allocate a significant portion of their educational expenditures to adult education. The positioning of these countries in the ranking evaluation of both indicators is presented in Fig. 3. We obtained a linear-regression dependence described by the equation $y = -0.103x + 12.32$ with a high level of interdependence (coefficient of determination $R^2 = 0.74$). However, the correlation coefficient has a negative value, $r = -0.8612$, indicating an inverse relationship—a higher level

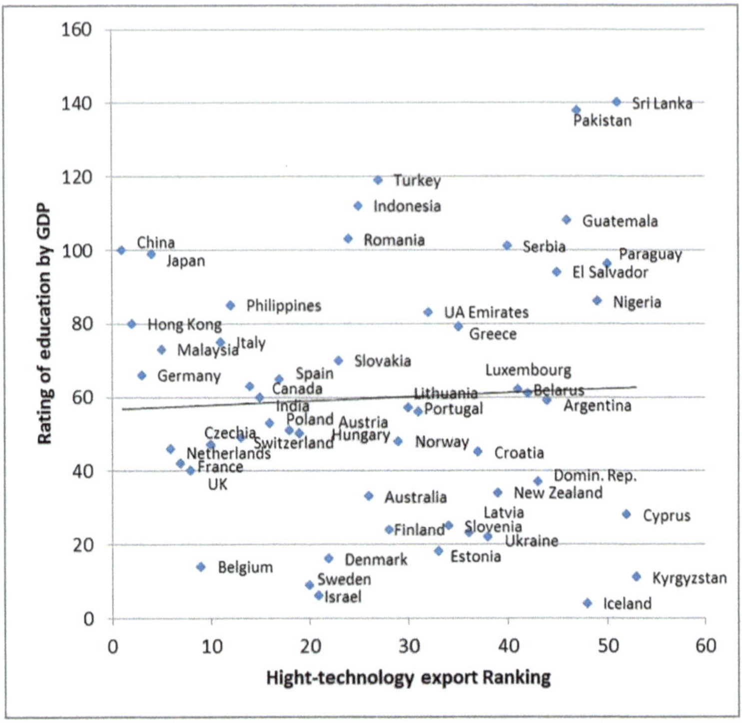

Fig. 2 Rating dependence between the level of spending on education and the high technology export index of countries. *Source* Built by the authors according to data [28, 29]

of a country's education spending does not guarantee a better position in terms of high-tech exports.

The Fisher Criterion (with 1 and 6 degrees of freedom) with a value of 11.79 and a hypothesis level of A = 0.05 (confidence probability of 0.95) is 0.0139 < 0.05, confirming the adequacy of the model. The test for static significance is confirmed by the Student's criterion with a probability value of 0.00047 and a calculated result of 0.0139 < 0.05. The explanation for such an inverse relationship can take into account the following:

- Most of the TOP 10 countries presented in the figure belong to the EU group (Germany, the Netherlands, France, Belgium) and adhere to a basic policy of improving the quality of education combined with minimizing unjustified spending on it. For instance, there is a goal to reduce the proportion of early school leavers (ELET) to below 9% by 2030 [31]. The Eurostat Report emphasizes that education and vocational training are key driving forces in the economies of countries, influencing employment, productivity, innovation, and competitiveness. These are essential prerequisites for supporting and increasing high-tech

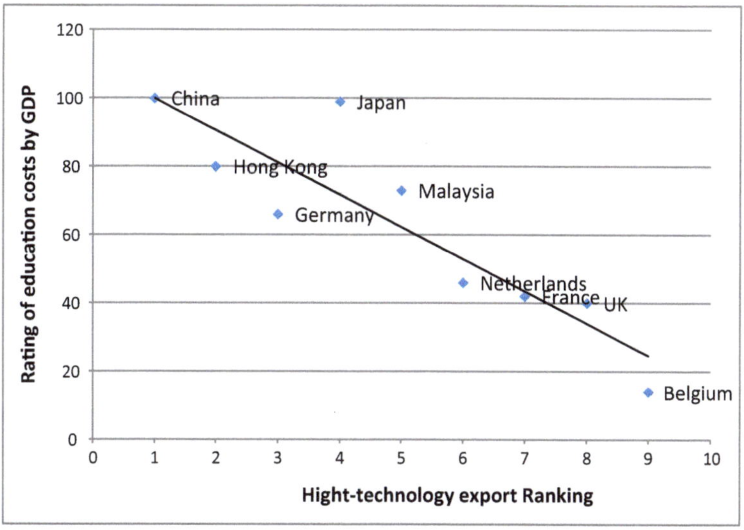

Fig. 3 Ranking dependence between levels of education spending and high-technology exports of countries (TOP 10). *Source* Built by the authors according to data [29, 31]

exports and other sustainable development goals in the EU strategy. Special attention is also given to improving the quality of higher education and continuous adult education as investments in human capital, whose quality and retraining depend on the dynamics of changing technologies, especially in high-tech sectors of the economy. Scientists who studied the export of medium and high-tech production in Europe [32] prove that the basis of the production of goods for high-tech export is technological innovation and intangible assets of enterprises, which were created by their own system of research and innovation. This system is directly influenced by the continuous learning process aimed at acquiring knowledge through professional, higher and postgraduate higher education. The formation of the chain "continuous education - technological innovations at the enterprise - production of competitive high-tech products for export" is included in the basis of financing of professional and higher education in the EU countries, distinguishing the costs of its acquisition from scientific and research activities, which are financed mainly at the expense of enterprises and directly depend on its qualities.

- The results of the analysis of China's State expenditures on education and economic growth (2009–2019) highlight a disproportionate relationship between the growth of education spending and GDP growth, particularly favouring the latter. There is a significant gap between education spending in rural and urban areas. The funding ratio between primary, secondary, and higher education in 1995 was 1:2.28:20.43, allowing for the development of quality higher education in China. However, by 2019, the funding ratio was 1:1.46:1.96, indicating China's attempt to balance the ratio between education levels. Nevertheless, experts argue that the lack of a significant increase in spending on higher education, relative to GDP, may lead to problems in technological development [33].

According to the Standard International Trade Classification (SITC) [34], high-technology exports include aerospace products, computers and office equipment, electronics and telecommunications, pharmaceuticals, scientific instruments, electrical machinery, chemistry, nonelectrical machinery, and weapons. Competitive, political, environmental, and security restrictions among countries trading in such goods can reduce the volume of their exports. In our opinion, the explanations provided by researchers are well-founded. They can serve as a basis for the conclusion that a country's entry into the global market of high-tech products requires higher education with significant knowledge creation potential.

Thus, the implementation of the concept of sustainable development in Ukraine cannot be achieved without its own industrial production that is resource-efficient and environmentally friendly. Therefore, we consider the statement made by us above that higher education plays a key role in ensuring the necessary transformations in the economy and society to be justified. Unfortunately, the massive destruction suffered by the country's industrial sector during the war has sharply reduced Ukraine's ability to maintain its positions in the international market for technology and technological equipment.

Moreover, the share of exports of technically complex products has been steadily decreasing over the past 10 years (from 13.7% in 2014 to 9% in 2021). In 2022, the total export of such products across all categories in terms of statistically highlighted groups fell by a third—from 6.1 billion USD to 4.2 billion USD [35]. Individual cases of successful relocation of industrial enterprises to the western regions of Ukraine are unable to significantly improve the situation in the industry.

Moreover, due to the war, a considerable portion of qualified workers in Ukraine, who were carriers of engineering and technical knowledge and skills, has been irreversibly lost. As a result, the structure of the labor market has changed and the unemployment rate has increased. Certainly, the unemployment rate in any country depends on the state of the economy in a specific historical period.

It varies across different categories of workers, ranging from age to professional skills or types of activities. In the conditions of evolutionary changes in the economy, the structure of employer demands in the labor market gradually changes, allowing job seekers to adapt relatively quickly and find employment. However, with the onset of the war, the labour market in Ukraine entered a zone of high turbulence due to migration processes and mobilization of people for military actions, as well as massive physical destruction of production systems of many economic entities, especially industrial enterprises. In just 2022, Ukraine's economy lost almost a third of its GDP, undergoing radical structural changes, leading to shifts in the demand structure for labor and intensifying structural unemployment. Structural unemployment arises when workers lack the necessary professional or qualification skills, or conversely, have an excess of them in their region or field of activity. Given the large number of internally displaced persons who lost their jobs due to the war, their professional specialization does not always enable them to find suitable employment in the new location.

This indicates a significant mismatch between the needs of employers and what workers can offer. To reduce structural unemployment, comprehensive measures are needed, including education reform. Given the key role of education in meeting the workforce needs of the business environment, its participants must directly participate in defining these needs. Some work in this direction is being carried out in Ukraine. For instance, within the USAID program "Dream and Act," a study on the Future of Work 2030 was conducted to identify areas vital for Ukraine's recovery [36]. In October 2022, the following sectors were identified:

- Security and defense industry
- Medical services and healthcare
- Construction
- Architecture and infrastructure
- Agribusiness and agriculture
- IT and digital technologies
- Energy
- Social services and psychological support
- Retail trade
- Transport and logistics
- Education and science.

It is evident that the compiled list of priority sectors is primarily relevant for the restoration of Ukraine's economy, ensuring its defense capability, and supporting the population during active military, political, and economic confrontation with the Russian Federation.

In this list of important sectors, education and science occupy the 11th (last) position. Although it is clear that without qualified personnel, all other priority sectors cannot develop as dynamically as the circumstances demand. The scale and uniqueness of technical and technological tasks that Ukraine will have to tackle in the near future make the need for specialists in engineering design, technological construction, information support of production and logistics processes, and positioning the value characteristics of manufactured products for entry into target consumer market segments, among other things, urgent. It is essential to ensure the balanced development of sectors in the economy that can provide not only faster but also greater economic returns on invested capital.

For example, the agro-industrial sector cannot generate significant added value if its processed products (e.g., organic) are not properly positioned in prospective (in terms of economic efficiency) markets and have limited sales. Marketing technologies and tools should complement and expand the technological capabilities of processing enterprises, as detailed in the work [37] regarding the food market. Hence, the professional training of all specialists of the necessary qualifications should also be balanced in terms of quantity and quality, meeting not only immediate but also future needs of the economy. Considering the duration of preparing future professionals, educational programs today should incorporate the learning outcomes required by the Ukrainian economy in 4–5 years.

As mentioned earlier, these outcomes should contribute to achieving sustainable development goals. Given that preparing qualified professionals requires significant expenses (with the amount varying by specialty) and that the state bears a substantial portion of these costs, it is crucial to rationally formulate orders for specific specialties. There is already some experience in justifying such needs in Ukraine. In April 2023, the Ministry of Economy developed a list of specialties for training professions using education vouchers for internally displaced persons who can obtain a new specialty at the state's expense.

While the approved list includes engineering specialties, it is important for the content of educational programs to align with the current demands of the labor market in Ukraine and the programs for post-war economic recovery. Professionals in these fields will experience increased demand, participating in justifying regional or cross-industry projects for socioeconomic development. They should not only consider the region's resource potential and the technological trends in a specific industry but also ensure the maximization of societal benefits from their implementation. In the conditions of rapid and multi-vector scientific and technical progress, the updating of knowledge in the main trend areas that can open up new opportunities for economic growth takes place in real time. Therefore, the education system should not only outline the basic outline of the chosen profession to those seeking educational services and fill it with the necessary knowledge for mastering the relevant competencies, but also form a vision of various options for their professional future.

And therefore—to teach them to select from the information space exactly the knowledge they need to improve their professional skill and competence demand in the selected segment of the labour market. In the context of sustainable development goals, higher education institutions must contribute to the expanded reproduction of the country's intellectual capital. Therefore, their functioning should be subordinated to the law of increasing returns, where each new learning cycle concludes with the release of specialists possessing competencies that enhance their professional performance, considering the dynamics of market changes.

This includes not only the changes that have occurred but also those expected based on the strategies laid out for the country's socio-economic development. In the process of shaping these goals, higher education should play a proactive role, indicating directions for prospective transformations—demonstrating transformational leadership. Transformational leadership is a term used to describe leadership characterized by strategic thinking and anticipating the future. It has a decentralized nature and relies on broad support from followers, to whom authority is delegated for active participation in the transformations occurring in socio-economic systems.

The term "transformational leaders" traditionally refers to individuals with a set of competencies capable of radically changing the organizational environment in which they work-starting from the mission, structure, functional purpose of individual elements, and their relationships, ending with a change in corporate culture [25–27]. In our opinion, attributes of strategic foresight and the formation of a conceptually new approach to building the model of economic system functioning in a highly dynamic environment with a new set of resource constraints should be present in higher education as an important societal institution.

However, for this, higher education institutions must not only lead in the processes of internal transformations, introducing progressive changes in the structure of educational products and the organization of the educational process, as viewed from the standpoint of maintaining the system's stability in the face of dynamic changes in its operating environment. In our view, higher education leadership should be considered through the lens of its transformative impact on this environment-through complementary integration of processes for generating new knowledge and disseminating it.

Higher education should show proactivity in creating new knowledge that can be valuable for business. Many of the world's leading universities work according to this principle. New knowledge, formalized in the form of an intangible asset, can become an economically attractive value proposition in the innovation market. Its acquisition (provided exclusive rights to use it are acquired) can become a key factor in business competitiveness and be the basis of market success. In this case, it can be argued that the research component of a higher school produces a resource for creating new points of economic growth, first by increasing the information asymmetry in the business environment with its value propositions, and then by offering an educational product to overcome it.

Thus, by transformative leadership in higher education, we mean its ability to ensure the expanded reproduction of the country's intellectual capital in line with global trends in technological development and socio-economic progress, taking into account its natural resource potential. This is achieved through the proactivity of higher education institutions in shaping and disseminating new knowledge, which serves as the foundation for defining the directions of social and economic transformations. The author's position is that this can only be realized based on an anticipatory approach in management, which shapes a proactive approach to working with the market. Anticipatory management is "aimed at anticipating all possible events in the internal and external environment based on the identification of weak signals in the functioning environment of entrepreneurial structures" [38].

Obviously, achieving proactivity should be the goal of all higher education institutions—competing among themselves in the quality of educational products, the efficiency of organizing the educational process, and its effectiveness in imparting competencies relevant to the "future" chosen specialties. The author's vision of the model of functioning of a higher education institution based on proactivity and transformational leadership and the role of strategic marketing in this process are presented in Fig. 4.

An anticipatory orientation allows higher education institutions, in market research of educational services, to give preference to forecasts of structural changes in the economy, as structural changes in the labor market usually follow changes in employer demands. The more competitive higher education institutions will be those that recognize the vectors of structural transformations in the global economic space at an early stage. This will provide them with sufficient agility

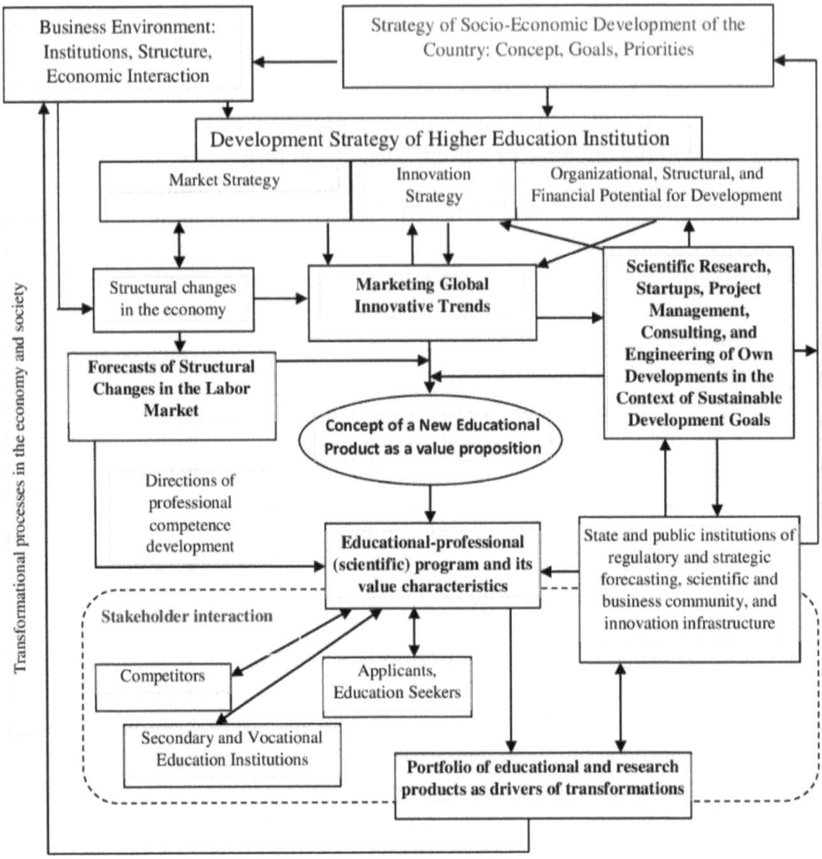

Fig. 4 Proactivity and leadership of higher education institutions in the processes of socio-economic system transformation

to incorporate the necessary knowledge components into the structure of current educational-professional programs or in the development of new ones.

They should expand the portfolio of graduates' competencies, preparing them for professional activities in the face of future structural shifts. Thus, it can be argued that higher education institutions will ensure the expanded reproduction of the country's intellectual capital. And an important role in determining priorities in the development of educational products of higher education institutions will be played by marketing analysis and strategizing value propositions in each component of the extended reproduction cycle (Fig. 5).

At each stage of the cycle of expanded reproduction of intellectual capital, innovations are introduced into the corresponding processes, adding value to the educational product for different categories of learners. Innovations can relate to both the structure of knowledge that forms the professional profile of the learner's competencies and the methods of organizing the educational process, which create better conditions

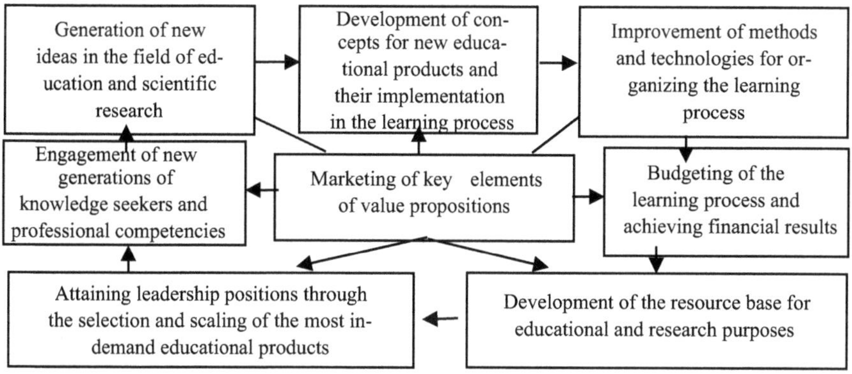

Fig. 5 Cycle of expanded reproduction of intellectual capital in higher education institutions

for achieving learning outcomes in various educational products. It is the complementarity of innovations in these components of the educational process that will ensure the growth of the educational product's value for the learner (new knowledge may require new methods of presentation in the learning process and new ways of assimilation for future professional use).

Proactivity in modern conditions is ensured by using artificial intelligence for planning and organizing the educational process—both for predicting technological trends that could radically change the structure of the labor market and for developing effective marketing tools to maintain demand for educational services. This will allow educational institutions to solidify certain educational directions, enrich them with new teaching technologies and professional competencies, expand the scope of scientific research, introducing those that align with the demands of new sectors of the economy.

By the way, the use of advanced information technologies in the educational process is a necessary condition for the qualitative development of higher education. It develops skills in learners to work in the conditions of global digitization of the entrepreneurial environment—where routine processes are maximally automated.

And in the aggregate, the skills of professional competence are recognized as crucial in the job market, enabling the algorithmization of most components of business processes, correctly selecting and improving decision support systems at their critical points [39].

To implement the functions of transformative leadership in higher education institutions, active research activities should be conducted, during which new knowledge is created and transformed into tangible or intangible products for practical application. It is important to conduct scientific research and carry out project research considering the natural resource potential of regions, focusing on the creation of high-tech and environmentally friendly processing complexes.

These complexes would generate greater added value and expand the possibilities of positive structural transformations in the regional economy in line with the priorities of sustainable development. Such proposals for innovative technological development of production will serve as an effective motivator for expanding collaboration between higher education institutions and representatives of the entrepreneurial sector [40]. Considering that modern conditions for providing educational services in Ukraine combine state orders and contractual forms of education, an increase in the number of individuals studying on a contract basis expands the financial opportunities for implementing the development strategy of higher education institutions (HEIs). The quantitative and qualitative characteristics of the student body depend on the ability of HEIs to use modern marketing tools to position new educational products. When choosing them, attention should be paid to the peculiarities of forming consumer value in innovations [30].

In the service sector, in the structure of the market offer, more attention should be given to its cognitive component, which relates to the sphere of perception [41]. The more attractive the substantive content of specific educational products is for target consumer groups (in terms of the economic potential of their practical application), and the more perfect (in terms of assimilation) the form of its presentation, the higher the likelihood of an increase in consumer demand for the services of a specific HEI [42]. As a result, the financial capabilities of this educational institution will grow from their consumption.

Consequently, the return on educational activities will increase. To increase knowledge creation potential and ensure the exclusivity of new knowledge, every modern higher education institution must conduct research within authorial scientific schools. The results of these studies are taken into account when forming new educational products.

The practical implementation of scientific developments (their commercialization in the business environment) provides additional income to the budget of HEIs. The success of commercializing RD results will enhance the university's rating assessment in the market for research and consulting services. As a result, the likelihood of a positive evaluation of grant applications and, accordingly, obtaining funds for conducting research in the following years will increase.

6 Conclusions

The implementation of social and environmental priorities of the Sustainable Development Concept in Ukraine requires significant structural changes in the national economy, as its current state does not provide the necessary level of economic results for technological transition and addressing social problems resulting from conflict. Substantial losses have occurred in the labour market, where there is a sharp shortage of professionals capable of developing and implementing sustainable development projects at a high professional level, both at the regional level (considering their resource specificity) and cross-industry (with adherence to the requirements of maximizing societal benefit from their implementation).

These and other factors complicate the strategic planning of socio-economic processes in Ukraine in line with the priorities of sustainable development, and their consideration is a mandatory condition for the success of transformative processes. The initial hypothesis for forming recommendations on identifying driving forces and priority points of transformative changes in the socio-economic life of Ukraine was based on the assertion that in the modern world, social progress and ecological sustainability depend on the quality of education.

This dependency has been statistically proven by other researchers during comprehensive ranking assessments of the economies of 200 countries. Building on the assumption that achieving sustainability is possible through a country's ability to engage in international trade with significant value propositions, we argued that high-tech production falls under such propositions. To confirm the hypothesis, linear regression and variance analysis were conducted to establish the interdependence of the quality of higher education with the ability of countries to export high-tech products, positioning Ukraine's rating in the evaluated set.

We pointed out the insufficient utilization of the potential of new knowledge in Ukraine for implementing technological transition according to the requirements of the sustainable development concept. We introduced the idea of transformative leadership in higher education to facilitate such a transition and argued for considering higher education leadership through its transformative impact on the business environment. Transformative leadership in higher education was defined as the ability to ensure the expanded reproduction of the country's intellectual capital in the context of global trends in technological development and socio-economic progress.

This is achieved through the proactivity of higher education institutions in shaping and disseminating new knowledge, which serves as the foundation for defining the directions of social and economic transformations.

We substantiated the tasks of such leadership for higher education institutions and emphasized its proactive nature. The content was structured, and the main directions of proactivity and leadership of higher education institutions in the processes of transforming socio-economic systems were delineated. Development of higher education directions and financing of research works according to the:

1. Hausman-Idalgo Economic Complexity Index (ECI), the logic of which is tied to hightech exports based on two criteria: "diversity" and "ubiquity". Diversity refers to the quantitative list of high-tech goods exported by the country, while ubiquity is the number of countries also exporting high-tech goods. The higher the list of goods and the lower their ubiquity in the country's exports, the higher the index.
2. Coverage of the population with digital education and inclusion of a sub-index of higher education digitization into the official Ukrainian Digital Transformation Index.
3. Expansion of grant funding for education acquisition and conducting research work through budgets at all levels.

Emphasized the priority of strategic marketing in shaping and positioning the value of new educational products. Implementing these recommendations will be of practical importance for reforming the system of higher education in Ukraine in relation to the priorities of sustainable development.

References

1. United Nations Sustainable Development Goals (2016) 17 goals to transform our world. http://www.un.org/sustainabledevelopment/climate-change-2. Last accessed 2024/11/21
2. Feldman M, Hadjimichael T, Kemeny T, Lanahan L (2016) The logic of economic development: a definition and model for investment. Environ Plan C Gov Policy 34(1):5–21. https://doi.org/10.1177/0263774X15614653. Last accessed 2024/11/21
3. Pietak L (2014) Review of theories and models of economic growth. Comp Econ Res Cent East Eur 17(1):45–60. ISSN 2082–6737. https://doi.org/10.2478/cer-2014-0003. Last accessed 2024/11/21
4. Pelsa I, Balina S (2022) Development of economic theory—from theories of economic growth and economic development to the paradigm of sustainable development. DIEM. https://www.researchgate.net/publication/367814319. Last accessed 2024/11/21
5. White RJ (2024) Imagining education beyond growth: a post-qualitative inquiry into the educational consequences of post-growth economic thought. Curric Perspect 44:539–549. https://doi.org/10.1007/s41297-024-00232-3. Last accessed 2024/11/21
6. Lesuisse P (2024) Education, public expenditure and economic growth under the prism of performance. HAL Id: hal-03685311. https://hal.science/hal-03685311v1. Last accessed 2024/12/21
7. The importance of human capital development in the modern world. What should be Ukraine's strategy (2021). Ukrainian Institute for the Future, p 13. https://uifuture.org/publications/vazhlyvist-rozvytku-lyudskogo-kapitalu-u-suchasnomu-sviti-yakoyu-maye-buty-strategiya-ukrayiny/. Last accessed 2024/12/21
8. Shinkaruk L, Bevz I, Baranovska I et al (2015) Structural transformations in the economy of Ukraine: dynamics, contradictions and impact on economic development. Scientific report. Institute of Economics and National Academy of Sciences of Ukraine, Kyiv
9. Sachs JD (2005) The end of poverty: economic possibilities for our time. Penguin Books, New York
10. Bashnianyn HI, Shevchyk BM (2015) «Pastka bidnosti» ta ekonomichnyi oriientalizm ["Poverty trap" and economic orientalism]. VisnykLvivskoikomertsiinoiakademii Seriiaekonomichna 49:185–193
11. Tabachuk A (2021) Directions of transformation of the economic system of Ukraine in today's conditions. Econ Soc 26:5–21. https://doi.org/10.32782/2524-0072/2021-26-29. Last accessed 2024/10/21
12. Havlovska N, Rudnichenko Y, Lisovskyi I (2019) Transformation processes in the system of providing economic security of investment activities of industrial enterprises. Baltic J Econ Stud 5(2):18–23
13. Bielinskyi AO (2023) Recurrence quantification analysis of energy market crises: a nonlinear approach to risk management [Electronic resource]. In: Proceedings of the selected and revised papers of 10th international conference on monitoring, modeling & management of emergent economy (M3E2-MLPEED 2022). CEUR workshop proceedings 3465, pp 110–131. https://ceur-ws.org/Vol-3465/paper14.pdf. Last accessed 2024/10/20
14. Hryhoruk P, Khrushch N, Grygoruk S (2021) Environmental safety assessment: a regional dimension. IOP conference series: earth and environmental science, vol 628. IOP Publishing, p 012026. https://doi.org/10.1088/1755-1315/628/1/012026

15. Kechichian ER, Takemoto S, Shin NY (2020) Industries competitiveness in the face of disasters. World Bank, Washington, DC. http://documents.worldbank.org/curated/en/682501604040773738/Resilient-Indu-stries-Competitiveness-in-the-Face-of-Disasters. Last accessed 2024/10/20
16. Zamazii O, Dupliak O, Karpenko V, Proskurovych O, Mazarchuk A (2021) Place of environmental management in Ukraine in the system of modeling management of sustainable development of the region. WSEAS Trans Environ Dev 17:253–261
17. World digital competitiveness ranking-2021. Technical report. https://imd.cld.bz/Digital-Ranking-Report-2021/354. Last accessed 2024/12/20
18. Voitko S, Sakalosh T, Mazanko T (2023) Circular economy represented in selected global economy indices. Innov Sustain 2:90–96
19. Herbst M (2020) Performance-based funding, higher education in Europe. In: The international encyclopedia of higher education systems and institutions, pp 2227–2231. https://doi.org/10.1007/978-94-017-8905-9_561. Last accessed 2024/12/20
20. Tomchuk O, Tserklevych V, Hurman O, Petrenko V, Chymosh K (2020) The efficiency of higher education institutions: evaluation and management of managers. Estud Econ Apl 38(4)
21. Kobets V, Yatsenko V, Buiak L (2021) Bridging business analysts competence gaps: labor market needs versus education standards. Commun Comput Inf Sci 1308:22–45
22. Kovtunets V, Londar S (2021) Evaluation of higher education services cost. Econ Ann XXI 178(7–8):57–69
23. On the distribution of state budget expenditures between higher education institutions based on indicators of their educational, scientific and international activities. Resolution of the Cabinet of Ministers of Ukraine dated 24.12.2019 No. 1146 (2019). https://zakon.rada.gov.ua/laws/show/1146-2019-%D0%BF#Text. Last accessed 2024/12/20
24. Vitrenko Y (2020) The plan for the restoration of the sphere of higher education of Ukraine as a management tool in the conditions of martial law. Education of Ukraine in the conditions of martial law: management, digitalization, European integration aspects: a collection of abstracts of reports of the 4th international scientific and practical conference, pp 80–83
25. Collins RT II, Algaze C, Posner BZ (2023) The leadership/management concept scale: differentiating between actions constituting leadership and management. Leadersh Organ Dev J 44(5):657–677
26. Klein G (2023) Transformational and transactional leadership, organizational support and environmental competition intensity as antecedents of intrapreneurial behaviors. Eur Res Manag Bus Econ 29(2)
27. Greimel NS, Kanbach DK, Chelaru M (2023) Virtual teams and transformational leadership: an integrative literature review and avenues for further research. J Innov Knowl 8(2)
28. The Global Economy (2024) Business and economic data for 200 countries. High tech exports—country rankings. https://www.theglobaleconomy.com/rankings/high_tech_exports/. Last accessed 2024/12/20
29. World Population Review (2023) Education rankings by country 2023. https://worldpopulationreview.com/country-rankings/education-rankings-by-country. Last accessed 2024/12/20
30. Stadnyk VV, Zamazii OV (2015) Innovative factors in the system of value-based management of an industrial enterprise. Actual Probl Econ 171(9):242–249
31. European Commission. Statistics explained. Quality education. Technical report. https://ec.europa.eu/eurostat/statistics-explained/index.php?title=SDG_4_Quality_education#Quality_education_in_the_EU:_overview_and_key_trends. Last accessed 2024/12/20
32. Leogrande A, Costantiello A, Laureti L (2022) The export of medium and high-tech products manufactured in Europe. https://doi.org/10.20944/preprints202208.0300.v1. https://www.preprints.org/manuscript/202208.0300/v1. Last accessed 2024/12/20
33. Liu Y, Tan Z, Ning X (2023) Public education expenditure and economic growth: based on panel data from 2009 to 2019 in China. World J Adv Res Rev 11:184–192
34. Standard international trade classification. https://unstats.un.org/unsd/trade/sitcrev4.htm. Last accessed 2024/12/20

35. Commodity structure of foreign trade of Ukraine. https://csrv2.ukrstat.gov.ua/operativ/operativ2024/zd/tsztt/arh_tsztt2024_u.html. Last accessed 2024/12/20
36. Future of work 2030: the impact of war and the profession for the recovery of Ukraine. http://careerhub.in.ua/future-of-work-rvplyv-vijny-ta-profesyidlya-vidnovlennya-Ukrayiny/. Last accessed 2024/12/20
37. Stadnyk V, Pchelianska G, Holovchuk Y, Dybchuk L (2023) The concept of marketing of balanced development and features of its implementation in the food market. Agric Resour Econ Int Sci E-J 6(3):80–96
38. Ashley WC, Morrison JL (1997) Anticipatory management: tools for better decision making. The Futurist 31(5):47–509
39. Kraus K, Kraus N, Nikiforov P, Pochenchuk G, Babukh I (2021) Information and digital development of higher education in the conditions of innovatyzation economy of Ukraine. WSEAS Trans Environ Dev 17:659–671. https://wseas.com/journals/ead/2021/b305115-652.pdf. Last accessed 2024/12/20
40. Stadnyk V, Izhevskiy P, Zamazii O, Goncharuk A, Melnichuk O (2018) Factors of enterprises' strategic selection of participation forms in integration formations. Probl Perspect Manag 16(2):90–101
41. Yepifanova I, Dzhedzhula V (2023) Assessment of the level of intellectual capital of industrial enterprises. In: 2023 13th international conference on advanced computer information technologies (ACIT). IEEE, pp 367–371. https://doi.org/10.1109/acit58437.2023.10275543
42. Rauer JN, Kroiss M, Kryvinska N, Engelhardt-Nowitzki C, Aburaia M (2021) Cross-university virtual teamwork as a means of internationalization at home. Int J Manag Educ 19(3):100512. https://doi.org/10.1016/j.ijme.2021.100512

Accounting Engineering as a Tool for Innovative Support of the Transformation of Socio-economic Systems of Enterprises and Their Economic Activity

Nataliia Kashchena, Oksana Chumak, Iryna Nesterenko, and Rasa Subačienė

Abstract The study is aimed at theoretical substantiation and development of methodological provisions and applied recommendations for the formation of a system of information support for managing the economic activity of trade enterprises using the technology of accounting engineering. The role of accounting engineering in the transformation of the information system of a trade enterprise, focused on ensuring the relevance, complexity and integration of accounting and analytical information necessary to support management decisions in the field of economic, environmental and social activities, is determined. Methodological provisions for the implementation of accounting engineering as an innovative tool for the formation of accounting and analytical data and optimization of information flows have been developed, which ensures the adoption of informed management decisions and contributes to the adaptability of the enterprise in the context of digital transformation of the economy. The accounting architecture of the accounting engineering technology is formed in the form of mega-accounts that are integrated into the accounting system and form the items of the derivative balance sheet. The accounting nomenclatures of financial and management accounting are defined, and their characteristics are given, taking into account the needs of information support for managing the

N. Kashchena · I. Nesterenko (✉)
State Biotechnological University, Kharkiv, Ukraine
e-mail: i.nesterenko@btu.kharkov.ua

N. Kashchena
e-mail: n.kashena@btu.kharkov.ua

O. Chumak · I. Nesterenko
State University "Kyiv Aviation Institute", Kyiv, Ukraine
e-mail: oksana.chumak@npp.nau.edu.ua

R. Subačienė
Vilnius University, Vilnius, Estonia
e-mail: rasa.subaciene@evaf.vu.lt

A. Semenov et al. (eds.), *Data-Centric Business and Applications*, Lecture Notes on Data Engineering and Communications Technologies 272,
https://doi.org/10.1007/978-3-032-06608-4_6

economic activity of trade enterprises. The study is aimed at theoretical substantiation and development of methodological provisions and applied recommendations for the formation of a system of information support for managing the economic activity of trade enterprises using the technology of accounting engineering. The role of accounting engineering in the transformation of the information system of a trade enterprise, focused on ensuring the relevance, complexity and integration of accounting and analytical information necessary to support management decisions in the field of economic, environmental and social activities, is determined. Methodological provisions for the implementation of accounting engineering as an innovative tool for the formation of accounting and analytical data and optimization of information flows have been developed, which ensures the adoption of informed management decisions and contributes to the adaptability of the enterprise in the context of digital transformation of the economy. The accounting architecture of the accounting engineering technology is formed in the form of mega-accounts that are integrated into the accounting system and form the items of the derivative balance sheet. The accounting nomenclatures of financial and management accounting are defined, and their characteristics are given, taking into account the needs of information support for managing the economic activity of trade enterprises.

Keywords Trade enterprises · Socio-economic systems · Transformation · Innovative support · Accounting engineering · Accounting and analytical data · Economic activity · Management · Digitalization

1 Introduction

1.1 Relevance

Innovative support for the transformation of socio-economic systems of business entities, increasing their economic activity and resilience to external and internal challenges is an integral part of sustainable development based on compliance with socio-ecological and economic standards to ensure the quality of life of current and future generations. Due to globalization, intensive digitalization and turbulence in the business environment, companies must constantly adapt to new challenges and focus on long-term efficiency, which ensures not only economic growth but also a harmonious combination of social and environmental responsibility. In this context, the integration of advanced management technologies and the latest methods of optimizing business processes, including financial and accounting ones, into corporate governance strategies, as well as information support for management actions, are becoming increasingly important.

As information influences on the functioning of trade enterprises increase, the efficiency of their economic activity and, consequently, economic activity increasingly depend on the quality of information used to make management decisions.

Expanding access to consumer markets, the development of new forms of trading activities and their digitalization increase the volume of information flows and lead to increased requirements for the efficiency of formation, sufficiency, quality of recording and analytical processing of accounting information system data, their further use in analysis and management. This, in turn, actualizes the creation of a qualitatively new format of information support for managing the economic activity of trade enterprises, capable of increasing economic activity and satisfying the information needs of management at all levels at all stages of the process of developing, making and monitoring the implementation of management decisions. As a result, the author recognizes the need to conceptualize the formation of information support for the management of economic activity of trade enterprises and recommendations aimed at improving accounting, analytical and control processes, determining the most effective methods of forming information arrays of data, establishing channels of information interaction and optimizing information flows in the management process.

The driver of solving these problems is an integrated approach to the formation of accounting and analytical information of a financial and non-financial nature, as well as its accumulation in a single circuit of the information support system for managing the economic activity of trade enterprises. In its implementation, the philosophy of management performs a general theoretical and methodological role in the adequate reflection of management activities [1] and directs it to continuous self-development. Ensuring the latter requires taking into account the philosophical basis for the formation of a subject-subject type of management relations, which in the future will ensure progress towards the formation of a system of real self-management, as well as the involvement of innovative models, technologies and tools that will optimize business processes and contribute to the economic activity of the enterprise. Taken together, this will strengthen the accounting and analytical function of management and ensure the formation of information adequate to management requests in the required context of accounting and analytical data to support strategic and operational decision-making.

Accounting engineering is an effective tool for integrating various financial and non-financial data, optimizing management activities and improving decision-making efficiency. It combines advanced methods of accounting, financial analysis and strategic planning with innovative technologies aimed at improving the efficiency of enterprise management in the face of dynamic changes. Accounting engineering allows to create models that integrate not only traditional accounting and analytical data but also additional information, which enables managers and executives to analyze and forecast financial and economic processes more deeply and helps companies to adapt to the ever-changing economic environment.

The experience of foreign companies that have used engineering technology in their management shows that such tools allow for continuous monitoring of the financial condition of the enterprise. The term "accounting engineering" refers to the adaptation and rethinking of the accounting profession in the context of modern technological advances, such as digitalization, artificial intelligence and the Fourth Industrial Revolution [2]. This concept emphasizes the need to integrate engineering approaches for the development of the accounting sphere in the context of rapid technological change. Therefore, accounting engineering is defined as a paradigm of modern accounting and analytical practice that integrates accounting theory with the economic goals of company management, primarily in terms of the relationship of mutual dependence with the dominant role of company goals and achievements and which acts as a tool ("catalyst") that allows to confirm the financial results expected by the board [3, p. 39].

In the scientific discourse, accounting engineering is considered as a multidimensional phenomenon that combines elements of creative accounting and regulatory maneuvering. Thus, P. Nicoleta interprets this phenomenon as synonymous with creative accounting, emphasizing that their difference lies solely in the intentions of use: within the legal framework, accounting engineering is aimed at optimizing financial performance without violating current standards [4]. In turn, L. Michalcik positions accounting engineering as a tool for authorized adjustment of financial results by stakeholders, which can create significant challenges for the economic environment. In his scientific work [5], he carries out a systematic analysis of this phenomenon, interpreting it as a set of accounting techniques that allow to vary the profitability of enterprises. The researcher constructs a theoretical model that reflects the potential of accounting engineering to achieve the projected financial parameters, focusing on the flexibility of accounting policies in regulating financial reporting. M. Young examines accounting engineering through the prism of its complexity and professional challenges. In his opinion, the effective application of this approach requires not only a deep awareness of the regulatory and legal aspects of financial accounting, but also a high level of competence of specialists [6]. He pays special attention to the problems of regulatory supervision and audit, emphasizing the difficulties of ensuring transparency of financial statements in cases of using accounting engineering to adjust economic indicators.

Domestic scientists consider accounting engineering as an accounting concept that includes the construction of a structured chart of accounts using new types of accounting architecture, aggregated and disaggregated objects in order to provide accounting support for the management of economic processes and determine the results of the enterprise, construction of derivative balance sheets [7, 8]. I. A. Yukhymenko-Nazaruk reveals accounting engineering as a set of management actions aimed at ensuring compliance of the enterprise's economic activity with the requirements that provide a different accounting reflection that meets the chosen strategy and management goals [9, p. 165]. The concept of "accounting engineering" is understood as a new system of methods for determining the financial condition of an enterprise and its stability margin, in particular as a tool for managing cash flows, reserves, enterprise risks, and most importantly, financial stability [10]. Engineering is a set of techniques and methods that a company uses to predict its own development in order to improve business efficiency, aimed at developing an optimal solution according to the specified parameters of the business entity and in accordance with its strategic goals [11, p. 138]. Also, we consider the proposed interpretation of financial engineering as an analytical process, which is the design, development and implementation of financial strategies that contribute to increasing the value of the enterprise and the efficiency of all business processes taking place within the socio-economic system [12, p. 124].

Taking into account the above and the results of the previous study of this issue [13], in the context of our study, we believe that accounting engineering should be considered as an innovative technology for transforming the information support system for managing the economic activity of an enterprise. It is based on the integration of the latest digital solutions and methodological tools for collecting, processing and analytical processing of relevant accounting and analytical data on the basis of the fair value of property, which allows optimizing accounting, analysis and planning processes to implement operational and strategic management impacts, ensuring the efficiency, adaptability and sustainability of the enterprise in a turbulent business environment.

Accounting engineering is changing the landscape of management information support, transforming traditional approaches to accounting, analysis and planning through the introduction of digital technologies, automation and analytical tools, which necessitates deepening scientific research to develop methodological foundations for its effective application in the context of the growing complexity of the business environment.

1.2 Goals and Objectives

Paying tribute to the achievements of leading scientists of our time, it should be emphasized that the methodological basis for solving the problem of improving the efficiency of the information support system for managing the economic activity of trade enterprises is the development and implementation of the relevant accounting engineering technology. Its structural dominant is the prevailing scientific paradigm of generating an information array of relevant accounting and analytical data in the context of management processes, supplemented by conceptual approaches to modeling the information space. At the same time, ensuring the effectiveness of management functions and their information support in the process of making tactical and strategic decisions appears as a key determinant of the efficiency of the enterprise management system. Therefore, when formulating the main concept and theoretical hypotheses of formation of the system of information support for management of economic activity of trade enterprises, it should be taken into account that this is an integrated system of strategic type, which operates as part of the information system of management of a trade enterprise in order to meet the information needs of management personnel at all levels, and through the integrated implementation of accounting, analysis and controlling functions ensures the creation of relevant databases for making management decisions.

The determining prerequisite for the effectiveness of the formation of a strategically oriented integrated system of information support for the management of economic activity of trade enterprises is the definition of information needs of management, reflecting the strategy and options for achieving the target parameters of its development (identification in quantitative and qualitative dimensions), their implementation and integration into the process of systematization of accounting and analytical information and the development of a feedback mechanism that indicates the degree of satisfaction of information users.

Knowledge of information needs, in turn, allows to identify the channels of interaction of the integrated information support system and its functional subsystems with the system of management of economic activity of enterprise and to ensure optimization of vertical information flows of data for making management decisions, namely: a downward flow of instructions and requirements for the formation of accounting and analytical information necessary for management purposes, and an upward flow of accounting and analytical information as a resource ready for use for management purposes.

In this context, this paper aims at scientific substantiation and development of methodological provisions and applied recommendations for the formation of a system of information support for the management of economic activity of trade enterprises using the technology of accounting engineering, focused on generating relevant accounting and analytical information of a retrospective and prospective nature in the context of economic, environmental and social spheres of activity.

The direction of scientific research involves the phased implementation of a set of scientific and practical tasks, namely:

- defining the concept of forming a strategically oriented integrated system of informational support for managing the economic activity of trade enterprises using accounting engineering technology;
- identifying the areas of business value-added formation (business units) and the constraints that may negatively impact the economic activity of trade enterprises;
- justifying the accounting engineering technology and modeling its functioning within the system of informational support for managing the economic activity of a trade enterprise.

The section concludes with a summary and the prospect of further research related to the methodological toolkit for assessing the economic activity of trade enterprises in the context of the digital transformation of business processes.

2 Theoretical and Methodological Foundations and Applied Recommendations for the Formation of an Information Support System for Managing the Economic Activity of Trade Enterprises

2.1 The Concept of Forming an Information Support System for Managing the Economic Activity of Trade Enterprises Based on Accounting Engineering

The issue of information support for management is not new and remains under close scrutiny by modern researchers. However, due to its multidimensional nature and the variety of manifestations in the functional areas of trade enterprises' economic activities, it is still underdeveloped and requires further in-depth exploration.

The core idea of the concept for forming an information support system for managing the economic activity of trade enterprises is formulated as the modeling of a unified information space for economic activity management, aimed at maximizing the satisfaction of users' informational needs.

In accordance with this concept, the following theoretical hypotheses have been proposed:

1. all elements of the information support system are interconnected and interact under the influence of various business environment factors;
2. the effectiveness of the information support system is determined by the innovative technology of forming accounting and analytical data through the application of accounting engineering.

The practical implementation of these hypotheses requires the justification of relevant fundamental theories and approaches, both philosophical and industry-specific. Among the key philosophical theories are management theory, information theory, reflection theory, cognition theory, and development theory. In the industry-specific domain, the most significant are accounting theory, analysis theory, and controlling theory.

The study has proven that an integral attribute of the functioning of any management information system, including the system of information support for managing economic activity, is the acquisition, transmission, and storage of various types of information. In this process, there is always an exchange of information between different system components, forming corresponding information flows. The simplest case is the transmission of information from the governing body to the executive body (task assignment). A more complex case is a closed-loop control system, in which information about task execution results is transmitted back to the governing body through the so-called "feedback" mechanism.

Therefore, information theory is not merely an applied science that employs probabilistic research methods but should be considered a branch of probability theory. Given the objectives of our study, it must necessarily be taken into account when forming the scientific basis for the concept of information support management of economic activity in commercial enterprises.

Among philosophical theories, the key ones for shaping the distinct integrity of the information support system are the theories of reflection, cognition, and development. "These theories do not contradict but rather complement one another, developing different quantitative and qualitative characteristics related to specific aspects of the management process. Through the synthesis of these theories, accounting and analytical support can become more efficient, useful, and cost-effective" [14, p. 220, 15, p. 294, 16, p. 324]. Thus, the synthesis of reflection and cognition theories characterizes the information complex, which exhibits features of managerial accounting. The combination of cognition and development theories expands the informational capabilities of management to the level of controlling, while the integration of development and reflection theories forms information for strategic management (strategic accounting) as a system for recording, summarizing, and providing data necessary for making strategic managerial decisions by the executives of an economic entity [14, p. 221].

The system of information support for managing the economic activity of trading enterprises, formed based on philosophical and industry-specific theories, acquires emergent properties through engineering technologies in accounting, analysis, and controlling. These technologies are founded on an in-depth detailing of information regarding the financial and resource potential of an enterprise and its utilization to enhance economic activity across functional dimensions. These include labor, intangible, material, and financial resources (both own and borrowed) at the level of core business processes in commercial, financial, human resource, innovation, investment, market, management, social, and environmental domains. This approach ensures the formation of large datasets containing relevant accounting and analytical information, thereby enhancing the effectiveness of the information support system for managerial decision-making.

Accounting engineering, as an advanced technology for information generation, enables the optimization of accounting and analytical processes. Based on the informational needs of management personnel, it facilitates the development of content for accounting, reporting, and analytical data, which serves as a foundation for substantiating, developing, implementing, and monitoring the execution of operational, tactical, and strategic management decisions. The instrumental profile of accounting engineering, as a key component in building a strategically oriented system of information support for managing the economic activity of a trading enterprise, has been developed based on sources [13, 17, 18] and is presented in Table 1.

The engineering approach, combined with the basic (system-structural, situational, functional, process, adaptive, value-oriented, informational) and scientific approaches formulated by the Financial Accounting Standards Board (FASB), namely: tax, legal, ethical, economic, behavioral, and structural, will enable the formation and representation of accounting and analytical data in various types of accounting and reporting in the required analytical perspectives.

Table 1 The instrumental profile of accounting engineering as a driver for the formation and effective functioning of an integrated strategic information system

Components of accounting engineering tools	Characteristic
Accounting engineering tools	Financial balances, derivative financial statements, business process reengineering, interactive methods of financial calculations, balanced valuation statement, intellectual capital valuation statement, net liabilities and capital valuation statement, zero balances
Technologies of accounting engineering procedures	Initial operator, corrective operations, adjusted balance, economic operations, specific intermediate, alternative, hypothetical operations, hypothetical final operator, control operations
Accounting and analytical support of engineering	Aggregated entries, structured chart of accounts, qualimetric components, tools for automating accounting and analytical processes, evaluation systems, algorithms, drivers, accounting and control points, accounting aggregates
Managed objects and processes	General management, ownership, reservation system, risks, innovations, financial results, reorganization, financial condition
Results of using accounting engineering tools	Generalized indicators by functional areas of economic activity and various types of value in accordance with the analysis requests of clients, managerial alternative solutions to current problems, information support for decision-making, control, automation of accounting and analytical support

2.2 Methodological Guidelines for the Formation of Information Arrays of Relevant Accounting and Analytical Data Based on the Theory of Constraints and Business Units

Economic activity is an indicator that reflects the business reputation of a trading enterprise in collaboration with stakeholders. Acting as a comprehensive characteristic of all business processes aimed at ensuring competitiveness and sustainable development through profitability and business value growth, it is determined by the level of staff qualifications, the condition, structure, and efficiency of the use of economic resources and capital of the enterprise. For objectivity and accuracy in identifying the level of economic activity of a trading enterprise, it is crucial to form the appropriate informational foundation within the accounting system.

The needs of adaptive management of the economic activity of a trading enterprise dictate the necessity of using scientifically grounded methods for collecting, analyzing, processing, and transmitting data across all business processes in the information management system, which can ensure a broad spectrum for forming accounting and analytical information to support decision-making for its development. The producer of such information is, by definition, the enterprise's accounting system. According to Evans [19], it provides more than 70% of the information needs of the management system. Thus, for effective management of the economic activity of trading enterprises, the accounting system must generate information across all its aspects (commercial, financial, human resources, market, managerial, investment, innovation, social, environmental) and spheres of managerial influence.

Modern conditions necessitate the management of the economic activity of trading enterprises based on the principles of, in particular, the Theory of Constraints, which takes into account assumptions about the existing constraints (internal and external) in business operations. The Theory of Constraints is a theory according to which every enterprise has at least one constraint (process, demand, capacity, etc., that limits the productivity of the enterprise and hinders the system from achieving its goal), which prevents it from earning money [20, p. 40]. After all, any economic system always has at least one constraint that limits the excessive growth of positive financial results.

Unlike other approaches, which operate locally, the Theory of Constraints analyzes and manages enterprises systemically, continuously improving their activities and forming competitive advantages with a focus on results [21, p. 201]. Therefore, the use of the Theory of Constraints in the accounting system ensures the identification of constraints that have the most negative impact on economic activity and prevents more significant constraints in future periods. The system of key constraints we have developed, whose presence can negatively affect the economic activity of trading enterprises, is presented in Table 2.

Theory of Constraints in Managing Economic Activity of an Enterprise allows to take into account the coherence of actions across all accounting nomenclatures, which are responsible for business units, department heads, guided by the available constraints and expected results for the reporting period. In Statement 4HH of the American Institute of Management Accountants [22], it is noted that in the system of the Theory of Constraints, the role of management accounting is to:

- ensure economic assessments of productivity, operating costs, and inventory;
- accumulate and integrate data from the Theory of Constraints system, total quality management, and related management models to ensure compatibility in the reporting system;
- confirm the identification of constraints;
- provide assessments of capacity costs and ensure investment analysis for potential capacity expansion;
- explain various assumptions regarding discrepancies in other types of strategic and tactical analysis;
- help operational managers make decisions to alleviate constraints and their impact;
- develop and maintain a management system using activity-based costing to complement the information provided and required by the Theory of Constraints;
- promote the development of comprehensive knowledge about incremental cost models and cost structure;

Table 2 The system of key constraints characteristic of trading enterprises

Key types of constraints	Key constraints for trading enterprises	
	Types	Characteristic
Capacity constraints	Inventory volume Production area volume Material and technical support	Inventory balances, goods in the sales floor, inventory turnover, merchandise calendar, energy efficiency, warehouse structure, number of retail outlets, logistics flows, volumes of retail and warehouse space, procurement planning, provision of fixed assets, IT support, etc.
Financial resource constraints	Credit loans Financial assistance Accounts payable and receivable Cash gaps	Lack of sufficient financial resources for operational activities, execution of investment projects, payment schedule, structure of financing sources for activities, settlement of obligations
Time resource constraints	Duration of the operational cycle Duration of the management cycle Store operating hours	Time lag of the inventory product cycle (logistics, storage, preparation, sales) Time lag in decision-making Time during which the buyer can make a purchase
Labor resource constraints	Human resources Intellectual resources Educational resources Cultural resources (corporate culture)	Staff turnover, low productivity, low competencies according to the job profile, structure and number of employees not matching the capacities, lack of internal regulations for personnel management, employment of people with disabilities, dress code, professional development of staff, etc.
Service constraints	Time-related (service, delivery, checkout) Security Communication Additional services	Timeliness and duration of order fulfillment, duration of customer service, availability of consulting, guarantees, conditions for the customer's stay in the sales area, additional service offerings, product packaging, communication channels with customers, customer safety, after-sales service, etc.
Managerial constraints	Organizational Control	Organizational culture, corporate governance, internal communications, internal standards, planning, decision-making efficiency, development strategy, internal control, compliance, audit
Market constraints	Macroeconomics Competition Customers Resources	Competitiveness, geographical location, competing retail outlets, pricing policy, product shortages, assortment structure, macroeconomic indicators

- provide data on productive marginal income and determine relevant constraints for all decisions;
- identify direct links between productivity and operating costs;
- report on the impact of constraints;
- ensure that the financial function has not become a constraint.

Business units should be understood as management areas where value-added business activities are formed, and which are key to achieving the main goal of a retail enterprise—delivering goods to the customer (through all stages of the product movement as defined by the operational cycle). For the traditional retail model, we have identified the following main business units:

- logistics (delivery, storage, internal movement of inventory);
- product sales (pre-sale preparation and sale);
- financial management;
- investment projects;
- material and technical supply.

The common aspect in using the management model by business units and the theory of constraints is that both concepts involve comparing revenues and expenses to define the performance indicator of economic activity (using various methodological approaches).

The application of the theory of constraints by retail enterprises involves identifying those business units that have constraints (resource, managerial, service, market) that need to be considered when ensuring the functioning of other business units or their individual processes. This approach allows using the established system of indicators defined by accounting policy for the respective business units.

The implementation of the conceptual principles of the theory of constraints in the enterprise's accounting system contributes to achieving the following management results:

- evaluate economic activity by business units in terms of product groups/nomenclature and business processes (supply, storage, sales);
- develop informational support for inventory management by business units (logistics and product sales), defining a range of performance indicators for their movement, sufficiency, demand, turnover, and efficiency of product flows, etc.;
- improve the financial structure in terms of cash flow management, accounts receivable and payable, equity capital, guarantees, and tax burden;
- provide economic justification for investment projects based on economic efficiency, possible sources of funding and their structure, resource budgets under a range of probable and real constraints in forecast conditions;
- develop a system of measures to prevent and improve those business processes that have constraints with significant negative consequences within business units.

2.3 Methodological Approach to Modeling the Accounting System Using Engineering Technology

The technology of accounting engineering involves the use of methodological approaches from other sciences to improve the information support system for managing the economic activity of an enterprise and expand its functionality. Within this system, the process of continuous and targeted preparation of indicators necessary for analysis, planning, control, and the preparation of operational, tactical, and strategic management decisions in all areas of the enterprise's operational, investment, and financial activities takes place [23, p. 24]. This information system is also used in the development of models, technologies, and engineering procedures. This information system is based on the development of models, technologies, and accounting engineering procedures aimed at integrating data from various subsystems of management, financial, and tax accounting. The use of engineering approaches allows for the creation of adaptive algorithms for data processing, ensuring the timeliness, relevance, and accuracy of accounting and analytical data.

The development of accounting engineering models involves the application of optimization concepts for accounting processes, automation of data collection, classification, and interpretation, as well as the implementation of digital technologies that enhance the effectiveness of management decisions. The technological toolkit includes methods of predictive analysis, scenario modeling, and controlling procedures that ensure system flexibility and its alignment with current challenges in the business environment. Accounting engineering procedures are aimed at standardizing and unifying information flows, which minimizes the risks of managerial errors and contributes to increasing the economic activity of the enterprise through effective resource distribution and control over their use.

To ensure the modeling of the accounting system using engineering technology, it is necessary to consider and implement a number of measures:

- analyze the financial structure and financial model of the enterprise;
- consider the possibility of implementing innovative methods of financial and operational management;
- evaluate the assortment structure of product groups and related products;
- assess the activities of retail enterprises with a focus on achieving sustainable development goals (retail, hypermarkets, etc.): distinguishing between economic, ecological, and social areas of activity and preparing transparent management reporting;
- analyze the impact of the identified results on the level of economic activity, which is an important indicator of performance.

Based on the above, accounting engineering is used to develop an algorithm for eliminating unprofitable activities, enhancing solvency, improving competitiveness, and increasing added value or capitalization.

Given that accounting policy is considered the methodological foundation for accounting and analytical support of economic activity management, under the implementation of accounting engineering, it is advisable to form sections for both financial and management accounting. Some enterprises develop two separate regulations: accounting policy for financial accounting and accounting policy for management accounting. The latter provides more opportunities for defining and establishing alternative provisions that regulate the use of accounting engineering mechanisms and tools within the enterprise. Thus, accounting engineering, by consolidating the assets of the enterprise and adjusting their value in the context of uncertainty and fair valuation, allows the determination of economic activity indicators in areas such as: financial, personnel, market, production, management, investment, and innovation. The established relationship between the accounting and analytical components of the information system through accounting engineering technology is presented in Fig. 1.

Considering that engineering tools form a strategically oriented system of information support for economic activity management, covering a wide range of relevant accounting and analytical data, we distinguish the following components:

- two autonomous working charts of accounts or one integrated chart for financial and management accounting;
- mega accounts (aggregated) and the corresponding account entries, according to the specifics of the activity;

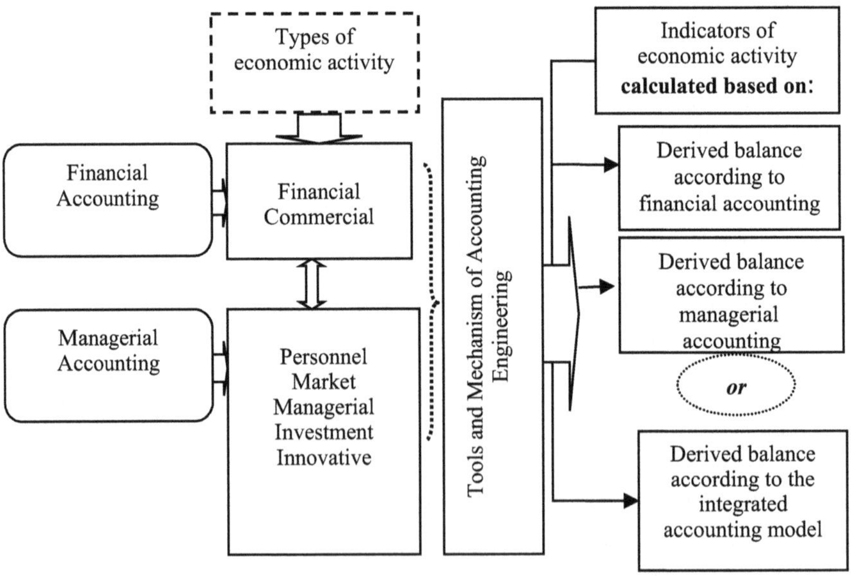

Fig. 1 Reference model of economic activity management for trade enterprises in the context of digital transformation of business processes

- a list of accounting nomenclatures and economic operations for financial and management accounting;
- integrated accounting registers for financial and management accounting;
- derived balances for financial and management accounting or an integrated derived balance;
- formation of a system of indicators for evaluating economic activity;
- results of using accounting engineering.

Given the external and internal factors and the purpose of applying engineering technology, its tools can influence the support of sustainable development processes, risk prevention, identification of reserves, contribute to the reengineering of economic operations or processes, consider inflationary processes, and predict the advisability of restructuring or other procedures.

The application of engineering tools in accounting covers several types of aggregated accounting operations: account aggregation and balances at the end of the period, assessment of high-risk business transactions, fair value assessment of assets, hypothetical (conditional) sale of assets, and determination of net assets and net liabilities. This algorithm of accounting engineering technology with innovative tools is designed to analyze projects or business processes (unit economics, in particular) based on risk levels and indicators of economic activity in the economic, social, and environmental spheres.

The functioning of accounting engineering technology in the system of information support for managing economic activity is structurally presented in relation (Fig. 2).

The accounting architecture of the information support system is based on the accounting policy (methods, tools, technologies), the working plan of mega accounts, accounting nomenclature, and the register of business transactions, the mechanism for constructing derivative balance reports, grouping balance sheet items into a system of economic activity indicators, evaluation, and monitoring of economic activity indicators.

Economic activity, as an aggregated evaluation indicator of efficiency, is determined by a system of representative indicators, the calculation base of which is the balance sheet, where the value of assets as of the reporting date is presented. The classic balance sheet, in the form of financial statements, expresses the value of assets at historical cost, while the derived balance sheets at fair (market) value allow the owner to understand the real state of assets and liabilities.

V. K. Makarovich has systematized the derived balance sheets used for managing economic processes and the resource potential of the enterprise in the following areas: property management, solvency management, reserve system management, and risk management [4, p. 132]. The real value of property objects, as per the derived balance sheet, is also used when performing analytical procedures to assess operational activities, calculate forecast data, build strategic management accounting, determine the ability to execute a development strategy, and implement new investment projects, etc.

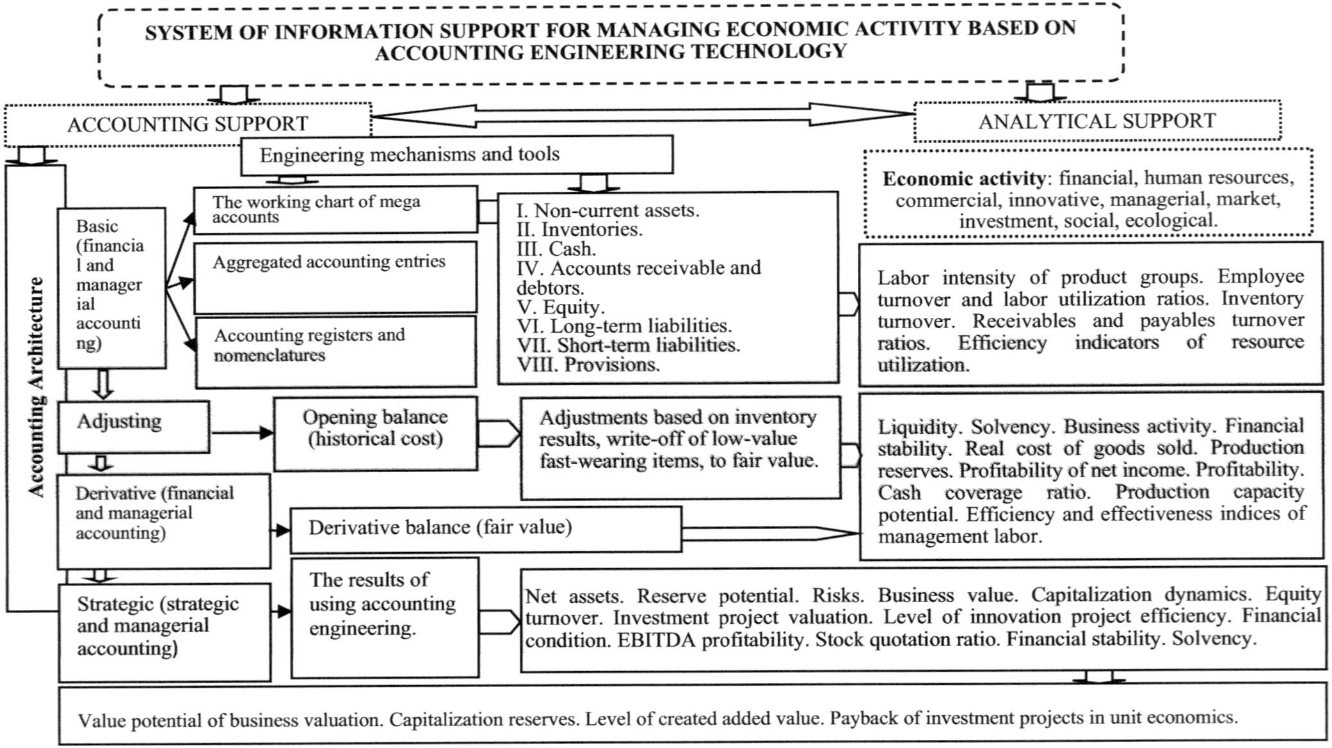

Fig. 2 The technology of accounting engineering in the information support system for managing the economic activity of trade enterprises

The preparation of the opening balance sheet is carried out based on the aggregated balances for mega accounts, and then corrected for "regulated" operations (losses, damaged inventory, defects, events after the balance sheet date, low-value rapid-wear items, auditor recommendations, balance sheet and off-balance sheet entries that affect the object of management) by making corrective journal entries followed by a fair valuation of balance sheet items. The use of different rules and norms for various types of accounting and other legal standards results in the disclosure of indicators in the reporting that reveal the essence of business processes from different economic-legal aspects, which define the quality of managerial actions in different ways. Therefore, enterprises are encouraged to choose criteria for recognizing and evaluating differences, which are reflected in the reporting and have an impact on managerial decisions [24]. At this stage of the engineering technology evaluation, it is assumed that all assets are subject to sale, and all liabilities to settlement.

For retail enterprises, mega accounts are recommended, which will later be reflected as items in the derived balance sheet, with corresponding accounting nomenclatures in financial and management accounting, as shown in Table 3.

At the beginning of applying the engineering technology, it is suggested that the business units for retail enterprises be conditionally denoted by the letter *B*, and further, each item of the derived balance sheet or mega account should be assigned to the business unit that consumes the relevant accounting nomenclatures of that account during its operations.

For retail enterprises, eight mega accounts and, accordingly, the aggregated items of the derived balance sheet are recommended, which consider the specifics of operations—inventory and cash turnover, accounts receivable and payable, logistics, and sales. Retail activities are characterized by a relatively small share of fixed assets, but today, the share of intangible assets is growing, mostly in the form of various software products (CRM, FinTech, accounting, etc.).

For enterprises that have a significant amount of retail space in their own buildings, it is advisable to allocate an accounting nomenclature for investment property, which may significantly change its fair value (in either direction) under certain external conditions, potentially having a substantial impact on the evaluation of the economic activity of the enterprise. The accounting nomenclatures of the first mega account "Non-current assets" are involved in supporting activities for the following business units: material and technical support (B_{MTS}), logistics (B_L), and investment project (B_{IP}).

During analytical procedures for evaluating economic activity, aggregated data from each mega account can be used to analyze the activities of the relevant business unit. For example, for analyzing the business unit of an investment project (B_{IP}), the data will be generated from the following mega accounts: I. Non-current assets, III. Cash, V. Equity, VI. Long-term liabilities, and VII. Short-term liabilities. In this format, the investment project manager receives the most detailed information about the project at fair value in the current period.

The second mega account "Inventory" supports business units such as goods sales (B_{GS}), material and technical support (B_{MTS}), and logistics (B_L), and is one of the most significant by weight in the balance sheet value of a retail enterprise, as it

Table 3 Accounting architecture of accounting engineering technology for retail enterprises

Mega account	Business unit (for networks, retail) ***B***	Accounting nomenclature (*fragments*)		Working plan of mega accounts ***XY*** (*fragments*)
		Financial accounting ***X***	Managerial accounting ***Y***	
I. Non-current assets	B_{MTS} material and technical support B_L logistics B_{IP} investment project	X_{I1} fixed assets, Low-value non-current material assets and capital investments X_{I2} intangible assets X_{I3} financial investments	Y_{I1} fixed assets, low-value non-current material assets Y_{I2} investment property Y_{I3} capital investments Y_{I4} intangible assets Y_{I5} financial investments	XY_{I12} office building XY_{I13} passenger cars XY_{I14} low-value non-current material assets in storage XY_{I21} investment property—Company-owned building
II. Inventories	B_{SG} sale of goods B_{MTS} material and technical support B_L logistics	X_{II1} goods in storage X_{II2} goods in trade X_{II3} transportation and procurement costs	Y_{II1} goods in storage. Region Y_{II2} goods in trade. Store number Y_{II3} transportation and procurement costs. Delivery	XY_{II11} goods in storage. Region. Uninvoiced Y_{II31} transportation and procurement costs. Import delivery
III. Cash and cash equivalents	B_{FM} financial management B_{IP} investment project	X_{III1} cash X_{III2} bank accounts X_{III3} other funds	Y_{III1} cash on hand Y_{III2} non-cash funds Y_{III3} foreign currency Y_{III4} cash equivalents Y_{III5} acquiring services	XY_{III11} cash. Source of financing XY_{III12} cash. Revenue XY_{III22} current account in US dollars XY_{III23} current account. Source of financing
IV. Accounts receivable and debtors	B_{SG} sale of goods	X_{IV1} settlements with buyers X_{IV2} settlements with employees/ other debtors X_{IV3} settlements with other debtors	Y_{IV1} settlements with buyers—Advances received Y_{IV2} settlements with buyers Y_{IV3} settlements with employees Y_{IV3} settlements for warranty obligations	XY_{IV11} settlements with buyers—Advances received XY_{IV12} settlements with buyers/export XY_{IV13} settlements with domestic buyers XY_{IV13} settlements with domestic buyers/offsetting
V. Equity	B_{FM} financial management B_{IP} investment project	X_{V1} registered capital X_{V2} additional capital X_{V3} profit X_{V2} other equity	Y_{V2} authorized capital contributed Y_{V2} profit Y_{V3} other equity	XY_{V11} source of financing XY_{V12} retained earnings XY_{V13} reinvested profit XY_{V13} revaluation of non-current assets

(continued)

Table 3 (continued)

Mega account	Business unit (for networks, retail) ***B***	Accounting nomenclature (*fragments*)		Working plan of mega accounts ***XY*** (*fragments*)
		Financial accounting ***X***	Managerial accounting ***Y***	
VI. Long-term liabilities	B_{FM} financial management B_{IP} investment project	X_{VI1} long-term loans X_{VI2} long-term lease liabilities X_{VI1} other long-term liabilities	Y_{VI1} long-term Bank loans Y_{VI2} long-term loans from owners Y_{VI2} long-term liabilities	XY_{VI11} financial assistance. Owners XY_{VI12} bank loans in national currency XY_{VI13} other long-term loans. Investment project XY_{VI13} financial lease liabilities/motor transport
VII. Short-term liabilities	B_{FM} financial management B_L logistics B_{IP} investment project	X_{VII1} short-term bank loans X_{VII2} current debt on long-term liabilities in national currency X_{VII3} settlements with domestic suppliers	Y_{VII1} short-term bank loans Y_{VII2} current debt on long-term bank liabilities in national currency Y_{VII3} settlements with suppliers for uninvoiced deliveries Y_{VII3} settlements with suppliers for advances received	XY_{VII11} short-term bank loans in national currency XY_{VII12} bank loans in national currency XY_{VII13} other long-term loans. Investment project XY_{VII13} financial lease liabilities/motor transport
VIII. Provisions	B_{SG} sale of goods B_{FM} financial management	VIII_{X1} provision for vacation payments VIII_{X2} provision for warranty liabilities VIII_{X3} provision for material incentives	$Y_{VIII\ 1}$ provision for employee payments Y_{VIII2} service claims Y_{VIII3} warranty payments (mandatory)	XY_{VII11} provision for incentive payments XY_{VII12} provision for vacation payments XY_{VII13} provision for warranty liabilities/goods

consumes the most cash to ensure continuous and profitable business operations. For retail enterprises that import goods or trade in food products, which often come from many different suppliers, the allocation of the logistics business unit under this mega account is important for further analytical procedures and decision-making regarding logistics operations. Depending on how the logistics operations are organized, which logistics channels and flows are used, the relevant accounts in the chart of accounts are opened, allowing managers to take into account accurate data for the previously outlined operational logistics issues.

As we can see, the accounting model, built according to business processes by business units, allows generating information support for the analysis of economic activity based on managerial queries aimed at making decisions that take into account the specifics of operations and future benchmarks. O.V. Fomina noted that the mismatch of inventory valuation methods in the theory of constraints and official accounting standards using the net cash income method (Throughput accounting) requires the organization of management accounting in a closed "circle" of accounts, which includes accounts for expenses, inventory, and financial results. The use of screen accounts in managerial and financial accounting allows for reconciliation based on the principle of equality between the aggregate valuations of expenses, inventory, and financial results [25, p. 923].

The third mega account "Cash" is key in the operations of retail enterprises, as trade is the sector with the largest cash mass in circulation during each reporting period. The accounts of management accounting and the chart of accounts are built in accordance with the selected accounting nomenclatures, defined by the specifics of the activity, forms of settlement with customers and suppliers, implemented financial technologies, conditions of cooperation with banking institutions, types of business operations that increase or decrease cash flows, etc. Scholars recognize the components of accounting engineering for managing cash flows as follows [11, 26]:

1. development and identification of business processes for managing cash flows as "objects of accounting";
2. cash account chart for double-entry bookkeeping;
3. accounting engineering tools;
4. accounting registers: "Cash Flow Reports," "Derivative Balances";
5. the result of accounting—determining the effectiveness of implementing accounting techniques in cash flow management.

Cash transactions in retail enterprises are among the key operations in the financial management system, and therefore require special attention. The present expands the scope of financial issues related to cash flow management, which should include the following:

- organizing cash settlements with customers (software, equipment, documentation, reporting, deposits to the bank);
- organizing and conducting internal settlements;
- conditions, forms, and organization of cashless payments and cooperation with banking institutions (acquiring, forms of payments, cash flow time on accounts, currency exchange);
- organizing the implementation and use of FinTech technologies;
- planning (budgeting), creating a payment schedule, and optimizing cash flows;
- analyzing the effectiveness of cash flows converted into inventory;
- effectiveness of cash flow management.

From the above, it follows that the accounting nomenclatures for the accounts in the accounting system, which reflect the movement of cash, must be defined by many factors and objectives of the management system. The chart of mega accounts, which allows collecting and accounting information about cash and can determine and calculate the indicators for cash flow management, is an important tool in accounting engineering. Timely and complete information about the cash of a retail enterprise will ensure the smooth operation of enterprises, for which the turnover indicators of cash are almost always quite high.

The fourth mega account "Accounts Receivable from Customers and Debtors" supports the business unit of product sales (B_{PS}). The main focus for this mega account is to timely identify and prevent the occurrence of receivables from customers or doubtful (uncollectible) debts. It is also important to manage settlements with employees regarding the occurrence of receivables, depending on the specifics of the activity and the conditions of the collective agreement. Thus, for retail enterprises, whose income is ensured by various numbers of customers, different types of goods, and various forms of settlement, it is advisable to control the conditions of occurrence and formats of debt repayment in the accounting system. This is possible with an effectively built system for documenting sales transactions and settlements with customers, controlling the final balances of debts through reconciliation with counterparties and internal documents, using claim procedures and correspondence, and comprehensive assessment of contractual activities. Therefore, accounts receivable is analyzed in terms of its complexity:

- recognition as an asset (financial), which has the potential (right) to receive future economic benefits;
- the "next" fair value estimate (net realizable value) and potential impairment of the receivables;
- assessment of risks of not receiving cash from debtors;
- impact on the safety of the financial condition and economic activity (in particular, as an indicator of the company's competitiveness);
- immobilization of working capital (temporarily withdrawn from circulation);
- assessment of the acceptable level for the normal operation of the enterprise.

Regarding the existing long-term accounts receivable, it may be caused by activities of other business units, and different approaches may be used in the accounting system. In particular, legal aspects of debt settlement terms in contracts, aspects of debt discounting, determination of the expected time lag between the transfer of ownership rights to the goods and its payment (time factor), recognition of credit characteristics and financing elements, terms of payment deferral, and the calculation of possible impairment reserves are considered. Well-grounded forecast information is used for evaluation, employing models for estimating impairment of receivables.

The fifth mega account "Equity" is supported by business units such as financial management (B_{FM}) and investment projects (B_{IP}). Reporting equity information is crucial for owners (investors)—users of the financial statements (regulated and voluntary reports), as it allows them to understand the scale of financial independence in the short-term and long-term periods and assess the level of business capitalization. Therefore, accounting aspects are decisive when recognizing it as an accounting object, in the classification, structuring as a source of funding, distribution, and reinvestment. At the same time, considering the significant legal aspect of this object, the organizational and legal form of the enterprise and the established system of corporate governance are the leading foundations. It is recommended to select analytical perspectives in accounting depending on the type of financing sources (internal or external), currency types, and directions for the use of equity capital. Since the technology of accounting engineering ensures the assessment of equity at fair value, such accounting methods, based on conducted analytical procedures, will allow decisions on:

- optimizing the structure of equity capital;
- determining the capitalized share of net profit;
- rationalizing sources of equity formation;
- determining the efficiency of equity capital distribution;
- optimizing the amount of equity in relation to attracted capital and existing assets (non-current and current assets);
- assessing the impact of obtained financial ratios;
- identifying the significant impact of factors on changes in capital size and capitalization.

Depending on the management request, it is determined whether it is necessary to introduce additional accounting nomenclature and sub-accounts into the chart of accounts regarding transactions that impact changes in capital by its types, the formation and direction of net profit, the formation of reserve capital, capital (share) withdrawal, and adjustments related to changes in the accounting policy of the enterprise.

The sixth mega account "Long-Term Liabilities" is supported by business units such as financial management (B_{FM}) and investment projects (B_{IP}). This includes debts to financial market entities, counterparties, and owners (class 5 of accounts). According to current regulations, long-term liabilities are presented in the balance sheet of the enterprise at present value using the discounting method [27], which allows equating the value of the enterprise's resources that will be used in the future to settle obligations to a common time denominator. A key issue today is the evaluation of long-term liabilities in terms of discounting debts, which is implemented by making changes to accounting policies or accounting estimates. However, according to paragraph 12 of Accounting Regulation (Standard) 6 "Corrections of Errors and Changes in Financial Statements," changes in accounting policies require the reflection of the impact of changes in accounting policies on past events and transactions [28]. Therefore, in the derived balance sheet, the enterprise can reflect the impact of changes by adjusting the balance of retained earnings at the beginning of the reporting year, performing a retrospective calculation of long-term debt.

Methodological techniques and methods of accounting policy also include determining the discount rate, calculating the discount amount, the adopted level of inflation in calculations, recognizing financial income or expenses that will lead to changes in the composition of financial performance (amortization of the discount). Adjustments in the derived balance sheet are also needed for the difference between the accounting or nominal value and the present value of liabilities to prevent the disclosure of distorted data to users of the reporting information. Therefore, it is recommended to technically introduce an appropriate sub-account into the chart of accounts for the new accounting object. If the enterprise has long-term repayable financial assistance in foreign currency, which is a common source of business financing today, it is advisable to also isolate a sub-account to reflect interest and separately discount such a loan. Organizational and methodological aspects of accounting support, which are important for decision-making, are determined by the terms of contracts with counterparties, documentation of transactions related to the incurrence of liabilities, the choice of accounting nomenclature and analytical breakdowns for sub-accounts in the chart of accounts, and control measures in the accounting system for timely transferring long-term liabilities to current ones and the timeliness of liability settlement, as well as ranking counterparties (lenders) by various criteria. A well-thought-out accounting policy regarding these aspects allows for preventing late settlement of obligations, timely reporting with up-to-date information about the status of the enterprise's credit portfolio, generating reports on non-resident loans, and coordinating decisions regarding the attraction and distribution of funds. It is also important to consider possible obligations with an undefined amount or time of settlement as of the balance sheet date. Current data on long-term liabilities form relevant informational support for analytical procedures related to determining the economic activity of the trading enterprise.

The seventh mega account "Short-Term Liabilities" is supported by business units such as financial management (B_{FM}), logistics (B_L), and investment projects (B_{IP}). The logistics business unit was involved because logistics operations are the most in-demand in terms of borrowed resources, as it includes many tightly time-bound operations of various types. This mega account reflects information about liabilities to financial market entities, counterparties, employees, and government bodies (class 6 of accounts), the presence of which, from the perspective of evaluating economic activity, reduces solvency and liquidity indicators. Regarding short-term liabilities, in addition to adhering to legal norms and contractual terms, attention should be paid to:

- documenting transactions and identifying analytical breakdowns for liabilities incurred;
- risks related to foreign currency liabilities (economic and legal aspects);
- organizing claims discipline for non-performance of contract terms and reflecting such amounts in the accounts;
- sources of financing for the settlement of such liabilities;
- the possibility of adjusting the credit balance of class 6 accounts in the derived balance sheet (in particular, through the use of clearing operations or debt offsets) and determining the causes of any existing debit balance.

Analytical accounts may be opened for the following types of short-term liabilities, which will have specific accounting treatment: bank loans, overdrafts, financial assistance (legal or natural persons), liabilities to employees and suppliers. Depending on the specifics, separate accounting nomenclature objects should highlight received advances and unbilled deliveries, promissory note settlements, financial leasing, and obligations for mandatory payments, etc. Since retail enterprises are characterized by the constant receipt of cash and non-cash funds from the sale of inventories (goods), there are situations where short-term liabilities are settled using funds from these operations, which should be foreseen and regulated by internal regulatory documents.

The eighth mega account "Provisions" is managed by business units such as goods sales (B_{GS}) and financial management (B_{FM}), given that retail enterprises may have

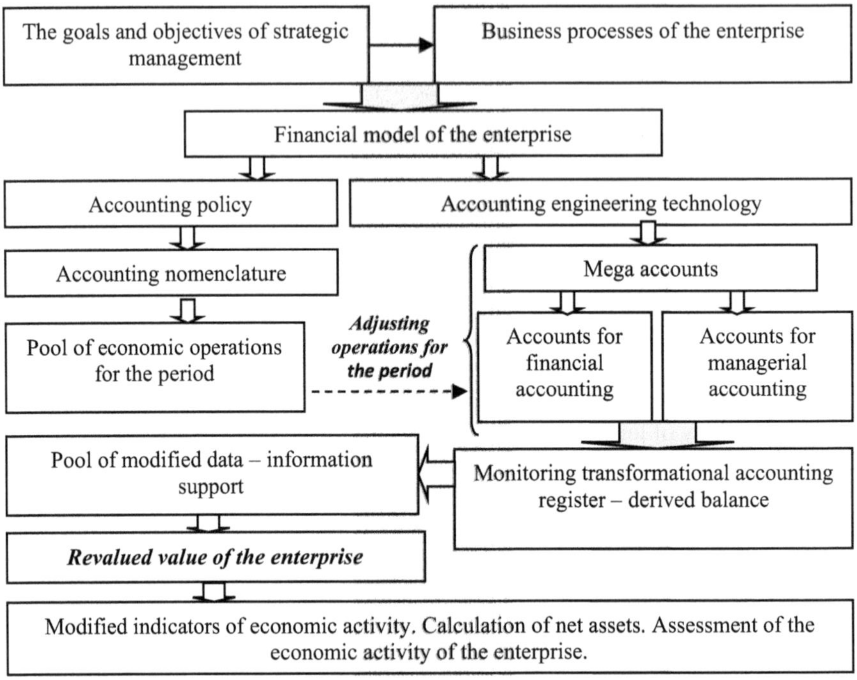

Fig. 3 Structuring of accounting modeling of information support when using accounting engineering technology

significant sums allocated for creating warranty provisions and employee payments. Here, there is a correlation between operating activity expenses in the ninth class of accounts and accrued provisions. The effectiveness of accounting engineering is also reflected in the optimal structuring of accounting modeling for informational support. A defined model structure of the accounting component of informational support for managing the economic activity of retail enterprises is shown in Fig. 3.

The considered technology includes the grouping of accounts and balances, the preparation of the opening and derived balance sheets, the use of the results obtained in analytical procedures and management. This allows increasing the effectiveness of economic activity in the future, preventing and identifying risks, evaluating investment projects, and predicting the impact of various external factors on operations [29, p. 411, 30, 31].

Referring to the structure of accounting modeling for informational support presented in the diagram, we can trace the correlational relationships between creative accounting, the evaluation of economic activity, and the tools of accounting engineering as the technology of the enterprise's accounting policy. This relationship allows for defining and justifying the appropriateness of using accounting engineering to achieve current or strategic goals, thereby reinforcing the significance of the accounting system in identifying economic effects from managerial decisions.

At the same time, the relationship between accounting engineering and creative accounting allows for expanding the latter's role in achieving the management objectives of the enterprise, which is one of the prerequisites for strengthening the role of accounting as a socio-economic institution in the functioning of economic systems. Since accounting engineering and creative accounting are branches of variable accounting, these technologies can have both threatening consequences and optimization opportunities, depending on the purpose of their use by the enterprise's accounting departments and management personnel, through potential free adjustments to the results of uniform economic operations.

3 Conclusions

Researching the theoretical-methodological basis and methodological tools for forming the information support system for managing the economic activity of trade enterprises using accounting engineering technology leads to the following conclusions:

1. The turbulence of the business environment forces trade enterprises to adapt to the challenges of globalization, the intensive digitalization of business processes, and the dynamic changes in economic realities, which leads to the need for transformations in their socio-economic systems. The innovative support for these transformations involves integrating advanced technologies, new management methods, and optimizing business processes, requiring the use of specialized tools and approaches, including engineering ones.
2. The appropriateness of implementing accounting engineering technology is driven by its ability to form an information support system for managing the economic activity of the enterprise, aimed at meeting the informational needs of management at all levels and ensuring the accuracy, relevance, and transparency of accounting and analytical information. It also involves its analytical processing and use in management to achieve sustainable development goals. This, in turn, allows for the effective evaluation of financial standing, optimization of business processes, making well-founded management decisions, and adapting to rapidly changing market conditions and the economic environment.
3. Accounting engineering is defined as an innovative technology for transforming the information support system for managing the economic activity of the enterprise. It is based on the integration of cutting-edge digital solutions and methodological tools for collecting, processing, and analytically processing relevant accounting and analytical data at fair value, allowing for the optimization of accounting, analysis, and planning processes to implement operational and strategic managerial influences. This ensures the effectiveness, adaptability, and resilience of the enterprise in a turbulent business environment. To qualitatively assess the economic activity of trade enterprises, it is recommended to implement accounting engineering technology in the accounting architecture of the trading

enterprise based on the theory of constraints and business units. This allows for generating informational arrays of relevant accounting and analytical data of both retrospective and prospective nature.

4. The application of engineering tools in accounting encompasses several types of aggregated accounting operations: aggregating accounts and balances at the end of the period, evaluating high-risk business operations, assessing assets at fair value, conditional (hypothetical) asset sales, and determining net assets and net liabilities. This algorithm for accounting engineering technology with innovative tools is designed for analyzing projects or business processes (unit economics, in particular) based on risk levels and indicators of economic activity in the context of economic, social, and environmental spheres of activity. To implement this, a platform for the functioning of accounting engineering technology within the system of information support for managing economic activity has been developed, structurally interconnected. It defines the foundation of a scientific-applied approach to the formation of information support for managing economic activity, which necessitates focusing on providing users with relevant information for managerial decision-making.

Promising directions for future scientific research include the development of conceptual and methodological foundations for assessing the economic activity of trade enterprises in the context of the digital transformation of business processes, as well as the formulation of adaptive strategies for their digital integration.

References

1. Melnychenko AA (2015) Teoretychni zasady suchasnoi filosofii upravlinnia [Theoretical foundations of modern management philosophy]. Visnyk NTUU «KPI». Filosofiia. Psykholohiia. Pedahohika 1(43):57–63
2. Tekbaş İ. The profession of the digital age: accounting engineering—IFAC. https://www.ifac.org/knowledge-gateway/discussion/profession-digital-age-accounting-engineering?utm_source=chatgpt.com. Accessed 30 Mar 2025
3. Michalczyk L (2012) Model of accounting engineering in view of earnings management in Poland, e-Finanse: financial internet quarterly, vol 8, no 3. University of Information Technology and Management, Rzeszów, pp 37–54
4. Pana Nicoleta D. Creative accounting—art. http://www.oeconomica.uab.ro/upload/lucrari/920071/52.pdf. Accessed 30 Mar 2025
5. Michalczyk L (2012) The role of accounting engineering in shaping the balance policy of a company. Financ Internet Q e-Financ 8(4):44–52
6. Young MR. Financial reporting after the subprime crisis: new challenges for audit committees. http://www.willkie.com/publications/2010/month/financial-reporting-after-the-subprime-crisis. Accessed 30 Mar 2025
7. Brukhanskyi RF (2014) Bukhhalterskyi inzhynirynh yak metod pobudovy stratehichno oriientovanoi systemy obliku [Accounting engineering as a method of building a strategically oriented accounting system]. Visnyk Volynskoho instytutu ekonomiky ta menedzhmentu 10:59–66
8. Herasymovych IA (2017) Mekhanizmy ta instrumenty bukhhalterskoho i finansovoho inzhynirynhu v upravlinni pidpryiemstvom [Mechanisms and tools of accounting and financial engineering in enterprise management]. Oblik i finansy (1):25–32

9. Yukhymenko-Nazaruk IA (2017) Bukhhalterskyi inzhynirynh v systemi kreatyvnoho obliku [Accounting engineering in the creative accounting system]. Rozvytok sotsialno-ekonomichnykh system v heoekonomichnomu prostori: teoriia, metodolohiia, orhanizatsiia obliku ta opodatkuvannia: Materialy Mizhnarodnoi naukovo-praktychnoi konferentsii. TNTU im. I. Puliuia. m. Ternopil, pp 165–166
10. Chumak OV, Nahorna IV (2017) Mekhanizm inzhynirynhu pry formuvanni informatsiinoho pidgruntia analizu finansovoi stiikosti pidpryiemstva [Engineering mechanism in forming the information basis for analyzing the financial stability of an enterprise]. Biznes-inform (12-2017):359–364
11. Zamula I, Prodanchuk M, Kovalchuk T, Myhalkiv A, Simakov K (2020) Engineering of business processes in accounting support of cash management. Agric Resour Econ 6(3):135–148. Accessed 30 Mar 2025
12. Sokhatska OM, Proboiv OA (2005) Vyrishennia problem korporatyvnoho upravlinnia shliakhom zastosuvannia tekhnolohii finansovoho inzhynirynhu [Solving corporate governance problems through the application of financial engineering technologies]. Zbirnyk naukovykh Prats Cherkaskoho derzhavnoho tekhnolohichnoho universytetu. Seriia: ekonomichni nauky, no 15. ChDTU, Cherkasy, pp 237–243
13. Kashchena N (2021) Accounting engineering in accounting and analytical support of management of economic activity of trade enterprises. Ukr J Appl Econ Technol 6(1):330–336. https://doi.org/10.36887/2415-8453-2021-1-39
14. Savchuk VK (2013) Teoretychni zasady formuvannia informatsiino-analitychnoho servisu upravlinnia [Theoretical foundations of the formation of an information and analytical management service]. Naukovyi visnyk Natsionalnoho universytetu bioresursiv i pryrodokorystuvannia Ukrainy. Ser.: Ekonomika, ahrarnyi menedzhment, biznes, vol 181, no 3, pp 218–223
15. Savchuk VK (2014) Heneza informatsiino-analitychnoi systemy upravlinnia [Genesis of the information and analytical management system]: zbirn. Mater. Sh mizhn. Nauk.-prakt. Konf. Prysviach. 70 richchiu kaf. Bukhhalterskoho obliku pidpryiemnytskoi diialnosti. KNEU, Kyiv, pp 293–297
16. Savchuk VK (2016) Oblikovo-analitychne zabezpechennia upravlinnia yak faktor pidvyshchennia yoho rezultatyvnosti [Accounting and analytical support of management as a factor in increasing its effectiveness]. Transformatsiini peretvorennia oblikovo-analitychnoho zabezpechenenyia upravlinnia v umovakh yevrointehratsiinykh protsesiv: mater. IV Mizhnar. nauk.-prakt. konf., m. Kyiv, 25 lystopada 2016 r. KNEU, Kyiv, pp 323–325
17. Nesterenko I, Kashchena N, Chmil H, Nesterenko O, Kovalevska N (2024) Devising a methodological approach to identifying the economic potential of production costs for eco-innovative products. East Eur J Enterp Technol 3(13(129)):6–15. https://doi.org/10.15587/1729-4061.2024.304805
18. Kashchena N, Chmil H, Nesterenko I, Lutsenko O, Kovalevska N (2024) Diagnostics as a tool for managing behavior and economic activity of retailers in the conditions of digital business transformation. In: Semenov A, Yepifanova I, Kajanová J (eds) Data-centric business and applications. Lecture notes on data engineering and communications technologies, vol 194. Springer, Cham, pp 149–173. https://doi.org/10.1007/978-3-031-53984-8_7
19. Evans MH. Discussion board articles—value based management. Creating value through excellence in financial management. http://exinfm.com/pdffiles/value_based_mgmt_articles.pdf. Accessed 30 Mar 2025
20. Holov SF (2012) Upravlinskyi oblik na osnovi teorii obmezhen [Management accounting based on the theory of constraints]. Bukhhalterskyi oblik i audyt (6):40–48
21. Krymska LO, Shevchenko OO (2012) Upravlinnia pidpryiemstvom na osnovi teorii obmezhen [Enterprise management based on the theory of constraints]. Biznes Inform (12):198–201
22. Statement of Management Accounting 4HH (1999) Theory of constraints (TOC) management system fundamentals. IMA, 60 p
23. Herasymovych IA (2015) Orhanizatsiine ta informatsiine zabezpechennia inzhynirynhu v systemi upravlinnia operatsiinoiu diialnistiu pidpryiemstva [Organizational and information

support for engineering in the enterprise's operational management system]. Bukhhalterskyi oblik, analiz ta audyt: problemy teorii, metodolohii, orhanizatsii 2(15):21–25
24. Chumak OV (2017) Uzghodzhennia pokaznykiv zvitnosti pry otsintsi upravlinnia diialnosti derzhavnykh pidpryiemstv [Harmonization of reporting indicators when assessing the management of state-owned enterprises]. Suchasni problemy upravlinnia pidpryiemstvamy: teoriia i praktyka: materialy Mizhnar. Nauk.-prakt. Konf., m. Kharkiv, 30–31 bereznia 2017, Kharkiv, pp 56–62
25. Fomina OV (2015) Upravlinskyi oblik za teoriieiu obmezhen [Management accounting according to the theory of constraints]. Hlobalni ta natsionalni problemy ekonomiky, № 3, pp 920–925
26. Lysak H, Morozova H, Gorokh O, Maliy O, Nesterenko I (2022) The system of financial control in the management of a small business enterprise: methods and tools of implementation. Rev Econ Financ 20(1):1034–1041. https://doi.org/10.55365/1923.x2022.20.115
27. Zoboviazannia. Natsionalne polozhennia (standart) bukhhalterskoho obliku 11 vid 31.01.2000 № 20 iz zminamy i dopovnenniamy. https://zakon.rada.gov.ua/laws/show/z0085-00#Text. Accessed 30 Mar 2025
28. Vypravlennia pomylok i zminy u finansovykh zvitakh. Natsionalne polozhennia (standart) bukhhalterskoho obliku 6 vid 28.05.1999 № 137 iz zminamy i dopovnenniamy. https://zakon.rada.gov.ua/laws/show/z0392-99#Text. Accessed 30 Mar 2025
29. Vasilyeva T, Samusevych Y, Babenko V, Bestuzheva S, Bondarenko S, Nesterenko I (2023) Environmental taxation: role in promotion of the pro-environmental behaviour. WSEAS Trans Bus Econ 20:410–427. https://doi.org/10.37394/23207.2023.20.37
30. Makarovych VK, Sydorchuk OH (2019) Bukhhalterskyi inzhynirynh v diialnosti pidpryiemstva [Accounting engineering in the activities of the enterprise]. Naukovyi visnyk Uzhhorodskoho universytetu. Seriia: Ekonomika, № 2, pp 129–134
31. Kashchena N, Nesterenko I, Chmil H, Kovalevska N, Velieva V, Lutsenko O (2023) Digitalization of biocluster management on basis of balanced scorecard. J Inf Technol Manag 15(4):80–96. https://doi.org/10.22059/jitm.2023.94711

Innovative Tools for Motivating Strategic Human Resource Development in IT Business

Inna Kuznetsova, Olha Balabash, and Yuliia Karpenko

Abstract The article offers a methodology of building a system of incentives on the basis of consideration of motivational features of labour activity with the aim of improvement of strategic management of human resources at IT enterprises. The study applies a number of methods: the method of dynamic series—to analyze the trend of development of the IT industry, M. Porter's method of preference rhombus—to analyze the determinants that affect the qualitative parameters and quantitative indicators of human resources of enterprises in the industry; the method of building a motivational profile—to identify common motivational types among employees of IT enterprises. The article allocates and analyses determinants that influence qualitative parameters and quantitative indicators of provision of IT industry enterprises with human resources, studies the nature of their influence and traces structural and logical interrelations between them. This study uses a methodology based on the typological model of labour motivation to build a motivational profile of IT company employees and identify common motivational types and the corresponding recommended methods of incentives for them. An infogram of correspondence of incentive methods to the motivational types of employees of IT enterprises is built and the results of the study are visualised. The practical significance of the infogram is to simplify the construction of an incentive system at IT enterprises by formalising the process of determining incentive methods in accordance with the motivational types of employees and the specifics of work in the IT sector.

Keywords Strategic management · Strategy · Enterprise resources · Human resources · Incentive system · Motivational profile · IT industry

I. Kuznetsova · O. Balabash (✉) · Y. Karpenko
Odesa National Economic University, Odesa, Ukraine
e-mail: balabash_olga987@ukr.net

I. Kuznetsova
e-mail: Inna9876543210@ukr.net

Y. Karpenko
e-mail: karpenko_julia_v@ukr.net

A. Semenov et al. (eds.), *Data-Centric Business and Applications*, Lecture Notes on Data Engineering and Communications Technologies 272,
https://doi.org/10.1007/978-3-032-06608-4_7

1 Introduction

Human resources occupy a special place in the totality of an enterprise's resources. The efficiency of human resource management in the strategic perspective determines the level of economic growth of the enterprise and its competitiveness in the market. Part of achieving these objectives is to establish an effective incentive system that should reflect the goals of a particular employee and the strategic plans of the organisation as a whole.

The IT industry is currently developing dynamically, and the need for skilled workers is therefore extremely acute. Today, the majority of employees in this sector are paid above the market average, enjoy decent working conditions and are provided with a social package. Despite the fact that IT companies are constantly improving their staff motivation and incentive systems, competition between companies for qualified and experienced staff is not decreasing. In view of this, we believe that the development of theoretical and practical foundations of strategic human resources management with a focus on taking into account the industry-specific features of labour activity in the IT sector will help to solve the identified problems.

Aim: to improve the methodology for building a system of incentives based on the motivational features of labour activity in order to improve the strategic management of human resources in IT business.

To achieve this goal, the following tasks were solved:

- -study of trends in the development of the IT industry to identify the determinants that affect the qualitative and quantitative parameters of human resources provision of IT enterprises;
- building a motivational profile to track typical industry features and motivational types of employees in the IT industry;
- Conducting an expert study based on a survey to identify and formalise the motivational features of labour activity at IT enterprises;
- improvement of the classification of methods of employee incentives by taking into account common motivational types of IT employees and motivational features of labour activity;
- developing an informogram of the incentive system of IT enterprisesbased on the motivational aspects of IT labour activity.

To achieve this goal, the following methods were used: the method of dynamic series—to analyse the trend of development of the IT industry, the volume of labour resources and the need for them; M. Porter's rhombus of advantages method—to analyse the available resources, information infrastructure, development of a cluster of supporting industries, market demand conditions and other determinants that affect the qualitative parameters and quantitative indicators of human resources provision of enterprises in the industry, to study the nature of their impact and to trace the structural and logical relations between them; the method of the method of building a motivational profile—to identify common motivational types among employees of IT enterprises; expert evaluation—to identify motivational features of labour activity at IT enterprises; systematisation and generalisation—to improve the classification of methods of non-material incentives for employees of IT enterprises; graph-analytical method, table method—to build an infographic of compliance of incentive methods with motivational types of employees of IT enterprises and visualisation of other research results.

The practical significance of the work lies in the possibility of using the developed infographics on the correspondence of incentive methods to the motivational types of employees for the improvement of strategic human resource management by forming an incentive system at IT enterprises.

2 Overview of Economic Indicators of the IT Industry in Ukraine

The IT industry is a significant sector of Ukraine's economy. In 2020, IT accounted for 8.3% of Ukraine's total exports. Today, the IT industry is developing rapidly. In 2020, it grew by 20.6% year-on-year, and over the past 5 years, it has increased by 2.5 times. This conclusion is based on the data from the Report 2021 on the state of the IT services market in Ukraine and the Governmentportal of the State Statistics Committee of Ukraine [1, 2]. According to experts, exports of IT services are expected to grow to USD 8.4 billion in 2025 (Fig. 1).

According to the study of the 2021 Tech Market Report, Ukraine ranks first in Europe in terms of exports of IT services. There are more than 110 centres of well-known international companies in Ukraine, and 21 companies of Ukrainian origin

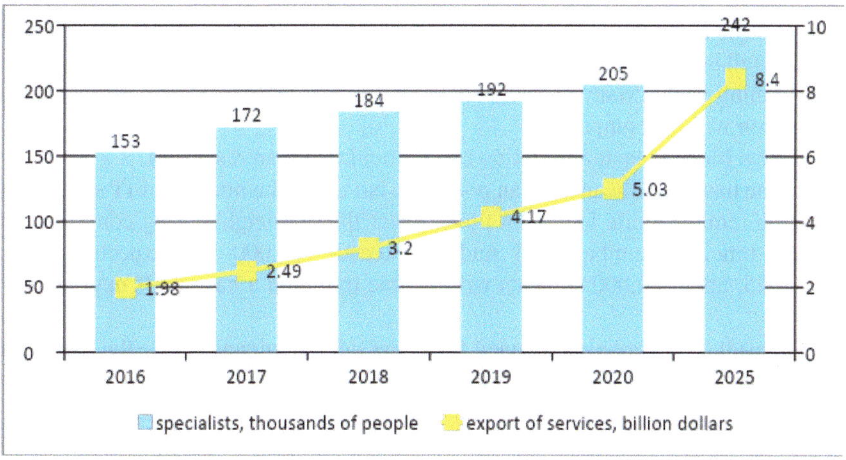

*2025 –forecast

Fig. 1 Development trends of the IT industry in Ukraine

with development offices in Ukraine are included in the Global Outsourcing 100 list. Ukraine has become a powerful technology hub capable of implementing projects of any complexity and geography.

Further analysis of the environment of functioning of enterprises in the industry will allow to identify a number of determinants, the detailing of which and the study of the nature of the impact will allow to trace the impact on the provision of enterprises in the IT industry with human resources.

In order to implement this, it is proposed to use M. Porter's rhombus of advantages. It includes an analysis by factor conditions (human resources, natural resources, finance, technology, infrastructure), a cluster of supporting industries, domestic market demand and competition in the domestic market, as well as randomness and government influence, as stated in the studies of Pokras [3]. This method allows to formulate a list of applied recommendations for the formation of a development strategy for primary industry enterprises, as stated in the studies by Balabash and Kuznetsova [4], Balabash [5].

The first block we considered is market demand conditions.

Firstly, it is important to note the growing demand for IT services. This was influenced by the forced transition to remote work of enterprises as a result of the COVID pandemic. In such circumstances, it was important to maintain the elements of control and management of projects and business processes remotely. This conclusion is based on data from the Government portal of the State Statistics Committee of Ukraine [6]. All this led to an increase in demand for the use of IT technology in various fields.

The main areas for which IT products are created are e-commerce, entrepreneurship and finance Fig. 2.

Ukraine has recently been actively developing and implementing IT technologies at the state level. The government is implementing digitalisation of the public sector. Starting from 1 October 2021, electronic document exchange between government agencies will be launched, and full digitalisation of the document flow will take place in the coming years. Many projects are being implemented by public authorities in cooperation with IT companies.

The next block is factor conditions, in particular human resources.

Ukraine has a significant human potential, but today the number of IT specialists per capita remains quite low—2.4 times lower than in neighbouring countries. At the same time, the number of IT students exceeds 100,000. It is expected that in 2022–2025, about 80,000 students will be able to obtain a bachelor's degree in IT [1, 2].

As a result of the growing demand for IT services, companies have the resources to fulfil more orders. They started to hire specialists more actively. According to the analysis, we can note a parallel increase in the number of specialists, but it is happening at a pace that does not meet the needs of the industry (Fig. 3).

Among the negative trends affecting the availability of labour resources for the enterprises of the industry under study is the "migration of qualified personnel abroad to the USA and Europe". This is due to higher wages and non-financial incentives for employees of global IT giants. The more developed infrastructure and better living

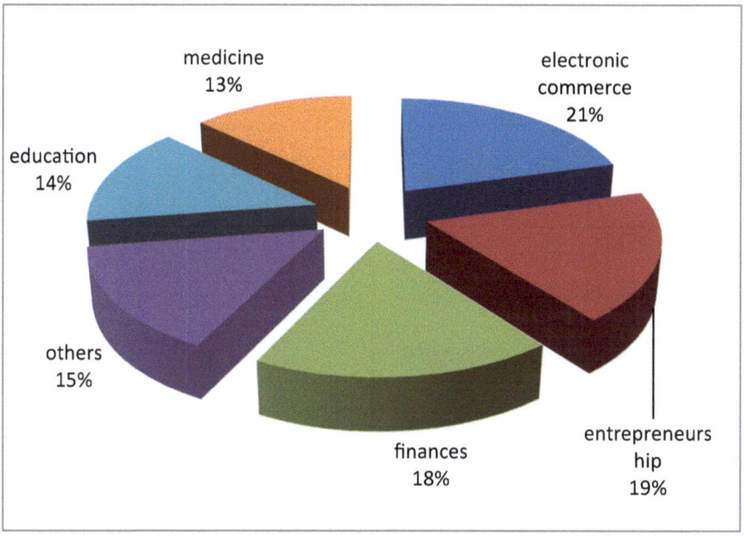

Fig. 2 Industries for which Ukrainian software is created

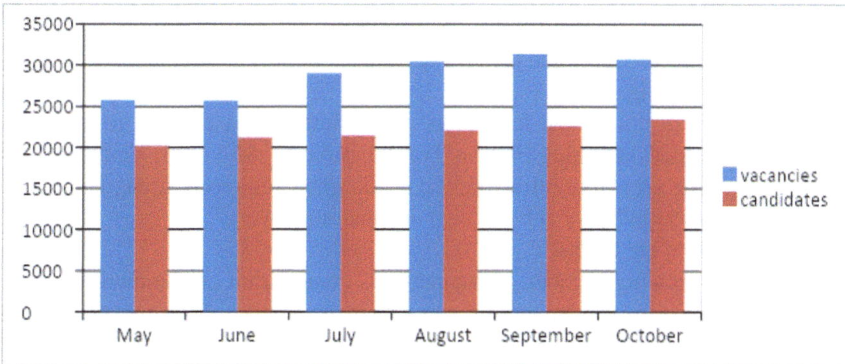

Fig. 3 The ratio of supply and demand for human resources in the IT industry, 2023

conditions in these countries also contribute to this. This negative factor was present before the war, and has intensified significantly since the russian invasion.

Infrastructure.

As the industry expands, there is a need to develop the necessary infrastructure. For IT, this includes offices, coworking spaces, open spaces, innovation parks, and R&D centres. Today, such projects are actively developing in large cities with a high concentration of IT companies, such as Kyiv, Kharkiv, and Lviv. They help to concentrate valuable resources for the industry in one place and conduct educational activities among specialists.

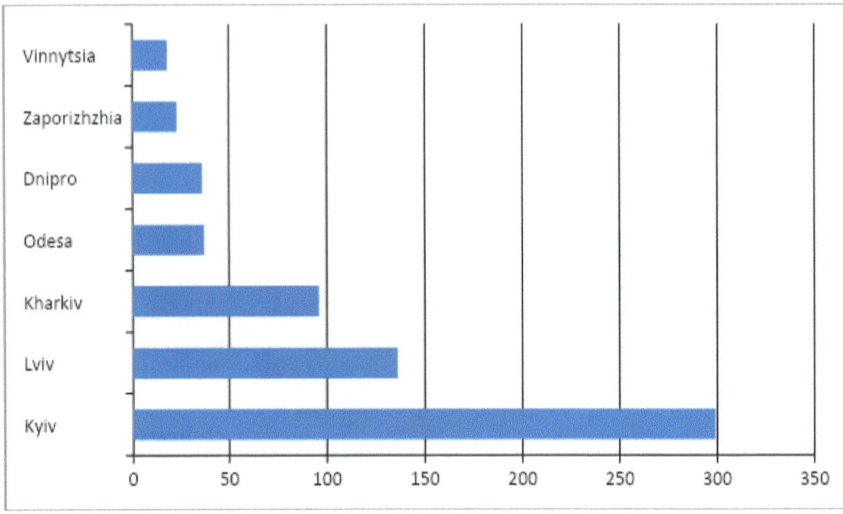

Fig. 4 Offices of service companies

Kyiv has the largest number of offices—1533, followed by Lviv (263) and Kharkiv (197).

More than 70% of Ukrainian tech professionals live in the top three cities: Kyiv, Kharkiv and Lviv (Fig. 4).

It is also worth noting the formation of IT clusters and centres in numerous medium-sized cities, including Poltava, Ivano-Frankivsk, Mykolaiv, Vinnytsia, and Kremenchuk. The growth of infrastructure in Ukraine's regions is strengthening the social impact of the industry at both local and national levels.

Competition on the market.

Despite the growing number of competitors, competition on the market can be characterised as medium intensity, as the growth in the number of companies is driven by the growing demand for their services both in the domestic and, most importantly, international markets.

According to the Report 2021 on the state of the IT services market in Ukraine [1, 2]. Ukrainian IT companies have a good reputation in the international market and are quite competitive. Ukraine is the first largest exporter of IT services in Europe. The high level of expertise and a flexible model of contractual interaction with private entrepreneurs are the industry's competitive advantages. More than 100 representatives of the Fortune 500 are loyal customers of the domestic IT business, and the Global Sourcing Association, a leading international organisation, has recognised Ukraine as the best country for IT services to the UK. In 2020, IAOP's Global Outsourcing 100 ranking of the world's best service providers included 13 Ukrainian companies and 8 companies with offices in Ukraine.

The cluster of supporting industries plays an important role in the development of the IT sector.

The development of the IT sector is not possible without a high level of communication and technology. Ukraine continues to develop Internet access networks. In 2020, the number of fixed Internet access lines (points) totalled 7265 thousand, up 18% or 1106 thousand compared to the previous year. In 2021, 4G mobile internet was launched for the first time in 7.4 thousand settlements with a population of 2 million people, and the mobile coverage area increased or an additional operator appeared in 8.4 thousand settlements for 7.1 million people. This conclusion is based on the data from the Government portal of the State Statistics Committee of Ukraine [1].

As noted above, IT products and services are widely used in many areas of the economy. A special place among them is occupied by retail and wholesale trade using the Internet, through various Internet platforms, etc., the so-called e-commerce. In Ukraine, the number of electronic orders of goods increases by about 25–30% annually, and the turnover in hryvnia by 40–60% (statistics) (Table 1).

The study of strategic human resource management is actively developing in modern scientific literature. In particular, Gulyk and Kobzev [7] note that strategic human resource management goes beyond traditional personnel management, focusing on long-term planning and coordination of personnel policy with the overall strategy of the enterprise. The author emphasizes that the main task of strategic management is to create competitive advantages through the effective use of human capital.

Heathfield [8] considered "human resources" as the people who manage the organisation, as well as each individual employee of the enterprise.

Pochynok [9], considering "human resources" at the micro level, emphasises that they are a special asset of the enterprise, an integral part of the enterprise's production resources, the formation and development of which requires investment".

The work of Litvinenko and Prykhodko [10] emphasizes the importance of attracting and retaining talented specialists as a key element of strategic human resource management. The author proposes a model of personnel management that includes the development of corporate culture, motivational mechanisms, training and professional development of employees, as well as the formation of effective communications within the organization.

Agreeing with the opinion of Karpenko [11]; Kuznetsova [12], Khmelyarchuk [13], Balabash [14, 15], we consider it important to note that the strategic level of management is ensured through the implementation of the operational and tactical level of management. Operational management of human resources is achieved through the implementation of the following stages: personnel accounting, planning, recruitment and adaptation of personnel, training and development of personnel, their motivation and incentives.

In the work of Mishina [16], devoted to the theoretical foundations of labor incentives, a hierarchical multi-criteria classification of methods of non-material personnel incentives was developed and the nature of their impact on the personnel of the enterprise was examinedas an effective tool of influence in the modern realities of the labor market.

Table 1 Analysis of the determinants of the development of the IT industry according to the method of M. Porter's rhombus of advantages

Group of factors	Factors	Contents	Character
1. **Factor conditions** *Labour resources*	Availability of skilled labour resources	Ukrainian universities provide 16,000 qualified IT specialists and 130,000 general engineering specialists annually. An extensive system of non-governmental education. The high quality of education is regularly recognised and confirmed by various rankings, both nationally and globally	Positive
	Outflow of qualified personnel	The migration of skilled personnel abroad to the US and Europe is driven by higher salaries and non-financial incentives for employees of global IT giants. The more developed infrastructure and better living conditions in these countries also contribute to this	Negative
Infrastructure	Is actively developing in large and medium-sized cities	Infrastructure for the industry includes offices, coworking spaces, open spaces, innovation parks, and R&D centres. Currently, such projects are actively developing in large cities with a high concentration of IT companies, such as Kyiv, Kharkiv, and Lviv. They help to concentrate valuable resources for the industry in one place and conduct educational activities among specialists. Formation of IT clusters in medium-sized cities, including Poltava, Ivano-Frankivsk, Mykolaiv, Vinnytsia, and Kremenchuk	Positive
2. **Cluster of supporting industries**	Development of related industries	The main areas for which IT products are created are e-commerce, business and finance (58% of orders). In Ukraine, the number of electronic orders of goods increases by about 25–30% annually, and the turnover in hryvnia by 40–60%	Positive
	Development of internet access networks	In 2020, the number of fixed internet access lines (points) totalled 7265 thousand, up 18% or 1106 thousand compared to 2018	Positive
	Active digitalisation of the public sector	Introduction of electronic document management. Launch of the applications Diia, eMalyatko, City, etc. Development of distance education and healthcare reform	Positive

(continued)

Table 1 (continued)

Group of factors	Factors	Contents	Character
3. **Market demand conditions**	Growing demand	IT technologies are being actively developed and implemented in many areas. This is facilitated by electronic document exchange between government agencies. Many projects are being implemented by public authorities in cooperation with IT companies. Development of distance education. In the non-governmental sector, the main areas for which IT products are created are e-commerce, entrepreneurship and finance	Positive
4. **Competition in the market**	Medium level of competition	The market includes around 4000 IT companies, with 1600 being service-oriented. The number of people working in this sector is also growing every year—in 2023, their number was 285,000. Ukrainian IT companies have a good reputation in the international market and are competitive with foreign companies due to the high level of expertise and flexible model of contractual interaction with clients	Positive

Therefore, contemporary research on strategic human resource management focuses on integrating HR policies into the overall corporate strategy, utilizing digital technologies in HR processes, developing corporate culture, and creating conditions for attracting and retaining talented employees.

The authors of the considered scientific studies distinguish two groups of incentive methods: economic (professional-qualification, compensatory, organizational) and non-economic (social-status, moral-psychological, disciplinary). They contribute to the stability of personnel, reducing turnover, and loyalty of personnel, which are important factors to consider in the management strategy. Agreeing with scientific statements about the advantages of these incentive methods, it is important to emphasize the need to study them and take into account the specifics of professional activity when developing a human resources management strategy.

3 Results

Despite the high level of wages in the industry and the significant efforts made by employers to develop working conditions and incentive systems, the problem of providing companies with skilled labour remains relevant.

Therefore, we consider it necessary to study the system of strategic human resources management on the example of specific enterprises in the industry, in order to take into account the industry-specific features of IT workers' labour activity and mechanisms for combining various forms of incentives in practice.

The commitment of employees to various forms of incentives, the effectiveness of their application in relation to specific representatives of the profession allows us to define such a tool as a motivational profile. Building a motivational profile of the staff allows developing a system of managerial influences on employees through the most effective incentive tools.

In the studies of Dzyuba [17], Kuznetsova [12], according to the methodology based on the typological model of labour motivation, the construction of a motivational profile is based on a combination of motivational types of a person and forms of motivation.

In the studies of Dzyuba [17], each person in terms of his or her motivation is a combination in some proportions of five pure motivational types: lumpen type (agrees to low pay, provided that others do not receive more; does not seek to improve qualifications, opposes it; low activity and responsibility, opposes the activity of others; strives to minimise efforts); instrumental type (interested in the cost of labour more than the content; the validity of the cost of labour is important, does not want "handouts"; the ability to provide for one's own life is important); professional type (interested in the content of work; does not agree to work that is not interesting to him/her, no matter how much he/she is paid for it; is interested in difficult tasks—the possibility of self-expression; considers freedom in operational actions important; professional recognition as the best in the profession is important); patriotic type (needs an idea that will drive them; public recognition of participation in success, recognition of indispensability in the company is important); economic type (voluntarily assumes responsibility; is characterised by an acute demand for freedom of action; does not tolerate control).

Using a standardized questionnaire and relying on the methodology substantiated by [17], we conducted a survey of employees of IT companies in Odesa. In total, 100 employees completed the survey, the results of processing these questionnaires are presented in Table 2.

The analysis showed that the professional type of motivation prevails among employees [18, 19]. Representatives of this type are focused on confirming their professional abilities and strive to achieve high results in their work [20]. Organizational, moral and material incentives are effective for them [20, 21]. The organizational component is the basis of their motivation, since they attach great importance to the conditions of work, its content and organization [20, 21]. Negative methods of influence, as well as a paternalistic approach, are unacceptable for them.

The second, less numerous type is the patriotic one. For such employees, moral incentives play a key role, along with non-monetary rewards, elements of paternalism, and the opportunity to participate in company management.

After the motivational profile of company employees has been established, it will be advisable to consider in more detail the incentive options within the established forms in order to study the industry-specific characteristics of IT employees' work.

Table 2 Raw data for building a motivational profile of employees of IT companies

Group of employees	Motivation index				
	Patriotic type	Economic type	Instrumental type	Professional	Lumpenized type
Men	0.22	0.21	0.20	0.26	0.11
Women	0.22	0.20	0.22	0.23	0.14
Managers	0.17	0.22	0.18	0.38	0.05
Main staff	0.24	0.20	0.20	0.24	0.13
They have a higher education	0.26	0.24	0.15	0.29	0.06
They do not have a higher education	0.22	0.20	0.19	0.24	0.15
Age up to 35 years	0.21	0.16	0.22	0.19	0.22
35–45 years	0.20	0.18	0.19	0.23	0.20
45–50 years	0.17	0.22	0.20	0.25	0.16
Older than 50 years	0.21	0.23	0.18	0.31	0.07
Work experience up to 5 years	0.24	0.16	0.18	0.19	0.23
5–10 years	0.19	0.22	0.22	0.22	0.17
More than 10 years	0.24	0.29	0.11	0.29	0.08
General index	0.21	0.21	0.19	0.25	0.14

For this purpose, we conducted a survey of employees to identify which incentive options are more acceptable and desirable for them. Based on the results of the survey, the answers were systematised, and some of them, which were similar in content, were combined.

The results of the survey are shown in Fig. 5, which grouped the answers to the questions about the incentives for IT employees and the percentage of their support.

After determining the key motivating factors, we distributed them among the identified motivational types of IT company employees. According to the specific characteristics of each motivational type, we determined the level of effectiveness of each type of stimulation: basic "+++"—as the most desirable and effective; acceptable—one that has a certain positive impact "±"; neutral "0"—which has no noticeable motivational effect; and unacceptable "–"—one that has a negative impact on motivation.

The distribution of incentive methods according to their impact on different motivational types of IT employees is presented in the Table 3.

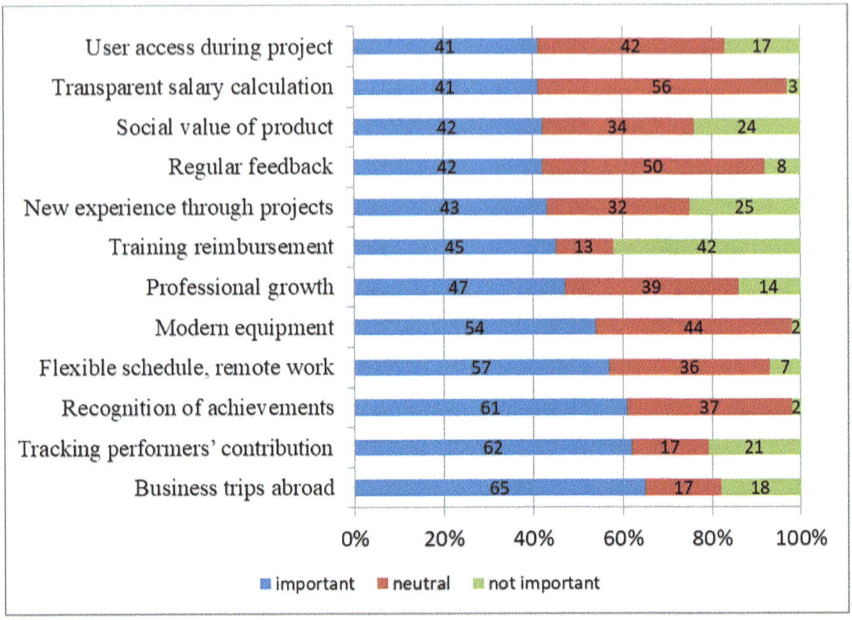

Fig. 5 Priority of incentive factors in the IT industry

Table 3 Infogram of the correspondence of methods of stimulation to motivational types of employees of enterprises in the IT industry

Methods of stimulation	Motivational type				
	Instrumental	Professional	Patriotic	Economic	Lumpenized
Negative	0	–	+/–	–	+++
Disciplinary	0	+++	+/–	+/–	0
Money	+++	+/–	0	+/–	0
Natural	+/–	0	+/–	0	+++
Moral	–	+/–	+	0	0
Paternalism	–	–	+/	–	+++
Qualifying	+/–	+++	+/	+/	0
Organizational	0	+++	0	+/	–
Enthusiasm	0	+++	+/	+/	0
Participation in management	0	+/–	+/	+++	–

The built infogram reflects recommendations on the use of various methods of incentivising employees depending on their motivational type and taking into account the specifics of work in the IT sector.

4 Conclusions

Based on the method of dynamic series, the article studies the trends in the development of the IT industry and traces the structural and logical relationships between the identified trends. The results are complemented by the use of M. Porter's rhombus of advantages method to analyse the available resources, information infrastructure, cluster of supporting industries, market demand conditions and other determinants that affect the qualitative parameters and quantitative indicators of human resources provision of enterprises in the industry. According to the results, it was found that:

- employees of the industry's enterprises are paid higher wages than the national and regional average, which contributes to the outflow of personnel from other industries and the prospects for young professionals to choose a profession;
- sufficient infrastructure development to ensure proper working conditions and skilled training for IT company employees not only in cities with a population of over a million, but also in cities of regional significance;
- the ability of Ukraine's education system to provide the industry with qualified personnel through an extensive system of both public and private educational institutions;
- the growing demand for IT services and products allows companies to accumulate significant resources that can be spent on the development of staff motivation systems;
- the development of a cluster of supporting industries allows expanding the scope and application of IT developments, ranging from e-commerce to social services, education, medicine, etc. This helps to increase staff interest in project implementation and motivation through a sense of personal contribution to the implementation of socially important projects at the national level.

At the same time, a number of determinants have been identified that have a negative impact, namely: the outflow of skilled personnel, migration of skilled personnel abroad to the United States and Europe due to higher wages and non-financial incentives for employees of global IT giants. The more developed infrastructure and better living conditions in these countries also contribute to this. This was already happening before the war, and has intensified since the Russian invasion.

Despite the fact that the existing determinants with a positive impact significantly outweigh the negative ones, competition between companies for qualified and experienced staff is not decreasing, as evidenced by the analysis of the dynamics of supply and demand for workers in the IT industry. Accordingly, the need for skilled workers is surprisingly acute. In view of this, we believe that the development of theoretical and practical foundations of strategic human resources management with a focus on taking into account the industry-specific features of labour activity in the IT sector will help to solve the identified problems.

Based on a methodology that is based on the typological model of labour motivation, the article builds a motivational profile of employees of IT companies and identifies common motivational types among the surveyed employees; identifies the inherent features of the identified motivational types and the corresponding recommended methods of stimulation for them. The analysis identified the characteristic features of each type of motivation and selected appropriate recommended stimulation methods.

Based on the conducted research, an infographic has been created to illustrate the alignment of incentive methods with the motivational types of IT company employees. It formalizes the process of selecting incentives based on an employee's motivational type, serving as a foundation for building an effective incentive system in the IT sector.

References

1. Governmentportal of the State Statistics Committee of Ukraine: use of information and communication technologies at enterprises: use of internet network, cloud calculation, robotics. Available: http://ukrstat.gov.ua (2024)
2. Ukraine: the Home of Great Devs: 2021 Tech Market Report. Available: https://beetroot.co/wp-content/uploads/sites/2/2021/03/Ukraine_-the-Home-Of-Great-Devs-2021-_-Ebook-v-2.0-2.pdf (2021)
3. Pokras O (2017) Analysis of the competitiveness of instrument making in Ukraine with the help of a diamond of national preferences. TAPR 4(5):31–36. https://doi.org/10.15587/2312-8372.2017.109114
4. Balabash O, Kuznetsova I, Levytska I, Zaika S, Skrypko T (2022) Strategic management of the enterprise: analysis of sectoral determinants. Independent J Manag Prod 13(3):196–214. https://doi.org/10.14807/ijmp.v13i3.1992
5. Balabash O (2021) Managing innovation in product retail: strategic aspects. In: Pasichnyk Y, Lyashenko V, Radchenko O, Sukach O (eds) Economic and legal principles of economic growth in the post-crisis period: collective monograph, pp 216–229
6. Governmentportal of the state statistics committee of Ukraine: Use of e-democracy tools by state authorities and local governments. Available: http://ukrstat.gov.ua (2020)
7. Gulyk TV, Kobziev AV (2018) Strategic management of personnel development as a source of increasing enterprise competitiveness. Econ Soc 18:25–258. Available: https://doi.org/10.32782/2524-0072/2018-18-49
8. Heathfield S (2016) What is a human resource? Available: https://www.thebalance.com/what-is-a-human-resource-1918144
9. Pochynok N (2010) Human resources in accounting theory and practice. Econ Anal 6:141–144
10. Lytvynenko OM, Prykhodko KV (2020) Problems of recruitment and retention of Generation Z in the realities of staff shortages in Ukraine. Nobel Bull 1(13):65–73. Available: https://doi.org/10.32342/2616-3853-2020-1-13-8
11. Karpenko Y, Pienova A, Melnychuk D, Kutsyk V, Nakonechna N, Kalinichenko S (2021) The main aspects of choosing the enterprises' operating strategy. Independent J Manag Prod 12(6):359–375. https://doi.org/10.14807/ijmp.v12i6.1758
12. Kuznetsova I, Balabash O, Ilin V, Poprozman N, Shushpanov D, Slavina N (2021) Content strategy in management of communications. Independent J Manag Prod 12(3):232–242. Available: https://doi.org/10.14807/ijmp.v12i3.1538
13. Khmelyarchuk M, Demko I, Kozmuk N, Balueva O, Balabash O (2019) The role of entrepreneurship in strategic management of the tourism industry. J Entrepreneurship Educ 22(2). Available: https://www.abacademies.org/articles/The-roleof-entrepreneurship-in-strategic-management-1528-2651-22-2-320.pdf

14. Balabash O, Kuznetsova I, Karpenko J, Dudziak O, Semenyshena N (2020) Management of biofuel production development on the basis of scenario planning. J Environ Res Eng Manag 76(3):35–46. https://doi.org/10.5755/j01.erem.76.3.25681
15. Balabash O, Kuznetsova I (2020) Technologies of strategic management of sustainable development of modern organizations. In: Kuznetsova I (ed) Disa Plyus, Kharkiv. Available: http://dspace.oneu.edu.ua/jspui/handle/123456789/11832
16. Mishina S, Mishin O (2019) Intangible labor incentives as a tool for strengthening personnel security at the enterprise. Econ Soc 20:582–592. https://doi.org/10.32782/2524-0072/2019-20-75
17. Dzyuba N (2017) Integrative approaches to the evaluation and development of management personnel in the context of the organization's strategic goals. Life Bus. Available: https://www.education.ua/blog/27873/
18. Urikova O, Ivanochko I, Kryvinska N, Strauss C, Zinterhof P (2013) Consideration of aspects affecting the evolvement of collaborative e-business in service organisations. Int J Serv Econ Manag 5(1/2):72. https://doi.org/10.1504/ijsem.2013.051863
19. Stoshikj M, Kryvinska N, Strauss C (2013) Efficient managing of complex programs with project management services. Glob J Flex Syst Manag 15(1):25–38. https://doi.org/10.1007/s40171-013-0051-8
20. Voynarenko M, Dzhedzhula V, Yepifanova I (2020) Modeling of the process of personnel motivation for innovation activity. WSEAS Trans Bus Econ 17:467–477. https://doi.org/10.37394/23207.2020.17.45
21. Yepifanova I, Dzhedzhula V (2023) Assessment of the level of intellectual capital of industrial enterprises. In: 2023 13th International conference on advanced computer information technologies (ACIT), pp 367–371. IEEE, USA. https://doi.org/10.1109/acit58437.2023.10275543

Smart Cities and Urban Resilience in Ukraine: Towards Sustainable Recovery and Digital Transformation

Oleh Kuzmak, Pierpaolo Pontrandolfo, Olena Kuzmak, and Tetiana Stepura

Abstract This chapter explores the evolution, current landscape, and future potential of smart city initiatives in Ukraine within the broader framework of the global Sustainable Development Agenda. A smart city is conceptualized as a modern urban ecosystem that leverages innovative digital technologies to enhance citizens' quality of life, optimize resource utilization, and improve infrastructure and urban governance. The study adopts a mixed-method approach, combining desk research, secondary data analysis, case studies of selected Ukrainian cities, and a review of international standards and policy frameworks. This enables the identification of key trends, achievements, and barriers in the development of smart urban systems. Particular attention is paid to transformative processes shaped by globalization, intensified economic interconnectivity, and the growing demand for sustainable organizational models. The chapter emphasizes the influence of strategic EU directives and regulatory frameworks in shaping Ukraine's urban modernization pathway. It systematically outlines the core elements of the smart and sustainable city concept, with a focus on creating inclusive, resilient, and environmentally responsible urban environments. The analysis further investigates the systemic characteristics of smart infrastructure, evaluating both its functional advantages and structural challenges, especially in integrating advanced technologies into outdated urban systems. While Ukrainian cities exhibit significant innovation potential, large-scale implementation remains constrained by financial, institutional, technological, and social factors. Against the

O. Kuzmak (✉) · O. Kuzmak
Faculty of Business and Law, Lutsk National Technical University, Lutsk, Ukraine
e-mail: oleh_kuzmak@lutsk-ntu.com.ua

O. Kuzmak
e-mail: o.kuzmak@lutsk-ntu.com.ua

P. Pontrandolfo
Department of Mechanics, Mathematics and Management, Politecnico Di Bari, Bari, BA, Italy
e-mail: pierpaolo.pontrandolfo@poliba.it

T. Stepura
Department of Business Economics and Investment, Lviv Polytechnic National University, Lviv, Ukraine
e-mail: tetiana.m.stepura@lpnu.ua

A. Semenov et al. (eds.), *Data-Centric Business and Applications*, Lecture Notes on Data Engineering and Communications Technologies 272,
https://doi.org/10.1007/978-3-032-06608-4_8

backdrop of the ongoing war, the study underscores the growing strategic importance of smart city initiatives for urban resilience, reconstruction, and long-term development. The findings provide practical recommendations for the development of a national smart city strategy and offer insights relevant to other countries undergoing crisis-driven or transitional urban transformation. The conclusions contribute to the global scholarly discourse on smart, sustainable, and future-oriented urbanism in regions affected by conflict or systemic change.

Keywords Smart cities · Sustainable urban development · Digital transformation · Urban resilience · Post-war reconstruction · E-governance · Smart infrastructure · Ukraine

1 Introduction

In the modern world, the concept of a smart city is rapidly evolving from a theoretical foundation to a practical model for addressing urban challenges. Under the influence of global urbanization, technological innovations, and environmental pressures, smart cities aim to improve the quality of life, optimize resource use, and enhance the resilience of urban systems through the integration and application of information and communication technologies (ICT), data-driven decision-making, and citizen participation to create a more efficient, inclusive, and sustainable urban environment. Globally, smart city initiatives are recognized as vital tools for achieving the Sustainable Development Goals (SDGs), particularly in areas such as climate resilience, mobility, energy efficiency, digital engagement, and responsive governance. These efforts have gained renewed relevance in the face of recent crises—particularly climate change, health emergencies, and armed conflicts—that have exposed the vulnerability of traditional urban systems and highlighted the need for innovation and adaptability in urban planning and policy.

For Ukraine, the paradigm of the smart city has become especially relevant in the context of ongoing transformations related to administrative decentralization, socio-economic reforms, and, critically, the war and the acute need for post-war reconstruction that has emerged as a result. These factors have revealed long-standing vulnerabilities in urban planning, infrastructure management, and public service delivery. At the same time, they have opened opportunities for redefining the future of Ukrainian cities through innovation, resilience, and strategic modernization. In recent years, Ukraine has made significant strides toward sustainable urban development, demonstrating considerable innovation potential supported by a growing digital economy, an active civic tech community, and the political will to modernize public administration. However, these efforts are often hindered by limited institutional capacity, uneven regional development, outdated infrastructure, and inefficient resource use. The decentralization reform laid the foundation for local governance, but gaps remain in the coherence of urban policies and the adoption of integrated approaches to planning. The military conflict has further complicated these challenges, leading to widespread destruction in many regions and shifting urban priorities toward recovery, resilience, and renewal.

Against this backdrop, smart city initiatives are increasingly seen as catalysts for sustainable post-war urban regeneration. Smart technologies, including digital management systems, intelligent mobility solutions, energy-efficient infrastructure, and open data platforms, offer new tools for rebuilding Ukrainian cities on more inclusive, adaptive, and future-oriented principles. Moreover, smart city strategies provide a pathway for aligning national urban development with international sustainable development standards, particularly the SDGs.

This section explores the evolution and current landscape of smart city development in Ukraine, emphasizing both achievements and structural barriers. It adopts a mixed-methods approach, combining desk research, case studies of selected Ukrainian cities, and an analysis of international policy benchmarks. Special attention is given to the systemic challenges technical, legal, economic, and sociocultural that hinder the widespread implementation of smart technologies. It also underscores the strategic importance of smart city initiatives in the context of Ukraine's recovery, proposing how digital transformation can support a sustainable future for cities. By placing the smart city agenda in Ukraine within the broader discourse on post-conflict urban resilience and digital innovations, this research contributes to understanding how smart cities can serve both as a recovery strategy and a model for long-term development. The findings aim to inform policy, planning, and scholarly discussions not only in Ukraine but also in other regions with transitional economies and crisis-affected areas facing similar urban challenges.

2 Research Methods

The research methodology employed in this study combines a comprehensive literature review, surveys, and qualitative case studies, incorporating a range of scientific research methods. The study employs analysis and synthesis to examine the essence and characteristics of the smart city concept, explore its implementation processes in Ukrainian cities, and identify the challenges associated with these processes. Additionally, the computational-constructive method was utilized to propose strategic directions for sustainable urban development and the concept of smart cities. The dialectical and abstract-logical methods were applied for theoretical generalization and conclusion formulation.

The authors conducted an extensive review of domestic and international studies that focus on the integration of Sustainable Development Goals (SDGs) into urban development strategies, as well as on the implementation of the smart city concept. To assess the current state and trends regarding the smart city concept in Ukraine and globally, the study utilizes data from global rankings, including the Smart City Index 2023 by IMD and Singapore University of Technology and Design (SUTD), the IESE Cities in Motion Index 2022 by the IESE Business School, the Smart City

Awards competition from the Kyiv Smart City Forum, and EU regulations that serve as the conceptual foundation for sustainable urban management.

Additionally, the study incorporates insights from the World Bank, the "Diia" digital platform, websites of Ukrainian cities, and the analytical report "Smart Infrastructure in Sustainable Urban Development: Global Experience and Prospects for Ukraine" prepared by the Razumkov Center with support from the Friedrich Naumann Foundation for Freedom in Ukraine. This diverse range of sources facilitated the identification of key issues faced by Ukrainian cities in the implementation of smart city initiatives.

The aim of this research is to provide a comprehensive analysis of the development of the smart city concept in Ukraine, evaluate trends, and assess achievements in relation to the SDGs. The study also focuses on uncovering the challenges Ukrainian cities face in the implementation of smart technologies and providing strategic recommendations for developing a smart city model in Ukraine. By exploring these dimensions, the study seeks to highlight the progress and potential of smart city development in Ukraine and offer actionable recommendations to overcome existing challenges and support sustainable urban development.

3 Smart City Development Trends: Global and Local Influences

Sustainable development, institutionalized through global initiatives following the UN Conference on Environment and Development (UNCED) in 1992, has become a key paradigm in urban policy worldwide. This approach combines environmental, social, and economic principles, aiming to balance the needs of current generations without compromising the ability of future generations to meet their own needs. Since adopting the 2030 Agenda for Sustainable Development, sustainable development has become a central theme in international urban discourse [1–3].

Among the 17 Sustainable Development Goals (SDGs), Goal 11, "Make cities and human settlements inclusive, safe, resilient, and sustainable", clearly highlights the role of urban areas in shaping humanity's future [3].

These global imperatives are further reinforced by a number of European regulatory frameworks, such as the Leipzig Charter on Sustainable European Cities, the European Charter of Local Self-Government, and the European Spatial Development Perspective (ESDP). Together, these frameworks emphasize the importance of democratic governance, integrated urban planning, stakeholder participation, and spatial cohesion [4, 5].

European political frameworks serve as an important benchmark for striving to provide conceptual principles for sustainable urban development management. Specifically, the Congress of Local and Regional Authorities of the Council of Europe Charter [6] defines the role and powers of local and regional authorities in the European context, contributing to the development of democracy and local self-government.

The European Spatial Planning Charter, also known as the Torremolinos Charter [7], outlines the principles of regional spatial planning necessary to ensure the rational use of land and balanced development. The European Charter of Local Self-Government [5] recognizes and protects the rights of local self-government, promoting the development of democracy at the local level. The Charter of European Cities on the Path to Sustainable Development [8] obliges representatives of both large and small European cities to initiate the preparation of local "Agendas 21" and develop long-term action plans for the transition to sustainable development.

The European Spatial Development Perspective [9] defines the principles of spatial development to ensure comprehensive and balanced planning at various levels. The EU Sustainable Development Strategy [10] places a particular emphasis on environmental sustainability and social justice. The Leipzig Charter on Sustainable European Cities helped create the concept of integrated urban development at the EU level and influenced the development of EU initiatives such as the Urban Agenda. It sets standards for the new European urban policy focused on sustainable development [4, 11].

These documents have become key in defining the principles of democracy and local self-government. For countries aspiring to EU integration, such as Ukraine, they offer both a normative direction and practical tools for aligning urban development with European standards. They are designed to foster interaction between different levels of governance and ensure coordinated and integrated territorial development planning, considering aspects such as transport, the environment, the economy, social development, innovation, and the integration of economic growth, social justice, and environmental sustainability to create a viable, competitive, and socially just society.

As indicated by the aforementioned documents, sustainable development is a key focus at the international organizational level [12]. Despite criticism regarding its functioning [13], resilience today is a fundamental principle of urban development policy. These initiatives, in particular, have brought urban and regional development to the forefront.

The concept of the smart city has gained increasing popularity each year. The rapid growth of urban populations has led to significant changes in settlement structures worldwide. For the first time in human history, in 2008, the majority of the global population began living in cities [14]. In European countries, more than three-quarters of citizens already live-in cities. This trend indicates a gradual decline in the attractiveness of rural life. Cities are not only centers of economic power but also exert significant influence on political processes, actively implementing innovative technologies [15]. According to forecasts, by 2050, global urbanization will exceed 70%, marking the transition of cities into both engines and pressure points of global development [16].

In the Ukrainian context, the level of urbanization is similarly high: as of early 2022, Ukraine had 461 cities, with 67% of its population residing in urban areas [17, 18]. However, the onset of full-scale war in 2022 fundamentally changed the trajectory of urban development. According to the Kyiv School of Economics, direct infrastructure losses exceeded $170 billion by the end of 2023 [19]. The most vulnerable sectors are housing and utilities, transport, energy, water supply, and communal services. More than 2.4 million Ukrainians have lost their homes, and entire city districts, such as Mariupol, Severodonetsk, and Bakhmut, have become uninhabitable [19, 20].

This situation presents not only challenges but also opportunities for strategically rethinking the urban model in Ukraine during the post-war period by embedding the principles of smartness, resilience, inclusivity, and sustainability into reconstruction strategies. Thus, the transition towards smart, sustainable cities can serve not only as a tool for restoring physical infrastructure but also for rebuilding human capital, social trust, and institutional capacity.

The concept of the smart city, as emphasized in the UN Habitat III documents, highlights four main pillars: "(1) Society: a city for its inhabitants. (2) Economy: a city that prospers jobs, funding, and economic growth. (3) Environment: a city that operates sustainably for current and future generations. (4) Governance: a city that is reliable in its policymaking capacity" [8]. Overall, the United Nations underscores the concept of "smart urban governance," which is closely aligned with the principles of the Sustainable Development Goals (SDGs).

The World Bank emphasizes that the concept of a smart city is centered on people, their quality of life, and life opportunities. In this context, the smart city is envisioned as an environment where data and technology support the integration of urban infrastructure, enhance service delivery efficiency, and facilitate citizen-oriented solutions. Digital technologies play a key role in improving urban planning and governance [21]. This vision reflects a broader shift from hardware-focused solutions to systemic approaches that integrate physical infrastructure with digital intelligence.

For a modern smart city to function effectively, interaction is required on three levels (Fig. 1) [22].

Technological infrastructure—encompassing a network of connected sensors, high-speed communications, and open data systems, as well as access to open data portals. Sensors continuously monitor various parameters such as traffic flow, energy consumption, air pollution, and other aspects of daily life, transmitting the collected information to relevant stakeholders.

Service applications—software that transforms data into actionable solutions. This stage includes the development of practical applications, such as mobile apps for traffic monitoring, energy efficiency management systems, and other innovative solutions.

Citizen engagement—public participation, behavioral change, and digital inclusion. Broad awareness and behavioral adaptation among citizens are key factors for the successful implementation of many programs.

According to the McKinsey Global Institute, smart cities have a more positive impact on various aspects of quality of life than previously anticipated, including social connectedness, civic participation, employment levels, and cost of living. Specifically in the areas of public health and safety, even in their early stages, smart cities have reduced mortality rates by 8–10%, accelerated emergency response times

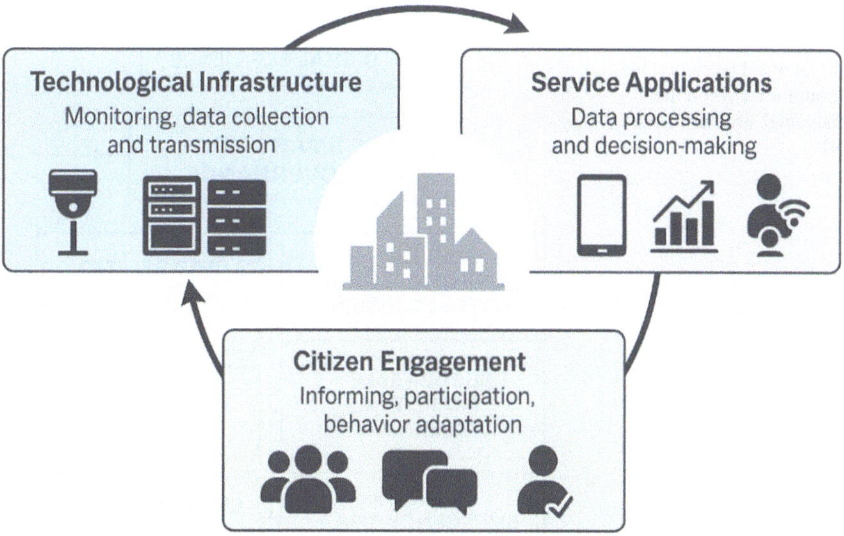

Fig. 1 Smart city functional model: a cyclical interaction of infrastructure, applications, and citizens. Developed by the author based on the conceptual structure from [22]

by 20–35%, lowered the disease burden by 8–15%, and cut greenhouse gas emissions by 10–15%, making these cities not only more resource-efficient but also more resilient [22].

"Smart" sustainable cities, also referred to as "innovative cities", utilize information and communication technologies (ICT) and other resources to enhance quality of life, improve the efficiency of urban operations and governance, provide infrastructure services, and strengthen competitiveness. These cities consider the needs of both current and future generations across economic, social, environmental, and cultural dimensions, ensuring alignment with contemporary demands [23]. In other words, improving quality of life and overall resilience requires the development of smart and sustainable cities based on ICT infrastructure, which includes big data, open data, the Internet of Things (IoT), data access and management, data security, mobile broadband access, and extensive sensor networks.

The application of ICT not only increases the efficiency of complex urban systems but also improves the quality and effectiveness of delivering essential services through various e-solutions. These technologies facilitate citizens' access to knowledge and opportunities, while also addressing environmental challenges and risks of emergencies through technology-driven measures.

In this context, smart approaches contribute to the achievement of SDG 11, which aims to make cities inclusive, safe, resilient, and sustainable. These approaches focus on the elements illustrated in Fig. 2.

In this context, the smart city framework is not limited to technological solutions alone. The United for Smart Sustainable Cities (U4SSC) initiative, launched by UN-Habitat, the International Telecommunication Union (ITU), and other UN

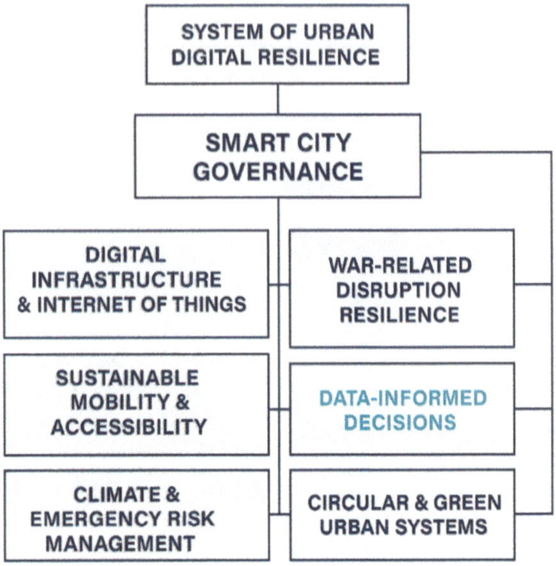

Fig. 2 Key components of ICT-enabled urban ecosystems. Developed by the author based on the conceptual structure from [24]

agencies, proposes a holistic model that encompasses digital infrastructure, environmental responsibility, social innovation, and urban resilience [25]. Today, approaches to smart and sustainable cities increasingly intersect with broader agendas, such as climate adaptation and the energy transition. The European Green Deal and the EU Mission for Climate-Neutral and Smart Cities serve as key reference points for long-term urban policy planning in Ukraine [26]. These initiatives promote decarbonization, circular economy models, and nature-based solutions elements that are critical for enhancing the resilience of Ukrainian cities to both environmental and geopolitical shocks. In essence, this entails the conscious use of advanced technologies and in-depth data analysis to support optimal decision-making and improve quality of life. Ukraine's participation in such frameworks holds strong potential to accelerate post-war recovery by aligning local practices with global standards.

Quality of life encompasses a broad range of aspects, from air quality breathed by residents to the sense of safety while walking through city streets. Providing more comprehensive real-time data enables local authorities and other entities to monitor events as they unfold. This allows an understanding of how demand patterns change, enabling quick and cost-effective responses. The application of smart technologies significantly alters the nature and economy of infrastructure. These technologies help reduce costs related to gathering information about usage patterns, and with the abundance of data points, local authorities, employers, and residents can identify new ways to optimize existing systems. The result is not only a more comfortable city but also a more productive environment for business operations.

Based on the analysis of global and European political documents and the growing convergence of digitization and sustainable initiatives, we propose a conceptual distinction between smart, sustainable, and smart sustainable cities. While each model shares common goals, they differ in origin, operational logic, and key objectives. These distinctions are further detailed in Table 1.

In this study, the following operational definitions are used:

Smart City—a city that leverages digital and communication technologies to enhance the efficiency of urban systems, the quality of services, and citizen participation while optimizing city management towards sustainable development. Key aspects of a smart city include (1) an integrated communication network, (2) effective city management, (3) urban planning and development, (4) future transportation systems, (5) environmental sustainability, (6) community engagement, and (7) safety and cybersecurity.

Sustainable City—a city that ensures long-term ecological balance, social inclusivity, and economic viability, minimizing its environmental footprint. In other words, it meets the current needs of its inhabitants without compromising the ability of future generations to meet their own needs. Key aspects of a sustainable city include (1) ecological sustainability, (2) economic sustainability, (3) social sustainability, (4) risk and crisis management, (5) inclusivity and community participation, and (6) infrastructure and the urban environment.

A sustainable city aims to balance economic, ecological, and social objectives, providing a healthy and thriving urban environment for both present and future generations. The implementation of the smart city concept can positively impact the city's resilience by reducing costs, optimizing resource usage, and improving residents' quality of life. A smart city creates conditions for sustainable development, enhancing the city's efficiency and competitiveness in a global context.

Smart Sustainable City—a comprehensive approach to urban development, this city strategically integrates smart technologies and sustainable development principles to improve urban efficiency and resilience. It promotes the effective use of resources, environmental preservation and ensures a comfortable and safe life for all its residents and visitors.

Table 1 Comparative characteristics of smart, sustainable, and smart sustainable cities

Characteristics	Smart city	Sustainable city	Smart sustainable city
Main focus	Efficiency, innovation, technology	Environmental balance, equity, economy	Integration of digital and sustainable aspects
Drivers	ICT, data, automation	Policy, regulation, community	Strategic governance, cross-sectoral synergy
Key dimensions	Infrastructure, mobility, e-governance	Ecology, society, economy	All combined
Outcomes expected	Service optimization, smart management	Long-term viability, quality of life	Resilience, inclusivity, adaptability
Governance style	Tech-centric, top-down	Participatory, bottom-up	Hybrid, adaptive and inclusive

Built by the authors

Table 2 Functional domains of a smart sustainable city and associated technologies

Functional domain	Examples of smart technologies
Smart governance	E-democracy tools, participatory budgeting platforms, open data portals
Smart economy	E-commerce platforms, fintech solutions, startup incubators
Smart mobility	Intelligent transport systems (ITS), electric vehicles (EVs), traffic sensors
Smart environment	Air and water quality monitoring, waste tracking systems, green IoT
Smart people	E-learning platforms, digital literacy programs, innovation networks
Smart living	E-health services, public safety sensors, smart housing systems
Smart energy	Smart grids, renewable energy integration, automated metering
Smart water	Water leakage detection, consumption monitoring, automated irrigation
Smart buildings	Energy-efficient architecture, BMS (building management systems), green roofs

Compiled by the authors

We propose two complementary matrices to go beyond typologies and operationalize the SSC concept (Table 2).

The post-war phase offers a unique window of opportunity for the implementation of smart, sustainable urban development at all levels of governance. By integrating global frameworks (e.g., SDGs, U4SSC), European strategies (e.g., the Green Deal, Urban Agenda), and national policies, Ukraine can contribute to the development of a coherent urban management model adapted to local realities yet globally oriented in its vision (Table 3).

In conclusion, the convergence of global, regional, and national influences creates a strong foundation for Ukraine's transition to smart, environmentally clean cities. This integrated approach will be essential not only for the restoration of damaged infrastructure but also for the modernization of urban governance systems and the socio-economic structure of urban areas. In the next section, we will examine specific examples that illustrate both the challenges and new practices in Ukrainian cities, serving as starting points for the application of the principles discussed here.

4 Smart City Achievements Worldwide and in Ukraine: Successful Examples and Innovation Implementation

In the context of rapid global urbanization, the development of an intelligently managed urban environment has become an urgent necessity. The growing urban population is increasing the demand for energy, water, efficient transport systems, and other essential services. The implementation of the smart city concept is considered a strategic response to these challenges. According to various estimates, more than 140 smart cities are currently operating worldwide, while alternative sources suggest that the principles of a smart city including technological modernization,

Table 3 Contribution of smart technologies to sustainable urban development goals

Smart domain	Technological solutions	Sustainability impact
Smart governance	Open data, digital services, feedback platforms	Transparency, civic engagement, accountability
Smart economy	Fintech platforms, green bonds, circular economy apps	Sustainable growth, responsible consumption, innovation support
Smart transport	GPS-based public transport, EVs, mobility apps	Reduced emissions, improved accessibility, traffic decongestion
Smart environment	Environmental sensors, waste tracking, GIS mapping	Pollution control, resource monitoring, disaster risk reduction
Smart people	Digital education tools, civic tech, community networks	Social capital development, inclusion, SDG awareness
Smart living	Telemedicine, e-learning, emergency alerts	Health improvement, education access, social inclusion
Smart energy	Smart meters, renewable integration, demand-side management	Lower carbon footprint, energy efficiency, cost savings
Smart water	Leakage sensors, consumption analytics, water reuse systems	Conservation of resources, service reliability, reduced water loss
Smart buildings	Automated control systems, green materials, energy audits	Decreased energy demand, indoor comfort, emissions reduction

Compiled by the authors

support for sustainable development, improved quality of life, economic dynamism, energy efficiency, and public safety are already being implemented or actively applied in over 2500 cities globally.

Two main approaches to smart city development are typically distinguished.

Greenfield development refers to the construction of cities from the ground up using integrated smart systems. Notable examples include Masdar City (UAE), Songdo (South Korea), and Tianjin Eco-City (China). These projects showcase cutting-edge innovations in water recycling, e-mobility, and environmental monitoring. Although this approach offers new horizons for technological advancement, building a smart city from scratch requires substantial financial and technical resources to ensure successful implementation.

Gradual modernization involves the integration of intelligent technologies within existing urban infrastructure. This is the dominant global model, allowing cities to adapt flexibly, maintain continuity, and incrementally improve services in transportation, energy, education, and healthcare. This approach supports scalable and cost-effective digital integration aligned with local capacities. It enables cities to avoid abrupt disruptions and ensures a smooth transition towards a smart city environment.

This section analyzes both successful international cases and the development of the smart city landscape in Ukraine by comparing approaches, identifying best practices, and evaluating progress based on global indices and practical outcomes. Among global leaders in smart urban innovation, several cities exemplify integrated approaches to technology-driven governance, mobility, and sustainability.

Singapore is recognized for its seamless coordination of public and private transportation systems. By implementing adaptive traffic lights that respond to real-time congestion data and smart parking systems that direct drivers to available spaces, the city has significantly reduced traffic jams. Additional measures such as road sensors, surveillance cameras, and interactive maps enable residents to access up-to-date traffic and environmental information while allowing city authorities to adjust routes and policies dynamically, especially during emergencies or public health events.

New York City stands out for its advanced public safety and environmental monitoring systems. The city has deployed widespread camera networks and acoustic sensors capable of detecting gunshots, instantly alerting law enforcement. The BigBelly smart waste management system reduces operational inefficiencies, while in healthcare, platforms like IBM Watson Health analyze wearable data to assist in diagnosis and personalized treatment.

London integrates urban mobility and safety through smart parking systems, subway environmental monitoring, and fire prevention networks. One notable project, SkyCycle, envisions a network of elevated cycling highways above rail lines to ensure safe and efficient commuting.

Barcelona exemplifies the holistic smart city model, with smart irrigation and waste management systems, intelligent public transport stops, and energy-saving lighting integrated into a centralized platform (Sentilo). Environmental and noise pollution levels are continuously monitored to inform policy decisions and enhance quality of life.

These cities illustrate diverse, yet coherent strategies in integrating digital solutions to improve resilience, operational efficiency, and citizen experience. Their success hinges on institutional coordination, citizen engagement, and long-term vision—lessons relevant for transitional urban contexts such as Ukraine.

Global assessments of smart city development offer valuable benchmarks for comparing urban innovation and sustainability. Among the most reputable, the IESE Cities in Motion Index (2024) and the IMD Smart City Index (2025) provide multidimensional evaluations based on economic vitality, digital infrastructure, human capital, environmental performance, and governance.

According to the IESE Cities in Motion Index, five of the top ten cities are in Europe—London (1st), Paris (3rd), Berlin (5th), Oslo (7th), and Amsterdam (8th). These cities score highly in areas such as international connectivity, urban planning, environmental sustainability, and public engagement. North America is represented by New York (2nd), San Francisco (9th), and Chicago (10th), while Tokyo (4th) and Singapore (6th) reflect Asia's growing prominence in urban innovation (Table 4) [27].

Of the top 50 global cities, 26 are situated in Europe, while 14 are located in North America. The remainder are concentrated in the Asia–Pacific region, including Tokyo, Singapore, Seoul, Beijing, Taipei, Hong Kong, and Shanghai, as well as major urban centers in Australia–Melbourne, Sydney, and Canberra. This distribution reflects the increasing global competition among cities in areas such as sustainability, digital transformation, and the enhancement of urban quality of life.

Table 4 Top 10 smart cities—IESE cities in motion index

Rank	City	Country	Continent	Key strengths
1	London	UK	Europe	Human capital, international profile, Urban planning
2	New York	USA	North America	Economy, Human capital, Mobility, Urban planning
3	Paris	France	Europe	International profile, Economy, Public transport
4	Tokyo	Japan	Asia	Technology, Mobility
5	Berlin	Germany	Europe	Urban planning, Sustainability
6	Singapore	Singapore	Asia	Technology, Governance
7	Oslo	Norway	Europe	Environment, Governance
8	Amsterdam	Netherlands	Europe	Mobility, Urban design
9	San Francisco	USA	North America	Technology, Innovation
10	Chicago	USA	North America	Economy, Transport

Compiled by the authors based on [27]

The IMD Smart City Index 2025, developed by the IMD World Competitiveness Center in partnership with Singapore University of Technology and Design, highlights Zurich, Oslo, and Geneva as the top three cities globally. These cities excel in citizen-centric digital services, governance transparency, and urban sustainability. Notably, London ranks 6th, Singapore 9th, while New York drops to 49th, and Barcelona to 92nd—indicating divergence in smart performance dimensions [28].

This variation illustrates the complexity of smart city development: while some cities focus on high-tech integration, others prioritize environmental targets, social inclusiveness, or participatory governance. Importantly, these indices emphasize coherence, strategy, and usability over mere technological quantity.

In summary, the evolution of the smart city concept spans four generations, differing in the degree of digital technology integration and the level of urban transformation. Southern European countries have already implemented approaches typical of the fourth generation, Smart City 4.0, which include high levels of digitization, sustainable development, and citizen engagement. At the same time, several Eastern European countries, including some EU members, are in the process of developing and implementing strategies characteristic of the third generation of smart cities. Meanwhile, in several cities in Eastern Europe, including Ukrainian cities, models of the first generation, Smart City 1.0, are still being implemented, primarily focusing on technological infrastructure upgrades without sufficient integration of social and management innovations [29].

For Ukraine, global benchmarks serve as strategic reference points for aligning local initiatives with the best international practices. Rather than directly replicating existing models, cities should adapt specific components such as data-driven mobility management from Singapore, safety innovations from New York City, or integrated civic platforms from Barcelona while taking into account the local institutional and socio-economic context.

While best international practices provide a valuable foundation, the Ukrainian case is unique: cities must recover and transform simultaneously, laying the groundwork for a digital and sustainable future amid adversity. Against this backdrop, an analysis of global experience, national digital infrastructure, and local municipal practices enables a deeper understanding of how smart city strategies can support urban renewal and long-term resilience in Ukraine.

Ukraine has clearly articulated a draft Sustainable Development Strategy through 2030, aimed at aligning the country's social and economic trajectory with internationally recognized sustainable development agendas. As noted by Anatolii Kutsevol, Deputy State Secretary of the Government of Ukraine, during the 2022 European Urban Resilience Forum in Athens, hundreds of experts and government representatives jointly developed an ambitious 10-year Recovery Plan for Ukraine. This plan encompasses over 850 projects and initiatives, with a total projected budget exceeding USD 750 billion. The plan is aligned with the 2030 Agenda for Sustainable Development [30].

Although several cities have demonstrated notable progress in digital transformation, implementation remains uneven across the country. The ongoing war has further complicated development efforts, yet it has also underscored the strategic importance of smart technologies for recovery and resilience.

The concept of a smart city in Ukraine began to take shape in 2015, gaining momentum as part of broader reforms following decentralization and in response to the destruction caused by the full-scale war. Ukraine's transition to a digital society has accelerated significantly over the past decade, laying the groundwork for smart city development. The emergence of nationwide digital platforms such as Diia, Helsi.me, and Smart City illustrates the state's commitment to building an interoperable, citizen-oriented digital infrastructure that can be adapted at the local level.

In 2019, Ukraine established the Ministry of Digital Transformation, which set ambitious goals for widespread digitalization. The government's strategic vision, articulated in the Digital Transformation Roadmap to 2024, includes: making 100% of key public services available online; providing 95% of settlements and social infrastructure with high-speed internet access; engaging 6 million Ukrainians in digital skills development; and achieving a 6% share of the information technology sector in the national GDP [31]. These initiatives aim to enhance Ukraine's competitiveness in the digital era and foster rapid economic development.

A flagship example of Ukraine's digital transformation is the Diia ecosystem [32], a cornerstone of e-governance. Launched by the Ministry of Digital Transformation in 2020, Diia represents a milestone in Ukraine's digital governance, encompassing a mobile application, online portal, digital education services, and a business hub. The platform provides access to over 70 public services, including digital ID cards, business registration, tax services, and vaccination certificates.

The Diia ecosystem also promotes digital inclusion and civic engagement. Tools for e-voting, petitions, and online education (e.g., Diia.Education) help bridge the digital divide and strengthen democratic participation. A particularly remarkable feature is the eBaby service, which enables parents to register the birth of a child and receive up to 10 public services from various government agencies in a single application. As of today, this service is available in all Ukrainian cities. With over 19 million users, Diia has become the foundation for citizen-state interaction and a symbol of Ukraine's digital resilience.

Helsi.me: E-Health for Resilience [33]. Helsi.me is Ukraine's national e-health platform. It allows citizens to schedule medical appointments, consult with health-care professionals remotely, manage prescriptions, and access telemedicine services. During wartime disruptions and the COVID-19 pandemic, Helsi.me played a vital role in preserving access to primary healthcare, thereby contributing significantly to the resilience of the national healthcare system.

These national platforms illustrate the importance of digital leadership at the state level in enabling local smart city transformations. By offering scalable, interoperable, and citizen-oriented tools, they reduce the burden on municipalities and ensure service continuity even during times of crisis.

In 2021, Kyiv became the first and only Ukrainian city to be included in the IMD Smart City Index (SCI), ranking 82nd globally (Fig. 3) [34], surpassing even Tokyo. This position, ahead of some well-known urban centers, reflected Kyiv's early success in areas such as e-governance, digital payments, and mobility services. Initiatives like online parking systems, e-ticketing, and digital documentation contributed to the city's international recognition.

The Smart City Index evaluates cities based on technological development across five key areas: health and safety, mobility, activities, opportunities, and governance. Each city is also classified according to its Human Development Index (HDI) and grouped accordingly for comparative analysis.

For this comparison, four cities from the same HDI group were selected: Kyiv, the capital of Ukraine; Warsaw, representing a neighboring EU country; Abu Dhabi, the group leader; and Rome, ranking at the lower end of the same group.

Despite Kyiv's inclusion, the results also highlighted critical challenges. Residents identified core issues such as corruption, air pollution, road congestion, housing affordability, and the limited availability of basic amenities. While Warsaw's main concerns included housing, pollution, and traffic, corruption was seen as less relevant in both Poland and Abu Dhabi. In contrast, for Rome, corruption remained a serious concern.

The analysis indicates that although digital tools are being implemented, institutional quality, the depth of integration, and citizen trust still require significant improvement.

Ukraine's partial representation in such international ranking's signals both progress and fragmentation. The absence of other Ukrainian cities in these indices suggests that while smart initiatives exist, they are often isolated, underfunded, or poorly integrated. The key challenge lies in transforming innovation into scalable, system-wide improvements that enhance urban resilience and equity.

Although Kyiv and other Ukrainian cities had been actively developing smart city strategies well before the armed conflict, the full-scale war since 2022 has

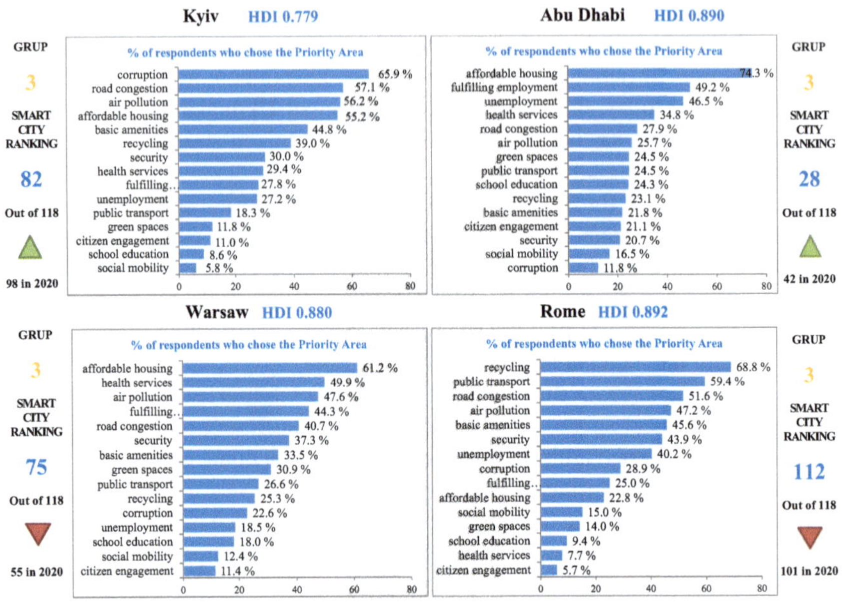

Fig. 3 Comparative smart city index rankings of selected cities with similar human development index (2021). Constructed by the author based on [34]

significantly reshaped both their priorities and implementation mechanisms. Amidst destruction and displacement, digital transformation has evolved into a vital tool not only for modernization but also for ensuring resilience, continuity, and security.

As the capital and hub of digital innovation, Kyiv leads national smart initiatives. With a digitalization level estimated at 75%, the city offers one of the most integrated e-service systems in Eastern Europe. The mobile application "Kyiv Digital", used by over 2 million residents, provides access to public transportation, e-parking, utility payments, citizen alerts, and participatory budgeting.

The city's "Safe City" program includes over 10,000 surveillance cameras equipped with facial recognition and license plate tracking technologies, reportedly contributing to a 37% reduction in street crime. Kyiv is also deploying air quality sensors, open data dashboards, and intelligent transportation systems. Despite infrastructure damage caused by attacks, the city continues to expand its digital infrastructure through cloud migration and enhanced cybersecurity measures.

Lviv, the cultural and innovation center of Western Ukraine, has become a model for sustainable urban development. With a digitalization level of 68%, the city emphasizes energy efficiency, urban mobility, and open governance. Initiatives include GPS-tracked public transport, real-time planning apps, open budget portals, and smart parking systems. Lviv is actively implementing the "Vision Zero" approach to eliminate traffic fatalities through smart traffic coordination.

Notably, Lviv was among the first to adopt the uMuni system, which has proven to be an effective tool for increasing energy efficiency, demonstrating significant results in reducing energy costs and optimizing consumption.

Lviv is also an active participant in the international "Covenant of Mayors for Climate and Energy" initiative, uniting thousands of local authorities committed to achieving the European Union's climate and energy goals. In July 2022, the city approved its Sustainable Energy and Climate Action Plan (SECAP) through 2030, aiming to reduce CO_2 emissions by 35% by enhancing energy efficiency in public institutions, residential buildings, public transport, street lighting, and other sectors [35].

During the war, Lviv has served as a logistics and humanitarian hub, employing digital tools to manage shelters and coordinate resource distribution.

Dnipro has invested in public safety, emergency responsiveness, and utility monitoring. Its E-contact program facilitates citizen reporting of issues such as broken streetlights or waste collection problems, while a centralized surveillance system enhances security across the city. The Safe City platform integrates video monitoring, air quality tracking, and public alert systems. The estimated level of digitalization stands at 61%.

Despite its proximity to the active war zone, Dnipro continues to implement intelligent infrastructure projects, including smart street lighting, online administrative services, and environmental monitoring panels.

A mid-sized city with strong administrative capacity, Vinnytsia demonstrates structured digital integration of public services. The city features smart benches, online administrative platforms, open urban planning tools, and transparent budget monitoring systems. Public transport is GPS-enabled and monitored via mobile applications. Its current digitalization level is 65%.

Vinnytsia's municipal strategy emphasizes citizen engagement and process automation. The city has also developed a local smart city roadmap aligned with the national digital policy and the Sustainable Development Goals (SDGs).

Before the war, Kharkiv was one of Ukraine's leading smart cities, with a digitalization index of 59%. The city implemented smart transportation systems, energy monitoring platforms, and the Mobile ID authentication tool becoming the first Ukrainian city to pilot this advanced security feature.

Following 2022, much of Kharkiv's infrastructure was damaged by shelling and occupation-related disruptions. Nevertheless, local authorities have worked to maintain essential digital services and are planning for recovery through smart reconstruction strategies.

Other cities, such as Ternopil, Poltava, and Lutsk, have launched localized smart initiatives, including smart bus stops, e-governance portals, and digital safety tools. However, resource constraints and outdated infrastructure frequently impede implementation.

An analytical comparison of these cities reveals significant disparities in project depth, integration levels, and citizen participation. The most pressing challenges include fragmented system architecture, insufficient cybersecurity protocols, and low platform interoperability. Moreover, the variability in public engagement affects service usability and the quality of user feedback.

Table 5 summarizes key smart city projects and digitalization levels across selected Ukrainian cities as of 2022.

Before the full-scale invasion, Kyiv regularly hosted the Smart City Awards as part of the annual Kyiv Smart City Forum. The competition recognized cities in seven categories, with evaluation criteria focusing on the originality and effectiveness of local digitalization initiatives, improvements to urban infrastructure and systems, cost-efficiency, and overall societal impact.

Figure 4 presents a consolidated overview of Ukraine's most innovative and technologically advanced cities based on the rankings from the Kyiv Smart City Forum 2020. The data provides comparative insights into the digitalization progress across cities of various population scales.

A comparative analysis reveals diverging smart city trajectories among leading Ukrainian cities. While Kyiv, Lviv, and Vinnytsia exhibit robust innovation ecosystems, Kharkiv and Dnipro have encountered infrastructural challenges and integration issues further exacerbated by wartime disruptions. Smaller cities, such as Ternopil and Poltava, have advanced through modular and localized smart solutions, although they often lack sufficient funding and technical expertise.

Table 5 Smart city development trajectories and digitalization challenges in selected Ukrainian cities (2022)

City	Key smart city projects launched (as of 2022)	E-government tools (as of 2022)	Estimated digitalization level (%, 2022)	Main challenges
Kyiv	Urban transport app, smart lighting, air quality monitoring	"Kyiv Digital", e-petitions, e-parking	75	Cybersecurity risks, incomplete system integration
Lviv	Smart parking, public Wi-Fi, Open Budget	eOffice, participatory budgeting	68	Limited funding, weak interoperability
Dnipro	Smart traffic, CCTV, emergency response apps	eCity portal, SMS-informing citizens	61	Low citizen engagement, legacy systems
Vinnytsia	Data portal, GPS for transport, smart benches	Transparent city system	65	Staff capacity, regulatory inconsistencies
Kharkiv	Pilot on digital urban planning, smart traffic lights	City dashboard (beta)	59	Post-conflict infrastructure recovery
Ternopil	Public service chatbot, smart bus stops	Local smart portal	54	Technical support and expertise

Assessed and compiled by the authors based on existing digital services, integration with the "Diia" platform, transparency tools, and expert evaluations [31, 36–39]

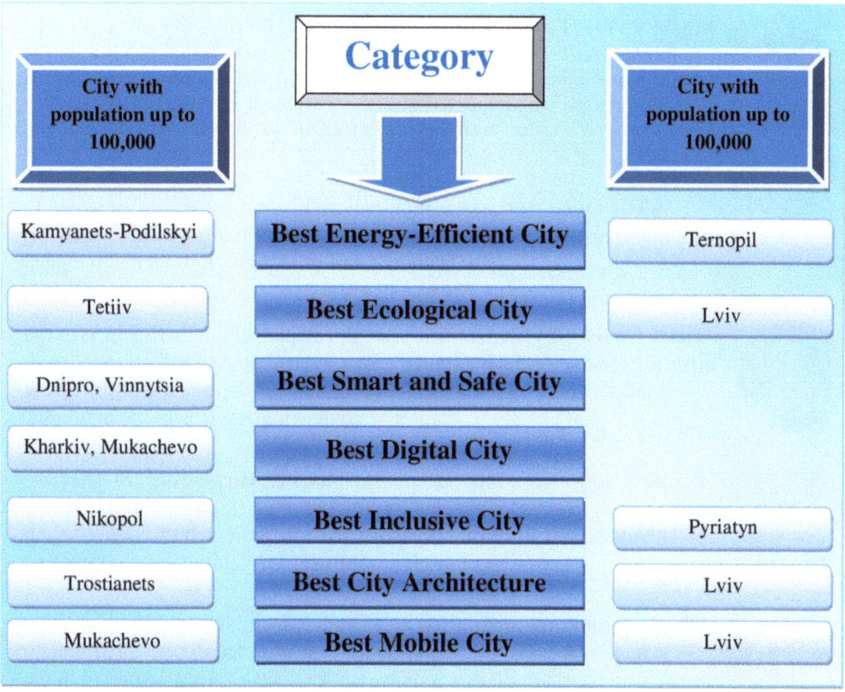

Fig. 4 Summary of top-performing Ukrainian cities by smart city category, according to the Kyiv smart city forum 2020. Constructed by the author based on [36]

This analysis illustrates that smart transformation in Ukraine is not only possible but is actively underway, albeit unevenly. For international observers, Ukraine presents a unique case of urban resilience amid conflict, offering valuable lessons in adaptability, modular innovation, and citizen-centered service provision.

Importantly, Ukraine has laid a solid foundation for smart city development, particularly through the implementation of national digital platforms and localized initiatives in major urban centers. While local contexts vary significantly, the trajectory of digital urban transformation reveals recurring thematic clusters where digital tools are systematically deployed. These clusters include public service delivery, governance transparency, energy and environmental monitoring, mobility, urban safety, and healthcare services.

To better illustrate the scope and diffusion of these initiatives, Fig. 5 provides a visual summary of the most prominent smart solutions currently being implemented in Ukrainian cities as of 2022.

The analysis of Ukrainian urban digitalization trajectories reveals three primary models of smart city implementation, each differing in terms of scalability, governance structure, and institutional dependence:

Fig. 5 The main innovative projects being implemented in Ukrainian cities in the field of smart development include. Constructed by the author based on [31, 36–39]

(1) *State-driven digital platforms with high scalability.* These include nationwide initiatives such as Diia, Helsi.me, and Diia.Business. Designed and maintained at the central level, they are highly interoperable and provide a unified interface for public service delivery. Their strength lies in consistent technical support, legal integration, and wide accessibility, enabling cities of all sizes to benefit from digital public infrastructure. These platforms also serve as tools for ensuring digital equity in the context of severe regional disparities and crises.

(2) *Municipal initiatives with localized governance.* Cities like Kyiv (Kyiv Digital), Lviv (Smart Lviv), and Vinnytsia (Transparent Office) demonstrate the value of context-specific smart city development. These initiatives offer tailored solutions in transport, governance, and participatory budgeting. However, their success depends heavily on local leadership, institutional capacity, and access to funding. Without national coordination, their scalability and interoperability often remain limited.
(3) *Modular or pilot smart projects.* Smaller cities such as Ternopil or Poltava have introduced discrete innovations such as smart benches, e-governance portals, or intelligent bus stops. These initiatives often rely on grant funding, local activism, or public–private partnerships. While they demonstrate ingenuity and adaptability, their expansion is typically constrained by infrastructure limitations and a lack of standardized technical frameworks.

This classification underscores the need for a balanced, multi-level strategy in Ukraine's smart city policy, combining top-down digital ecosystems with bottom-up innovation capacity and local adaptation.

5 Challenges and Barriers to Sustainable Smart City Development in Ukraine

Despite noticeable progress in digitalization and urban innovation, the implementation of smart city initiatives in Ukraine remains fragmented and inconsistent. The development of smart, sustainable urban systems is hindered by several systemic challenges that affect the institutional, financial, technical, and social dimensions of governance. This section provides a critical assessment of these barriers and proposes preliminary directions for overcoming them.

A classification of the most pressing barriers is presented in Table 6. This typology helps to systematically identify the cross-cutting challenges faced by Ukrainian cities and aligns them with levels of governance and specific urban outcomes.

5.1 Lack of Strategic Coordination and Unified Framework

One of the key obstacles to effective smart city development in Ukraine is the absence of a coherent national policy and standardized methodological framework. While certain municipalities, such as Kyiv and Lviv, have launched local initiatives, these often lack alignment with national priorities or international standards (e.g., U4SSC, ISO 37120). In many cities, local authorities operate without a strategic vision or coordination mechanisms, resulting in duplicated efforts and inefficient use of resources.

Table 6 Typology of barriers to smart city implementation in Ukraine

Type of barrier	Description	Level	Examples or consequences
Institutional	Lack of national strategy and coordination mechanisms	National/Local	Fragmentation of policies, duplication of efforts
Financial	Budget constraints, limited access to investment	Local	Delays in digital infrastructure upgrades, dependence on donor funding
Human resources/Skills	Shortage of IT personnel, skill gaps among municipal staff	Local/Regional	Low implementation capacity, reliance on outsourcing
Technological/ Infrastructure	Outdated communication networks, insufficient broadband and 5G infrastructure	Physical/Network	Inability to deploy IoT and real-time monitoring systems
Social/Digital Divide	Low digital literacy, uneven access across regions and demographics	Community	Underutilization of services, inequality in access to digital tools
Legal/Regulatory	Absence of standards, unclear legal framework for data usage	National	Cybersecurity risks, delayed EU integration

Compiled by the authors based on sources [20, 21, 24, 25, 40]

To address this, a multi-level governance model is needed. The central government should establish a national smart city framework with adaptable guidelines, unified data standards, and legal support mechanisms. This would enable harmonization across municipalities, promote interoperability of systems, and facilitate benchmarking with EU member states.

5.2 *Insufficient Funding and Investment Mechanisms*

Financial constraints significantly limit the scalability of smart initiatives. Local budgets are often oriented toward essential services, leaving limited resources for digital infrastructure, e-services, or innovation projects. The war has further exacerbated this challenge, with funding redirected toward emergency needs and recovery efforts.

As emphasized in the World Bank's Rapid Damage and Needs Assessment [20, 21], the post-war urban agenda requires not only reconstruction financing but also the institutional alignment of national and local priorities for resilient digital recovery. Furthermore, the United for Smart Sustainable Cities (U4SSC) initiative, supported by UN-Habitat and ITU, identifies coordination, regulation, and KPI-based benchmarking as critical enablers of successful smart city transitions [25].

Another major barrier relates to the absence of a unified classification of challenges. As outlined in Table 5, barriers span multiple domains and must be addressed through integrated strategies and partnerships.

Attracting private capital, fostering public–private partnerships (PPPs), and engaging international donors are essential to addressing this barrier. Special attention should be given to post-war recovery funds and resilience-focused investment programs. Smart city projects, if aligned with broader national recovery strategies, could become a priority area for international development banks and EU financial instruments.

5.3 Human Capital and Digital Skills Gaps

A major challenge, especially in smaller cities, is the shortage of skilled personnel in IT, data analytics, and systems integration. Talented professionals often choose the private sector due to higher wages, leaving public institutions with limited human capacity to plan, implement, and maintain smart city systems.

Moreover, municipal staff frequently lack the necessary competencies to manage digital transformation projects. This issue is compounded by the absence of specialized training programs and inter-municipal knowledge exchange platforms. Establishing training hubs in cooperation with academic institutions could help build the required skill base and ensure institutional sustainability.

5.4 Infrastructure Deficiencies and Technological Constraints

The deployment of smart technologies requires a robust physical and digital infrastructure [41]. However, in many Ukrainian cities, outdated communication networks, poor internet connectivity, and limited access to reliable power supply hinder the deployment of sensor networks, high-speed data transmission, or IoT applications.

A particular barrier is the underdevelopment of 5G infrastructure, which is essential for real-time data processing and large-scale device connectivity. Investments in next-generation broadband networks should be prioritized within national digital strategies, especially in the context of post-conflict reconstruction.

5.5 *Digital Inclusion and Public Engagement*

Low levels of digital literacy, particularly among elderly populations and rural communities, present a challenge to the adoption and usability of smart services. Gender, generational, and urban–rural digital divides persist, undermining the inclusiveness of smart solutions.

A special emphasis on digital inclusion is made in the UN-Habitat Issue Paper on Smart Cities [24], which highlights that for countries undergoing transformation or affected by conflict, ensuring equitable access to digital services is not only a development priority but also a condition for citizen participation in governance.

To bridge this gap, awareness campaigns, community-based training programs, and inclusive design principles should be incorporated into all stages of project planning and implementation. Digital tools must be user-centric and accessible to all social groups.

5.6 *Legacy Urban Systems and Environmental Constraints*

Many Ukrainian cities still operate within the spatial and infrastructural logic of the Soviet era. Aging residential buildings, inefficient energy systems, congested road networks, and high levels of pollution challenge the effectiveness of smart interventions. Urban models that were not designed for private vehicle use now face significant mobility bottlenecks.

While these limitations complicate implementation, they also create an opportunity to embed sustainability principles into the reconstruction process. Smart technologies can support the modernization of transport, energy, and housing systems in an integrated and future-oriented manner.

5.7 *Strategic Foundations for Smart City Development*

Effective implementation of the smart city concept in Ukraine requires an integrated model based on three strategic pillars (Fig. 6):

City Infrastructure and Services: Ensuring the quality and accessibility of transport, energy, water supply, education, and healthcare systems.

Institutional Enablers: Establishing supportive legal frameworks, governance capacities, and multi-stakeholder collaboration mechanisms.

Digital Transformation: Leveraging ICTs, open data, and smart platforms to optimize decision-making and service delivery.

These pillars should be supported by participatory governance, data-driven planning, and continuous monitoring of implementation outcomes.

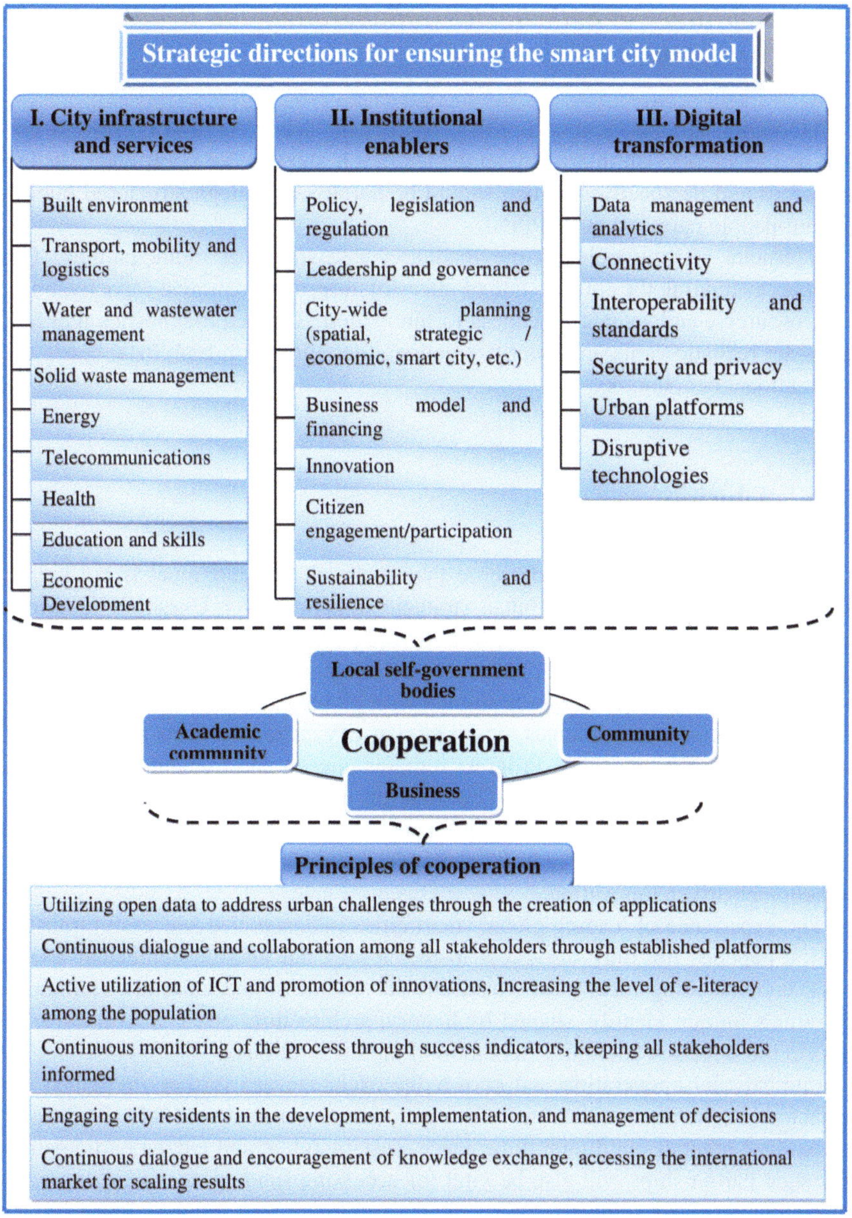

Fig. 6 Strategic directions for ensuring the smart city model. Built by the authors

5.8 *Importance of Stakeholder Collaboration and Readiness Assessment*

Collaboration among local authorities, businesses, academia, and citizens is critical to the success of smart city strategies. Each stakeholder contributes distinct resources and perspectives, which should be harnessed through structured public–private-people partnerships (PPPPs).

A crucial starting point is the readiness assessment of municipalities. This involves evaluating the current state of infrastructure, administrative capacity, financial resources, and societal openness to innovation. Tailored diagnostics will help cities set realistic goals and prioritize actions that align with both local needs and international benchmarks.

6 Conclusion

The findings of this study underscore that smart city development is not merely a technological enhancement but a strategic imperative for Ukraine's long-term urban recovery and resilience. In the context of post-war reconstruction, smart city frameworks present a unique opportunity to reimagine Ukrainian cities as inclusive, efficient, and adaptive ecosystems aligned with global sustainability standards.

Despite promising initiatives, no Ukrainian city currently meets the full operational criteria of a smart city. Progress remains uneven and frequently constrained by fragmented governance, weak coordination mechanisms, and resource limitations. To evolve from isolated digital projects to integrated urban strategies, Ukraine must prioritize multi-level policy coherence, institutional capacity-building, and stakeholder-driven innovation.

The experience of leading global smart cities indicates that successful transformation depends not only on access to technology but also on contextual awareness, understanding local needs, institutional readiness, and cultural specificities. Ukraine's unique identity, shaped by history, architecture, and civic values, must inform the selection and implementation of digital solutions.

International cooperation and knowledge exchange can further strengthen this trajectory. Rather than replicating existing models, Ukraine should selectively adopt global best practices in digital governance, sustainability, and citizen participation. Strategic partnerships will also be vital for accessing funding, technical expertise, and implementation standards.

The implementation of a national smart city strategy must begin with a clear-eyed assessment of local digital readiness, infrastructure resilience, and administrative capacity. This will enable cities to formulate tailored development roadmaps that reflect both their challenges and opportunities.

By fostering intersectoral collaboration, leveraging national platforms (e.g., Diia, Helsi.me), and aligning local initiatives with the Sustainable Development Goals

(SDGs), Ukrainian cities can position themselves as future-oriented centers of innovation, well-being, and economic dynamism.

Beyond national relevance, Ukraine's experience holds global significance. It stands as the only European country actively pursuing smart city transformation under conditions of ongoing armed conflict, institutional disruption, and humanitarian crisis. Unlike post-crisis recovery models, Ukrainian cities are advancing digital governance under real-time pressure, often in parallel with emergency response and reconstruction.

This situation creates a unique testing ground for crisis innovations. Ukraine illustrates how digital services can ensure continuity, how state platforms can reinforce local resilience, and how citizen-centered innovation can thrive even amid adversity.

Ultimately, the smart city model offers Ukraine not just a recovery pathway but a long-term vision grounded in resilience, human-centric services, and sustainability. With strategic focus and coordinated action, Ukrainian cities can emerge from the crisis stronger, smarter, and more inclusive than ever before.

Acknowledgments This publication was produced with the financial support of the European Union within the framework of the Jean Monnet Module project "European values and best practices in urban and territorial development towards sustainable develop-ment" (101174676-EVDOC-ERASMUS-JMO-2024-HEI-TCH-RSCH).

References

1. United Nations (1992) Rio De Janeiro declaration on environment and development. Planet Earth Summit. Rio de Janeiro. Retrieved from: https://www.un.org/en/development/desa/population/migration/generalassembly/docs/globalcompact/A_CONF.151_26_Vol.I_Declaration.pdf. Accessed 15 Mar 2025
2. United Nations (2015) Transforming our world: the 2030 agenda for sustainable development. Resolution adopted by the general assembly on 25 Sept 2015. (A/RES/70/1). Retrieved from: https://sdgs.un.org/2030agenda Accessed 15 Mar 2025
3. United Nations (2017) New urban agenda—Habitat III. https://habitat3.org/the-new-urban-agenda/. Accessed 15 Mar 2025
4. The Leipzig charter on sustainable European cities. Retrieved from: https://territorialagenda.eu/wp-content/uploads/leipzig_charter_2007.pdf. Accessed 17 Mar 2025
5. Council of Europe (1985) European charter of local self-government. Strasbourg. European treaty series—No. 122. Retrieved from: https://rm.coe.int/european-charter-of-local-self-government-eng/1680a87cc3. Accessed 20 Mar 2025
6. Council of Europe (2022) The congress of local and regional authorities of the council of Europe. Retrieved from: https://rm.coe.int/general-leaflet-of-the-congress-of-local-and-regional-authorities-of-t/1680719f4d. Accessed 25 Mar 2025
7. Council of Europe (1983) Resolution No. 2 on The European regional/spatial planning charter: Torremolinos charter: Adopted on 20 May 1983 at Torremolinos (Spain): European conference of ministers responsible for regional planning. Retrieved from: https://rm.coe.int/6th-european-conference-of-ministers-responsible-for-regional-planning/168076dd93. Accessed 20 Mar 2025
8. Charter of European cities & towns towards sustainability (1994). Retrieved from: https://sustainablecities.eu/fileadmin/repository/Aalborg_Charter/Aalborg_Charter_English.pdf. Accessed 20 Mar 2025

9. European Commission (1999) ESDP—European spatial development perspective. Towards balanced and sustainable development of the territory of the European Union. Office for official publications of the European communities, Luxembourg, 87 pp. Retrieved from: https://territorialagenda.eu/wp-content/uploads/ESDP.pdf. Accessed 22 Mar 2025
10. Commission of the European Communities (2001). A sustainable europe for a better world: A European union strategy for sustainable development. Retrieved from: https://eur-lex.europa.eu/LexUriServ/LexUriServ.do?uri=COM:2001:0264:FIN:EN:PDF. Accessed 22 Mar 2025
11. European ministers responsible for urban matters. (2020). The new Leipzig charter—The transformative power of cities for the common good. Retrieved from: https://www.bmi.bund.de/SharedDocs/downloads/EN/eu-presidency/gemeinsame-erklaerungen/new-leipzig-charta-2020.pdf?__blob=publicationFile&v=8. Accessed 22 Mar 2025
12. Espey J, Parnell S, Revi A (2023) The transformative potential of a global urban agenda and its lessons in a time of crisis. NPJ Urb Sustain 3:15. https://doi.org/10.1038/s42949-023-00087-z
13. Baravikova A, Coppola A, Terenzi A (2021) Operationalizing urban resilience: insights from the science-policy interface in the European Union. Eur Plan Stud 29(2):241–258. https://doi.org/10.1080/09654313.2020.1729346
14. Dirks S, Keeling M (2009) A vision of smarter cities: how cities can lead the way into a prosperous and sustainable future. Executive Report. IBM Institute for Business Value. Retrieved from: https://www.foresightfordevelopment.org/sobipro/55/1042-a-vision-of-smarter-cities-how-cities-can-lead-the-way-into-a-prosperous-and-sustainable-future. Accessed 15 Feb 2025
15. United Nations (2010) World urbanization prospects: The 2009 revision. Department of Economic and Social Affairs/Population Division, New York. Retrieved from: https://ru.scribd.com/document/78681125/WorldUrbanization-Prospects-2009-Revision-United-Nations-2010. Accessed 28 Mar 2025
16. Kuzmak O, Kuzmak O, Stepura T, Lorvi I (2025) Digital transformation to the efficiency and sustainability of cities. In: Semenov A, Yepifanova I, Kajanová J (eds) Data-centric business and applications. Lecture notes on data engineering and communications technologies, vol 240. Springer, Cham. https://doi.org/10.1007/978-3-031-81557-7_13
17. State Statistics Service of Ukraine (2022) Number of present population of Ukraine, as of Jan 1 2022. Retrieved from http://db.ukrcensus.gov.ua/PXWEB2007/ukr/publ_new1/2022/zb_Chuselnist.pdf. Accessed 28 Mar 2025
18. Worldometers (2025) Population of Ukraine (2025 and historical). Retrieved from https://www.worldometers.info/world-population/ukraine-population/. Accessed 08 Apr 2025
19. Kyiv School of Economics (KSE) (2025) Report on the direct infrastructure losses in Ukraine caused by Russia's armed aggression (as of Nov 2024), Feb 2025. Retrieved from https://kse.ua/wp-content/uploads/2025/02/KSE_Damages_Report-November-2024-UA.pdf. Accessed 25 Mar 2025
20. World Bank (2023) Ukraine rapid damage and needs assessment February 2022–February 2023. World Bank Group, Washington, D.C. Retrieved from https://ukraine.un.org/sites/default/files/2023-03/P1801740d1177f03c0ab180057556615497.pdf. Accessed 22 Mar 2025
21. World Bank (2025) Global smart city partnership program. Retrieved from: https://www.worldbank.org/en/programs/global-smart-city-partnership-program. Accessed 28 Mar 2025
22. McKinsey Global Institute (2018) Smart cities: digital solutions for a more livable future. Retrieved from: https://www.mckinsey.com/~/media/mckinsey/business%20functions/operations/our%20insights/Smart%20cities%20digital%20solutions%20for%20a%20more%20livable%20future/mgi-Smart-cities-full-report.pdf. Accessed 30 Mar 2025
23. International Telecommunication Union (2014) Overview of key performance indicators in Smart sustainable cities. Retrieved from: https://www.itu.int/en/ITU-T/focusgroups/ssc/Documents/Approved_Deliverables/TS-Overview-KPI.docx. Accessed 28 Mar 2025
24. United Nations (2015) Habitat III issue paper: 21-Smart cities. New York, 31 Mar 2015. https://habitat3.org/wp-content/uploads/Habitat-III-Issue-Paper-21_Smart-Cities-2.0.pdf. Accessed 12 Mar 2025

25. United for Smart Sustainable Cities (U4SSC) (2020) Guidelines on tools and mechanisms to finance smart sustainable cities projects. Retrieved from: https://www.itu.int/en/ITU-T/ssc/ Accessed 28 Mar 2025
26. European Commission (2019) The European green deal. https://eur-lex.europa.eu/legal-content/EN/TXT/?uri=CELEX:52019DC0640
27. London is the world's smartest metropolis, according to IESE Cities in Motion Index. Smart City, 09 Feb 2024. Retrieved from: https://www.thesmartcityjournal.com/en/cities/london-is-the-world-s-smartest-metropolis-according-to-iese-cities-in-motion-index Accessed 25 Mar 2025
28. IMD (2025) IMD Smart City Index 2025. The housing affordability challenge: A growing concern. World Competitiveness Center. Retrieved from: https://imd.widen.net/s/psdrsvpbk7/imd_smart_city_2025_report. Accessed 30 Mar 2025
29. Pontrandolfo P, Polinkevych O, Scozzi B, Kuzmak O (2023) Smart cities for sustainable development of local communities. Comp Econ Res. Cent East Eur 26(3):53–85. https://doi.org/10.18778/1508-2008.26.21
30. United Nations Development Programme (2022) Urban development and post-war reconstruction of Ukrainian cities. Summary of the European Urban Resilience Forum 2022 in Athens, Greece. UNDP. Retrieved from: https://www.undp.org/ukraine/blog/urban-development-and-post-war-recovery-ukrainian-cities. Accessed 25 Mar 2025
31. Goals until 2024. Ministry of digital transformation of Ukraine. Retrieved from: https://thedigital.gov.ua/ministry. Accessed 02 Apr 2025
32. Diia—The state portal of public services. Retrieved from: https://diia.gov.ua/. Accessed 13 Apr 2025
33. Helsi—Medical information system. Retrieved from: https://helsi.me/. Accessed 13 Apr 2025
34. Smart City Index 2021. A tool for action, an instrument for better lives for all citizens. Retrieved from: https://www.imd.org/smart-city-observatory/home/2021/#index. Accessed 02 Apr 2025
35. UA-Energy. (2022) Угода мерів—перезавантаження: міста за збереження клімату та енергії. Retrieved from: https://ua-energy.org/uk/posts/uhoda-meriv-perezavantazhennia-mista-za-zberezhennia-klimatu-ta-enerhii Accessed 12 Apr 2025
36. Kyiv City (2020) The annual Smart City Awards ceremony was held at Kyiv Smart City Forum 2020. Retrieved from: https://kyivsmartcity.com/news-na-kyiv-smart-city-forum-2020-proveli-shhorichnu-ceremoniyu-nagorodzhennya-smart-city-awards/. Accessed 02 Apr 2025
37. Razumkov Centre (2021) Smart infrastructure in sustainable urban development: global experience and prospects for Ukraine. Friedrich Naumann Foundation. Retrieved from: https://razumkov.org.ua/uploads/other/2021-SMART-%D0%A1YTI-SITE.pdf. Accessed 04 Apr 2025
38. Kyiv City State Administration (2022) Kyiv 2022. Annual Reports. Retrieved from: https://media-stg.kyivcity.gov.ua/kyivcity/sites/22/richnui%20zvit/2022/Zvit_Kyiv_2022_ENG_peregliad.pdf Accessed 05 Apr 2025
39. Cities of free people. Results of transparency and accountability rankings—2021. Retrieved from: https://transparentcities.in.ua/transparency-rating Accessed 15 Mar 2025
40. Kuzmak O, Kuzmak O (2023) Transformational processes of ensuring sustainable urban development: the realities of Ukraine. IOP Conf Ser Earth Environ Sci 1254:012079. https://doi.org/10.1088/1755-1315/1254/1/012079
41. Poniszewska-Maranda A, Kaczmarek D, Kryvinska N, Xhafa F (2018) Studying usability of AI in the IoT systems/paradigm through embedding NN techniques into mobile smart service system. Computing 101(11):1661–1685. https://doi.org/10.1007/s00607-018-0680-z

Processing of Audio Recordings and Feature Extraction for Machine Learning

Andrii Kobylin and Oleksandr Ryazantsev

Abstract This study presents a comprehensive framework for the automated analysis of emotional tone in video lectures, combining modern audio-processing libraries, feature-extraction techniques, and supervised classification methods. We evaluated open-source tools—Librosa for spectral and cepstral analysis, Praat for phonetic parameters, and PyDub for flexible audio manipulation—to extract over 100 acoustic features, including Mel-Frequency Cepstral Coefficients (MFCCs), chroma statistics, signal-strength metrics (maximum, mean, and average loudness), and prosodic cues such as pitch and intensity. Unsupervised clustering with K-Means on both spectral and signal-strength feature subsets achieved initial cluster accuracies of up to 90% for non-emotional and 79% for emotional segments. Building on these results, we trained a Support Vector Machine (SVM) with an Radial Basis Function (RBF) kernel on selected feature subsets, obtaining a baseline classification accuracy of 83% on unseen data, which improved to 89% upon inclusion of additional MFCC coefficients. To streamline the labor-intensive annotation process, we developed a web-based tool using Wavesurfer with customizable hotkeys, reducing manual tagging time. Our findings demonstrate that combining spectral, chromatic, and signal-strength features with SVM classification provides a robust approach for detecting emotional fragments in lectures. These results have important implications for enhancing pedagogical feedback systems and student engagement, and point toward future extensions involving deep-learning models and real-time analysis.

Keywords Audio processing · Spectral analysis · Feature extraction · Machine learning · Speech analysis · Emotional analysis

A. Kobylin (✉) · O. Ryazantsev
Volodymyr Dahl East Ukrainian National University, Kyiv, Ukraine
e-mail: asp-122-23-73@snu.edu.ua

O. Ryazantsev
e-mail: a_ryazantsev@snu.edu.ua

A. Semenov et al. (eds.), *Data-Centric Business and Applications*, Lecture Notes on Data Engineering and Communications Technologies 272,
https://doi.org/10.1007/978-3-032-06608-4_9

1 Introduction

At present, lectures are one of the primary formats for transferring knowledge from lecturers to students, intended for mastering theoretical material. The quality of a lecture directly influences how well students absorb knowledge. The emotions experienced by the lecturer during the lecture delivery affect the lecture's overall quality. Research findings [1, 2] indicate that a lecturer's emotions are as crucial to students' perception of information as their expertise and teaching style. The success of learning and the quality of acquired practical skills depend on how well the lecture is delivered. Additionally, the emotional engagement of lecturers influences the popularity and ranking of educational institutions [3].

A high level of student education promotes rapid career advancement and enhances the university's reputation. As a result, the number of applicants wishing to study at the institution increases, which, in turn, raises tuition fees and boosts the institution's income. Emotions manifest in various ways: hand gestures, facial expressions, voice tone, and more. In this work, we focus on analyzing the lecturer's voice tone.

The emotional tone of lectures plays a critical role in the pedagogical process because emotions directly impact students' ability to assimilate and recall information. Emotionally charged lectures can enhance students' motivation and engagement, contributing to a deeper understanding of the material.

Conversely, lectures with a monotonous or insufficiently emotional delivery can reduce students' interest and hinder their learning process. Given this, studying the emotional tone of lectures may help develop methods and tools that can assess and improve teaching quality.

Therefore, our objective is to process and analyze video lectures, synchronize video with subtitles, and develop tools for segmenting the video into parts with higher and lower emotional intensity. Using machine learning methods, we aim to classify emotional fragments and create a web application for annotating audio fragments to facilitate the analysis of large volumes of material.

2 Computed and Experimental Data

To study the emotional tone exhibited by the speaker, we selected the lecture course "The Making of Modern Ukraine" by Professor Timothy Snyder of Yale University [4]. Timothy Snyder has been engaged in academic activities for over 24 years and has published 15 books [5]. His open-access lectures on YouTube have garnered millions of views, indicating their popularity and the audience's interest.

The course "The Making of Modern Ukraine" consists of 23 lectures, each lasting between 46 and 55 min, recorded in mp4 format. The lecture series covers important topics such as the history of Ukraine, its geographical position, the development of relations with European countries, especially Poland, pivotal events of 2014 and

2022, and the country's cultural heritage. The course emphasizes the uniqueness of Ukrainian national identity and the specifics of the formation of Ukrainian statehood, addressing pressing questions raised after the war began: why Ukraine exists as a nation, what it has been, and what future awaits Ukraine after victory.

3 Methods of Audio Analysis

To solve the task of audio analysis [6] in lectures, it was decided to use machine learning (ML) and artificial intelligence (AI) methods [7, 8] for signal processing and information analysis. Modern ML models are widely used for detecting the emotional tone of audio recordings [9], speech recognition [10], sound classification [11], and more. However, the effectiveness of such models largely depends on two factors:

1. Proper signal preprocessing and accurate feature extraction.
2. Selection and training of the artificial intelligence model.

3.1 Signal Preprocessing and Accurate Feature Extraction

To ensure proper feature extraction from the audio signal, the following tools were selected:

- Librosa: an open-source Python library for audio analysis, which enables spectral analysis, calculation of Mel-Frequency Cepstral Coefficients (MFCC), and other acoustic features [12].
- PyDub: a tool for audio processing that allows converting, segmenting, and adjusting audio file parameters [13].
- Praat: software for phonetic analysis of speech, which provides parameters such as pitch, intensity, speech duration, and other aspects of vocal signals [14].
- Specialized web application: developed by the author of this study for annotating long video and audio recordings [15, 16].

The review of these tools covers their functional capabilities, efficiency, advantages, and limitations in the context of feature extraction for further use in ML models. The focus is on identifying the most relevant features for the tasks of emotional tone analysis of audio recordings, particularly loudness, spectral features, and voice signal characteristics [17]. Additionally, it is necessary to select the required features for segments that will later be used in machine learning. Clustering methods will be applied to mathematically justify the number of emotional tone categories.

For clarity, a 10-s segment from 3:15 to 3:25 was selected as an example, shown in Fig. 1. In this segment, the lecturer says the following phrase:

> If you do the reading and you look at the maps, it's going to be very straightforward. If you don't do the reading and you don't look at the maps, you're going to be baffled.

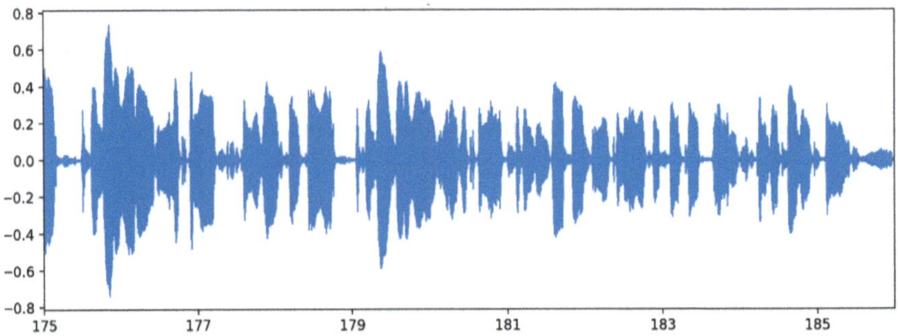

Fig. 1 The audio signal segment of the lecture shows the dependence of the normalized signal amplitude on time in seconds

3.1.1 Examples of Using the Python Library Librosa for Audio File Processing

Feature Extraction: MFCC. One of the most common methods for audio analysis is the extraction of Mel-Frequency Cepstral Coefficients (MFCC) (Fig. 2), which are a crucial feature for many ML applications, such as speech recognition.

MFCC is one of the most commonly used features in speech and sound analysis. They effectively represent the spectral characteristics of an audio signal in a way that approximates human hearing. Changes in the spectral "fingerprint" of a voice—rises or falls in frequency, decreases or increases in spectral dynamics—often correlate

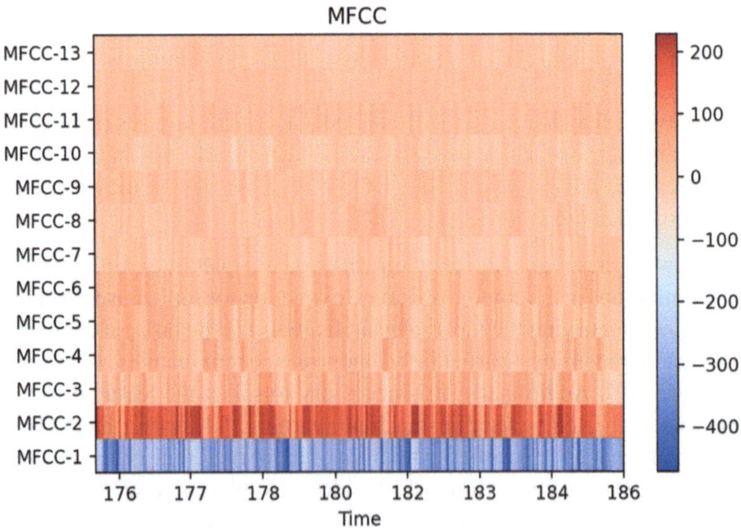

Fig. 2 Distribution of mel-frequency cepstral coefficients (MFCC) over time

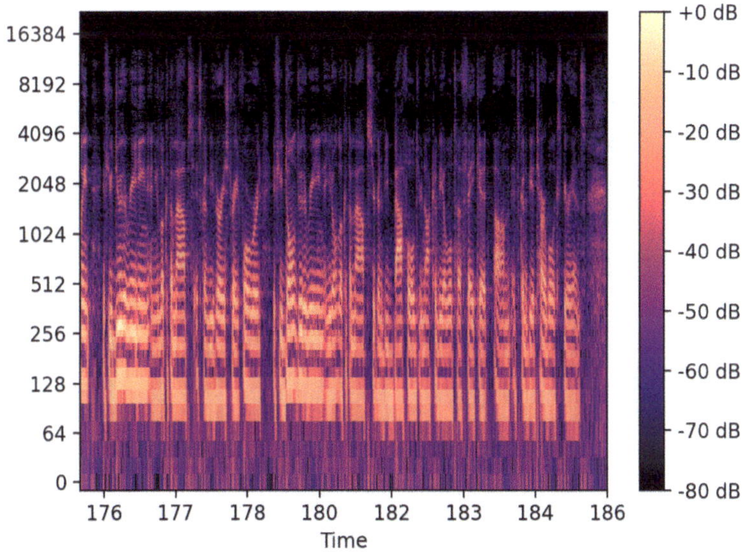

Fig. 3 Frequency spectrum. Distribution of frequency and signal amplitude over time

with different emotional states. For example, a more "sharp" spectrum may indicate anger or excitement, while a softer spectrum may indicate calmness.

Frequency Spectrum. A spectrogram of the frequency spectrum (Fig. 3) shows how the signal's frequency varies over time. It is useful for analyzing sound characteristics.

The frequency spectrum allows us to observe how the energy of the voice is distributed across different frequencies. This is important because voice timbre, formants (characteristic resonance frequencies of the vocal tract), and harmonics can provide information about the speaker's emotional state, gender, age, and other characteristics.

Pitch of the Audio Signal. Pitch refers to the perceived height of a sound, as well as its representation either by a letter or verbally (Fig. 4).

Chroma features reflect the distribution of energy across musical semitones in the voice. Although speech is not a song, the vocal tract can generate harmonics, and variations in these harmonics may be related to the speaker's emotional state. An uneven distribution of frequency components, their fluctuations, peaks, and dips can indicate the presence of emotional expressiveness.

3.1.2 Examples of Using PyDub

Converting to Another Format. PyDub allows easy conversion of audio files from one format to another, such as converting from.mp4 to.wav for further analysis.

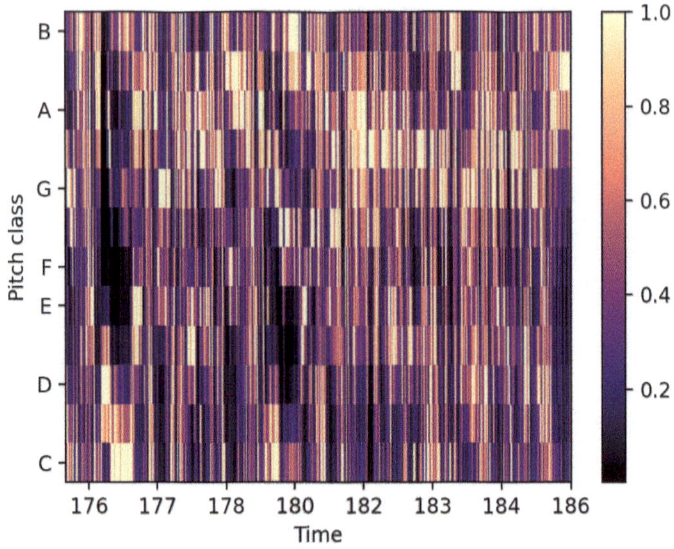

Fig. 4 Chromagram. Distribution of pitch over time

```
mp4_file = AudioSegment.from_file("lecture1.mp4", "mp4")
mp4_file.export("lecture1.mp3", format="mp3", bitrate="192k")
```

Reducing Volume by 3 dB. Using PyDub, the audio volume can be reduced by 3 dB, which is useful for normalizing audio levels before processing.

```
first_10_seconds = mp4_version[:ten_seconds]
beginning = first_10_seconds - 3
```

3.1.3 Examples of Using Praat

Praat is software designed for the phonetic analysis of speech, developed to study and analyze sound signals. It provides a wide range of tools for detailed examination of speech features (Fig. 5).

At the top of the figure, the waveform of the audio signal is displayed. At the bottom, a spectrogram is shown with additional features:

1. Intensity (green line): Represents the amplitude of the sound.
2. Formants (red dots): Concentrations of acoustic energy around specific frequencies of the speech wave.
3. Pitch (blue line): A perceptual property that allows sounds to be organized on a tonal scale (A, B, C, etc.).

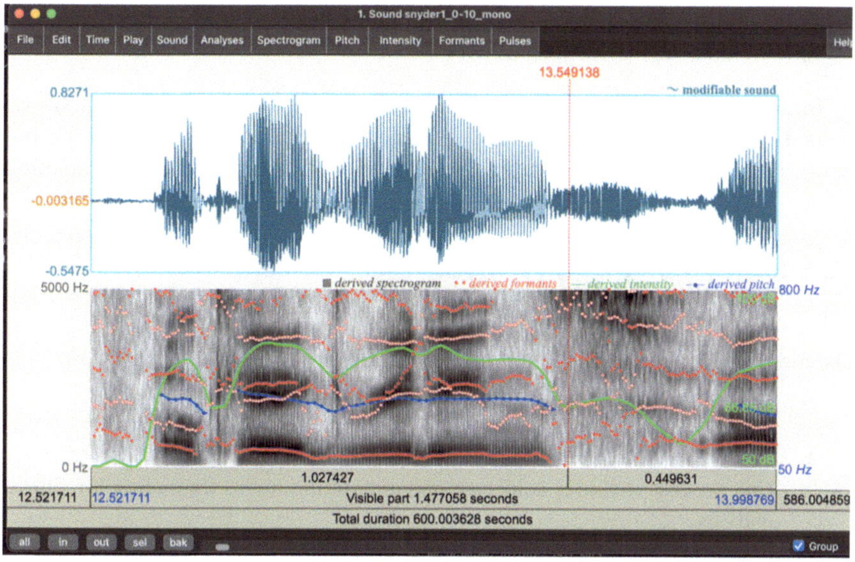

Fig. 5 Audio analysis window in the Praat software

Praat is also convenient for annotating audio files by marking time intervals and adding descriptions. Additionally, it includes built-in tools for generating MFCC diagrams. Praat can be used to verify results obtained from Librosa.

3.1.4 Web-Based Annotation Tool

During the work, the convenience of existing tools was compared, and it was decided to develop a custom tool (Fig. 6) with the following features:

1. Fast selection and tagging of segments with the required labels.
2. Availability of "hotkeys" for annotation.

Fig. 6 Tool for annotating emotions in an audio file within a web browser

3. Extensibility: Interactive display of additional characteristics for the entire file or individual segments, such as MFCC.

To annotate an audio file with a duration of 46 min and 4 s, a custom tool was created using the Wavesurfer library, written in JavaScript, enabling it to run in a browser. Additionally, hotkey settings were developed to facilitate faster annotation.

3.2 Data Analysis and Model Creation

The method of extracting audio recording features involves several stages:

1. Data Collection and Preparation:
   a. Utilizing a web application for annotating fragments of video lectures.
   b. This process includes selecting and annotating audio fragments for further analysis.
2. Audio Data Analysis:
   a. Using Python tools for processing audio files.
   b. Leveraging the Librosa library for loading and processing audio files and extracting features, as well as NumPy and PyPlot for mathematical computations and visualization.
3. Analysis of Extracted Features:
   a. Classification and clustering of data points based on the extracted features.

3.2.1 Data Collection and Preparation

For emotion classification, two categories were selected:

- Emotional
- Non-emotional.

Advantages of Using Two Categories

1. Simplified Problem Definition:
   a. Using only two categories significantly reduces the complexity of the task. Instead of recognizing a wide range of emotional states, the model only determines whether a segment is emotionally charged or not.
   b. This simplifies the choice of machine learning methods, lowers the requirements for training data volumes, and shortens the preparation time.
2. Reduced Data Annotation Effort:

   a. Annotating data is easier with two categories. Experts do not need to detail every emotion (e.g., joy, sadness, surprise); they simply divide the data into "emotional" and "non-emotional" segments.
   b. This accelerates the process of data collection and preparation.

3. Easier Interpretation of Results:

   a. Users of the system can more easily understand the model's conclusions. For example, if the system is designed for quick detection of emotionally charged sections of a lecture, two categories provide a clear signal: "Pay attention here" or "Neutral segment."

Disadvantages of Using Two Categories

1. Loss of detail and reduction in the depth of information available for more thorough analysis.
2. Potential imbalance between classes.

Using only two categories simplifies the process of classifying emotional features and is useful in the early stages of research or for specific technical applications.

In the first lecture by Timothy Snyder, 180 segments were identified (90 emotional, 90 non-emotional). In the next chapters we will see their further analysis.

3.2.2 Feature Extraction

Finding the right features is crucial, as each feature reflects a specific aspect of the acoustic and musical properties of the audio signal that may correlate with the emotional tone of speech:

1. Segment Duration (duration): The length of an audio fragment can affect the perception of emotional tone. For instance, longer pauses or stretched intonational fragments might indicate certain emotional states such as contemplation, sadness, or hesitation.
2. Signal Strength (max/average/mean_audio_strength): The intensity (loudness) of the audio signal directly reflects the speech's intensity. An emotionally excited or joyful speaker might speak louder, while a subdued, monotone voice could suggest neutrality or sadness.
3. Chroma Features (chroma_mean, chroma_min, chroma_max, chroma_std): Chroma features represent the distribution of energy across musical semitones in the voice. Although speech is not music, the vocal tract can generate harmonics, and changes in these harmonics may be associated with emotional states. Uneven distribution of frequency components, their fluctuations, peaks, and dips can signal the presence of emotional expressiveness.
4. Mel-Frequency Cepstral Coefficients (MFCC: mfcc_mean, mfcc_min, mfcc_max, mfcc_std): MFCC is one of the most widely used features in speech and sound analysis. They effectively capture the spectral characteristics of an audio

signal in a way that approximates human hearing. Changes in the spectral "fingerprint" of a voice—rises or falls in frequency, decreases or increases in spectral dynamics—often correlate with various emotional states. For example, a sharper spectrum may indicate anger or excitement, while a softer spectrum suggests calmness.

5. Statistical Characteristics (minimum, maximum, mean, standard deviation): Using statistical measures allows capturing not only the average value of a feature but also its variations and extreme values. This adds depth to the model's understanding of how stable or variable the signal is over time. Emotional speech often features abrupt changes in intonation and timbre, reflected in increased standard deviation or prominent maximum and minimum values.

In summary, combining these features allows capturing various aspects of the speech signal—from amplitude and spectral properties to variability and stability of the sound. This provides a more comprehensive picture for an ML model, increasing the likelihood of accurately classifying the emotional tone of audio fragments.

The following features were extracted for each segment with a duration ranging from 1.5 to 4 s:

- Segment Duration (duration)
- Maximum Signal Strength (max_audio_strength)
- Average Signal Strength (average_audio_strength)
- Mean Signal Strength (mean_audio_strength)
- Chroma Features:
  - Mean (chroma_mean)
  - Minimum (chroma_min)
  - Maximum (chroma_max)
  - Standard Deviation (chroma_std)
- MFCC Features (first 13 coefficients):
  - Mean (mfcc_mean)
  - Minimum (mfcc_min)
  - Maximum (mfcc_max)
  - Standard Deviation (mfcc_std).

In total, 106 features were extracted for the analysis of the emotional tone of the segments.

3.2.3 Analysis of Extracted Features, Classification, and Clustering

Calculating the correlation between individual audio signal features and emotional tone helps identify which features significantly impact classification results [18].

The correlation between the features and emotional tone was computed using Pearson's correlation coefficient [19] with the SciPy library (Fig. 7).

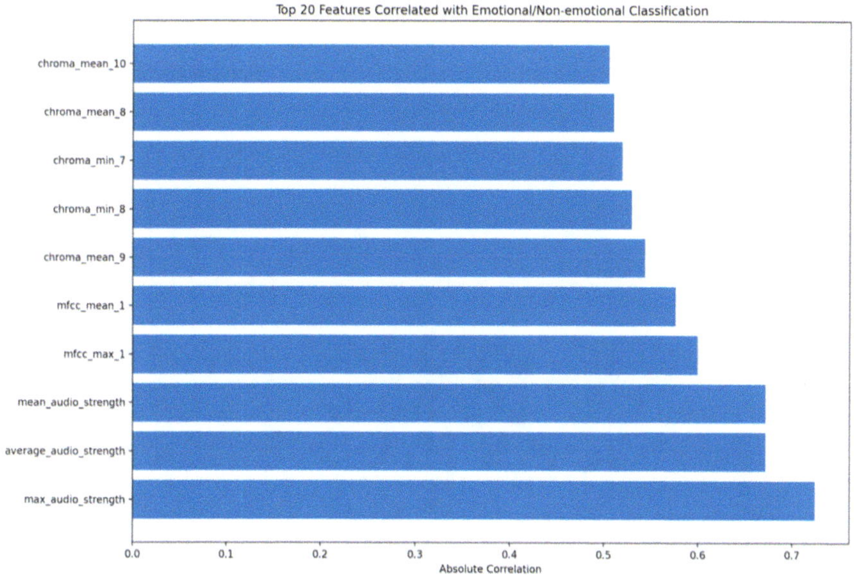

Fig. 7 Correlation between features and emotional tone

The following features demonstrated the highest correlation with emotional tone:

These features have a high correlation with the presence or absence of emotional tone and are likely to be useful for the machine learning model.

We will perform clustering on some of these features using the K-Means method. Clustering helps to preliminarily explore the data, identify patterns, and prepare a foundation for further classification model training or other analysis methods.

In the first case, clustering is applied to chroma_mean_1, chroma_mean_2, and chroma_mean_3 (Fig. 8).

The following results were obtained:

- Clustering accuracy for segments with the tag "emotional": 55.6%. A total of 99 segments were assigned to the cluster, with 44 incorrectly clustered (marked with "X" in the figure).
- Clustering accuracy for segments with the tag "non-emotional": 56.8%. A total of 81 segments were assigned to the cluster, with 35 incorrectly clustered (marked with "X" in the figure).

In the second case, clustering will be performed for max_audio_strength, average_audio_strength, and mean_audio_strength (Fig. 9).

The following results were obtained:

- Clustering accuracy for segments with the tag "emotional": 79.0%. A total of 105 segments were assigned to the cluster, with 22 incorrectly clustered (marked with "X" in the figure).

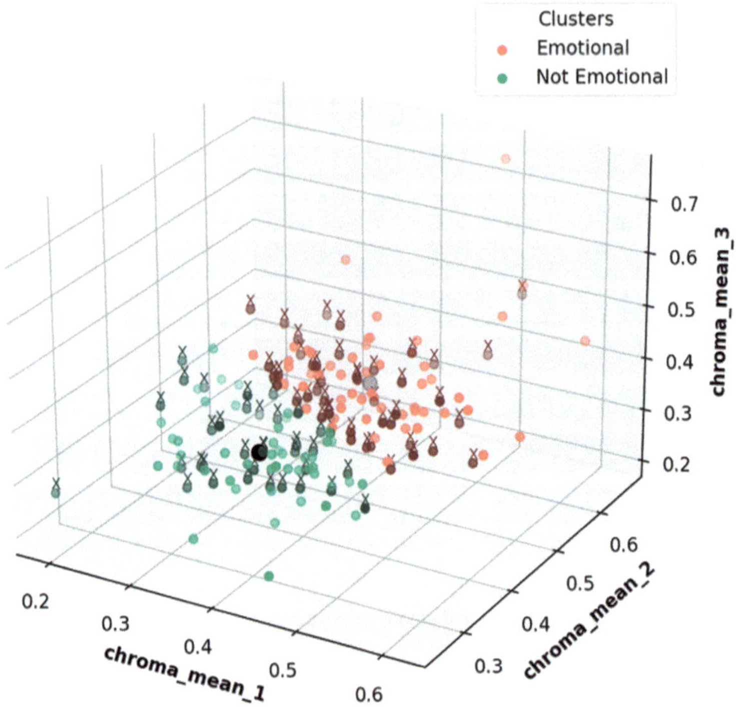

Fig. 8 K-Means clustering of features chroma_mean_1, chroma_mean_2, chroma_mean_3

- Clustering accuracy for segments with the tag "non-emotional": 90.7%. A total of 75 segments were assigned to the cluster, with 7 incorrectly clustered (marked with "X" in the figure).

The second set of parameters for clustering is much more accurate, so we will select these parameters for training the model. There are some errors, but the overall accuracy is good. These features can be used for model training.

3.2.4 SVM Model Training

For training, the SVM model [20] and its implementation in the Scikit-learn package were selected. This model is suitable for the task because:

1. Handling High-Dimensional Data: In our case, up to 106 features will be used. SVM performs well in high-dimensional feature spaces, which is often encountered in audio processing tasks.
2. Non-Linear Classification: The RBF (Radial Basis Function) kernel will be used, allowing SVM to create non-linear decision boundaries. This is beneficial for complex relationships between audio features and segment types.

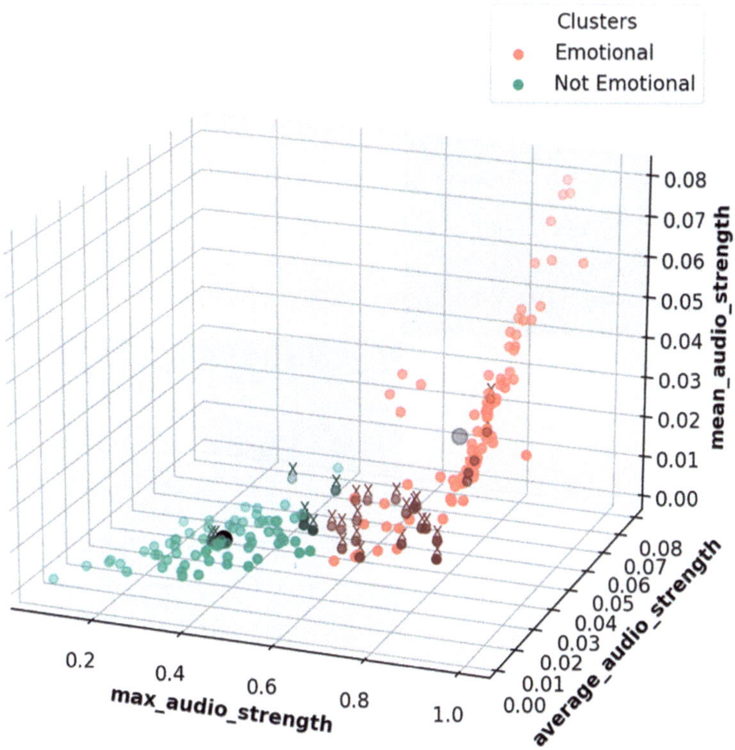

Fig. 9 K-Means clustering of features max_audio_strength, average_audio_strength, mean_audio_strength

3. Overfitting Prevention: SVM aims to maximize the margin between classes, which helps reduce overfitting, especially when working with relatively small datasets.
4. Efficiency with Imbalanced Classes: SVM handles imbalanced classes effectively, making it a good choice for this type of classification problem.
5. Good Generalization Ability: SVM typically generalizes well to new data, which is crucial for reliable classification of audio segments.
6. Interpretability: While SVM is not as interpretable as some other methods, it still allows for analyzing feature importance and support vectors.

The initial dataset was split into a training set (80%) and a test set (20%). Using three features: max_audio_strength, average_audio_strength, and mean_audio_strength, the following results were obtained:

- Training accuracy: 0.84
- Test accuracy: 0.83.

The confusion matrix in Fig. 10 illustrates the performance of the algorithm.

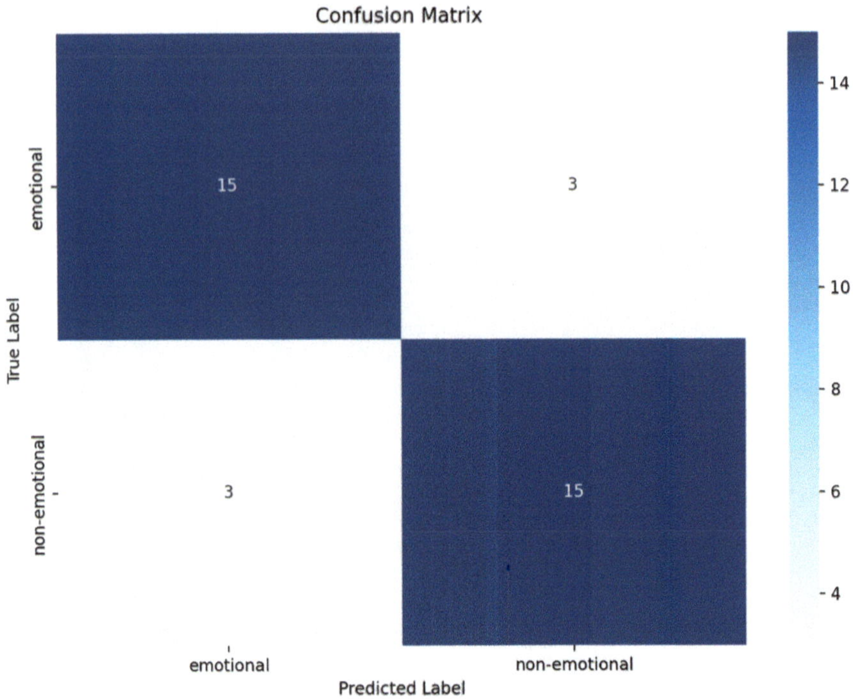

Fig. 10 Confusion matrix for the SVM method using three features: max_audio_strength, average_audio_strength, mean_audio_strength

1. Correct Predictions:
   a. 15 emotional samples were correctly classified as emotional.
   b. 15 non-emotional samples were correctly classified as non-emotional.
2. Incorrect Predictions:
   a. 3 emotional samples were incorrectly classified as non-emotional.
   b. 3 non-emotional samples were incorrectly classified as emotional.
3. Overall Performance:
   a. The model correctly classified 30 out of 36 cases.
   b. Accuracy is approximately 83.3% (30/36).
   c. Error rate is approximately 16.7% (6/36).
4. Balance:
   a. The confusion matrix indicates a well-balanced model because:
   b. The number of errors is equal in both directions (3 each).
   c. The number of correct predictions is also equal for both classes (15 each).

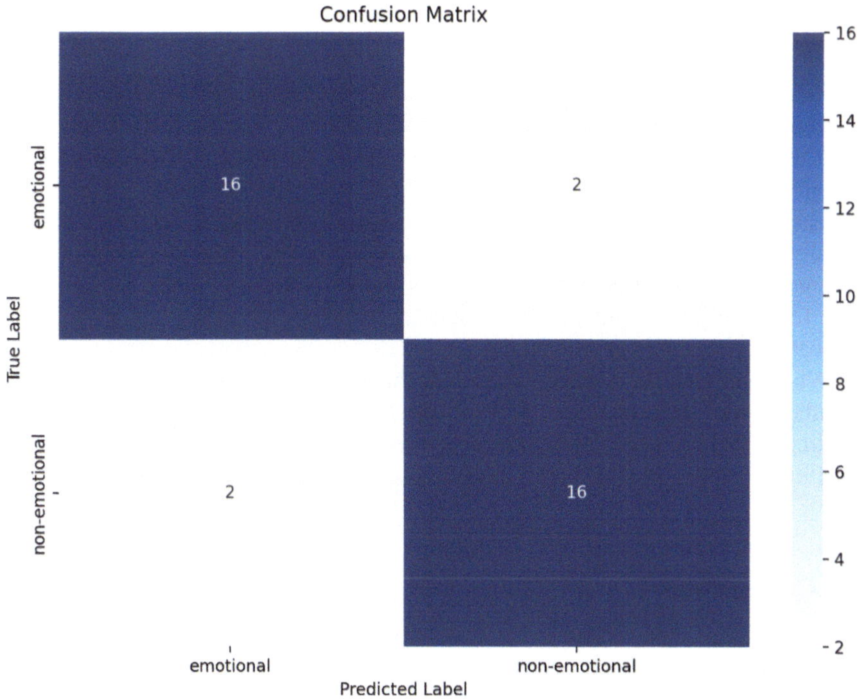

Fig. 11 Confusion matrix for the SVM method using 16 features

This suggests that the model performs well and is equally effective for both classes, without significant bias toward either class.

Thirteen additional MFCC coefficients were added for training the SVM model. This resulted in improved accuracy, which can be observed in the confusion matrix in Fig. 11.

When comparing the confusion matrices in Figs. 10 and 11, we can observe that:

- The new model shows better results: accuracy increased by 5.6%.
- The number of errors decreased from 6 to 4 cases.
- Both models remain well-balanced.
- In both cases, the number of correct predictions is equal for both classes.
- Correct classification of emotional samples: increased from 15 to 16.
- Correct classification of non-emotional samples: increased from 15 to 16.
- Incorrect classifications: reduced from 3 to 2 for each class.

Overall Conclusion: the new model demonstrates significant improvement compared to the previous one, while maintaining good balance between classes. This indicates successful enhancement of the classification algorithm.

3.3 Results Analysis

The study evaluated the accuracy of clustering and classification of emotional and non-emotional audio segments extracted from lectures. Various feature sets were used for analysis, including Mel-Frequency Cepstral Coefficients (MFCC), chroma features, and signal strength indicators.

The results of K-Means clustering indicated that the feature set related to signal strength (max_audio_strength, average_audio_strength, mean_audio_strength) significantly outperformed other features, achieving 79 and 90.7% accuracy for the respective clusters. This suggests that signal strength is a more relevant indicator for detecting emotional tone.

When applying the Support Vector Machine (SVM) method, the classification results showed high accuracy: 83% on the test dataset using the initial features. Adding 13 MFCC coefficients increased the accuracy to 89%, indicating that the additional features provide better modeling of emotional characteristics.

4 Conclusion

Modern tools for audio processing and feature extraction for machine learning, including Librosa and Praat, have been analyzed and tested to extract over 100 acoustic features. It has been demonstrated how these tools can be applied for audio annotation and the extraction of key features such as Mel-Frequency Cepstral Coefficients (MFCC), spectral features, chroma features, and others.

Thus, using a combination of these tools improves the accuracy and reliability of feature extraction, which is crucial for subsequent ML model training, especially in tasks related to speech emotion analysis. The inclusion of spectral features, prosodic characteristics, and acoustic properties can significantly enhance the performance of recognition and classification algorithms for audio signals.

Unsupervised clustering with K-Means on both spectral and signal-strength feature subsets achieved initial cluster accuracies of up to 90% for non-emotional and 79% for emotional segments.

Additionally, a new annotation tool was developed, which accelerated the annotation of long files (> 30 min).

This study demonstrates that combining modern audio processing tools and machine learning methods enables effective analysis of the emotional tone of lectures. Using features such as signal strength and MFCC, combined with the SVM algorithm, ensures high accuracy in classifying emotional fragments obtaining a baseline classification accuracy of 83% on unseen data, which improved to 89% upon inclusion of additional MFCC coefficients.

The developed web application for audio fragment annotation significantly reduced the time required for data preparation, which is essential for analyzing large datasets. Moreover, the integration of additional features showed potential for further accuracy improvements.

Future research can focus on expanding the range of features, exploring new approaches to clustering and classification, and adapting the system for analyzing other types of audio content, including not only lectures but also other forms of public speaking.

This opens up opportunities for creating universal tools that will improve the quality of the educational process and enhance interaction between lecturers and students.

Given modern technologies and machine learning methods, we aim to develop, in the future, a system capable of automatically analyzing and assessing the emotional tone of lectures. This opens up broad prospects for improving teaching methods and increasing the efficiency of the learning process in universities and other educational institutions.

References

1. Becker E, Goetz T, Morger V, Ranellucci J (2014) The importance of teachers' emotions and instructional behavior for their students' emotions—An experience sampling analysis. Teach Teach Educ 43:15–26. https://doi.org/10.1016/j.tate.2014.05.002
2. Kouhsari M, Huang X, Wang C, Lee J (2024) The association between school climate, teacher emotions, and adaptive instruction. Front Educ 9. https://doi.org/10.3389/feduc.2024.1447440
3. Ikonne O (2024) The relationship between institutional pedagogy and the brand identity formation of higher education institutions. Int J Manag Human 11:2394–0913. https://doi.org/10.35940/ijmh.B1761.11021024
4. Timothy Snyder: The making of modern Ukraine. https://www.youtube.com/playlist?list=PLh9mgdi4rNewfxO7LhBoz_1Mx1MaO6sw_
5. Timothy Snyder career. https://history.yale.edu/people/timothy-snyder
6. Lerch A (2022) An introduction to audio content analysis: Music information retrieval tasks and applications, 2nd edn. Wiley, IEEE Press
7. Purwins H et al (2019) Deep learning for audio signal processing. IEEE J Sel Top Sig Process 13:206–219. https://doi.org/10.1109/JSTSP.2019.2908700
8. Sharma G et al (2020) Trends in audio signal feature extraction methods. Appl Acoust 158:107020. https://doi.org/10.1016/j.apacoust.2019.107020
9. Luitel S et al (2022) Audio sentiment analysis using spectrogram and bag-of-visual-words. In: 2022 IEEE 23rd International conference on information reuse and integration for data science (IRI), pp 200–205. https://doi.org/10.1109/IRI54793.2022.00052
10. Mishra A et al (2024) Speech recognition using machine learning techniques. In: 2024 2nd International conference on disruptive technologies (ICDT), 1142–1146. https://doi.org/10.1109/ICDT61202.2024.10489508
11. McLoughlin I et al (2015) Robust sound event classification using deep neural networks. IEEE/ACM Trans Audio Speech Lang Process 23:540–552. https://doi.org/10.1109/TASLP.2015.2389618
12. https://librosa.org/
13. https://pydub.com/
14. https://www.fon.hum.uva.nl/praat/

15. Kobylin A, Khoroshun G, Kobylin O (2023) Detection of emotional tone in video lectures using machine learning methods. In: Collection of scientific and practical papers of the 9th all-Ukrainian forum "IT-Idea 2023", pp 71–74 https://cse.snu.edu.ua/wp-content/uploads/2024/03/Збірник-науково-практичних-праць-ІТ-Ідея-2023.pdf
16. Kobylin A, Khoroshun G, Ryazantsev O (2024) Tools for audio processing and feature extraction for machine learning. In: Collection of scientific and practical papers of the 10th all-Ukrainian forum "IT-Idea 2024", pp 91–94 https://cse.snu.edu.ua/wp-content/uploads/2025/01/Збірник_науково_практичних_праць_ІТ_Ідея_2024.pdf
17. Rahaman ME et al (2019) Performance analysis of isolated speech recognition technique using MFCC and cross-correlation. In: 2019 10th International conference on computing, communication and networking technologies (ICCCNT), pp 1–4. https://doi.org/10.1109/icccnt45670.2019.8944534
18. Kovtun V, Kovtun O, Semenov A (2022) Entropy-argumentative concept of computational phonetic analysis of speech taking into account dialect and individuality of phonation. Entropy 24(7):1006. https://doi.org/10.3390/e24071006
19. Wackerly D, Mendenhall W, Scheaffer R (2008) Mathematical statistics with applications, 7th edn. Duxbury Press
20. Cristianini N, Shawe-Taylor J (2000) An introduction to support vector machines and other kernel-based learning methods. Cambridge University Press, Cambridge

University Rankings and Students' Perception of Educational Quality as Drivers of Reputation and Enrollment Strategy in Service Marketing

Olena Bilovodska, Olga Prygara, Natalia Rudenko, Liudmyla Yarosh-Dmytrenko, Kseniia Bliumska-Danko, and Ihor Ponomarenko

Abstract The quality of educational services is a key factor in evaluating the effectiveness of higher education institutions, especially in the context of global competition, technological change, and shifting labor market demands. As active participants in the educational process, students offer valuable insights into quality, making their perceptions crucial for shaping institutional reputation and guiding choice in an increasingly competitive education market. This paper investigates the influence of university rankings on students' perception of educational quality, using a case study conducted at the Faculty of Economics, Taras Shevchenko National University of Kyiv. Through a quantitative survey, the study explores the factors that shape student decision-making and evaluates the extent to which global ranking indicators align with student priorities. The study confirms that university rankings strongly affect

O. Bilovodska (✉) · O. Prygara · N. Rudenko · L. Yarosh-Dmytrenko
Taras Shevchenko National University of Kyiv, Kyiv, Ukraine
e-mail: o.bilovodska@knu.ua

O. Prygara
e-mail: olgaprygara@knu.ua

N. Rudenko
e-mail: natalia.rudenko@knu.ua

L. Yarosh-Dmytrenko
e-mail: liudmilayarosh-dmytrenko@knu.ua

K. Bliumska-Danko
Sumy National Agrarian University, Sumy, Ukraine

Paris-Saclay University, Sceaux Cedex I, France

K. Bliumska-Danko
e-mail: kseniia.bliumska-danko@universite-paris-saclay.fr

I. Ponomarenko
State University of Trade and Economics, Kyiv, Ukraine
e-mail: i.ponomarenko@knute.edu.ua

A. Semenov et al. (eds.), *Data-Centric Business and Applications*, Lecture Notes on Data Engineering and Communications Technologies 272,
https://doi.org/10.1007/978-3-032-06608-4_10

students' perception of quality but reveal discrepancies between ranking criteria and students' priorities. While rankings like QS and THE focus on research output, international prestige, and citations, students prioritize practical education, employability, and personal development. This indicates that current ranking methodologies may not fully capture what students consider essential in assessing educational quality. So, the results reveal a significant gap between the metrics emphasized by major university rankings. The paper highlights the marketing function of rankings and discusses their impact on enrollment strategy. It concludes by offering recommendations for universities to enhance their ranking performance while remaining responsive to student needs.

Keywords Ranking system · Educational service · Educational quality · Student perception · Higher education · Marketing in education · Institutional reputation · Internal marketing · Service marketing · Brand identity competitive advantage

1 Introduction

The quality of educational services is one of the key indicators used to assess the effectiveness of higher education institutions (HEIs) in the modern world. As higher education systems undergo rapid transformation driven by technological advancement, globalization, and increasing competition for students and resources, the need for transparent and reliable measures of educational quality has become increasingly urgent. In this evolving landscape, educational quality is no longer viewed solely in terms of academic outcomes but also through the lens of student satisfaction, graduate employability, and alignment with labor market demands. Student assessment of educational services has become a critical component of this evaluative process, as students are not only recipients but also active participants and stakeholders in the educational experience. Their perspectives offer valuable insights into the relevance, accessibility, and overall effectiveness of the education they receive. Understanding how students perceive quality is especially important in a globalized education market where prospective students are faced with a growing number of choices and where institutional reputation plays a key role in decision-making.

In this context, international and national university rankings have become essential benchmarks for assessing the effectiveness of HEIs and have emerged as highly influential tools. Prominent rankings such as the QS World University Rankings, Times Higher Education (THE), the Academic Ranking of World Universities (ARWU), and Webometrics have gained global recognition and are widely used by various stakeholders, including students, parents, policymakers, and university administrators to compare institutional performance. They often serve as proxies for educational excellence and play a significant role in shaping the perception of educational quality. These rankings typically evaluate institutions based on criteria such as academic reputation, research output, faculty quality, international diversity.

As a result, they offer a comparative tool for measuring educational quality and institutional performance.

However, while these rankings are influential, they do not always capture the full complexity of what constitutes educational quality from a student's perspective. Students are increasingly looking for institutions that offer more than just academic prestige; they also prioritize factors such as career prospects, industry connections, and opportunities for personal growth.

2 Literature Review and Theoretical Background

University rankings evaluate a range of factors, including research output, teaching quality, internationalization, and institutional prestige. However, they tend to emphasize metrics that may not directly reflect the day-to-day learning experiences of students.

Recent research has raised important questions about the extent to which university rankings accurately reflect the quality of education, particularly from the student perspective. While rankings are widely used by prospective students and institutional stakeholders as indicators of excellence, studies have shown that their methodologies often emphasize research productivity over teaching effectiveness or student satisfaction [4]. The simplified metrics used in university rankings contribute to a skewed perception of academic quality, overlooking essential dimensions such as teaching effectiveness, and broader educational outcomes [14]. The international ranking systems mostly rely on research productivity and do not comprehensively reflect student satisfaction [24]. These observations highlight the need to critically evaluate the influence of rankings on how students perceive quality and to consider more student-centered indicators in assessing university performance.

In addition to the critique of university rankings, a substantial body of research emphasizes the multifaceted nature of educational service quality and the importance of aligning it with student expectations. Service quality in higher education has been studied from various perspectives, focusing on dimensions such as accessibility, completeness of services, and responsive teaching [6]. Scholars argue that students, as primary consumers of education, evaluate quality based on their overall experience—ranging from academic services and administrative support to campus infrastructure and professional outcomes [11, 28].

Literature reviews confirm that integrating student-centered quality models, such as SERVQUAL, can significantly enhance both satisfaction and institutional reputation [2]. These insights highlight the need for higher education institutions to balance formal ranking criteria with deeper, service-based quality frameworks.

In today's academic environment, university rankings also function as powerful marketing mechanisms. High rankings contribute to a university's global visibility, attract international students and faculty, and foster partnerships with industry and other educational institutions. They influence not only student choice but also institutional funding, government policy, and strategic planning. As a result, universities invest considerable effort in improving their standing in global league tables.

Various international and national rankings—such as QS World University Rankings [22], Times Higher Education (THE) [30], Academic Ranking of World Universities (ARWU) [27], and Webometrics [7]—evaluate institutions based on different criteria. Among the core indicators are teaching quality, academic publications, level of international cooperation, graduate employability, and financial sustainability of the university. University rankings play a pivotal role for HEIs by performing several important functions. First and foremost, they are powerful marketing tools: high positions in rankings enhance a university's prestige and attract a greater number of prospective students. At the same time, rankings foster international competition, enabling institutions to expand partnerships and secure grant funding. For potential students university rankings serve as a key point of reference that influences their decision-making process regarding the choice of institution. Moreover, participation in rankings encourages universities to improve, motivating them to raise teaching standards, intensify research activities, and implement innovative practices.

The concept of service quality assessment is fundamental in service marketing. A service is a product with distinct characteristics, such as intangibility, inseparability from the provider, variability in quality, and inability to be stored [21, 31]. In the context of education, service delivery involves active interaction between the student and the teacher, implying their co-creation in the learning process. A substantial body of research has examined students' perceptions of educational quality in higher education [2, 6, 8, 11, 16, 19, 20, 23]. The quality of services largely depends on the engagement of the learner and their active participation. Typically, consumers do not receive an immediate tangible benefit from educational services, and their value is realized through the practical application of acquired knowledge and future professional growth.

In the academic literature, the concept of educational service quality is approached from various perspectives. Harvey and Green [8] suggest that quality in higher education can be interpreted through multiple dimensions, including the transformation of knowledge and competencies, alignment with labor market demands, and the innovativeness of the educational process. Building on multidimensional interpretations of educational quality, Parasuraman et al. [21] proposed the SERVQUAL model, which evaluates service quality through five key dimensions—tangibility, reliability, responsiveness, assurance, and empathy—widely applied in higher education to assess the student experience. Student evaluation of learning quality is a multifaceted process encompassing several aspects, such as the academic component (teaching, learning materials), administrative services, infrastructure, and the overall atmosphere within the university [12]. Student evaluation of educational quality is inherently multidimensional, encompassing not only academic instruction but also the accessibility and efficiency of administrative services, campus infrastructure, and the overall institutional environment [6]. Students increasingly expect universities to provide a cohesive experience that integrates high-quality teaching with timely support and functional learning spaces. Alemu [2] found notable gaps between student expectations and actual service delivery across the SERVQUAL dimensions—tangibility, reliability, responsiveness, assurance, and empathy—highlighting that students evaluate quality through both academic and support-related encounters. This indicates that learning satisfaction depends on a broader ecosystem of services rather than on academic factors alone. Kayyali [14] further reinforces this view by framing academic excellence as a combination of teaching quality, infrastructural resources, student engagement, and institutional impact, thereby urging universities to adopt more holistic strategies in their quality improvement efforts.

University rankings frequently influence student decisions regarding institution selection. However, perceptions of educational quality are also shaped by other factors. Research indicates that student perceptions may vary depending on factors such as faculty support, access to modern technologies, opportunities for international exchange and internships, and the overall university environment [29]. Additionally, student satisfaction is closely tied to the extent to which their expectations are met throughout the learning process. Institutions that actively engage students in feedback mechanisms and adapt their educational programs based on student input are more likely to sustain high levels of student satisfaction.

University rankings play a significant role in shaping perceptions of educational quality among students and other stakeholders, including employers, government bodies, and peer institutions [9, 10]. Rankings typically reflect universities' achievements across areas such as research, innovation, international collaboration, and teaching quality. However, they often emphasize indicators that may not directly capture the student experience. As such, rankings do not always correlate with actual levels of student satisfaction regarding educational services. Some studies [17, 18] indicate that a high ranking does not necessarily guarantee high perceived educational quality from the student perspective. This discrepancy may be attributed to the differences in evaluating academic achievements versus educational experiences. Hazelkorn [10] examines the impact of university rankings on the global higher education system and explores how rankings have become a critical factor in shaping university policies, as well as influencing decision-making processes among students, faculty, and governments.

Salmi [25] investigates the creation of rankings and the factors influencing universities' positions in global league tables. Marginson [17, 18] critically evaluates ranking methodologies and argues that they often overemphasize research-related indicators, which do not adequately reflect teaching quality or student engagement. Altbach [3] discuss rankings as tools for comparing both academic and instructional performance of universities. Nonetheless, many studies critique the methodology of university rankings for their disproportionate focus on research performance. Marginson [18], for example, highlights that rankings such as the Academic Ranking of World Universities (ARWU) heavily rely on indicators related to scientific publications, citations, and research achievements, while neglecting important dimensions such as teaching quality, pedagogical practices, and student involvement in the learning process. Hazelkorn [10] emphasizes that rankings exert pressure on universities to increase the number of research publications and strive for academic prestige, rather than focusing on student development and the provision of practical skills for the labor market. Altbach [3] points out that large research universities with greater resources have a competitive advantage in rankings due to their ability to attract top researchers and secure research funding. In contrast, institutions that prioritize teaching and student support often rank lower, despite offering considerable value in terms of educational delivery. Clarke [5] notes that the criteria used by most university rankings are narrowly focused and fail to capture the complex multidimensional nature of educational quality. Similarly, Salmi [25] argues that rankings may mislead students by prioritizing research accomplishments, which may not be the most critical factor in their educational journey.

The reviewed literature highlights both the influence and limitations of university rankings as indicators of educational quality. While rankings serve as powerful marketing tools and strategic instruments that shape institutional reputation and student enrollment decisions, their methodologies often prioritize research output over teaching quality and student satisfaction. This creates a misalignment between what rankings measure and what students value in their educational experience. As a result, there is a growing consensus among scholars on the need to integrate student-centered quality frameworks—such as SERVQUAL—into institutional evaluation practices. These models emphasize responsiveness, reliability, and empathy, offering a more holistic understanding of service quality in higher education.

These critical perspectives highlight the need for more balanced evaluation metrics that consider not only research excellence but also teaching quality, student support, and the overall educational experience provided by the institution. To remain competitive and responsive in a globalized education market, universities are thus encouraged to balance performance metrics with strategies that prioritize student expectations, engagement, and long-term satisfaction.

Also, the literature reveals that university rankings, while influential in shaping institutional reputation and guiding student choices, often fail to reflect the full complexity of educational quality as experienced by students. Their heavy reliance on research metrics neglects key aspects such as teaching effectiveness, student engagement, and support services, which are central to the student experience. In contrast, service marketing perspectives emphasize the importance of aligning institutional offerings with student expectations, suggesting that quality in higher education should be assessed through more comprehensive, student-centered models. As universities seek to enhance both their standing in rankings and their appeal to prospective students, an integrated approach that combines formal performance indicators with service quality principles becomes essential for sustainable development and strategic enrollment growth.

So, this article aims to explore the influence of university rankings on students' perception of the quality of education. It analyzes how students interpret and prioritize the criteria used in rankings and whether these align with their actual expectations and experiences. Additionally, the paper investigates how rankings impact institutional behavior and provides recommendations for higher education institutions seeking to enhance both their perceived and actual educational quality.

3 Methods

To investigate how university rankings influence students' perceptions of educational quality, this study employed a quantitative survey-based approach. The research was conducted among students of the Faculty of Economics at Taras Shevchenko National University of Kyiv using Google Forms. To form the sample, the quota sampling technique was used.

A structured questionnaire was designed to capture student priorities when selecting a university and their views on what constitutes quality in education. The survey included multiple-choice and Likert-scale questions aimed at identifying which factors students consider most important in evaluating educational quality.

The questionnaire consisted of questions, among them:

1. Do you agree with the statement: a university ranking indicates a high quality of educational service?
2. Please rank the factors that were most important to you when choosing a university for your studies (from 1—most important to 5—least important);
3. Please evaluate the factors that influence the quality of educational services (from 1—most important to 8—least important);
4. When evaluating the quality of teaching of academic subjects at the university, the most significant factors for you are (from 1—most important to 6—least important);
5. Which university rankings are you familiar with?

Specific attention was paid to how students perceive the importance of institutional rankings, reputation, academic programs, faculty qualifications, employability, and other quality-related indicators. The sample consisted of respondents, selected through purposive sampling to target students actively engaged in higher education decision-making. Responses were analyzed using descriptive statistics to determine the frequency and ranking of preferences. The findings were categorized by first- and second-priority choices to better reflect the relative weight students assign to various quality indicators.

This methodology enabled a comparison of students' actual preferences with the criteria commonly used in global university rankings, highlighting areas of alignment and divergence between institutional metrics and student expectations.

4 Results

The survey findings provide a clear overview of how university rankings influence students' perceptions of educational quality.

First of all, the results indicate that students agree that the rating of a higher education institution reflects the high quality of educational services—25% of respondents fully support this thesis, and 71.6% partially endorse it. Overall, this signifies a high

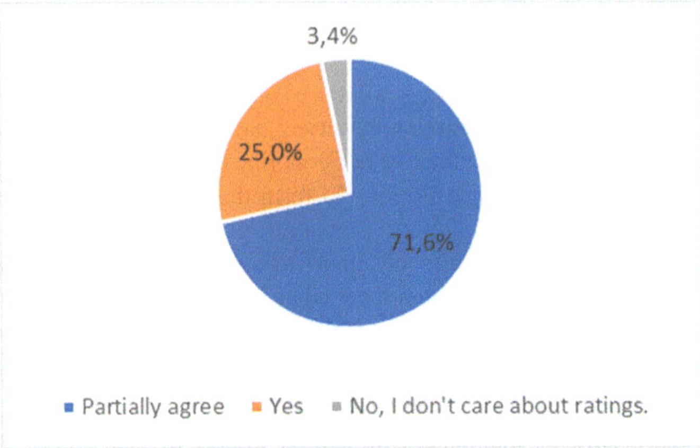

Fig. 1 Results on the question "Do you agree with the statement: the rating of a higher education institution indicates the high quality of educational services"

level of trust among students in ranking systems as a tool for assessing education quality and their influence on the choice of HEI. However, the vast majority of respondents also point out that when selecting a university, they do not rely solely on ranking indicators (Fig. 1).

Also, the findings indicate that the importance of rankings and the reputation of higher education institutions is recognized as a first- and second-priority factor by 28.13% and 21.88% of respondents, respectively. At the same time, the description of academic programs and specializations, along with the employment opportunities, is also considered a primary criterion in choosing a university (23.44% marked it as their first priority). Additionally, recommendations and reviews play a significant role, cited as a first-priority factor by 20.31% of respondents and as a second-priority factor by 23.44%. Although infrastructure and reviews/recommendations were less frequently ranked as top priorities, they still hold value—particularly as secondary or mid-level considerations. Notably, infrastructure was ranked 3rd and 4th by 31.25% and 26.56% of respondents, respectively, indicating its moderate importance. Meanwhile, reviews and recommendations were evenly distributed across the rankings, suggesting their supplementary role in decision-making. Overall, students tend to prioritize factors that directly affect their academic and professional outcomes, while external opinions and institutional facilities serve more as supportive considerations (Table 1).

At the same time, the qualifications of teaching staff (34.38%), the practical orientation of education (31.25%), and opportunities for professional development and employment (32.81%) are identified as decisive indicators of the quality of educational services. By contrast, factors such as teaching materials, faculty publications and academic support/consultations are considered less decisive, receiving relatively lower rankings. Similarly, infrastructure and student support services were

Table 1 Importance of factors when choosing a higher education institution (from 1—most important to 5—least important)

Importance of the factor	University ranking and reputation	University ranking and reputation, %	Description of academic programs and specializations	Description of academic programs and specializations, %	Employment opportunities	Employment opportunities, %	University infrastructure	University infrastructure, %	Reviews and recommendations	Reviews and recommendations, %
1	18	28.13	15	23.44	15	23.44	9	14.06	13	20.31
2	14	21.88	18	28.13	13	20.31	4	6.25	15	23.44
3	10	15.63	12	18.75	10	15.63	20	31.25	10	15.63
4	11	17.19	10	15.63	9	14.06	17	26.56	13	20.31
5	10	15.63	7	10.94	16	25.00	9	14.06	10	15.63

less frequently chosen as top priorities but remained moderately important in mid-tier rankings.

Overall, the findings highlight that students value education that combines high teaching quality, practical relevance, and employability, while more traditional academic metrics or supportive services play a secondary, yet still meaningful, role (Table 2).

Teaching materials, faculty publications, and the level of academic support are generally not considered first-priority factors in the university selection process. However, these are the very criteria that receive significant emphasis in university rankings.

University rankings differ significantly in terms of the indicators they use, the weight assigned to those indicators, and their overall methodological approach. For instance, the QS World University Rankings emphasize academic and employer reputation (together accounting for 50% of the total score), followed by faculty-to-student ratio, research citations, and internationalization [22]. Times Higher Education (THE) employs a broader range of 13 indicators grouped into five categories: teaching, research, citations, international outlook, and industry income [30]. ARWU (Academic Ranking of World Universities), also known as the Shanghai Ranking, relies entirely on objective research-based metrics such as Nobel Prizes, Fields Medals, highly cited researchers, and publications [27]. Meanwhile, the Webometrics Ranking focuses on digital visibility, openness, and excellence based on web presence and academic output [7].

Despite their global influence, these rankings have been widely criticized for their overemphasis on research productivity and institutional prestige at the expense of student-centered indicators. Teaching quality, student engagement, and satisfaction are rarely measured directly. QS and THE attempt to reflect aspects of teaching through proxies like faculty/student ratios or teaching reputation, but these are often indirect and may not correspond with actual student experience. ARWU and Webometrics, by contrast, do not account for teaching at all, making their relevance to prospective students more limited. This disparity between what rankings measure and what students prioritize—such as employability, practical learning, and support services—suggests a need for more holistic and balanced evaluation frameworks in higher education (Table 3).

The comparative table presents notable methodological differences among the four leading global university rankings—QS, THE, ARWU, and Webometrics. It is particularly important to consider the composition of indicators and the degree to which these rankings account for the quality of teaching. QS and THE are generally perceived as the most balanced systems, as they combine reputational surveys, quantitative data, and measures of internationalization. However, their assessment of teaching quality is largely indirect, relying on metrics such as faculty-to-student ratio or institutional income, rather than direct evaluations of student satisfaction or teaching effectiveness. In contrast, the ARWU (Shanghai Ranking) employs a methodology that focuses exclusively on research outputs and academic excellence. It does not include any indicators related to teaching or student experience, making it most relevant for assessing research performance rather than educational quality.

Table 2 Importance of factors influencing the quality of educational services (from 1—most important to 8—least important)

Importance of the factor	Qualifications of teaching staff	Qualifications of teaching staff, %	Teaching materials and faculty publications	Teaching materials and faculty publications, %	Practical orientation of education	Practical orientation of education, %	Availability of student support and guidance
1	22	34.38	10	15.63	20	31.25	6
2	8	12.50	12	18.75	11	17.19	7
3	6	9.38	11	17.19	7	10.94	8
4	3	4.69	4	6.25	2	3.13	13
5	4	6.25	4	6.25	2	3.13	6
6	2	3.13	4	6.25	4	6.25	3
7	5	7.81	5	7.81	2	3.13	7
8	9	14.06	7	10.94	11	17.19	9

Availability of student support and guidance, %	University infrastructure and a comfortable learning environment	University infrastructure and a comfortable learning environment, %	Level of academic support and faculty consultation	Level of academic support and faculty consultation, %	Opportunities for professional development and employment	Opportunities for professional development and employment, %
9.38	2	3.13	6	9.38	21	32.81
10.94	6	9.38	10	15.63	3	4.69
12.50	12	18.75	8	12.50	7	10.94
20.31	8	12.50	6	9.38	3	4.69
9.38	5	7.81	11	17.19	2	3.13
4.69	13	20.31	7	10.94	1	1.56
10.94	6	9.38	7	10.94	9	14.06
14.06	7	10.94	6	9.38	10	15.63

Table 3 Methodological features of university rankings

Ranking	Key indicators	Methodological features	Teaching quality
QS world university rankings	Academic and employer reputation, faculty/student ratio, citations, international students and staff	Balanced with a strong focus on reputation and global presence	Partially via faculty/student ratio, no direct student feedback
THE (Times Higher Education)	Teaching quality, research volume and quality, international outlook, industry income	Comprehensive 13-indicator model using both subjective and objective data	Included via teaching pillar (income, Ph.D. ratios), but indirectly
ARWU (academic ranking of world universities)	Nobel prizes, highly cited researchers, publications in nature/science, and indexed articles	Purely objective, focused on elite research output and awards	Not considered
Webometrics	Web visibility, openness (citations of top authors), research excellence	Measures digital presence and scholarly openness, not academic quality	Not considered

The Webometrics Ranking adopts a different perspective by evaluating the digital visibility and openness of universities. It emphasizes web-based impact and access to academic content rather than academic or pedagogical performance. While this approach provides insight into the online presence of institutions, it does not reflect the quality of teaching or student engagement.

So, each ranking system has its strengths and weaknesses, which are worth considering when interpreting the results (Table 4).

In conclusion, none of the analyzed rankings offer a comprehensive and objective evaluation of teaching quality. This limitation should be taken into account when using these rankings for shaping institutional image or informing prospective students' choices.

5 Discussion

University rankings play a significant role in shaping the reputation of higher education institutions. Higher positions in rankings generally enhance the perceived credibility and prestige of a university, which positively influences its ability to attract prospective students. This correlation underscores the importance of universities maintaining or improving their rankings as part of their broader strategic goals.

However, university rankings alone do not entirely determine students' choices. Students tend to base their decisions not only on official rankings but also on their

Table 4 Comparative analyze of university rankings

Ranking	Advantages	Disadvantages
QS world university rankings	The QS rankings place strong emphasis on the reputation of universities among academics and employers, providing a clear indication of global prestige It also considers the number of international students and staff, which highlights a university's global reach and openness	While QS considers academic and employer reputation, it does not include direct feedback from students about their educational experiences, which may overlook student satisfaction The emphasis on reputation can overshadow actual teaching quality, making it less reliable for evaluating the day-to-day learning experience
THE (Times Higher Education)	THE utilizes a comprehensive model with 13 indicators, combining both subjective and objective data, offering a more balanced view of university performance The rankings include aspects that are directly linked to teaching quality, such as the student-to-PhD ratio, making it a more holistic measure of academic strength	Despite being comprehensive, THE teaching quality aspect in THE rankings is measured indirectly through factors like income and PhD ratios, which do not always correlate with actual teaching effectiveness This can lead to an overemphasis on certain institutional features while undervaluing others that directly affect student experience
ARWU (academic ranking of world universities)	It is focused on research output and achievements, making it ideal for evaluating universities with a strong research focus The methodology is entirely objective, minimizing bias and providing a clear view of research excellence	It does not consider teaching quality at all, which is a significant disadvantage for prospective students who are focused on the educational experience rather than research output The ranking is heavily skewed toward elite research achievements, offering limited insight into student life or educational quality
Webometrics	It focuses on universities' digital presence and scholarly openness, important in today's digital and open-access academic environment	It focuses on digital visibility and research openness, which are important but not directly related to the quality of teaching or the overall student experience The ranking does not account for aspects like university infrastructure or student support, which are critical to a student's overall educational experience

perception of the quality of education offered. This perceived quality becomes a decisive factor in the university selection process. In fact, the subjective evaluation of educational quality often outweighs even the objective measures found in rankings, as students want to feel that the education they receive will be both academically and professionally beneficial.

In terms of enrollment strategies, universities need to carefully integrate both academic reputation and service quality perception into their promotional efforts. This means that beyond focusing on ranking improvements, universities should actively communicate their strengths in terms of teaching quality, student satisfaction, and global influence. This holistic approach in university marketing is essential for competing effectively in the increasingly competitive higher education market.

First, universities should emphasize the practical orientation of their academic programs, as students place high value on real-world applicability, career readiness, and opportunities for professional development. This includes promoting internships, dual education formats, and partnerships with employers.

Second, it is crucial to provide clear and detailed descriptions of academic programs, highlighting their relevance to current labor market needs and showcasing success stories of graduates. Student decision-making is heavily influenced by transparent information and evidence of career outcomes.

Third, while university rankings and reputation remain influential, institutions should focus on building trust through value-driven communication rather than relying solely on league tables. Sharing authentic student experiences, testimonials, and institutional achievements can significantly enhance public perception.

Fourth, adopting a personalized communication strategy is essential. Targeted marketing campaigns, educational fairs, personalized emails, and active engagement on digital platforms allow institutions to reach prospective students more effectively and maintain long-term interest.

Fifth, enhancing the quality of student services and support systems can boost both student satisfaction and institutional appeal. Efficient application processes, prompt communication, and a supportive learning environment are critical touchpoints for prospective applicants.

Lastly, universities should embrace digital marketing strategies, increasing their visibility on social media platforms such as Instagram, YouTube, and TikTok, and using engaging content formats that resonate with younger audiences. Highlighting vibrant campus life, student achievements, and innovative learning environments can strengthen the university's brand identity.

For universities aiming to attract international students, global rankings become even more critical. These rankings act as indicators of academic excellence and institutional potential on a worldwide scale, helping universities build an international reputation. As such, universities must strive for strong performances in global rankings, while also considering the diverse needs and expectations of international students.

To sum up, universities must recognize that rankings alone are not sufficient to drive student enrollment. A comprehensive strategy that includes improving the quality of education, enhancing student satisfaction, and actively participating in global research is essential. This approach will not only boost rankings but will also foster a positive perception of the university, ultimately leading to greater success in student recruitment.

The findings of this study confirm that university rankings significantly influence students' perception of institutional quality, but also highlight important discrepancies between ranking criteria and students' actual priorities. While rankings such as QS and THE offer some indicators related to teaching and reputation, the emphasis on research output, international prestige, and citation metrics does not necessarily reflect the aspects of quality that matter most to students—namely, the practical orientation of education, employability, and personal development opportunities. This suggests that current ranking methodologies may only partially capture what students consider essential in evaluating the quality of education. A substantial number of researchers, in their studies, distinguish and substantiate the indicators that are crucial for ranking metrics. Selten et al. [26] demonstrate that the variables used to construct the rankings primarily measure two underlying factors: a university's reputation and its research performance. Rafique et al. [24] indicate that university ranking should include indicators like governance and digital presence, as these are missing in global ranking indicators.

Furthermore, the data demonstrate that students place relatively less emphasis on indicators such as faculty publications or academic support systems—criteria that carry considerable weight in most global rankings. This disconnect reveals the need for ranking systems to evolve towards more student-centered models that account for the lived academic experience, including teaching effectiveness, access to hands-on learning, mentorship, and support services.

Universities aiming to improve both their actual performance and public image [1, 13, 15] should strive to balance their efforts between enhancing research visibility and meeting the concrete educational needs of their students.

6 Conclusion

The research findings demonstrate that university rankings and institutional reputation are important factors for students when choosing a place of study. These criteria rank among the top decision-making factors, following only the quality of academic programs, employment opportunities, and faculty qualifications. At the same time, students show greater interest in the practical orientation of education and opportunities for professional development, rather than in faculty publications or the level of academic support—areas that carry considerable weight in rankings but have less influence on student choices.

To enhance their positions in university rankings while simultaneously increasing their attractiveness to prospective students, universities should adopt a multifaceted strategy. This includes expanding international cooperation by attracting foreign students and faculty and engaging in global research collaborations. Improving the quality of teaching is also essential and can be achieved by revising academic programs to emphasize practical skills and incorporating modern educational technologies. Additionally, strengthening relationships with the business sector through internship opportunities, dual education pathways, and broader cooperation with employers can significantly contribute to institutional development.

Alongside these efforts, universities are also shaping comprehensive student enrollment strategies that reflect both the formal requirements of ranking systems and the real expectations of prospective students. These strategies involve targeted marketing campaigns, personalized communication, participation in educational fairs, and active engagement on digital platforms to build a strong institutional brand and maintain long-term interest among potential applicants. Moreover, institutions increasingly rely on data-driven approaches to understand student behavior and preferences, allowing them to tailor outreach efforts, enhance student experience from first contact, and foster loyalty throughout the applicant journey.

Such enrollment strategies are essential in today's competitive educational environment, where students act as informed consumers, carefully comparing institutions based on perceived value, career prospects, and alignment with personal goals. By aligning their recruitment efforts with students' expectations—while also addressing key ranking indicators—universities can effectively position themselves as attractive, student-centered institutions, thereby improving both their market presence and overall performance.

Finally, universities should actively engage in ranking assessments by providing accurate and timely data to ranking organizations and boosting their public visibility through strategic public relations efforts.

In doing so, universities can effectively use rankings as a marketing tool—one that not only aligns with formal ranking criteria but also addresses the actual needs of students and the labor market.

References

1. Adnan WH, Saabar SS, Ibnu IN, Rahman SFA, Yahaya M (2025) Shaping perceptions: sustaining the corporate image of public universities in Malaysia. Int J Res Innov Soc Sci 9(3):274–289. https://doi.org/10.47772/IJRISS.2025.903SEDU0016
2. Alemu A (2023) Assessing service quality in tertiary education using adapted SERVQUAL scale. Cogent Educ 10(2):2259733. https://doi.org/10.1080/2331186X.2023.2259733
3. Altbach PG (2015) The dilemmas of ranking. Int High Educ 80:2–3. https://ejournals.bc.edu/index.php/ihe/article/view/8727
4. Chiyón I, Quevedo AV, Vegas S, Mosquera JC (2023) Reflecting on how university rankings measure undergraduate academic excellence. In: 21st LACCEI international multi-conference for engineering, education, and technology, pp 1–8. https://doi.org/10.18687/LACCEI2023.1.1.268
5. Clarke M (2002) Some guidelines for academic quality rankings. High Educ Eur 27(4):443–459. https://doi.org/10.1080/0379772022000071922
6. Clemons R, Jance M (2024) Defining quality in higher education and identifying opportunities for improvement. SAGE Open 14(3):1–11. https://doi.org/10.1177/21582440241271155
7. Cybermetrics Lab (2024) Ranking web of universities: methodology. Retrieved Apr 9, 2025, from https://webometrics.info/en/Methodology

8. Harvey L, Green D (1993) Defining quality. Assess Eval High Educ 18(1):9–34
9. Hazelkorn E (2008) Learning to live with league tables and rankings: the experience of institutional leaders. High Educ Pol 21(2):193–215. https://doi.org/10.1057/hep.2008.1
10. Hazelkorn E (2015) Rankings and the global reputation race. Palgrave Macmillan
11. Hill FM (1995) Managing service quality in higher education: the role of the student as primary consumer. Qual Assur Educ 3(3):10–21. https://www.emerald.com/insight/content/doi/10.1108/09684889510093497/full/html
12. Hill Y, Lomas L, MacGregor J (2003) Students' perceptions of quality in higher education. Qual Assur Educ 11(1):15–20. https://doi.org/10.1108/09684880310462047
13. Horodovenko V, Bilovodska O, Vatras V, Kanzafarova I, Melnyk R (2021) Improving the quality of higher education based on the positive image of HEE in the social and information space: legal and management aspects. Estud Econ Apl 39(7). Special issue: Impact of current trends in social commerce, economics, and business analytics. https://doi.org/10.25115/eea.v39i7.4996
14. Kayyali M (2023) The relationship between rankings and academic quality. Int J Manag Sci Innov Technol 4(3):1–11. https://ssrn.com/abstract=4497493
15. Kazaishvili A, Khmiadashvili L (2022) The impact of the communicative behaviour in social media on the university brand image. Int J Manag Knowl Learn 11:135–143. https://doi.org/10.53615/2232-5697.11.135-143
16. Latif KF, Latif I, Farooq Sahibzada U, Ullah M (2017) In search of quality: measuring higher education service quality (HiEduQual). Total Qual Manag Bus Excell 30(7–8):768–791. https://doi.org/10.1080/14783363.2017.1338133
17. Marginson S (2007) Global university rankings: implications in general and for Australia. J High Educ Policy Manag 29(2):131–142. https://doi.org/10.1080/13600800701351660
18. Marginson S (2014) University rankings and social science. Eur J Educ 49(1):45–59. https://doi.org/10.1111/ejed.12061
19. Nadiri H, Kandampully J, Hussain K (2009) Students' perceptions of service quality in higher education. Total Qual Manag Bus Excell 20(5):523–535. https://doi.org/10.1080/14783360902863713
20. Nguyen HV, Vu TD, Saleem M, Yaseen A (2024) The influence of service quality on student satisfaction and student loyalty in Vietnam: the moderating role of the university image. J Trade Sci 12(1):37–59. https://doi.org/10.1108/JTS-12-2023-0032
21. Parasuraman A, Zeithaml VA, Berry LL (1988) SERVQUAL: a multiple-item scale for measuring consumer perceptions of service quality. J Retail 64(1):12–40
22. QS Quacquarelli Symonds (2024) QS world university rankings 2024 methodology. Retrieved Apr 9, 2025, from https://support.qs.com/hc/en-gb/articles/4405955370898-QS-World-University-Rankings
23. Quinn A, Lemay G, Larsen P, Johnson DM (2009) Service quality in higher education. Total Qual Manag Bus Excell 20(2):139–152. https://doi.org/10.1080/14783360802622805
24. Rafique T, Awan MU, Shafiq M, Mahmood K (2023) Exploring the role of ranking systems towards university performance improvement: a focus group-based study. Heliyon 9(10):e20904. https://doi.org/10.1016/j.heliyon.2023.e20904
25. Salmi J (2013) The race for excellence: a marathon not a sprint. Int High Educ 72:14–15
26. Selten F, Neylon C, Huang C-K, Groth P (2020) A longitudinal analysis of university rankings. Quant Sci Stud 1(3):1109–1135. https://doi.org/10.1162/qss_a_00052
27. ShanghaiRanking Consultancy (2024) Academic ranking of world universities 2024 methodology. Retrieved Apr 9, 2025, from https://www.shanghairanking.com/methodology/arwu/2024
28. Stankovska G, Ziberi F, Dimitrovski D (2024) Service quality and student satisfaction in higher education. In: Education in developing, emerging, and developed countries: different worlds, common challenges. BCES conference books, vol 22. Bulgarian Comparative Education Society, Sofia, pp 153–160
29. Telford R, Masson R (2005) The congruence of quality values in higher education. Qual Assur Educ 13(2):107–119

30. Times Higher Education (2024) World university rankings 2024 methodology. Retrieved Apr 9, 2025, from https://www.timeshighereducation.com/world-university-rankings/world-university-rankings-2024-methodology
31. Zhurylo V, Prygara O (2018) Consumer behaviour formation in the customized service markets: a case of Ukraine. Balt J Econ Stud 4(1):156–163. https://doi.org/10.30525/2256-0742/2018-4-1-156-162

Innovative Post-war Recovery Plan for Ukraine: Demographic Challenges, Growth Drivers, Cybersecurity

Svitlana Luchyk, Yulia Manachynska, and Marharyta Luchyk

Abstract The post-war reconstruction of Ukraine necessitates the restoration and rebuilding of its national economy, severely impacted by the conflict. The reconstruction and recovery processes require substantial financial, material, and human resources, all of which are limited. The deepening demographic crisis and intensified external migration, particularly among young people, have highlighted the urgency of digitally transforming the national economy. This transformation involves high-tech production, innovative approaches, and modernizing all economic sectors using information and communication technologies. It also includes automating and optimizing routine processes where human involvement is limited or unnecessary. This article aims to assess the current state of human and labor resources in Ukraine, determine the role of digital transformation in post-war economic recovery, and develop recommendations for staffing economic sectors with highly qualified personnel, considering labor market challenges. The scientific research was conducted using general scientific methods, including analysis and synthesis, analogy and modeling, abstraction and concretization, as well as system and graphical analysis. The increasing proportion of pre-retirement and retirement-age individuals in the workforce necessitates raising the age threshold for employees and job seekers. The state and businesses must act proactively to leverage the value of aging workers, lest labor productivity decline and negatively impact economic growth. To meet domestic labor market needs, it is crucial to engage people with disabilities, adapt veterans, and encourage the inclusion of internally displaced persons. Additionally, efforts should focus on attracting foreign specialists and labor migrants willing to contribute to value creation in Ukraine. Improving education quality and expanding

S. Luchyk (✉)
Kharkiv National University of Internal Affairs, Kharkiv, Ukraine
e-mail: kaf-ist@univd.edu.ua

Y. Manachynska
Chernivtsi Institute of Trade and Economics of SUTE, Chernivtsi, Ukraine
e-mail: yu_manachynska@ukr.net

M. Luchyk
JSC Credit Agricole Bank, Kyiv, Ukraine
e-mail: luchik-margarita@ukr.net

A. Semenov et al. (eds.), *Data-Centric Business and Applications*, Lecture Notes on Data Engineering and Communications Technologies 272,
https://doi.org/10.1007/978-3-032-06608-4_11

training programs in modern digital and information-communication technologies, artificial intelligence, cybersecurity, and information protection are essential for the population, particularly older age groups.

Keywords Post-war reconstruction · Innovations · Digital transformation · Growth drivers · Population aging · Labor shortage · Skills · Competencies · Cybersecurity

1 Introduction

On November 19, 2024, it marked 1000 days since the start of Russia's full-scale war against Ukraine, resulting in significant losses among both military personnel and civilians, the emergence of refugees and displaced people, and the destruction of critical infrastructure. According to estimates from the KSE Institute as of January 2024, the total amount of documented direct losses inflicted on Ukraine's infrastructure due to Russia's full-scale invasion had risen to $155 billion (at replacement cost). By November 2024, this figure increased to nearly $170 billion. The amount of indirect losses incurred by Ukraine due to income losses in various sectors of the economy was estimated at $265.6 billion as of June 2023. Among these, the largest losses were suffered by the manufacturing sector (19.4% or $51.5 billion) and the agricultural sector (15.2% or $40.3 billion).

The ongoing destruction caused by Russia during combat operations and shelling across the entire territory of Ukraine has necessitated the recovery of its national economy. This primarily involves the restoration of destroyed or damaged energy infrastructure, hospitals, schools, and residential buildings. The process of economic recovery requires significant resources, both human and financial, the availability of which is limited. Therefore, cutting-edge technologies are being actively used to automate and optimize processes in various areas of activity. These advanced technologies enable the rapid transformation of key economic sectors, with big data, robotics, augmented reality technologies, and artificial intelligence forming the foundation for their development.

The widespread use of information and digital technologies in the country's economy requires technologically skilled professionals. However, does Ukraine currently have such resources available in the required quantity? The issue of labor and, in general, human resources is currently very pressing in the country. According to the UN, Ukraine's population has decreased by 10 million people, or approximately a quarter, since the beginning of Russia's full-scale invasion due to the outflow of refugees, a decline in birth rates, and war-related mortality. As a result of these processes, there has been a significant distortion in the structure of Ukraine's population, particularly with a sharp decrease in the share of youth and an increase in the proportion of elderly individuals. Elderly people may encounter specific difficulties when using new computer and digital services and content due to a lack of specialized knowledge and skills. Therefore, the question that is especially critical today is: to what extent is the Ukrainian state and its available human potential capable of restoring the domestic economy while simultaneously undertaking its reconstruction?

2 Literature Review

The issue of post-war recovery and the reconstruction of the Ukrainian state and its economy is currently being widely discussed among international and domestic political, civil, and scientific communities and organizations. These are the state authorities and their Ukraine Recovery Plan, the Ukraine Facility Plan—a plan for comprehensive recovery and changes for the period 2024–2027, international organizations, in particular, the World Bank, business structures, public organizations, scientists and others. The scientific and practical interest lies in the works of both foreign and domestic experts, which include analyses of the experiences of European and other countries worldwide and recommendations for economic recovery after wars and military conflicts. Notably, the works of Reinert (2019) stand out, as they explore the strategic choices and specific features of European countries' post-war economic reconstruction strategies [1]. Views on the process of post-war reconstruction in Ukraine and the necessity of supporting it through foreign investment and financial aid are expressed by Bregman, Valero and others [2, 3]. The issue of post-war reconstruction is actively being discussed by Ukrainian scholars. Notable contributions come from M. Skripnychenko, L. Kuznetsova, O. Bilotserkivets, I. Pidoricheva, K. Kravchenko, L. Niemets, K. Sekhida, S. Khodakevych, O. Zhuravlov, A. Strizhak, V. Yarmolenko, and others. These scholars have studied various strategic approaches implemented by foreign countries to support their economies under challenging conditions. The analytical materials from the Institute for Analytics and Advocacy are of practical importance for forming a plan for Ukraine's post-war recovery. These materials outline key reform areas that can unlock or expand domestic financial resources in Ukraine and contribute to its progress toward EU membership. They also include decisions on digitization and the development of digital infrastructure to enhance the internal capacity of the Ukrainian economy [4].

The choice of the strategic course for Ukraine's post-war revival, focusing on innovation and technological renewal of economic sectors, while considering the existing negative demographic situation in the country, requires the study of the experience of foreign countries in this direction. Significant interest is generated by the scientific publications of Gao et al. who explored the relationship between population aging, technological innovations, and qualitative economic development [5], Wang and Zhao studied the impact of the population aging process on the digital transformation of enterprises, particularly highlighting the negative effects of this factor on non-governmental and small- and medium-sized enterprises [6]. Xueyan and Xiaohui empirically tested the impact of population aging on the digital transformation of small and medium-sized enterprises, noting the regional heterogeneity of this phenomenon [7]. The research by Jing and Li revealed a significant inverted

U-shaped relationship between population aging and technological innovations in developing market countries, affecting both the labor supply and investments in human capital [8].

The shortage of highly qualified personnel, the significant share of the aging population, and their lack of necessary knowledge and skills to work with information and communication technologies and digital technologies are issues accompanying the processes of innovative growth and digital transformation of the economy not only in Ukraine. The importance of personnel competencies in the context of a digital economy is highlighted, and a three-module competency model for personnel in the context of digital transformation is proposed. This model identifies skills such as: working with digital technologies, creative thinking and behavior, and professional skills in production and management activities, as presented in the work by Belolipetskaya et al. [9]. Research on the links between digitalization and the skills required for working with artificial intelligence was conducted by Matei et al. [10]. Cognitive and non-cognitive skills of the workforce in digital sectors and how they are rewarded were studied by Grundke et al. [11], as well as Ostmeier and Strobel [12].

The importance of educating various population groups on cybersecurity issues in modern conditions has been determined by several researchers. For instance, Gajapathy and Patil [13] and Vazquez [14] explored the serious impact of cyber risks on a country's economy, the awareness of these risks, the varying levels of awareness among people regarding cybersecurity based on age, gender, education, and employment, the activities of the government, corporations, institutions, and society in the context of protecting individuals from data breaches, and the proactivity of businesses in addressing the cybersecurity needs of older employees. The mentioned scientific contributions are useful for shaping the vision of Ukraine's post-war recovery, taking into account the existing problems and challenges.

3 Statement of the Problem

Reconstruction and recovery are two interconnected yet distinct concepts. The term "reconstruction" refers to the process of restoring or rebuilding what has been destroyed or neglected. In the context of wartime, reconstruction typically involves the restoration or construction of critical infrastructure essential for ensuring the security and functionality of the state and society. In contrast, "recovery" describes the formulation of a vision and strategy for the future development of the state, regions, and communities. Essentially, recovery provides the foundation for reconstruction. Therefore, post-war recovery should focus not only on restoring what was destroyed or damaged during the conflict but also on advancing the development of the state and its economy under fundamentally new conditions, leveraging the latest technologies, platforms, and methods. Stepaneko and Kurilets define post-war recovery as "a set of organizational, legal, financial, and other measures aimed at accelerating reconstruction and creating conditions to improve the livelihoods of the population and the operations of business entities in a given area, while also addressing the consequences of war, including its humanitarian, social, economic, and other impacts." [15, p. 38].

Post-war recovery in Ukraine will necessitate substantial human resources, as previously outlined. Consequently, a critical factor in the state's strategic decisions regarding the post-war reconstruction of Ukraine and its economy will involve a comprehensive evaluation of the current demographic landscape, the condition of the existing human capital, the level of educational attainment, and the population's preparedness for socio-economic transformation.

4 Results and Discussions

During wartime, the demographic situation inevitably deteriorates, characterized by increased mortality rates, decreased birth rates, and intensified population migration, all of which are natural reactions to conflict. As of January 1, 2022, Ukraine's population was approximately 41.2 million. By the end of 2024, according to Ukraine's Deputy Prime Minister, O. Chernyshov, the population had decreased to 32 million. However, the broader Ukrainian nation numbers around 60 million, with half of this population living abroad. This group includes refugees, individuals who left the country prior to the war, and the diaspora. For instance, about 5 million Ukrainians currently reside in Europe, having left the country due to the conflict. According to M. Skiba, an education expert at the Ukrainian Institute for the Future (UIF), Ukraine's population is expected to be approximately 30 million in 2025, with an estimated range of 29–30.5 million. However, these figures are only approximate.

In 2024, Ukraine recorded 495,090 deaths, which is 2.8 times higher than the number of registered births, totaling 176,679. Currently, for every newborn, there are approximately three deaths. In comparison to the pre-COVID years of 2018–2020, when there were two deaths for every child born, this ratio serves as a key indicator of a demographic crisis in any country. For Ukraine, these figures suggest that the population reproduction rate has reached a critical threshold. Prior to the full-scale invasion, the total fertility rate for women of childbearing age was 1.16 children per woman; by 2024, this had decreased to 0.9 [16].

Thus, the rapid decline in the birth rate, coupled with large-scale external migration, has triggered a demographic shift in Ukraine, resulting in an increasingly elderly population. According to the United Nations High Commissioner for Refugees, as of mid-February, approximately 6.9 million Ukrainians are currently abroad, with 6.3 million residing in Europe. This migration has contributed to the rapid aging of the population. Within a few years, the average age of Ukrainians has risen from 41 to 45. Comparative data from the United Nations on Ukrainian demographics for the period 2021–2023 further highlight the disappearance of the younger generation, particularly those in their early twenties. In 2021, there were 200,000 young men in this age group, but by 2023, their number had decreased to approximately 70,000. Similarly, the number of young women in this age group has fallen from 200,000 to 50,000. Another emerging challenge is the mass exodus of 10th-and 11th-grade students from Ukraine before completing their education [17, p. 15].

The trends related to population aging are observed in many EU countries. In 2021, the share of the population aged 65 and older in EU countries was approximately 21%. The burden of the 60 + age group on the working-age population (20–64 years old) was 31%. In Ukraine, the corresponding figures for the same period were 17% and 27%, respectively, which were somewhat lower. However, this trend is rapidly worsening in Ukraine, and by 2023, individuals aged 65 and older made up about 18.56% of the population, indicating that the aging process is advancing. It is well-established that population aging leads to a shortage of the labor force in the job market.

The official labor market in Ukraine in 2021 comprised 11.5 million workers. By 2023, this number had decreased to 9 million. As of 2024, experts assert that Ukraine's economy has gradually recovered to about 90%, with 88% of companies reopening new vacancies. Consequently, business activity has resumed, resulting in an increasing demand for personnel. According to data from the State Employment Service of Ukraine, the greatest shortage of workers is expected in skilled trades involving tools, accounting for approximately 21% (e.g., professions such as seamstresses, painters, electricians, and vehicle repairers). The shortage of workers in equipment and machinery maintenance and operation is estimated at about 18% (e.g., professions such as machine operators, tractor drivers, and crane operators). The shortage of workers in the retail and service sectors is estimated at around 15% (e.g., professions such as salespeople, cooks, bartenders, baristas, waiters, and security guards). In categories such as teachers, educators, pharmacists, general practitioners (family doctors), engineers, and practical psychologists, the shortage is expected to be approximately 14%.

As we can observe, the domestic economy requires not just workers, but skilled and technologically literate individuals. The current reconstruction processes in the country, as well as the programs for post-war recovery of the national economy, are focused primarily on innovation, high-tech industries, and scientific production to transition the national economy from a raw material orientation to a digital one. At present, high-tech exports constitute approximately 5.74% of Ukraine's manufactured exports, not just industrial goods exports. Furthermore, between 2012 and 2021, Ukraine's high-tech exports experienced fluctuations, but the global context shows that high-tech exports are a significant component of many economies. For instance, Ukraine's high-tech exports were valued at $1.28 billion in 2021, which is significantly lower than those of leading European countries like Germany, Poland, Slovakia, and Romania. Additionally, 58.5% of Ukraine's total goods exports consist of raw material products supplied to foreign markets [18]. The full-scale invasion by the aggressor, the destruction of production assets and infrastructure, as well as the emigration of highly skilled professionals, have significantly hindered the development of the country's high-tech sectors.

The domestic information and communication technology (ICT) sector, despite the ongoing war, remains the largest exporter of services for the state, contributing

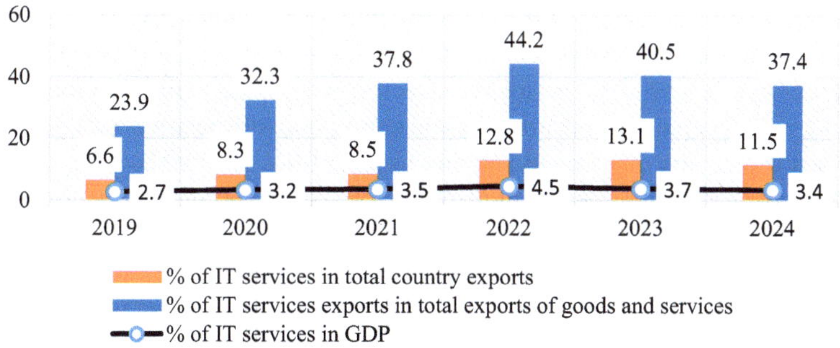

Fig. 1 IT export as a share of Ukraine's GDP, total exports, and export of goods and services, %. *Source* National Bank of Ukraine, 2024

foreign currency revenues to the country's economy. In 2022, its share of the national GDP was 4.5%. In the subsequent years, this share gradually decreased (Fig. 1).

According to data from the National Bank of Ukraine, in 2024, the IT industry contributed $6.45 billion to the Ukrainian economy through the export of its services. This represents a 4.2% decrease, or $280 million, compared to the 2023 figure. In 2021, the contribution was $6.9 billion. Overall, within the general structure of Ukraine's exports, the information technology (IT) sector occupies second place, with a share of 11.5%, following only the export of food products. In 2023, this share was 13.2%. Within the export of goods and services, computer services accounted for 37.4% of the total volume of exported services. Between 2019 and 2024, IT service exports experienced a growth of 54.5%.

The analysis of human resources within IT companies has revealed a significant outflow of specialists abroad. According to data from the DOU portal, the total number of employees in the 50 largest IT companies decreased by 20,000 over the three years of the war. The largest outflow of specialists occurred in the second year of the full-scale war, with a reduction of 10,700 people compared to 7000 in the first year. This period also marked a peak in the legalization of individuals abroad. In the third year of the war, the outflow notably decreased to 2400 specialists. The Ukrainian teams of NIX and Grid Dynamics Group were reduced by half. Among product companies, the gaming development sector was particularly affected. The relocation of specialists abroad became the primary reason for the departure of professionals from Ukrainian teams. Most companies actively opened offices abroad, with over 30 offices of the largest Ukrainian IT companies opening in Poland. Additionally, offices were established in Bulgaria (12), Latin American countries (12), Romania (11), Portugal (8), and India (7) during the three years of the war.

In 2025, the IT sector remains highly active, with an increasing number of job vacancies. Demand is rising for both technical and non-technical positions. In January and March of 2025, the number of vacancies in the IT sector surpassed 6000 per month. The technologies offered by the sector encompass cloud computing, big data analytics, the Internet of Things, cybersecurity, artificial intelligence, virtual reality, as well as 3D and 4D printing, modeling, and more. The active integration of these technologies into key sectors of the economy will lend an innovative character to the modernization and recovery processes, thereby accelerating their development.

According to experts, the agricultural sector, processing industries, the defense-industrial complex, and the information and communication technology (ICT) sector are poised to become the primary driving forces of the national economy. Furthermore, Ukraine has significant potential to develop sectors such as healthcare, research and development (R&D), biotechnology, and digital industries.

In Ukraine, substantial progress has already been made in establishing the legal and regulatory framework for the digital transformation of the national economy. Supporting the development and creation of new digital technologies is one of the primary priorities of the Ukrainian government, as outlined in regulatory acts such as the Law of Ukraine "On Stimulating the Development of the Digital Economy in Ukraine," the Cabinet of Ministers of Ukraine Order "On Approval of the Concept for the Development of Digital Competencies and the Approval of the Action Plan for Its Implementation," and the Cabinet of Ministers of Ukraine Order "On Approval of the Concept for the Development of Artificial Intelligence in Ukraine." In Ukraine, several strategic documents have been approved, including the Innovation Development Strategy until 2030, approved by the Cabinet of Ministers of Ukraine on July 10, 2019, No. 526; the National Economic Strategy until 2030, approved by the Cabinet of Ministers of Ukraine on March 3, 2021, No. 179; and the Strategy for the Digital Development of Innovative Activity in Ukraine until 2030, approved by the Cabinet of Ministers of Ukraine on December 31, 2024, No. 1351-p.

According to the 2024 European Innovation Scoreboard, Ukraine achieved a score equivalent to 32.5% of the EU average, classifying the country as an "Emerging Innovator." The strengths of Ukraine's innovation ecosystem include high broadband penetration, the export of knowledge-intensive services, and significant employment in knowledge-intensive activities. However, improvements are required in areas such as the introduction of innovative products by small and medium-sized enterprises, innovation in business processes, and the development of application design.

According to estimates provided by the World Intellectual Property Organization (WIPO) in the 2024 Global Innovation Index, Ukraine ranks 60th out of 133 countries surveyed globally. This marks a decline in its ranking position, down five spots compared to the 2023 index and three spots compared to the 2022 index. The decrease in Ukraine's innovation performance is primarily attributed to the prolonged nature of the full-scale war with the Russian aggressor, extensive damage to critical infrastructure, and large-scale migration of youth abroad.

Positive trends in innovation development in Ukraine, as identified by WIPO in 2024, include:

- An increase in the production of innovative products relative to the level of investment in innovation as a percentage of GDP, where Ukraine ranks 6th.
- Accelerated innovation-driven economic growth over the long term from 2014 to 2024.
- Leadership in the sub-index of utility models by origin.

- Growth in employment among women with higher education, where Ukraine ranks 2nd.
- An increase in spending on software, where Ukraine ranks 4th.
- Improvement in the online creativity criterion, where Ukraine ranks 39th.
- Relative stabilization of the "market complexity" component, despite its negative state, where Ukraine ranks 85th.

The deterioration of Ukraine's innovation standing in global rankings can be attributed to several negative factors, including a decrease in foreign direct investment inflows, stagnation in infrastructure development, and a decline in key indicators related to institutions, human capital, and scientific research. Additionally, trends such as a reduction in higher education coverage and a decrease in the duration of schooling, as noted by observers, are atypical for Ukraine.

Despite these challenges, including the ongoing war, significant losses, and widespread destruction, Ukraine continues to make strides in the innovation sector. The country has made notable progress in expanding and refining the legislative framework for the digital transformation of its national economy. Notably, Ukraine has become the first country to digitize a majority of official and government documents. The adoption of digital tools and the corresponding infrastructure has proven instrumental in combating corruption and enhancing the efficiency of resource utilization.

However, the scale of recovery efforts, coupled with insufficient financial resources, inefficient allocation of available funds, and a high level of corruption, presents the primary challenges to Ukraine's post-war recovery. To effectively address these recovery challenges, digital solutions must focus on the following areas [19]:

1. Building an integrated national digital platform that will enable the tracking of financial flows, from "donor to construction" and from the state budget to local budgets, with detailed breakdowns for any projects. This platform should operate on the principle of "everyone sees everything."
2. The integration of data from various digital registries and platforms—both existing (such as RPM, Prozorro, Prozorro.Sale, Diia, spending.gov.ua, etc.) and newly created—is essential. There must be a clear connection between the recovery programs and the allocated funds at all levels of government, from national to local, with projects identified by a unique code.
3. The implementation of a modular approach to the deployment of digital systems will facilitate the integration of various recovery initiatives and promote transparent public governance through a unified digital platform. For example, this approach could include modules for prioritizing applications related to business recovery, social infrastructure projects in communities, non-financial assistance, humanitarian aid, and more.

The introduction of digital solutions, automation, and robotics is reshaping the skills and competencies required for employment and success in the labor market. Companies now seek professionals who not only comprehend these processes and technologies but also possess the ability to apply them effectively to address complex challenges and enhance operational efficiency. Among the key technological skills, the following are particularly noteworthy: computer skills; the ability to use and manage artificial intelligence; organization, analysis, and categorization of large datasets; programming using modern programming languages such as TypeScript, Python, JavaScript, C#, Java, and PHP; knowledge of audiovisual editing and design; the ability to develop applications, websites, and social media accounts, applying user experience (UX) and user interface (UI) design principles; skills in digital marketing tools. Among the most in-demand "soft" skills today are: analytical thinking; resilience, flexibility, and adaptability; leadership and social influence; creative thinking; motivation and self-awareness.

It is evident that such skills are predominantly possessed by the younger and middle-aged generations, with companies competing for their talent. However, as previously mentioned, the proportion of young people in the workforce is rapidly shrinking. It is also important to note that young individuals are highly selective when choosing a workplace, placing significant emphasis on factors such as working conditions, wages, various forms of motivation and incentives, as well as opportunities for career advancement. According to the "Rating Lab" survey, a significant majority of young people in Ukraine (73%) prefer self-employment and entrepreneurship. The survey organizers attribute this trend to the prevailing mentality within Ukrainian society, where self-employment has long been a common aspiration. Furthermore, the current socio-economic environment necessitates a certain degree of mobility, which may encourage young people to pursue remote work. For those Ukrainians who still choose to work within a company, the most important factor is a regular and stable income. This preference underscores the instability of modern life, insufficient financial resources to pursue self-employment, and uncertainty regarding one's own abilities. Notably, 38% of respondents indicated the need for fixed working hours, while 36% cited a lack of additional knowledge, skills, and competencies as barriers to self-employment. Among the most frequently mentioned reasons for hesitancy to engage in self-employment, young people often cited the absence of real support from more experienced colleagues. In terms of preferred sectors, the majority of Ukrainian youth aspire to pursue careers in trade, IT, and online commerce. Other sectors, such as culture, design, and tourism, are also highly sought after. Manufacturing ranks fifth, with only 7% expressing interest in this field. Notably, the youngest respondents tend to lean towards IT and online commerce, while older respondents are more inclined towards trade, design, and manufacturing. Design appears to be the most attractive business opportunity for unemployed youth.

Despite the availability of vacancies (with 221,000 positions listed as of February 5, 2025) and the growing demand for workers in the labor market, unemployment remains a significant issue in Ukraine. According to the State Employment Service of Ukraine, the number of registered unemployed individuals was 186,500 in December 2022, and this figure decreased to 96,100 by December 2023. However, as of February 5, 2025, 123,000 Ukrainians were still seeking employment, with 100,000 officially registered as unemployed. According to the National Bank of Ukraine's (NBU) forecast, the unemployment rate in Ukraine is expected to be 11.6% in 2025 and

10.6% in 2026. The unemployment rate in the economy will gradually decrease due to the growing demand for labor, but it is expected to remain higher than pre-war levels, primarily due to the limited supply of qualified labor, which will be impacted by the consequences of the war.

There is also unemployment among young people. While the level is noticeably lower compared to the average European indicator, it still exists. This is largely due to the fact that Ukrainian students often begin their employment immediately after completing secondary education, and by the third or fourth year of university, they are already working in various jobs, not necessarily related to their field of study. This often compromises the quality of education, but at the same time, it allows young people to gain practical experience. The State Employment Service of Ukraine actively supports initiatives aimed at improving youth employment and integrating them into the labor market. One such initiative involves providing compensation to employers who hire young specialists under 25 years old with less than one year of work experience, or under 35 years old for their first job. The compensation covers 50% of the minimum wage and is paid every odd month for six months. This approach significantly reduces wage expenses at the start of employment and serves as a strong incentive for small and medium-sized businesses to expand their workforce and create new job opportunities, which positively impacts the economy as a whole. It also provides a powerful incentive for young people to be officially employed, offering them the opportunity to secure stable employment and gain experience.

The current labor market situation indicates that not only young people need to acquire knowledge, skills, and abilities related to the use of modern technologies in various fields of activity. People of pre-retirement and retirement age must also possess technological skills, be flexible, and be mobile in order to remain in their professions and continue working. Today, companies are often forced to lower or even abandon age-related restrictions for employees and hire more retirees. In 2023, the share of employees aged 51–60 years (pre-retirement age) was 18.3%, while retirees (over retirement age) accounted for 10.3% of the total employed population. Since 2018, the number of working pensioners in Ukraine has been steadily increasing. As of early 2024, their number stood at approximately 2.8 million, which is only slightly fewer than before the 2011 reform (Fig. 2).

The vast majority of working pensioners (25–26% in recent years) are retirees by age or length of service, and they are the ones contributing to the overall increase in the number of working pensioners. The percentage of working pensioners in relation to the total labor force has consistently been between 13 and 15%, and in relation to the total employed population, it has been 14–17%. According to the national sample survey of the socio-economic condition of households, conducted with the support of UNICEF (United Nations Children's Fund) in Ukraine, in 2023, the percentage of working pensioners in the labor force increased to 19%, and among the employed population, it rose to 21%. This means that currently, about one in five employed individuals is a pensioner.

The leading role in forming new competencies and digital skills is played by education and training. Information and communication technologies enable more personalized learning, compensate for disabilities, and provide new opportunities for

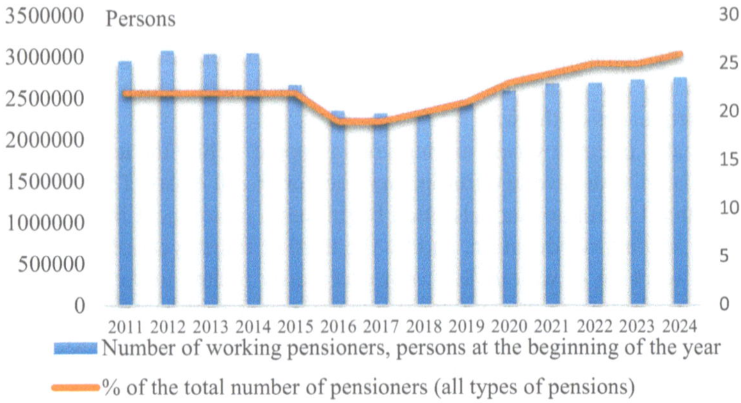

Fig. 2 Number of working pensioners in Ukraine from 2011 to 2024. *Source:* State Statistics Service of Ukraine

accessing information and services. The digital competence of specialists includes the following components: information literacy and media literacy, communication and collaboration, creation of digital content (including programming), security (including protection of personal data in the digital environment and cybersecurity), lifelong learning, as well as problem-solving. According to the Ministry of Digital Transformation of Ukraine's "Digital Literacy Research in Ukraine," as of 2023, digital skills are possessed by:

- 93% of the adult population of Ukraine aged 18–70 years possess digital skills, marking an increase of 8% since 2019.
- 95% of teenagers aged 10–17 years possess digital skills.
- 99% of people with hearing impairments aged 18–59 years have digital skills, showing an increase of 15% over the past four years.
- 38.0% of the adult population have acquired advanced digital skills, which represents an increase of 12.5% since 2019.

The pandemic and the war, which led to the replacement of traditional learning formats with remote or online education, have resulted in a decline in the level and quality of education. To overcome these issues, peace in the country, combined efforts from the state, education, business, and, of course, time are needed. Ukraine invests a relatively large share of its gross domestic product in education, specifically around 7% in 2024. This is more than most European Union countries, where the average is approximately 4.7% of GDP, or about 10% of total government spending. Nevertheless, as noted in the Strategy for the Digital Development of Innovative Activities in Ukraine, investments in the country do not always translate into quality education, as confirmed by the international PISA study on education quality. At the same time, Ukraine is among the top ten countries where the founders of startups with a capitalization of over $1 billion were born, which not only indicates a high level of talent but also a significant level of human capital outflow [18].

In 2025, education will undergo significant transformation under the influence of technology. The primary goal of education will be to prepare adaptable students who are ready to learn throughout their lives and continuously develop professionally. The growing demand for personalized learning solutions and the integration of modern pedagogical methods using artificial intelligence and data analytics are the main factors driving the development of the educational services market. It is expected that the global market for educational technologies, valued at $163.49 billion in 2024, will grow by approximately 13.3% from 2025 to 2030.

Improving employee qualifications and retraining has also become one of the main directions of employers' strategies. More companies are ready to invest in employee training, finance retraining programs, and create support initiatives. Salary growth, flexible working conditions, including remote work, and care for employees' well-being are becoming key methods for attracting and retaining talent. The State Employment Service of Ukraine offers unemployed citizens opportunities for professional training to enhance their competitiveness in the labor market (Table 1). The data from Table 1 indicate that from January to November 2024, 62% of 347,167 unemployed individuals were employed, including 62% youth, 55% women, 52% internally displaced persons, 28% people with disabilities, and 23% participants in combat operations.

The State Employment Center of Ukraine [https://niss.gov.ua/doslidzhennya/sotsialna-polityka/osvitno-kvalifikatsiyni-dysproportsiyi-rynku-pratsi-ukrayiny] offers various professional training programs to the unemployed. For example, individuals who have registered as unemployed can receive a certificate that entitles them to professional training at educational institutions licensed to provide such services, or through individual training at the workplace (in production) or an internship with the employer. The certificate covers the cost of training up to 10 living wages for working-age persons, which, as of now, amounts to 30,280 UAH. Expenses for

Table 1 Number of unemployed individuals who received services from the state employment service in January–November 2024

Categories of the unemployed	Unemployed, persons	%		
		Employed	Received training vouchers	Completed vocational training
Total	347, 167	62	6.2	10.7
Women	266, 796	55	5.9	11.5
Youth	91, 268	62	2	11.4
Persons with disabilities	27, 218	28	4.4	8.1
Internally displaced persons	67, 580	52	6.8	8.4
Participants in combat operations	11, 219	23	5.8	5.1

Source: [20]

accommodation and travel to the educational institution during the training period are also reimbursed. From January to November 2024, 37,111 people, or 10.7% of the total number of unemployed, underwent professional training. Specifically, 8.1% of people with disabilities and 8.4% of internally displaced persons participated in professional training.

One of the proposals from the State Employment Service of Ukraine is a training voucher, which allows certain categories of citizens (such as individuals aged 45+ , veterans, internally displaced persons, people with disabilities, etc.) to study one of 155 professions and specialties, the list of which is approved by the Ministry of Economy of Ukraine. The value of the voucher cannot exceed 10 times the living wage for working-age persons, which as of now is 30,280 UAH. If the cost of training exceeds this amount, the individual must pay the difference themselves. From January to November 2024, 21,473 unemployed people (or 6.2%) received training vouchers. Specifically, among unemployed youth, 2% received vouchers, 6.8% of internally displaced persons, and 4.4% of people with disabilities.

Participants in combat operations (PCO) and individuals with disabilities caused by the war are offered free training at vocational education centers of the State Employment Service. In eight modern vocational training centers, participants can acquire one of 97 trades or undergo professional development through over 400 educational programs. According to the data, 570 participants in combat operations (or 5.1%) took advantage of this offer and underwent free professional training, while 651 individuals (or 5.8%) received training vouchers.

In November 2024, the Ukrainian government launched a new project allowing Ukrainian women to acquire over 30 professions in fields where they were previously underrepresented. Among the professions that women will be able to pursue are: tram driver, trolleybus driver, forklift operator, CNC machine operator, milling machine operator, renewable energy systems installation and maintenance specialist, and more. The Employment Service covers the cost of education, which cannot exceed 10 times the minimum subsistence level for working-age individuals (currently 30,280 UAH). If the cost of training exceeds this amount, the difference can be paid by the employer or the participant in the project. According to the data in Table 1, 15,794 women (5.9% of the total) received training vouchers, and 30,669 women (11.5% of the total) underwent professional training.

In the context of the ongoing full-scale hybrid war in Ukraine, the development of cybersecurity and information protection skills among the population has become increasingly critical. Over the past eight years, Ukraine's cybersecurity market has expanded fourfold, reaching a value of $138 million in 2024. Projections indicate that the market will continue to grow, with an expected 50% increase over the next five years, bringing the total value to approximately $209 million.

Ukraine currently accounts for less than 1% of the global cybersecurity market share. Despite this, the country is considered a trendsetter within the cybersecurity sector, particularly excelling in research and development (R&D) focused on the innovation of cybersecurity technologies. While the global cybersecurity market is predominantly driven by the security services segment, which is projected to comprise 53% of the market in 2024, Ukraine's cybersecurity market is primarily oriented towards cybersecurity solutions, which constitute 57% of the domestic market.

The ongoing conflict has led to an escalation in both the frequency and complexity of cyberattacks targeting Ukraine. Government and business entities are increasingly subjected to various forms of cyberattacks, including ransomware, AI-driven cyber operations, information theft, and vulnerabilities in peripheral devices. These cyberattacks, such as ransomware and data breaches, disrupt organizational operations, induce downtime, and often incur substantial recovery costs. Beyond the immediate financial implications, affected companies may also face diminished revenue, reputational damage, and a loss of customer trust.

In 2024, a record number of data breaches were documented, affecting over 353 million individuals, leading to widespread identity theft and fraud. In numerous instances of cyber incidents, organizations are compelled to allocate additional resources for breach notifications, IT investigations, and recovery efforts, while also facing regulatory penalties. Moreover, businesses experience financial losses due to reduced productivity, operational disruptions, and declines in stock prices.

Phishing and social engineering have emerged as some of the most significant threats in cybersecurity. These manipulative techniques exploit human psychology to deceive individuals into divulging sensitive information or engaging in actions that compromise security. Such attacks frequently involve fraudulent emails, counterfeit websites, or the impersonation of trusted entities, all aimed at exploiting vulnerabilities in human behavior.

One of the significant cybersecurity threats in Ukraine is malware, including computer viruses, spyware, and other malicious scripts designed to infiltrate computers, mobile devices, and entire networks with the intent to cause harm. Malware infections can result in data theft, financial fraud, and severe malfunctions of affected devices. Furthermore, the compromise of user accounts, including the theft of usernames and passwords, exposes sensitive information stored on computers and other systems. Cybercriminals continuously evolve their tactics to breach computer systems, underscoring the importance of robust cybersecurity measures. Protection against these threats is achievable through a comprehensive understanding of the risks and the implementation of appropriate skills and strategies in cybersecurity and information protection [21, 22].

Cybercriminals continually seek victims for their illicit activities, with older individuals, including employees, often being primary targets. These individuals are more susceptible to exploitation due to factors such as naivety, excessive trust, and the increasingly sophisticated nature of cyberattacks. Criminals typically engage in meticulous research of their victims, utilizing information available on social media platforms, such as LinkedIn, to tailor their strategies. Older adults are generally more inclined to trust strangers, particularly when presented with emotionally charged messages, such as requests for assistance concerning a family member. Fraudsters frequently exploit psychological manipulation, creating a sense of urgency that coerces the victim into making hasty decisions. Furthermore, older individuals tend to accumulate savings over their lifetimes, rendering them an attractive target for cybercriminals.

To enhance digital literacy among the older population and promote the adoption of fundamental cyber hygiene practices, educational initiatives, courses, and programs are gaining traction in Ukraine. Numerous major cities have established "Third Age Universities" catering to individuals aged 60 and above. For instance, the "PROTON" Third Age University, operated by the Let's Help charity foundation, is one of the largest institutions of its kind, providing inclusive educational opportunities for elderly individuals. The university offers weekly lessons focusing on payment security and cyber hygiene, equipping students with essential skills to safely navigate the digital landscape and safeguard their personal information from cyber threats.

In Ukraine, 55 state universities offer educational programs in cybersecurity, specializing in the field of "Cybersecurity and Information Protection" (specialization 125), while at least 16 private IT schools provide specialized courses in the domain. However, experts highlight a notable quality disparity between state and private educational institutions. The ratio of state to private programs, approximately 3.5:1, underscores that the quantity of programs does not necessarily correlate with their quality. Additionally, the lack of practical cybersecurity skills, which many universities are unable to impart, results in a deficiency of hands-on experience and specialized knowledge crucial for professional practice. As a consequence, numerous graduates are relegated to general IT positions, such as front-end or back-end development. In response to these gaps, private enterprises are stepping in to offer a broad spectrum of practical programs and specialized courses, delivering cybersecurity training that aligns more closely with industry standards and professional requirements.

Therefore, investments in emerging technologies within the fields of cybersecurity, information protection, and employee training will enable organizations to better mitigate cyber risks, safeguarding their operations, data, and reputation against future cyber threats.

5 Conclusion

The post-war recovery of Ukraine involves the restoration and reconstruction of the national economy, which was severely impacted by the conflict. This rebuilding process requires substantial resources—financial, material, and human—all of which are, unfortunately, limited. The exacerbation of the demographic crisis, along with the growing external migration of the Ukrainian population, particularly among the youth, has heightened the need for the digital transformation of the national economy. This transformation necessitates the adoption of high-tech production methods, innovation-driven approaches, and the modernization of all economic sectors through the integration of information and communication technologies (ICT), automation, and the optimization of routine processes in areas where human involvement can be reduced or eliminated. Technologies such as artificial intelligence (AI) offer significant potential for replacing human labor, thereby enhancing resource management efficiency, streamlining production processes, and supporting data-driven decision-making through the analysis of large datasets.

Modern technologies should optimize processes in many industries; however, in our opinion, the drivers of national economic growth should be the information and communication technology sector, the agricultural sector, the manufacturing industry, and the military-industrial complex. Their further development is capable of increasing the export of goods and services, ensuring the country's food and military security.

Threats to the digital transformation of the country's economy may arise from the intensification of the aging process of the Ukrainian population, caused by significant losses of young and able-bodied people due to the full-scale military aggression, declining birth rates, and large-scale youth migration abroad. The increasing proportion of the pre-retirement and retirement age categories in the population structure necessitates employers to raise the age limit for employees and job seekers. The state and businesses must act proactively, or else they risk losing the potential value that aging workers can bring, which will lead to decreased labor productivity and a negative impact on economic growth. Workers of all categories, regardless of age, must possess the skills and abilities to work with modern digital and information technologies, including artificial intelligence. Cybersecurity and information protection skills will also become mandatory for personnel. Therefore, employers must continue to implement various training programs for employees to master the technologies required for their jobs. Lifelong learning and the continuous updating of competencies and skills should become the main driving force of society.

To meet the needs of the domestic labor market, it is necessary to actively involve people with disabilities, adapt veterans, and stimulate the inclusion of internally displaced persons. According to Deputy Minister of Social Policy Daria Marchak at the HR Wisdom Summit, a summit of HR professionals from leading Ukrainian companies, "There are 700,000 internally displaced persons (IDPs) of working age who are not employed. The number of IDPs who could be retrained, offered better jobs, and used to solve business problems has not yet been determined… In addition, there are about one million people with disabilities who are not working but could be employed. Considering that we have one million potential veterans, a significant portion of whom will be returning with serious injuries, we are starting to think that this will be the new normal for society. It is also time to open Ukraine's borders to the global labor market and more actively engage foreign professionals who are willing to generate added value within the country. Without openness to international labor markets, Ukraine has limited prospects for rapid economic recovery and sustainable growth. While the risks associated with an uncontrolled influx of migrants are considerable, Ukraine has historically derived strength from its openness to diverse cultures, religions, ideas, and people.

While placing high hopes on the digital transformation of the national economy and the technological modernization of all its sectors, it is essential to consider the additional challenges and threats that may arise in the process. Organizations and professionals must strengthen their cybersecurity efforts, as threats to cybersecurity and the number of data security incidents continue to grow. Artificial intelligence is already being utilized by cybercriminals, and the number of cyberattacks involving its use increases each year. The most dangerous threat lies in the context of social engineering, where generative AI can assist malicious actors in bypassing authentication systems, hacking corporate email passwords, spreading disinformation, and carrying out various forms of fraud. Therefore, businesses must actively implement comprehensive cybersecurity measures. Key steps include the deployment of multi-layered security systems comprising antivirus software, firewalls, and threat detection systems; regular software updates and the use of reliable authentication mechanisms such as multi-factor authentication (MFA); data backups for rapid recovery; and employee training in cybersecurity fundamentals to minimize the risk of human error-related attacks.

References

1. Reinert ES (2007) How rich countries got rich... and why poor countries stay poor. Constable. London. https://www.academia.edu/108092576/How_Rich_Countries_Got_Rich_and_Why_Poor_Countries_Stay_Poor
2. Bregman R (2022) Why Europe needs Ukraine. The Atlantic. https://www.theatlantic.com/ideas/archive/2022/04/why-europe-needs-ukraine/629476/
3. Valero J (2022) Europe expects to pay bulk of Ukraine's reconstruction costs. Bloomberg. https://www.bloomberg.com/nP,ews/articles/2022-04-19/europe-expects-to-pay-bulk-of-ukraine-s-reconstruction-costs
4. Exploring Additional Funding Paths for Revival (2024) Ukraine's Essay book a collection of inspirational concepts (2024) Ilona Sologoub, VoxUkraine scientific editor The Institute of Analytics and Advocacy. https://iaa.org.ua/en/portfolio/exploring-additional-paths-for-ukraines-revival-a-collection-of-inspirational-concepts/
5. Gao W, Chen Y, Xu S, Lyulyov O, Pimonenko T (2023) The role of population aging in high-quality economic development: mediating role of technological innovation. Sage J. https://doi.org/10.1177/21582440231202385
6. Wang P, Zhao T (2024) Impact of population aging on corporate digital transformation strategies: an empirical investigation based on population census data from various provinces in China. Financ Res Lett 67. https://doi.org/10.1016/j.frl.2024.105838
7. Xie X, Xu X (2024) The impact of population aging on SME digital transformation: evidence from China. PLOS ONE 19. https://doi.org/10.1371/journal.pone.0300660
8. Jing Z, Li Z (2023) The effect of population ageing on technological innovation in emerging market countries. Technol Forecast Soc Chang 200. https://doi.org/10.1016/j.techfore.2023.123096
9. Belolipetskaya A, Golovina T, Polyanin A, Vertakova Y (2020) Transformation of the personnel competency model in the context of the transition to the digital economy. In: Topical problems of green architecture, Civil and environmental engineering 2019 (TPACEE 2019). E3S web of conferences, vol 164, pp 09005. https://doi.org/10.1051/e3sconf/202016409005
10. Maer Matei MM, Mocanu C, Zamfir AM, Nastasa A (2023) Implications of digitalization on skill needs in a sustainable economy. Amfiteatru Economic, 25(Special №17):1115–1130. https://doi.org/10.24818/EA/2023/S17/1115
11. Grundke R, Marcolin L, Squicciarini M (2018) Which skills for the digital era? Returns to skills analysis. OECD Science, Technology and Industry Working Papers, 2018/09, OECD Publishing, Paris. https://doi.org/10.1787/9a9479b5-en

12. Ostmeier E, Strobel M (2021) Building skills in the context of digital transformation: how industry digital maturity drives proactive skill development. J Bus Res 139:718–730. https://doi.org/10.1016/j.jbusres.2021.09.020
13. Gajapathy V, Patil RM (2021) Demographics and cyber risk awareness—is there an association? J Appl Manag-Jidnyasa 13(1):41–50. file:///C:/Users/luchy/Downloads/admin,+Jidnyasa,+Vol+13,+Issue+1,+2021–46–55.pdf
14. Vazquez JCB (2019) The influence of demographic factors on the cybersecurity awareness level of individuals in an academic environment. Graduate program in information systems and technology. York University Toronto, ontARIO http://hdl.handle.net/10315/36700
15. Stepanenko NV, Kurilets OO (2023) Essence, concept and content of post-war recovery: theoretical and legal aspect. Leg Bull 34–41. https://doi.org/10.31732/2708-339X-2023-10-34-41
16. Osadcha Y (2025) In Ukraine, there is a record low in fertility: one less child per woman. Lb.ua. https://lb.ua/health/2025/03/04/663811_ukraini_antirekord_z.html
17. Pishchulina O (2023) Assessment of the demographic factor for economic growth and post-war recovery. Analytical report. May 2023. Razumkov Center. chrome-extension://efaidnbmnnnibpcajpcglclefindmkaj/https://razumkov.org.ua/images/2023/05/22/2023-MATRA-I-KVARTAL-7.pdf
18. Strategy for the digital development of innovative activity in Ukraine for the period until 2030 approved by the order of the Cabinet of Ministers of Ukraine dated December 31, 2024 №1351-r. https://zakon.rada.gov.ua/laws/show/1351-2024-%D1%80#Text
19. Shapoval N, Fedoseenko M, Grybanovsky O, Tereshchenko O (2022) Post-war reconstruction of Ukraine. New markets and digital solutions. Kyiv school of economics. chrome-extension://efaidnbmnnnibpcajpcglclefindmkaj/https://kse.ua/wp-content/uploads/2022/09/Digital-instruments-in-Ukrainian-recovery.pdf
20. Educational and qualification disparities in the Ukrainian labor market: current realities and prospects (2025) NISS. https://niss.gov.ua/doslidzhennya/sotsialna-polityka/osvitno-kvalifikatsiyni-dysproportsiyi-rynku-pratsi-ukrayiny
21. Kovtun V, Grochla K, Kharchenko V, Haq MA, Semenov A (2023) Stochastic forecasting of variable small data as a basis for analyzing an early stage of a cyber epidemic. Sci Rep 13(1). https://doi.org/10.1038/s41598-023-49007-2
22. Sajid F, Hassan MA, Khan AA, Rizwan M, Kryvinska N, Vincent K, Khan IU (2022) Secure and efficient data storage operations by using intelligent classification technique and RSA algorithm in IoT-based cloud computing. Sci Program 2022:1–10. https://doi.org/10.1155/2022/2195646

Microprocessor Device for Predictive Maintenance Systems for Electrical Machines

Volodymyr Kucheruk, Pavlo Kulakov, Roman Lishchuk, Tetiana Neskorodieva, Andrii Semenov, Serhii Kontseba, Wiktoria Mankovska, and Svitlana Skurtol

Abstract The article presents the microprocessor device for predictive maintenance systems for electrical machines, the principle of operation of which is based on measuring the quality factor of windings. It is proposed to use the quality factor as an informative parameter for diagnosing the winding condition, in particular, detecting short-circuited coils. Complete and simplified equivalent winding diagrams were constructed, and analytical expressions for the quality factor were obtained, taking into account losses in the magnetic core and insulation. The method for determining the quality factor based on the phase shift between voltages has been developed, which is implemented in the form of a structural diagram of a sensor with a microcontroller. Simulations were conducted that confirmed the high sensitivity of the quality factor to the number of short-circuited coils, unlike inductance. The

V. Kucheruk · P. Kulakov (✉) · R. Lishchuk · T. Neskorodieva · S. Kontseba · W. Mankovska · S. Skurtol
Uman National University, Uman, Ukraine
e-mail: p.kulakov@udau.edu.ua

V. Kucheruk
e-mail: kucheruk@udau.edu.ua

R. Lishchuk
e-mail: roman.lishchuk@udau.edu.ua

T. Neskorodieva
e-mail: t.neskorodeva@udau.edu.ua

S. Kontseba
e-mail: kontseba@udau.edu.ua

W. Mankovska
e-mail: mankovska@udau.edu.ua

S. Skurtol
e-mail: skurtol@udau.edu.ua

A. Semenov
Vinnytsia National Technical University, Vinnytsia, Ukraine
e-mail: semenov.a.o@vntu.edu.ua

A. Semenov et al. (eds.), *Data-Centric Business and Applications*, Lecture Notes on Data Engineering and Communications Technologies 272,
https://doi.org/10.1007/978-3-032-06608-4_12

proposed approach allows for accurate diagnostics of windings as part of predictive maintenance systems, meeting the requirements of digitalization of industrial processes.

Keywords Winding quality factor · Predictive maintenance · Sensor · Microprocessor device · Electric machine · Technical condition of windings

1 Introduction

Windings of electrical machines are critically important elements that directly affect their reliability, efficiency, and uninterrupted operation. Damage to winding insulation is one of the most common causes of motor failures, which can lead to equipment downtime, production losses, and high costs for repair or replacement of units.

Modern industrial systems operate in difficult conditions—at elevated temperatures, humidity, vibrations and exposure to aggressive environments [1]. Under such conditions, the winding insulation gradually ages, loses its dielectric properties, and becomes vulnerable to breakdown. By implementing effective diagnostic methods—such as frequency impedance spectroscopy, thermal imaging, time or phase correlation methods—it is possible to detect potential defects at an early stage, reduce the risk of accidents and plan maintenance in advance.

Diagnostics becomes particularly relevant in the context of the transition to predictive maintenance systems [2, 3], where data on the condition of windings are integrated into the overall digital production infrastructure. Thus, winding diagnostics is not only technically necessary, but also economically feasible, as it helps increase the life of machines, reduce operating costs, and ensure energy efficiency.

Predictive Maintenance (PdM) is a modern approach to equipment maintenance that involves using sensor data, analytics, and artificial intelligence algorithms to predict when a system failure will occur. Its relevance is constantly growing due to the following key factors. The first factor is to increase reliability and continuity of production. In industry, equipment downtime leads to significant financial losses [4]. Predictive Maintenance allows to detect signs of potential failures before they actually occur, which allows to plan repairs without stopping the technological process. The second factor is a significant reduction in maintenance costs. Unlike preventive maintenance, predictive maintenance is performed only when it is really needed. This reduces the cost of unnecessary repairs, reduces the cost of spare parts and reduces the burden on service departments. The third factor is the rapid development of Industry 4.0 and the Internet of Things (IoT). The widespread adoption of IoT and digital technologies allows for the collection of large amounts of data from sensors in real time. It is this data that is the basis for the machine learning and analytics used in PdM. This makes PdM a key element of "smart manufacturing". The fourth factor is increasing security. Diagnostics and prediction of wear of critical equipment elements helps prevent accidents, explosions, overheating, etc. This is important in the energy, aviation, and oil and gas industries, where a failure can threaten people's lives. The fifth factor is the development of artificial intelligence. The use of neural networks, machine learning, and deep learning allows PdM systems to recognize complex signs of equipment degradation, even in non-standard situations. The above takes maintenance to a new level of precision and flexibility.

Predictive maintenance is not just a trend, but a strategically important component of modern production process management, which helps increase efficiency, reduce costs, and strengthen the competitiveness of the enterprise. Its relevance will only grow with the further development of digital technologies and automation.

The quality of electrical machines is largely determined by the quality of the windings [5, 6]. Based on the results of statistical studies of the causes of electrical machines failure [7, 8], most failures in the functioning of new windings occur due to defects that were not detected in a timely manner. Failures in electrical machine components are distributed as follows: 50–95% of failures occur due to damage to the stator winding (interturn insulation breakdown); bearings 10–40%; other components and parts 5–10%.

2 Research Results

Windings of electrical machines are manufactured on automated equipment; the high speed of processing winding wires and the frequent subsequent deformation of the winding lead to a high load on the insulation. The most typical type of winding defect is the presence of short-circuited coils. Thus, determining the degree of influence of the number of short-circuited coils on the winding parameters is of practical value.

To determine the winding parameters, the equivalent circuit of the winding of an electric machine, as an inductive element, must be depicted in the form of a complete equivalent circuit [9, 10], shown in Fig. 1. It was created based on the following physical considerations. The winding's own capacitance, equivalent to the distributed inter-turn capacitance of the winding, is connected to the input terminals. The winding has several types of losses, which are taken into account by the corresponding equivalent resistances [11, 12].

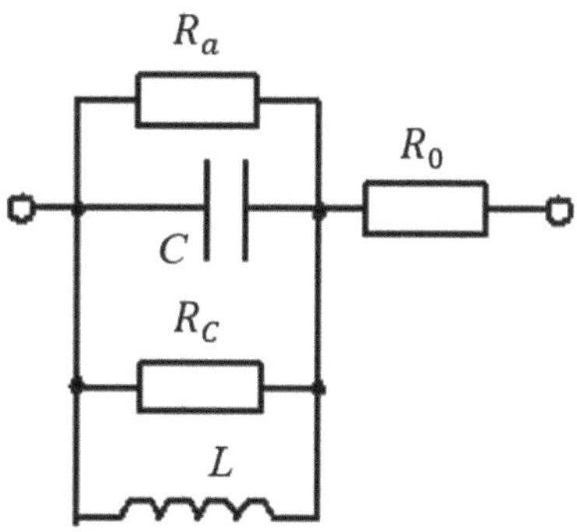

Fig. 1 Complete equivalent circuit of the winding of an electric machine

Winding losses are proportional to the square of the current. Therefore, they can be taken into account in the equivalent circuit by connecting the resistance R_0 in series in the general circuit. Losses in the magnetic circuit are determined by induction, which, in turn, is determined by the magnitude of the voltage that creates the magnetic flux coupled to the coils of the winding. Therefore, the resistance R_c, which takes into account losses in the magnetic core, is connected in parallel to the inductance in the equivalent circuit. The insulation losses due to the winding capacitance, $tg\ \delta$, and the voltage applied to the capacitance are represented by the resistance R_a, which is connected in parallel to the capacitance. As a result, a complete equivalent circuit of the winding of an electric machine is formed, which allows us to investigate the value of the quality factor Q of the winding from a quantitative perspective.

Let clarify the concept of quality factor. The quality factor of a winding is the ratio of the reactive energy stored in the winding to the active energy consumed during the oscillation period of the circular frequency ω (respectively, the reactive power to the active or the reactive component of the impedance to the active component) [13, 14].

Based on the equivalent winding diagram (Fig. 1) and the definition of the concept of the quality factor of the winding, we can obtain an expression for the quality factor Q:

$$Q(\omega) = \frac{\omega L}{R_0}\left[\left(1 - \omega^2 CL\right) + \frac{\omega^2 L^2}{\left(1 - \omega^2 CL\right) \cdot r} \cdot \left(\frac{1}{r} + \frac{1}{R_0}\right)\right]^{-1}, \tag{1}$$

where R_0, L, C—respectively, the active resistance, inductance and self-capacitance of the winding; $r = \frac{R_c R_a}{(R_c + R_a)}$, R_c and R_a—resistance losses in the magnetic core and insulation; $R_c = \frac{U^2}{P_c}$; $R_a = \frac{U^2}{P_a}$; U—winding voltage; P_c, P_a—respectively, the power losses in the magnetic core and in the insulation; ω—cyclic frequency.

For normal operation of the windings, it is necessary that the operating frequency of the winding be significantly lower than the resonant frequency, i.e. $\omega << \omega_0 = \frac{1}{\sqrt{LC}}$.

Otherwise, the inductance and quality factor of the winding will be significantly reduced due to the influence of its own compensating capacitance. Therefore, it is necessary to fulfill the condition $\omega^2 CL << 1$. When this condition is met, the winding can be represented using a simplified equivalent circuit (Fig. 2).

Based on the circuit configuration, its equivalent inductance L_{e1} and quality factor Q_1 are determined:

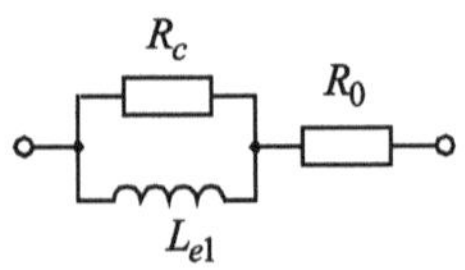

Fig. 2 Simplified equivalent winding circuit

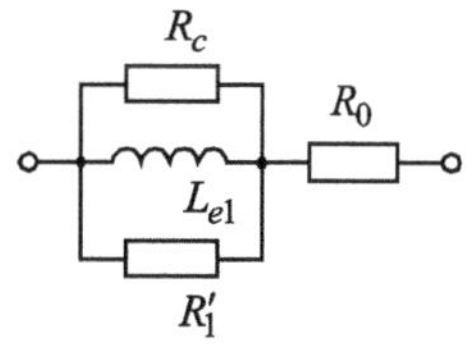

Fig. 3 Equivalent circuit of a winding with short-circuited coils

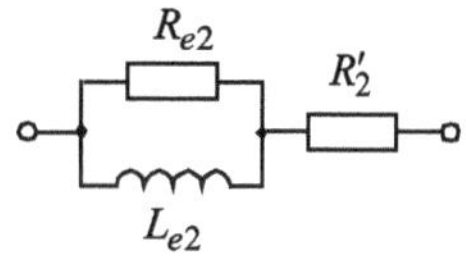

Fig. 4 Simplified equivalent circuit of a winding with short-circuited coils

$$L_{e1} = \frac{LQ_c^2}{\left(1 + Q_c^2\right)}; \tag{2}$$

$$Q_1 = \frac{Q_c}{1 + (Q_k Q_c)^{-1} + \frac{Q_c}{Q_k}}, \tag{3}$$

where $Q_k = \frac{\omega L}{R_0}$—winding quality factor; $Q_c = \frac{R_c}{\omega L}$—magnetic core quality factor.

In the presence of short-circuited coils, the winding is considered as a transformer with a short-circuited secondary winding (Fig. 3), where R_1' is the resistance of the secondary winding reduced to the primary coils.

Representing the parallel resistances R_1' and R_c by one equivalent resistance R_{e1}, we obtain a circuit (Fig. 4), similar in structure to the equivalent circuit shown in Fig. 4.

The parameters of this circuit are expressed as follows

$$R_{e2} = \frac{R_1' R_c}{R_1' + R_c}; \; L_{e2} = \frac{LQ_c^2}{\left(1 + Q_c^2\right)\left(1 + \frac{2Q_k Q_c k}{1+Q_c^2}\right)} = \frac{LQ_c^2}{1 + Q_c^2 + 2Q_k Q_c k}; \tag{4}$$

$$Q_2 = \frac{Q_c}{\left(1 + \frac{1}{Q_k Q_c} + \frac{Q_c}{Q_k\left(1 + kQ_k Q_c\right)}\right)\left(1 + kQ_k Q_c\right)} \times \frac{1}{1 + \frac{kQ_c Q_k\left(1 + kQ_c Q_k + Q_c^2\right)}{1 + kQ_c Q_k + Q_c^2 + Q_c Q_k\left(1 + kQ_c Q_k\right)}}. \tag{5}$$

After simplifying expression (5), we obtain

$$Q_2 = \frac{Q_c}{k^2 Q_c Q_k + k\left[2 + Q_k Q_c + Q_c^2\right] + (Q_k Q_c)^{-1} + \frac{Q_c}{Q_k} + 1}. \tag{6}$$

When $k = 0$, expression (6) is transformed into expression (4). In expressions (4), (6) it is taken into account that if R_0 and R_1' are expressed in terms of the resistance of one coil r_0, then $R_0 = Wr_0$; $R_2' = \frac{r_0 (W - W_k)^2}{W_k}$, where W and W_k are the number of coils of the winding and short-circuited coils, respectively. Taking into account that $W_k << W$, we obtain:

$$R'_2 \cong \frac{r_0 W^2}{W_k} = \frac{R_0}{k}, \quad (7)$$

where $k = \frac{W_k}{W}$—relative number of short-circuited coils. Then

$$R_{e2} = \frac{R_c}{\left(1 + kQ_kQ_c\right)}, \quad (8)$$

because $\frac{R_c}{R} = Q_kQ_c$.

From expression (6) we have found the relative number of short-circuited coils k, it is determined by the expression

$$k = \frac{1}{2} \cdot \frac{Q_c\sqrt{Q_2\left(Q_2Q_k^2 + 2Q_2Q_cQ_k + Q_2Q_c^2 + 4Q_k\right)} - Q_2\left(Q_2Q_k + Q_c^2 + 2\right)}{Q_cQ_kQ_2}. \quad (9)$$

Figure 5 shows the dependence $Q_1 = f(Q_k, Q_c)$.

Figure 6 shows the dependences $Q_2 = f(Q_k, Q_c)$ and $Q_1 - Q_2 = f(Q_k, Q_c)$ for different relative numbers of short-circuited coils k.

Analysis of Figs. 5 and 6 shows a significant influence of the relative number of short-circuited coils k on the quality factor of the winding.

For testing practice, testing the stator winding after installation is of great value, since it is during installation that the winding insulation can be damaged. Let investigate the possibility of using the simplified formula $Q \cong Q_k$ to determine the quality factor of a winding with a magnetic core.

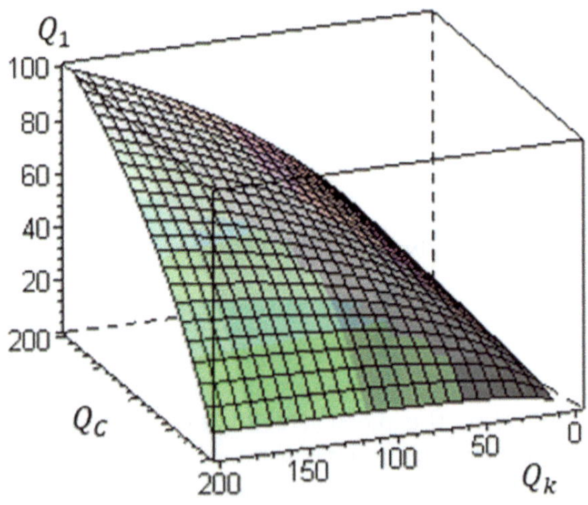

Fig. 5 Three-dimensional dependence $Q_1 = f(Q_k, Q_c)$

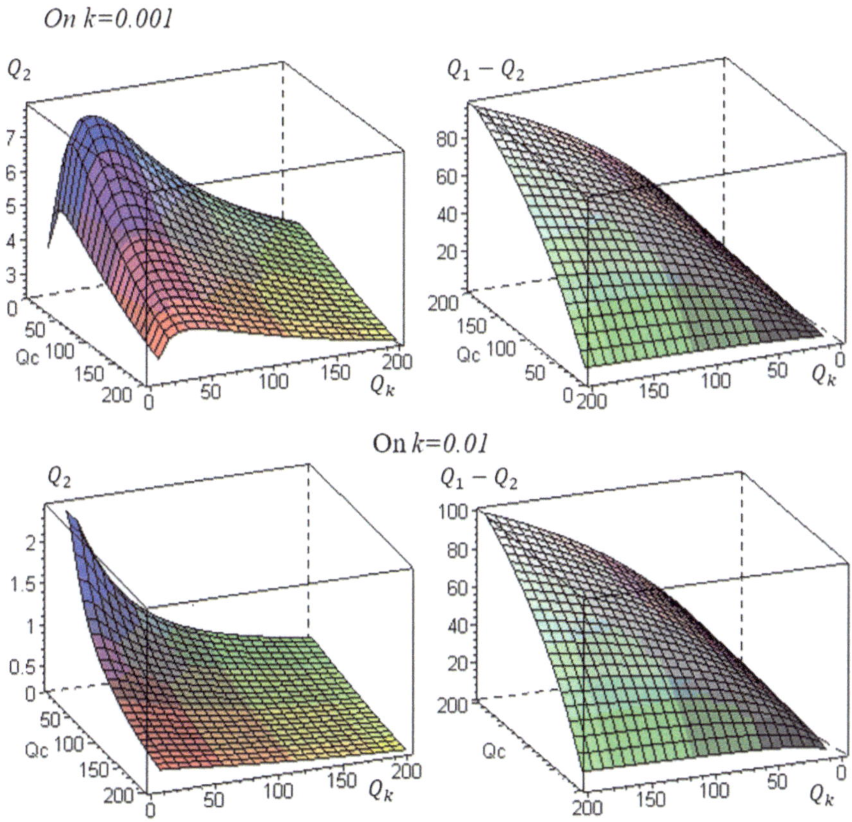

Fig. 6 Three-dimensional dependences for different relative numbers of short-circuited coils k. $Q_2 = f(Q_k, Q_c)$ and $Q_1 - Q_2 = f(Q_k, Q_c)$

Figure 7 shows the calculated dependences of Q_1 and Q_k on the operating cyclic frequency ω ($\omega = 2\pi f_0$). It should be noted that the nature of the change in the function $Q_1(\omega)$ significantly depends on the material of the magnetic core.

Figure 8 shows the results of the calculations of the relative deviation

Fig. 7 Dependencies $Q_k(\omega)$ and $Q_1(\omega)$

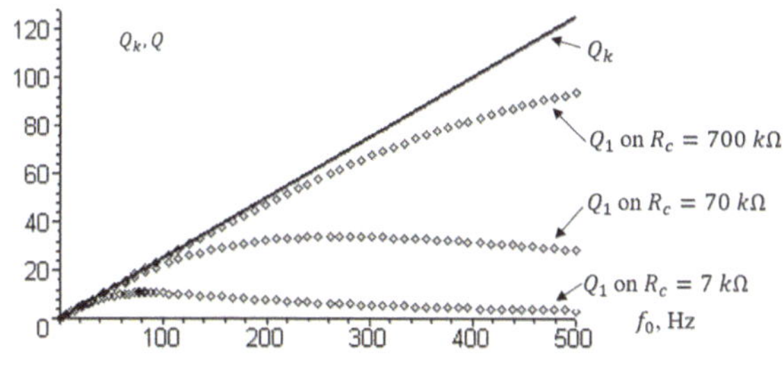

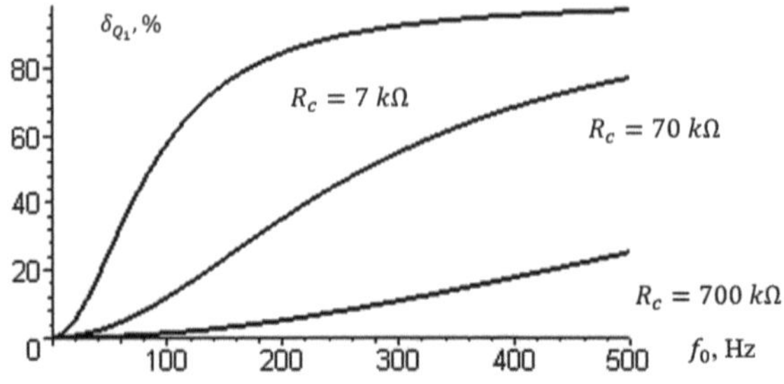

Fig. 8 Relative deviation $\delta_{Q_1}(\omega)$

$$\delta_{Q_1}(\omega) = \frac{|Q_1(\omega) - Q_k(\omega)|}{Q_1(\omega)} \cdot 100\%. \tag{10}$$

The simulation results show that with a low power loss in the magnetic core P_c in the frequency range $f_0 = 0$–50 Hz, it can be assumed that $Q \cong Q_k$. Under such conditions, the influence of the quality factor of the magnetic core Q_c on the resulting quality factor Q can be neglected and the equivalent circuit for replacing the winding with the magnetic core becomes identical to the equivalent circuit for replacing the winding without the magnetic core.

When taking into account the influence of the magnetic core on the resulting quality factor, it is possible to perform an equivalent transformation of the winding circuit with short-circuited coils, representing it in the form of the simplest inductor, shown in Fig. 9. The parameters of the inductor are determined by the expressions

$$R_{e3} = R_2' + \frac{\omega^2 L_{e2}^2 R_{e2}}{\left(R_{e2}^2 + (\omega L_{e2})^2\right)}; \tag{11}$$

$$L_{e3} = \frac{R_{e2}^2 L_{e2}}{\left(R_{e2}^2 + (\omega L_{e2})^2\right)}. \tag{12}$$

Moreover, $\lim\limits_{\omega \to 0} R_{e3}(\omega) = R_2'$, $\lim\limits_{\omega \to 0} L_{e3}(\omega) = L_{e2}$.

To theoretically justify the choice of the quality factor of the winding as an informative parameter when testing it, we present the following facts. By definition, the quality factor is the ratio of reactive power P_P to active power losses P_A. Increasing the number of short-circuited coils leads to the fulfillment of the following relations

$$\frac{dP_P}{dk} < 0; \quad \frac{dP_A}{dk} > 0. \tag{13}$$

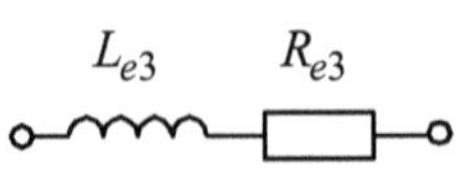

Fig. 9 Equivalent circuit of a winding with short-circuited coils in the form of the simplest inductor

The first inequality arises from the fact that as the number of short-circuited coils increases, the inductance of the winding decreases, and the second inequality arises from the actual increase in losses in the winding. Then, since the quality factor of the winding decreases with increasing number of short-circuited coils, we obtain

$$\frac{dQ_1}{dk} = \frac{d}{dk}\left(\frac{P_P}{P_A}\right) = -\frac{1}{P_A}\left(\left|\frac{dP_P}{dk}\right| + Q_1\frac{dP_A}{dk}\right). \tag{14}$$

Taking into account that $P_A = \frac{P_P}{Q}$, we obtain

$$\frac{\Delta Q}{Q} = -\left(\left|\frac{\Delta P_P}{P_P}\right| + \frac{\Delta P_A}{P_A}\right). \tag{15}$$

Analysis of expression (15) shows that the sensitivity to Q_1 is defined as the sum of the sensitivities to P_P and P_A.

Let's calculate the sensitivities $\frac{\partial L_{e2}(k)}{\partial k}$ and $\frac{\partial Q_2(k)}{\partial k}$.

$$\Phi_L(k) = \frac{\partial L_{e2}}{\partial k} = -2\frac{(1+Q_c)^2 Q_k Q_c}{\left(1+Q_c^2+2Q_kQ_ck\right)^2}, \tag{16}$$

$$\Phi_Q(k) = \frac{\partial Q_2(k)}{\partial k} = \frac{-Q_kQ_c\left(Q_kQ_c+1+Q_c^2\right)\left(2Q_kQ_ck+Q_kQ_c+Q_c^2+2\right)}{\left(Q_k^2Q_c^2k^2+\left(2Q_kQ_c+Q_c^2Q_k^2+Q_c^3Q_k\right)k+Q_cQ_k+1+Q_c^2\right)^2}. \tag{17}$$

The results of the calculations of the relative deviations $\frac{L_{e2}(k)}{L_{e2}(0)}$ and $\frac{Q_2(k)}{Q_2(0)}$ are presented in Fig. 10, and of the relative deviations $\frac{\Phi_L(k)}{\Phi_L(0)}$ and $\frac{\Phi_Q(k)}{\Phi_Q(0)}$ are presented in Fig. 11.

Analysis of the obtained results allows us to state that the change in the inductance of the winding, depending on the number of short-circuited coils, has a small value. This circumstance makes the determination of short-circuited coils by inductance inefficient. The value of the quality factor is much more sensitive to the number of shorted coils than the value of the inductance. The quality factor sensitivity function

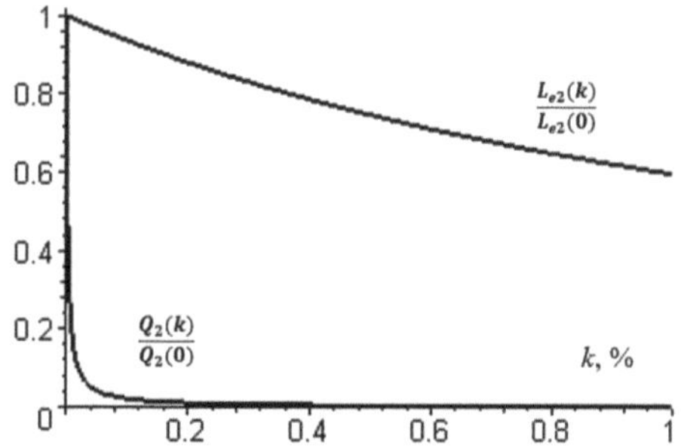

Fig. 10 Relative deviations $\frac{L_{e2}(k)}{L_{e2}(0)}$ and $\frac{Q_2(k)}{Q_2(0)}$

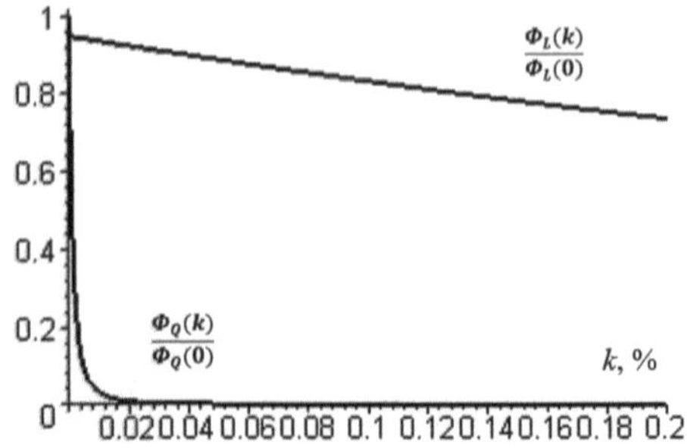

Fig. 11 Relative deviations $\frac{\Phi_L(k)}{\Phi_L(0)}$ and $\frac{\Phi_Q(k)}{\Phi_Q(0)}$

with a small number of short-circuited coils is nonlinear and has a large slope. The high sensitivity of the quality factor to the number of shorted coils allows for the identification of a small number of shorted coils, which is difficult to achieve using other methods.

Let use the phase shift method to measure the quality factor of the windings of electrical machines. Let consider the sensor circuit (Fig. 12), which contains the following elements: R is a separating resistor; C_0 is an exemplary capacitor; Z_x is a complex resistance of the investigated inductor; R_x, L_x is an active resistance and inductance of the investigated inductor, respectively.

Let analyze the proposed scheme using the method of complex amplitudes [15, 16]. The equivalent resistance of a circuit is defined by the expression

$$\dot{Z}_E = R + R_X - jX_{C_0} + jX_{C_X} = (R + R_X) + j\left(X_{L_X} - X_{C_0}\right)$$
$$= \sqrt{(R + R_X)^2 + \left(X_{L_X} - X_{C_0}\right)^2} \cdot \exp\left(j \cdot arctg\frac{X_{L_X} - X_{C_0}}{R + R_X}\right), \quad (18)$$

and the equivalent resistance of the investigated inductor is

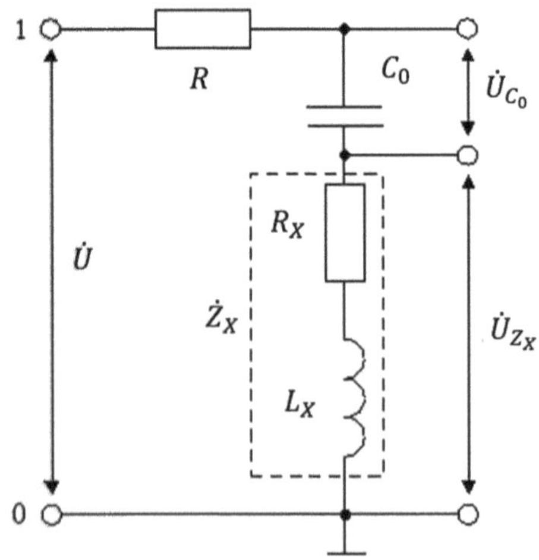

Fig. 12 Quality factor sensor scheme

$$\dot{Z}_X = R + jX_{L_X} = \sqrt{{R_X}^2 + {X_{L_X}}^2} \cdot \exp\left(j \cdot arctg\frac{X_{L_X}}{R_X}\right). \tag{19}$$

Let's find the voltage ratio $\dot{U} = \dot{U}_{Z_X} + \dot{U}_{C_0}$ and $\dot{U}_{Z_X}$

$$\frac{\dot{U}_{Z_X} + \dot{U}_{C_0}}{\dot{U}_{Z_X}} = 1 + \frac{\dot{U}_{C_0}}{\dot{U}_{Z_X}}. \tag{20}$$

The voltage $\dot{U}_{Z_X}$ is equal to

$$\begin{aligned}\dot{U}_{Z_X} = \dot{E}\frac{\dot{Z}_X}{\dot{Z}_E} &= \frac{E_m \exp(j0^o)\sqrt{R_X^2 + X_{L_X}^2}\exp\left(j\, arctg\frac{X_{L_X}}{R_X}\right)}{\sqrt{(R+R_X)^2 + \left(X_{L_X} - X_C\right)^2}\exp\left(j\, arctg\frac{X_{L_X} - X_{C_0}}{R+R_X}\right)} \\ &= \frac{E_m\sqrt{R_X^2 + X_{L_X}^2}}{\sqrt{(R+R_X)^2 + \left(X_{L_X} - X_C\right)^2}}\exp\left(j\left(arctg\frac{X_{L_X}}{R_X} - arctg\frac{X_{L_X} - X_{C_0}}{R+R_X}\right)\right).\end{aligned} \tag{21}$$

The voltage $\dot{U}_{C_0}$ is equal to

$$\begin{aligned}\dot{U}_{C_0} &= \frac{E_m \exp(j0^o) X_{C_0}\exp\left(-\frac{\pi}{2}\right)}{\sqrt{(R+R_X)^2 + \left(X_{L_X} - X_C\right)^2}\exp\left(j\, arctg\frac{X_{L_X} - X_{C_0}}{R+R_X}\right)} \\ &= \frac{E_m X_{C_0}}{\sqrt{(R+R_X)^2 + \left(X_{L_X} + X_C\right)^2}}\exp\left(-j\left(arctg\frac{X_{L_X} - X_{C_0}}{R+R_X} + \frac{\pi}{2}\right)\right).\end{aligned} \tag{22}$$

The phase difference ϕ_X between the voltages $\dot{U}_{C_0}$ and $\dot{U}_{Z_X}$ is defined as

$$\begin{aligned}\phi_X = \phi_1 - \phi_2 &= -arctg\frac{X_{L_X} - X_{C_0}}{R+R_X} - \frac{\pi}{2} - arctg\frac{X_{L_X}}{R_X} + arctg\frac{X_{L_X} - X_{C_0}}{R+R_X} \\ &= -arctg\frac{X_{L_X}}{R_X} - \frac{\pi}{2} = -arctg Q - \frac{\pi}{2}.\end{aligned} \tag{23}$$

From here we obtain

$$Q = ctg(\phi_X). \tag{24}$$

Thus, by measuring the phase shift ϕ_X between the voltages $\dot{U}_{C_0}$ and $\dot{U}_{Z_X}$ [17, 18], we can determine the quality factor of the inductor, and therefore the quality factor Q of the electric machine winding. Figure 13 shows the static characteristic of the quality factor sensor, and Fig. 14 shows the sensitivity function of the static characteristic to changes in the quality factor.

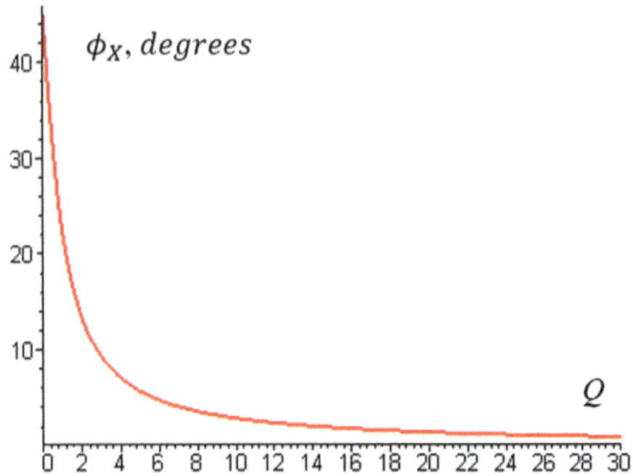

Fig. 13 Static characteristic of the quality factor sensor

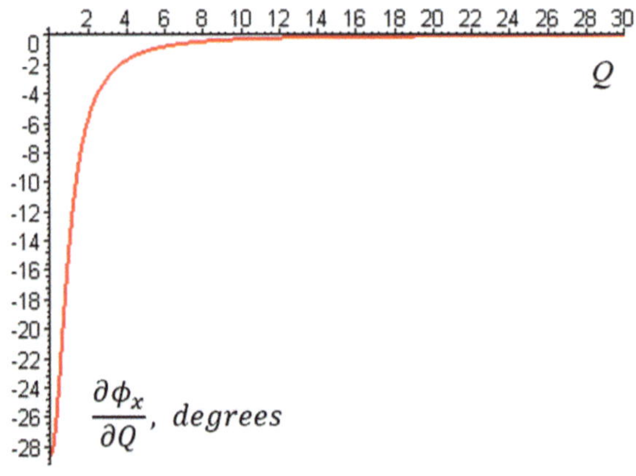

Fig. 14 The sensitivity function of the static characteristic to changes in the quality factor

The analysis shows that this method is suitable for measuring the quality factor at a generator frequency of 50 Hz within the range Q ∈ [0, 1…15], which corresponds to the nominal values of the quality factor of the windings of electrical machines.

Figure 15 shows a generalized structure scheme of the quality factor measuring device, which implements measurement of the phase shift of two voltages [19, 20].

Digital phase meters measure phase shifts by the value of the time interval between the moments when the corresponding sinusoid fronts pass through zero values (phase to time interval conversion) [21, 22]. Pulse generators F1 and F2 generate control pulses at the moments when the instantaneous voltage values pass through zero values, the time interval between which t_X is measured by the microcontroller MC, in our case Microchip ATTiny2313 [23, 24]. The time interval t_X is directly proportional to the measured phase shift ϕ_X. The dependence $\phi_X = f(t_X)$ is determined by the expression

$$\phi_X = \phi_1 - \phi_2 = \omega t_1 - \omega t_2 = 2\pi f_q t_X, \tag{25}$$

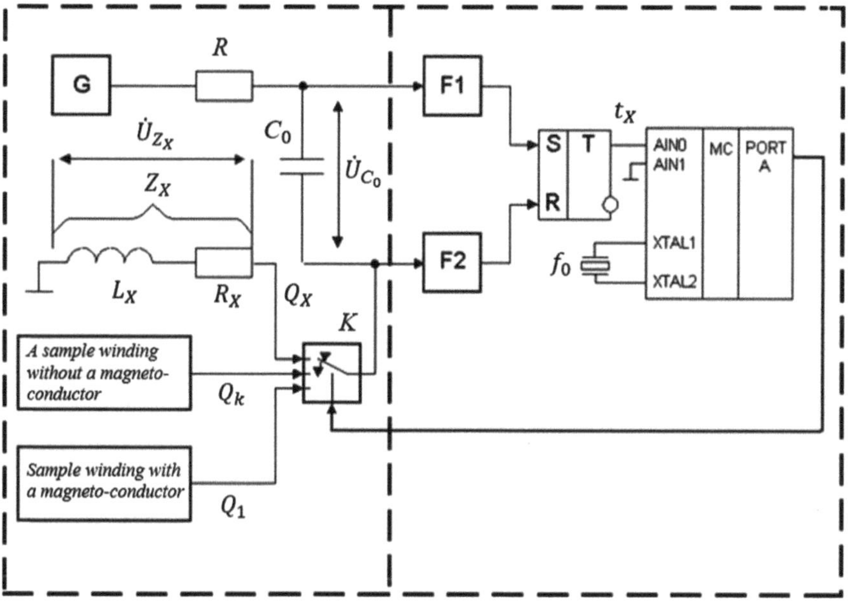

Fig. 15 Generalized structure scheme of the quality factor measuring device

where f_q is a frequency of the generator G output voltage.

The duration of the time interval t_X, proportional to ϕ_X, is determined by the expression

$$t_X = \frac{\phi_X}{2\pi f_q} = \frac{arcctg(Q)}{2\pi f_q}. \tag{26}$$

The number of pulses at the output of the microcontroller (MC) counter is equal to

$$N_X = \frac{t_X}{T_0} = t_X f_0 = \frac{f_0 \cdot arcctg(Q)}{2\pi f_q}. \tag{27}$$

Equation (27) is the general equation of the measuring transformation of quality factor. The microcontroller MC controls the switching of the analog switch K, which is used to switch to measuring the quality factor of the tested winding Q_x, the exemplary winding without a magnetic core Q_k, and the exemplary winding with a magnetic core Q_1. The quality factor of the magnetic core is determined by the expression

$$Q_c = \frac{\left(Q_1 Q_k + \sqrt{Q_1^2 Q_k^2 - 4Q_1^2 + 4Q_1 Q_k}\right)}{(2 \cdot (Q_k - Q_1))}. \tag{28}$$

The relative number of short-circuited coils k is determined using expression (10), with $Q_2 = Q_x$.

Figure 16 shows the family of static characteristics of the quality factor measuring device at different values of the quantization frequency f_0 at $f_q = 50$ Hz.

The sensitivity function of the static characteristic, which is determined by expression (27), is expressed by the formula

$$S = -\frac{f_0}{2\pi f_q(1 + Q^2)}. \quad (29)$$

The surface $S(Q,\ f_0)$ of the sensitivity functions defined by expression (29) is shown in Fig. 17. Its analysis shows that the range of quality factor measuring is 0.1–15.

Analysis of the presented results shows that to increase the accuracy of the measurement of Q and k, the choice of the value of the quantization frequency f_0 is important.

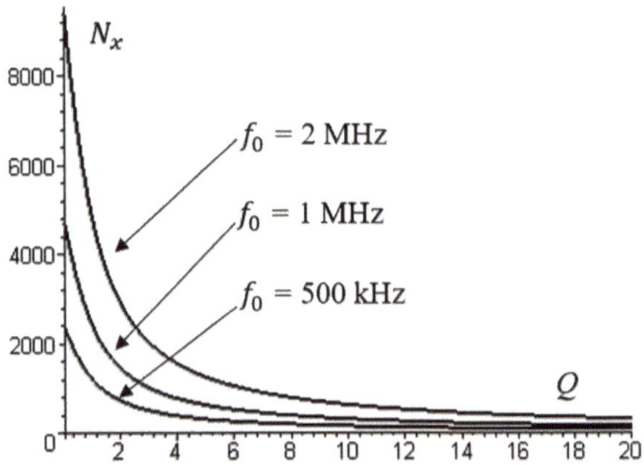

Fig. 16 Family of static characteristics of the quality factor measuring device

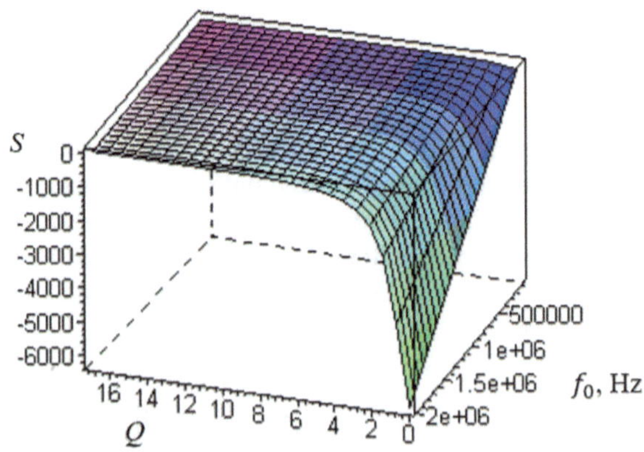

Fig. 17 Surface $S(Q,\ f_0)$ of sensitivity functions

3 Conclusions

The quality factor of a winding is a sensitive and informative parameter for assessing its technical condition, in particular the presence of short-circuited coils. Unlike inductance, the quality factor exhibits a much higher dependence on the number of short-circuited coils, especially at their small value [25, 26].

Mathematical models have been created that allow determining the relative number of short-circuited coils based on the quality factor, taking into account losses in the magnetic core and insulation. Expressions are proposed for calculating the quality factor in various winding configurations.

It has been proven that the phase shift method between two voltages is an effective way to measure the quality factor, and therefore—indirectly diagnose the condition of the windings. The use of a microprocessor-based measuring device allows for high-precision measurement of the quality factor in the range of 0.1–15.

The generalized structural scheme of the microprocessor-based measuring device for predictive maintenance systems of electrical machines has been developed, which allows for comparative measurement of the quality factor of the studied and sample windings. Implementing measurements based on phase shift matching ensures reliability and adaptability to various conditions.

The results of modeling and calculations confirmed the possibility of using the quality factor as the main criterion for assessing the degree of damage to windings, in particular during post-assembly tests of electrical machines.

The proposed approach helps to increase the efficiency of diagnostics and can be integrated into predictive maintenance systems, which meets the modern requirements of digitalization of industrial processes.

References

1. Lopes AC, Figueiredo RJ, de Seixas JM (2018) Maintenance 4.0: intelligent and predictive maintenance system architecture. In: Proceedings of 2018 IEEE 23rd international conference on emerging technologies and factory automation (ETFA), Turin, Italy, pp 1–8
2. Righetto SB et al (2021) Predictive maintenance 4.0: concept, architecture and electrical power systems applications. In: Proceedings CIRED 2021—the 26th international conference and exhibition on electricity distribution, pp 1–5
3. Peyghami S, Palensky P, Fotuhi-Firuzabad M, Blaabjerg F (2020) System-level design for reliability and maintenance scheduling in modern power electronic-based power systems. IEEE Open Access J Power Energy 8:123–134
4. Kucheruk VY, Palamarchuk EA, Kulakov PI, Gnes TV (2014) The statistical model of mechanical milking duration of farmyard milking installation. East-Eur J Enterp Technol 2(4(68):31–37. https://doi.org/10.15587/1729-4061.2014.23120
5. Kowalski M, Nowak A (2022) Methods of diagnosing the insulation of electric machines windings. Energies 15(22):8465 [Online]. Available: https://www.mdpi.com/1996-1073/15/22/8465
6. Kucheruk V, Kulakov PI, Kurytnik IP, Hrabovskyi O (2016) Invariant embedding method for rotor parameters identification of induction motors. Przegląd elektrotechniczny, R. 92 NR 7/2016, pp 136–139. https://doi.org/10.15199/48.2016.07.30

7. Hassan AS, Salam MA (2020) A comprehensive analysis of windings electrical and mechanical faults using frequency response analysis. Energies 13(1):105 [Online]. Available: https://www.mdpi.com/1996-1073/13/1/105
8. Rossi M, Bianchi A (2022) Electrical machines winding technology: latest advancements for performance improvement. Machines 10(7):563 [Online]. Available: https://www.mdpi.com/2075-1702/10/7/563
9. Gieras JF (2016) Electrical machines: fundamentals of electromechanical energy conversion. CRC Press, Boca Raton, FL, USA
10. Gholamian SA, Ardebili M (2020) Simplified algorithm for assessment equivalent circuit parameters of induction motors. IET Electr Power Appl 14(1):1–8
11. Ojaghlu P, Vahedi A, Totoonchian F (2018) Magnetic equivalent circuit modelling of ring winding axial flux machine. IET Electr Power Appl 12(3):293–300
12. Zidat F et al (2020) Induction motor equivalent circuit identification from finite element results. J Appl Electromagn Mech 59(3):345–356
13. Feng H, Cui X, Si J, Gao C, Hu Y (2019) Equivalent circuit model of novel solid rotor induction motor with toroidal winding applying composite multilayer theory. Appl Sci 9(16). Art. no 3288
14. Kong D, Zhao J, Mao L, Yin Y, Zhang J (2025) High-quality factor coil winding method accounting for proximity effect resistance and turn-to-turn mutual inductance. J Power Electron 25(3):123–134
15. Scuiller F (2020) General, compact and easy-to-compute winding factor formulation. IET Electr Power Appl 14(8):1243–1250
16. Ulaby FT, Ravaioli U (2015) Fundamentals of applied electromagnetics, 7th edn. Pearson, Boston, MA
17. Semenov AO, Savytskyi AY, Bisikalo OV, Kulakov PI (2018) Mathematical modeling of the two-stage chaotic colpitis oscillator. In: 2018 14th international conference on advanced trends in radioelecrtronics, telecommunications and computer engineering (TCSET), Lviv-Slavske, Ukraine, pp 835–839. https://doi.org/10.1109/TCSET.2018.8336327
18. Kucheruk V, Kulakov P, Mankovska W, Moskvichova J, Duk M (2018) Resistance-temperature detector besed on RL-diode generator of deterministic-chaotic oscillations. In: Proceedings of SPIE 10808, photonics applications in astronomy, communications, industry, and high energy physics experiments 2018, 108086H, 1 Oct 2018. https://doi.org/10.1117/12.2501664
19. Zangwill A (2013) Modern electrodynamics. Cambridge University Press, Cambridge, UK
20. Caruso T et al (2011) Time interval error effects in sine wave phase difference measurements. IEEE Trans Instrum Meas 60(7):2297–2303 [Online]. Available: https://doi.org/10.1109/TIM.2010.2103910
21. Kucheruk V, Kulakov P, Storozhuk N (2017) Measurement of the number servings of milk and control of water content in milk on stall milking machines. In: Szewczyk R, Kaliczyńska M (eds) Recent advances in systems, control and information technology. SCIT 2016. Advances in intelligent systems and computing, vol 543. Springer, Cham. https://doi.org/10.1007/978-3-319-48923-0_46
22. Ferrero A, Salicone S (2005) Measurement of phase difference between two signals using interpolated DFT and sine fitting. IEEE Trans Instrum Meas 54(2):715–719 [Online]. https://doi.org/10.1109/TIM.2004.838949
23. Carbone P, Petri D (1998) Effect of quantization and timing jitter in phase difference measurements. IEEE Trans Instrum Meas 47(5):1279–1283 [Online]. Available: https://doi.org/10.1109/19.744300
24. Microchip Technology Inc. Microchip Technology Inc. website. Access mode: https://ww1.microchip.com/downloads/en/DeviceDoc/Atmel-2543-AVR-ATtiny2313_Datasheet.pdf
25. Rudyk A, Semenov A, Kryvinska N, Semenova O (2022) Study of phase and amplitude-phase methods for measuring a reactive element quality factor. Measurement 187:110271. https://doi.org/10.1016/j.measurement.2021.110271
26. Rudyk AV, Semenov AO, Kryvinska N, Semenova OO (2021) Measuring quality factors of the radio-frequency system components using equivalent circuits. J Comput Electron 20(5):1977–1991. https://doi.org/10.1007/s10825-021-01770-z

Development of an Early Warning System for Crisis Phenomena in Banking Entities Using Neural Network Technologies

Mariya Kirzhetska, Stepan Vorobets, Oksana Musiiovska, Iryna Yepifanova, and Yuriy Kirzhetskyy

Abstract The study is dedicated to the development of an early warning system for detecting crisis phenomena within banking institutions. Particular attention is given to the application of neural network technologies, specifically Kohonen self-organizing maps, for analyzing financial reporting data from Ukrainian banks. The research demonstrates that banks which have been liquidated tend to cluster into neighboring groups on the Kohonen map, indicating a high potential for forecasting the onset of financial instability before its direct manifestation. Through the use of advanced machine learning techniques, the proposed methodology allows for the timely identification of risks associated with a bank's financial health, thus providing an opportunity to take preventive measures. This approach is particularly valuable for external stakeholders—such as investors, depositors, and regulatory bodies—who rely on publicly available financial information to assess the credibility and solvency of banking institutions. The developed tools not only enhance the precision of early crisis detection but also contribute to strengthening financial system stability by improving the overall risk monitoring processes. The findings emphasize the necessity of integrating innovative technologies into the risk management systems of banks and provide practical recommendations for their implementation in the Ukrainian banking sector.

M. Kirzhetska · S. Vorobets · O. Musiiovska
Lviv Polytechnic National University, Lviv, Ukraine
e-mail: mariya.s.kirzhetska@lpnu.ua

S. Vorobets
e-mail: stepan.y.vorobets@lpnu.ua

O. Musiiovska
e-mail: oksana.b.musiiovska@lpnu.ua

I. Yepifanova (✉)
Vinnytsia National Technical University, Vinnytsia, Ukraine
e-mail: yepifanova@vntu.edu.ua

Y. Kirzhetskyy
Lviv State University of Internal Affairs, Lviv, Ukraine

A. Semenov et al. (eds.), *Data-Centric Business and Applications*, Lecture Notes on Data Engineering and Communications Technologies 272,
https://doi.org/10.1007/978-3-032-06608-4_13

Keywords Early warning systems · Financial reporting indicators · Banking entities · Kohonen self-organizing maps

1 Introduction

Over the past decade, Ukraine has experienced a significant reduction in the number of banking service providers. This process began during the crisis period, when—due to financial instability, military actions, and the large-scale cleansing of the banking system by the National Bank of Ukraine—more than 100 banks were withdrawn from the market by the end of 2015. The downward trend in the number of banks continued in subsequent years. As of the end of 2024, more than half of the banks operating in 2012 had exited the market. The scale of bank closures over such a short period provides a solid informational basis for analyzing the prerequisites and identifying predictors that could have indicated potential problems within financial institutions before their actual occurrence. First and foremost, this will enable the development of an early warning system by expanding the methods for assessing financial indicators of businesses, including those in the banking sector.

Given the significant number of insolvency cases among Ukrainian banks, we aim to identify the earliest-reacting markers of forthcoming critical events affecting a bank's operations by analyzing changes in reporting indicators. Our dataset comprises 537 observations, covering all banks that operated in Ukraine during the 2016–2022 period. The data were obtained from financial statements published on the official website of the National Bank of Ukraine [1]. These are annual financial reports, including all indicators from the income statement, as well as asset and liability indicators from the balance sheet at the bank group level. We did not exclude outliers from the dataset, as we believe they may contain valuable information about the factors contributing to the occurrence of critical events in the functioning of individual banking institutions.

The assessment of a bank's performance based on financial statement indicators should begin with their normalization. The key reference indicators used for normalization were: net assets, total liabilities, total expenses, and total income. This approach allows all items in the financial statements to be brought to a common scale, thereby minimizing differences between large and small banks. As evidenced by the reporting data, this disparity is quite significant (Fig. 1).

At the initial stage of the study, factor analysis methods, particularly the principal component analysis (PCA), were chosen as the theoretical foundation. Their application as the theoretical basis is justified by the following [2]:

- firstly, it allows for a significant reduction in the dimensionality of the problem without losing the informativeness of the description of the studied objects;
- secondly, it enables the visualization of input data through the use of specialized software tools, specifically Statistica for Windows[3].

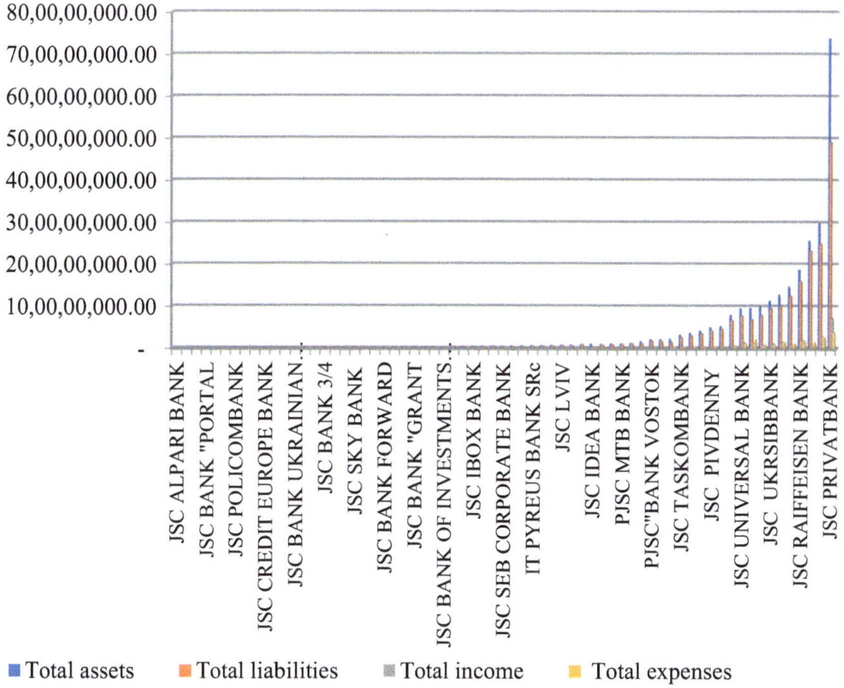

Fig. 1 Financial performance of banks without normalization of input indicators. (*Source:*based on NBU data)

Based on the analysis of the correlation matrix of the full set of input indicators, a preliminary filtering was conducted. At the stage of preliminary analysis, the method of correlation analysis was applied to the set of normalized financial indicators. This method enables the identification of multicollinearity and reveals strong correlations between specific pairs of financial indicators. Partial results of this analysis, presented in the form of a correlation matrix, are shown in Fig. 2.

Based on the results of this analysis, the following steps were taken:

- preliminary filtering of the array of studied input information;
- prior to performing factor analysis, it is advisable to ensure that the dataset is normally distributed and that, at the defined significance level ($\alpha = 0.05$), the correlation matrix of the studied indicators is not diagonal.

These calculations (Fig. 2) indicate that the investigated matrix of pairwise correlation coefficients of the input indicator system is statistically significant. When analyzing the correlation matrix of the selected indicator system (R) for diagonal structure, the χ^2 statistic was used [3] (Fig. 3):

$$\chi 2 = -n - \frac{1}{6}(2m + 11)\ln\left|\check{R}\right| \quad (1)$$

Correlations
Marked correlations are significant at p<0.05
N= 537 (Cfsewise deletion of missing data)

	Var 1	Var 2	Var 3	Var 4	Var 5	Var 6	Var 7	Var 8	Var 9	...
Var 1	1.00	- 0.17	0.08	- 0.09	0.12	- 0.04	- 0.03	- 0.05	- 0.03	...
Var 2	- 0.17	1.00	- 0.36	- 0.20	- 0.22	- 0.01	-0.02	0.04	0.06	...
Var 3	0.08	- 0.36	1.00	-0.18	0.01	- 0.03	0	0.01	0.03	...
Var 4	- 0.09	- 0.20	-0.18	1.00	0.27	- 0.04	0.01	- 0.07	- 0.04	...
Var 5	0.12	- 0.22	0.01	0.27	1.00	- 0.03	-0.07	- 0.05	- 0.06	...
Var 6	- 0.04	- 0.01	- 0.03	- 0.04	- 0.03	1.00	0.12	0.19	- 0.01	...
Var 7	- 0.03	-0.02	0	-0.04	-0.04	0.12	1.00	0.16	0.03	...
Var 8	-0.05	0.04	0.01	-0.07	- 0.05	0.19	0.16	1.00	0.35	...
Var 9	-0.03	0.06	0.03	- 0.04	- 0.06	- 0.01	0.03	0.35	1.00	...
...	...	...	...	...	...	...	...	...	...	...

Fig. 2 Fragment of input indicator system evaluation using correlation analysis methods. *Source* by authors

where n—is the sample size; m—is the number of selected indicators; |Ř|—is the determinant of the correlation matrix, defined as the product of the estimated eigenvalues of the correlation matrix: $|\check{R}| = \gamma_1, \gamma_2 \ldots \gamma_m$.

Based on the construction of confidence intervals for the eigenvalues λ_i of the correlation matrix Ř with a confidence level of $\infty = 0.95$, calculated using formula (1), it was found that the estimated eigenvalues of the principal components do not intersect. This indicates the statistical significance of the factor analysis results.

To construct the confidence interval for the eigenvalues γ_i, the statistic $\sqrt{N-1}(\gamma_i - \gamma_j)$,, was used, which is normally distributed as $N \rightarrow \infty$ with parameters (0.2 ∞2):

$$\frac{\gamma_i}{1+\delta\sqrt{\frac{2}{N-1}}} \leq \frac{\gamma_j}{1+\delta\sqrt{\frac{2}{N-1}}} \tag{2}$$

where γ_1 is the point estimate of the eigenvalue; γ_1;δ–q is the quantile of the standard normal distribution; 1–α/2 confidence level; and $\alpha = 1\text{-}\infty$.

The contribution of each principal component to the total variance of the entire set of input indicators is presented in Fig. 4. As shown by the results of factor analysis, six principal components explain only about 61% of the total variance of the input data. This highlights the need to apply more advanced analytical methods that are based on formalizing nonlinear relationships between financial indicators.

Based on the above assessment, the use of the principal component method as a research tool appears to be inappropriate. The task is to represent each individual bank as a point in a two-dimensional space in order to construct a plane in this space that minimizes the mean squared deviation of all points from that plane (1).

Indic. No	Variable	Factor Loadings					
		Factor 4	Factor 5	Factor 6	Factor 7	Factor 8	Factor 9
1	Amount of cash and cash equivalents	-0.143	0.314	0.2598	0.027	0.685	0.372
2	Amount of financial assets and due from other banks	0.164	-0.238	-0.094	-0.021	0.229	0.112
3	Amount of loans and advances to customers	-0.4822	0.547	0.025	0.032	-0.290	-0.274
4	Amount of other assets	0.276	-0.542	0.184	0.273	-0.249	-0,005
5	Amount of PPE and IA	0.067	-0.295	0.153	0.037	0.137	0.183
6	Amount of due to banks	0.0750	0.178	-0.038	-0.235	-0.481	0.736
7	Amount of customer accounts	-0.2045	-0.112	0.001	0.002	0.065	-0.016
8	Amount from Financial liabilities	-0.401	-0.179	0.103	0.088	0.084	-0.066
9	Amount from Provisions for liabilities	0.4779	0.293	0.096	0.074	0.036	-0.214
10	Amount from Other liabilities	-0.265	-0.189	-0.036	-0.010	0.041	0.048
11	Amount from Subordinated debt	0.067	-0.307	-0.531	-0.191	0.122	-0.147
12	Amount from Interest income	-0.0498	0.143	-0.393	0.86998	-0.018	0.203
13	Amount from Net interest income/(Net interest expense)	-0.1996	-0.174	0.673	0.181	-0.145	-0.056
14	Amount from Interest expense	0.141	-0.005	-0.069	-0.028	-0.032	-0.010
15	Amount from Charge to provisions	0.688	0.3498	0.177	0.065	0.043	-0.108
16	Amount of GNA	0.161	-0.179	0.188	0.071	-0.042	0.045
Expl.Var		1.433	1.332	1.118	0.983	0.973	0.934
Prp Totl		0.896	0.083	0.0698	0.061	0.061	0.058

Fig. 3 Fragment of the factor loadings matrix and the corresponding eigenvalue estimates. *Source*: by authors

$$\frac{\sum_s \left(r^\alpha - r_s^\alpha\right)}{\sum_s \left(r^\alpha - \breve{r^\alpha}\right)} = E_L \rightarrow \min \tag{3}$$

However, in our case, the mean squared deviation explained by the first two principal components accounts for less than one-third of the total variance of the input financial indicators. Therefore, even the optimal linear compression option does not allow for effective visualization of banks' financial conditions.

For this reason, the aim of this study is to develop an early warning system capable of identifying predictors and signals of potential problems in banks' financial performance before their actual occurrence, based on more flexible tools for assessing banks' financial condition compared to traditional methods.

Value	Eigenvalues			
	Extraction: Principal component			
	Eigenvalues	% Total variance	Cumulative Eigenvalues	Cumulative, %
1	2.491355	15.57097	2.49136	15.57097
2	1.759368	10.99605	4.25072	26.56702
3	1.621937	10.13711	5.87266	36.70412
4	1.43338	8.95863	7.30604	45.66275
5	1.331609	8.32256	8.63765	53.98531
6	1.117764	6.98603	9.75541	60.97134
7	0.983041	6.14401	10.73846	67.11535
8	0.972978	6.08111	11.71143	73.19646
9	0.933721	5.83576	12.64516	79.03222

Fig. 4 Eigenvalues of the principal components and the proportion of total variance explained by each across the full set of input financial indicators

2 Literature Review and Methodical Approaches

Early warning models have become a key tool for analyzing financial distress in banks. The prediction of bank failures began with the application of Altman's linear discriminant analysis to identify key financial indicators for assessing banks' stability [4]. Later, Martin [4] refined this approach by using logistic modeling. The authors assumed that severe banking problems primarily have internal origins [4, 5].

Research conducted in [6, 7] confirmed that the factors used by statistical early warning systems to predict bank failure remain stable over long periods of time. Therefore, they can be used to forecast bank bankruptcies even when there is little or no indication of trouble in the years immediately preceding the event.

The studies by Alam et al. [8] demonstrated that both fuzzy clustering and self-organizing neural networks are promising classification tools for identifying potentially bankrupt banks.

Hlazunov and Verchenko in 2020, developed an early warning model for micro- and macro-level analysis of individual and aggregated vulnerability factors of banks in Ukraine. This model is based on the methodology of aggregation into a baseline bank-level model, serving as a foundation for introducing the countercyclical capital buffer in Ukraine [9, 10].

The literature also contains numerous examples of the use of SOM in broader economic and financial contexts [11, 12]. In their work, Sarlin and Peltonen [13] developed a financial stability map of European banks aimed at forecasting financial crises. The authors demonstrated the attractive functionality of SOM, in particular its ability to visualize trajectories showing how individual countries moved across the map's plane during specific time periods [13].

One of the most significant works for our research is the work of Prof. Zarutska [14, 15], who investigates the practical tools for reforming the banking supervision system in Ukraine based on structural–functional analysis, employing methods for processing large volumes of information such as neural networks and Kohonen self-organizing maps [16, 17].

Our study contributes to the existing literature by developing an early warning system that allows the identification of predictors and signals of potential problems in the financial activities of banks before their actual occurrence, through the use of Kohonen maps to construct a dynamic coordinate system for the distribution of core financial reporting indicators.

3 Methodology and Data

Clustering algorithms serve as a convenient tool for grouping observations into homogeneous categories based on predefined characteristics [18]. The literature review confirms the applicability of cluster analysis in building early warning systems for detecting crisis phenomena in banks prior to their actual manifestation.

An important advantage of such algorithms is the minimization of the influence of the researcher's subjective judgment on the research results [18]. In this study, Kohonen Self-Organizing Maps (SOM) were used as one of the algorithms from the family of artificial neural networks. This is a two-layer neural network consisting of an input layer and an output layer. The formalized description of the Kohonen network and its training algorithm for solving the task of clustering banks based on financial reporting data assumes the following: let $X = (X_1, X_2, \ldots, X_i, \ldots, X_n)$ be a set of input vectors of size N ($i = 1, \ldots, N$), where each element of this set (X_i) represents a set of financial indicators for a specific bank — $X_i = (x_1, x_2, \ldots, x_i, \ldots, x_p)$. In our case, the dataset consists of annual financial reports of banks operating in Ukraine from 2016 to 2023 ($N = 537$), with 16 financial indicators per bank ($p = 16$).

The Kohonen map represents a set of interconnected neurons. There is a certain distance between these neurons, and it can be calculated accordingly. The simplest method for computing this distance is the Euclidean distance between two points located in a P-dimensional space.

Let us denote the set of such neurons as $Wm = (W_1{}^m, W_2{}^m, \ldots, W_j{}^m, \ldots, W_M{}^m), j = 1,M$. Each neuron W_{jm} is represented as a vector of numbers called neuron weights:

$$Wj^m = (w_1^m, w_2^m, \ldots, w_k^m, \ldots, w_P^m)$$

In a Kohonen network, the neurons are arranged on a two-dimensional grid. The set of neurons forms the corresponding network architecture.

Among various possible architectures, our study utilizes a hexagonal topology. The unique feature of this topology is that each node represents a hexagon, connected to six neighboring nodes, which are also hexagonal in shape. The only exception is the boundary neurons, which have fewer connections with neighboring nodes.

Each individual neuron in the network serves as a formalized representation of a group of banks that are similar to one another in terms of their financial indicators.

The learning algorithm of the Kohonen network begins with the initialization of the weights for all neurons in the network. In the Self-Organizing Map (SOM) algorithm, the weights w_j are typically assigned small random values, which is considered one of the drawbacks of this approach. This is because each time the network with a predefined topology is trained, it must start from different initial positions, which can affect convergence and results.

The second stage of the training process involves setting the current training time (iteration). In our example, we begin with $t = 1$.

At the next step, one of the elements of the input dataset is selected—X_i, representing the financial indicators of a bank for a particular time period. Then, from the set of weight vectors W_m, the element $W_j{}^{m*}$ is determined—the one whose metric is closest to Xi (i.e., the Best Matching Unit—BMU) according to a predefined distance function.

Following this, adjustments to all weights of the current neuron $W_j{}^{m*}$ are made using the following logic:

$$W_j^{m\mathcal{H}} = w_j^{m*} + a(t)N_{i,j}(t)\left(x_i - w_j\right) \quad (4)$$

where $W_j{}^{m}{}_{H}$ is the adjusted weight value, $a(t)$ is the learning rate parameter that depends on the training step and decreases progressively with each iteration. It starts at a moderately high value and gradually declines as the iteration process continues.

In our case, the learning rate decreased from 0.5 to 0.05 during the initial tuning phase and from 0.05 to zero during the fine-tuning phase.

The neighborhood function $N_{i,j}$ $(t, d(m,m^*))$, adjusts the update of weight coefficients depending on the distance from a neuron to the best-matching unit. It is a training parameter of the network that depends on step t and the distance d (m,m^*) between neurons in the Kohonen network geometry, and is defined by the following formula:

$$N(t) = \exp\left(-D^2/2\sigma(t)^2\right) \quad (5)$$

where D^2 is the distance on the map between neuron j and the best-matching unit (BMU); $\sigma(t)$ is the radius parameter.

Analogous to the learning rate parameter $\sigma(t)$, it decreases with each iteration. In our case, it starts at 2.5 during the initial tuning phase and decreases to 1, while during the fine-tuning phase it remains constant at 1. At the next step, the iteration counter is incremented by 1 ($t = t + 1$). If the maximum number of iterations is reached, the training algorithm terminates.

After all the above steps are completed, the next input data element (financial indicators of the next bank) is selected, and the network weights are adjusted according to the logic described above. To simplify, in formula (5), once the input vector of banking variables is presented to the model, the output layer weights are updated such that the weights of the best-matching unit are adjusted most closely to the input vector, while neighboring neurons are adjusted to a lesser degree depending on their distance. The farther a neuron is from the best-matching unit, the smaller the adjustment.

Thus, after many iterations, the resulting two-dimensional map acquires a topological structure that reflects the original input data.

Another parameter choice involves selecting the map size. We chose a 7×7 square, totaling 49 neurons. The choice of map size was determined by the size of the data sample, which in our case comprises 537 observations. MATLAB software was used for the calculations. The architecture of the Kohonen Self-Organizing Map in MATLAB is presented in Fig. 5.

The architecture of the Kohonen SOM consists of two layers—the input layer and the competitive layer (see Fig. 5). The input layer is represented by a matrix of input variables. The competitive layer uses a competitive activation function. The 'ndist' block computes the negative Euclidean distance m between the input vector P and the rows of the weight matrix W. The output of the activation function b—is the sum of the calculated distance and the bias vector S.

The competitive activation layer analyzes the values of the vector elements and assigns zero values to all neurons except for the winning neuron, which receives the highest activation output. This activation behavior can be described as follows:

$$a_i^1 = \left\{ \begin{matrix} 1, i = i^*, i^* = arg(maxn_i^1 \\ 0, i \neq i^* \end{matrix} \right\} \tag{6}$$

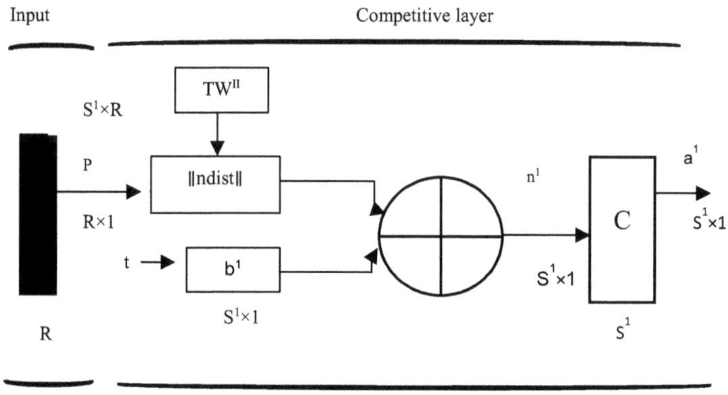

Fig. 5 Architecture of the Kohonen self-organizing map [16]

The index of the activated neuron determines the group (cluster) to which the closest input vector belongs:

W—an S-by-R weight matrix (or an S-by-1 bias vector);

P—an R-by-Q input matrix (or a matrix of ones (1, Q));

A—an S-by-Q output matrix.

Thus, the financial condition of Ukrainian banks can be described with acceptable accuracy using only two generalized financial indicators—namely, the two coordinates on the two-dimensional Kohonen map. Each bank, based on its financial reporting data, is represented by a specific neuron on the topological map. Neighboring neurons on the map correspond to banks with similar financial conditions. The further apart banks are on the map, the more their financial profiles differ from one another.

Let us consider some key features of this methodology in the context of our study. The inputs to this network are vectors of dimension (1×16), where each element of the vector corresponds to one of the 16 financial indicators of a bank, and the output neurons are the neurons of the Kohonen grid. The entire set of Ukrainian banks will be distributed among n clusters, which are identified by the corresponding neurons of the Kohonen topological map. The topological map itself represents a comparative integral assessment of the financial condition of banks in a 2-dimensional coordinate system based on the set of financial indicators provided as input to the neural network. One of the challenging decisions in applying this methodology was determining the number of clusters and, consequently, the size of the Kohonen topological map. The final decision regarding the map size was made based on the results of successive testing of different topological variants and training outcomes of such networks. As a result, the following topological variant was chosen: a hexagonal architecture with a size of 7×7, meaning the set of banks is distributed across 49 clusters. Considering that the input data set for the period from 2016 to 2022 consists of 537 elements, this neural network structure will allow for the accurate identification of bank clusters.

The neighbor connection map of this topology, shown in Fig. 6, illustrates that each hexagon corresponds to a neuron in the SOM. A dark color indicates a large distance between clusters, which, in turn, reflects significant differences in their weight values or can be interpreted as boundaries between clusters. A yellow color indicates minimal differences between neighboring clusters, with relatively similar weight values. Orange represents moderate distances between clusters, which can be interpreted as transitional zones.

Analyzing this visualization (Fig. 6), the following can be noted: for the majority of banks, there are no significant differences in the values of their financial indicators. However, a substantial SOM Neighbor Distance is observed between banks in clusters 13 and 14, as well as between clusters 14 and 21. A somewhat smaller difference is also present between clusters 7 and 14.

The visualization of input element planes (financial indicators), presented as SOM Weight Planes in Fig. 7, appears to be highly informative. This visualization tool consists of 16 individual maps, each corresponding to one of the financial input indicators under consideration. Each map represents the weight values of a particular financial indicator across the individual nodes of the topological map. The color

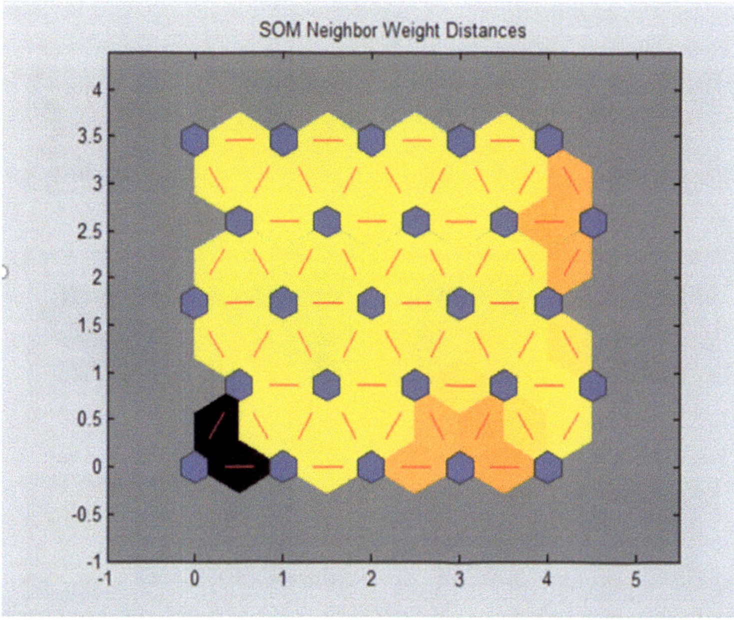

Fig. 6 SOM neighbor distances

scale ranges from black (minimum weight values for the respective SOM node) to yellow (maximum values), allowing for a visual assessment of each indicator's variation patterns. Indicators that exhibit similar color patterns may point to specific interdependencies among the financial characteristics of the banks.

As shown in Chart 12 of Fig. 7, there is a clearly visible continuous yellow area, indicating high weight values for this financial indicator (banks' interest income) across nearly all clusters. In contrast, the group of financial indicators (Chart 2 in Fig. 7) from 6 to 11 exhibits low weight values for the vast majority of clusters. This group primarily includes financial indicators listed in the 'bank liabilities' section of the balance sheet. The analysis of weight maps enabled the classification of banks based on common characteristics and revealed patterns in their financial indicators over the analyzed period.

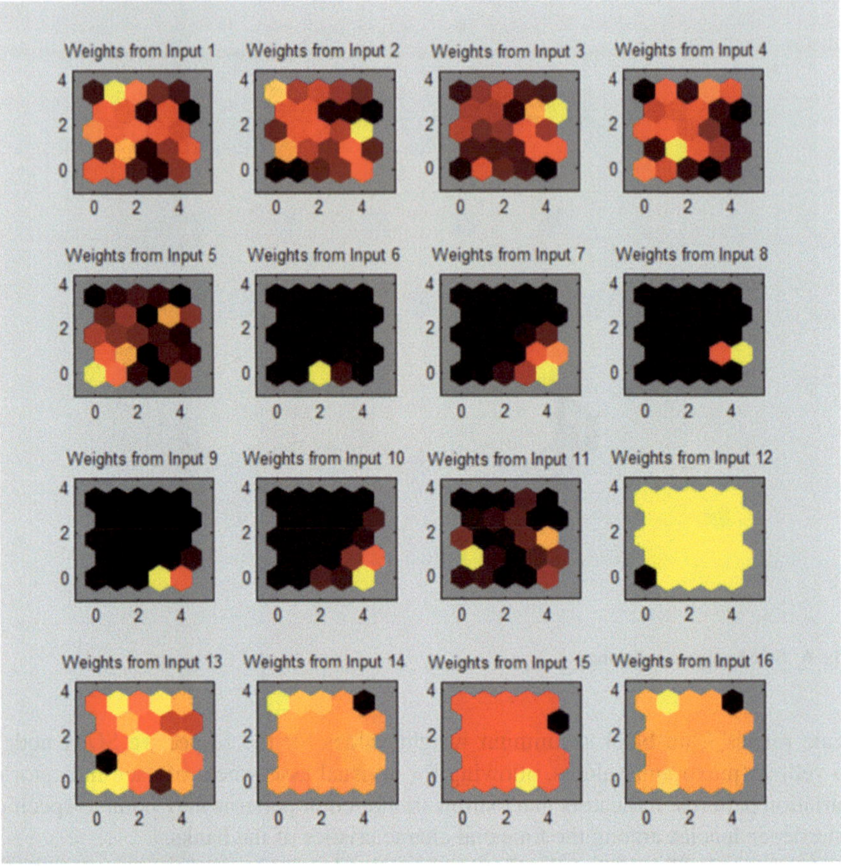

Fig. 7 SOM weight planes

The number of banks assigned to each specific cluster is presented in the form of the SOM Sample Hits plot (Fig. 8). The relative quantity of input vectors represented by each neuron in the SOM is reflected by the size of the figure that identifies it.

The following should be noted in this visualization (Fig. 8): the vast majority of banks were assigned to middle clusters, with nearly half of them concentrated in clusters 8–20. As shown in Fig. 6 (SOM Neighbor Distances), the overall financial condition of these banks can be considered relatively uniform.

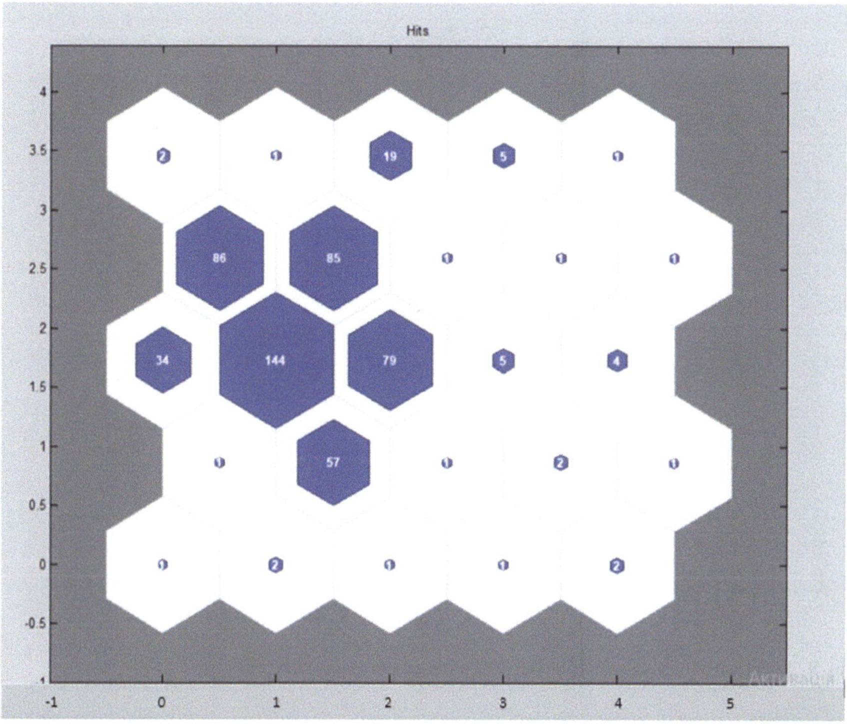

Fig. 8 SOM sample hits

The adequacy of representing the input financial reporting data of banks through mapping them onto the topological map can be visually evaluated using SOM Weight Positions (Figs. 9–10). The input vectors of the financial indicators are shown as green dots, the blue-gray dots represent the weight vectors of SOM neurons, and the red lines denote connections between neighboring neurons. The SOM architecture (Fig. 9) reflects the network state after 400 iterations of training. As seen, the nodes of the topological map are not yet sufficiently dispersed across the input vector space. Only after 2800 iterations was a significantly improved representation of the input elements by the neural network nodes achieved (Fig. 10).

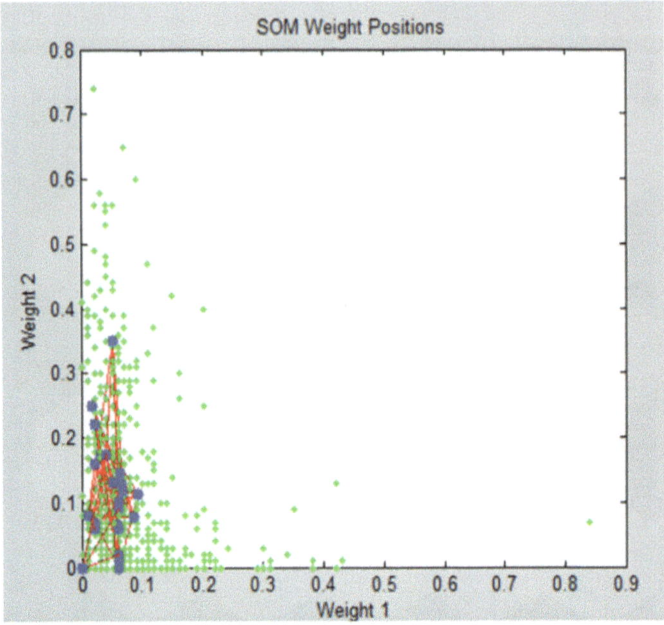

Fig. 9 SOM weight positions after the 400th iteration of network training

The trained network is tested by identifying the clusters of banks that remained operational in 2022. For this purpose, we identify the clusters of these banks based on their financial performance. The clustering results are presented in Table 1.

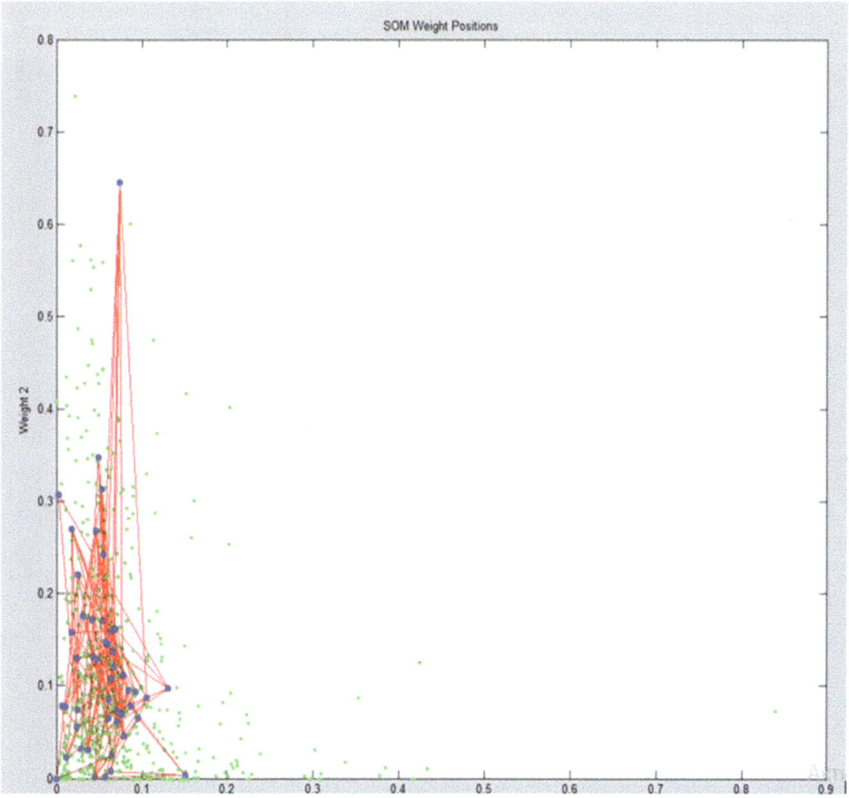

Fig. 10 SOM weight positions after the 2800th iteration of network training

One of the potential directions in the proposed analysis methodology is the study of trends in the integral metrics (cluster numbers) of the financial condition of individual banking entities over a specific time interval. This analysis was implemented using the monthly financial reporting data for 2024 of PUMB (Fig. 11).

This dynamic, as defined through cluster-based evaluation, is well explained by the trend of the normalized input data from the bank's financial statements during this period, which is presented in Fig. 12.

Table 1 Clustering results of Ukrainian banks in 2022

Name of the bank of Ukraine	Number of cluster members	Name of the bank of Ukraine	Number of cluster members	Name of the bank of Ukraine	Number of cluster members	Name of the bank of Ukraine	Number of cluster members
Cluster 12		JSC MTB Bank		Cluster19		JSC Creditwest Bank	
JSC Bank Alliance	5	JSC Kredobank		JSC MIB	2	JSC Crystal bank	
JSC Cominbank		JSC Credit Agricole Bank		JSC SKY Bank		JSC MotorBank	
JSC Meta Bank		JSC Credit Europe Bank		Cluster 26		JSC Akord Bank	
JSC RBC Bank		JSC Polycom Bank		JSC BANK 3/4	1	JSC Pravex Bank	
JSC Ukrbudinvest Bank		JSC Raiffeisen Bank		Cluster 34		JSC Unex Bank	
Cluster 18		JSC Sense Bank		JSC Alpari Bank	1	JSC Bank Lviv	
JSC Clearing House	25	JSC Taskombank		Cluster27			
JSC Bank PIVDENNIY		JSC Universal Bank		JSC IBOX Bank	3	Cluster47	
JSC A—BANK		JSCBank Vostok		JSC Concord Bank		JSC Procredit Bank	3
JSC Radabank		Cluster 20		JSC OKSI Bank		JSC Ukrgasbank	
JSC Grant Bank		JSC Cominvest Bank	1	Cluster 36		JSC Ukreximbank	

(continued)

Table 1 (continued)

Name of the bank of Ukraine	Number of cluster members	Name of the bank of Ukraine	Number of cluster members	Name of the bank of Ukraine	Number of cluster members	Name of the bank of Ukraine	Number of cluster members
JSC Bank of Invest and Savings		Cluster 24		JSC Citi Bank	1	Cluster48	
JSC Forward Bank		JSC CB PrivatBank	9	Cluster37		JSC First Investment Bank	2
JSCEPB		JSC AltBank		JSC Deutsche Bank	2	JSC Ukrainian Bank for Reconstruction and Development	
JSC Idea Bank		JSC Asvio Bank		JSC SEBCorporate Bank			
JSCING Bank Ukraine		JSC Bank Portal		Cluster 46			
JSC OTP BANK		JSC Avangard Bank		JSC Industrial Bank	12		
JSC Oschadbank		JSC Trust Capital Bank		JSC BM Bank			
JSC Pyreus Bank		JSC BTA Bank		JSC Ukrainian Capital			
JSC Poltava-Bank		JSC UkrsibBank		JSC Credit Dnipro			
JSC PUMB		JSC Bank Family		JSC Bank Globe			

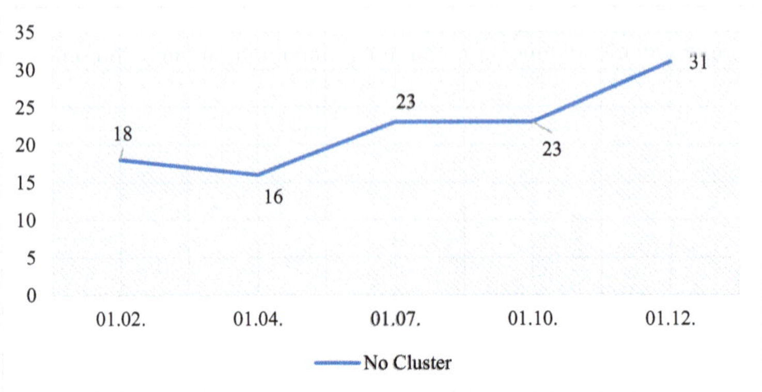

Fig. 11 Evaluation of the trend in integral metrics (clusters) of PUMB's financial condition in 2024

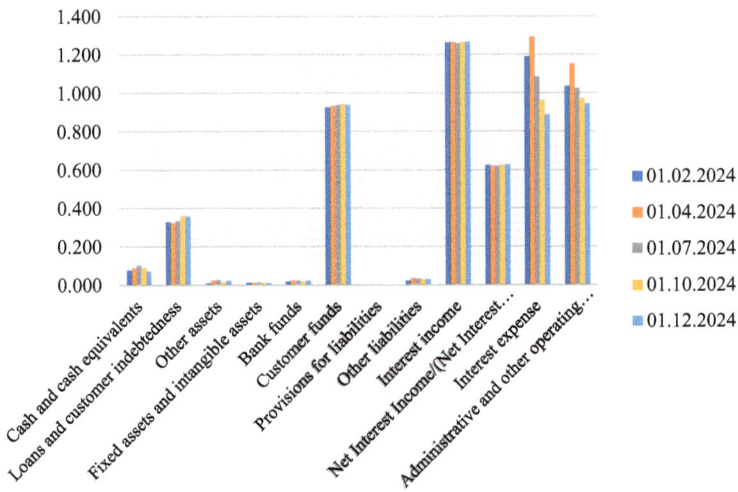

Fig. 12 Dynamics of normalized input data from monthly financial statements of PUMB in 2024

4 Conclusions

This study makes a significant contribution to the development of approaches for constructing early warning systems for crisis phenomena in banking entities using neural network technologies, specifically Kohonen self-organizing maps (SOM). This enables the identification of predictors and signals of potential financial problems in banking entities before their actual occurrence.

The proposed approach for developing an early warning system based on SOM involves clustering banks through the analysis of financial reporting data and offers the following advantages:

(1) Flexibility compared to traditional methods: SOM allows the identification of nonlinear relationships among financial indicators, contributing to the construction of more accurate models for assessing banks' financial condition;
(2) Visualization of financial solvency: the ability to present banks' financial characteristics as a topological map facilitates simplified analysis and monitoring of the situation in the banking sector.

At the same time, several debatable aspects of the proposed methodology should be noted:

(1) Selection of input indicators—although the study uses publicly available financial reporting indicators of banks, this list could be expanded by including additional derived indicators, which would provide a more detailed picture of banks' financial stability;
(2) Number of clusters—the lack of a theoretically grounded approach to determining the optimal number of clusters remains an open issue that requires further research.

Despite these discussion points, the method of clustering banks using SOM proves effective, provided the analysis results—such as SOM Neighbor Distances, SOM Weight Planes, SOM Sample Hits, and SOM Weight Positions—are correctly interpreted.

The assessment of bank clustering results in Ukraine over the analyzed period revealed a significant similarity in the financial indicators of most solvent banks that were actively providing banking services in 2024. These banks were primarily concentrated in clusters 12 through 27. Additionally, the analysis showed that banks liquidated in 2022 were assigned by the neural network to adjacent clusters 31 and 32, indicating that placement into these clusters may serve as a signal of potential financial distress.

Thus, the use of SOM for analyzing the financial performance of banks not only enables their clustering based on financial indicators, but also provides the potential for predicting the risk of liquidation. This makes the method a promising tool in the field of banking supervision and financial analysis, particularly for external users of publicly available information.

References

1. National Bank of Ukraine (n.d.) Financial statements of Ukrainian banks [Electronic resource]. Retrieved from http://www.bank.gov.ua/control/uk/publish/category?cat_id=7693080
2. Built In (n.d.) Principal component analysis [Electronic resource]. Retrieved from https://builtin.com/data-science/step-step-explanation-principal-component-analysis
3. Cramer D (2002) Quantitative data analysis with SPSS for windows: a guide for social scientists. Retrieved from https://www.researchgate.net/publication/275141290
4. Martin D (1977) Early warning of bank failure: a logit regression approach. J Bank Finance 1(3):249–276. https://doi.org/10.1016/0378-4266(77)90022-X

5. Pazarbasioglu C, Hardy DC (1999) Determinants and leading indicators of banking crises: further evidence. IMF Staff Pap 46(3):1–1. https://doi.org/10.2307/3867642
6. Kimmel RK, Thornton JH, Bennett SE (2016) Can statistics-based early warning systems detect problem banks before markets? N Amn J Econ Financ 37:190–216
7. Kakadiya R, Khan T, Diwan A, Mahadeva R (2024) Transformer models for predicting bank loan defaults: a next-generation risk management. In 2024 IEEE 6th international conference on cybernetics, cognition and machine learning applications (ICCCMLA), pp 26–31. https://doi.org/10.1109/ICCCMLA63077.2024.10871798
8. Alam P, Booth DE, Lee K, Thordarson T (2000) The use of fuzzy clustering algorithm and self-organizing neural networks for identifying potentially failing banks: an experimental study. Expert Syst Appl 18:185–199
9. Hlazunov A, Verchenko O (2020) Predicting bank defaults in Ukraine: a macro-micro perspective. Visnyk Natl Bank Ukr 250:33–44. https://doi.org/10.26531/vnbu2020.250.03
10. Kirzhetska M, Kirzhetskyy Y (2023) Model of interaction between the government and business towards legalization of unorganized imports. Soc Leg Stud 6(2):25–31. https://doi.org/10.32518/sals2.2023.25
11. Rashkovan V, Pokidin D (2016) Ukrainian banks' business models clustering: application of Kohonen neural networks. Visnyk Natl Bank Ukr 238:13–38. https://doi.org/10.26531/vnbu2016.238.013
12. Mingoti SA, Lima JO (2006) Comparing SOM neural network with Fuzzy c-means, K-means and traditional hierarchical clustering algorithms. Eur J Oper Res 174:1742–1759. https://doi.org/10.1016/j.ejor.2005.03.039
13. Sarlin P, Peltonen TA (2011) Mapping the state of financial stability. ECB working paper no. 1382. https://ssrn.com/abstract=1927847 or https://doi.org/10.2139/ssrn.1927847
14. Zarutska OP (2010) Studying peculiarities of Ukrainian banking system development by means of applying Kohonen map. Actual Probl Econ 5:255–262
15. Zarutska O, Novikova L, Pavlov R, Pavlova T, Levkovich O (2022) Evaluation of Ukrainian banks' business models by the structural and functional groups analysis method. Financ Credit Act Probl Theory Pract 4(45):8–20
16. Kohonen T (1982) Self-organized formation of topologically correct feature maps. Biol Cybern 43(1):59–69
17. Kohonen T (2013) Essentials of the self-organizing map. Neural Netw 37:52–65. https://doi.org/10.1016/j.neunet.2012.09.018
18. Shakhovska N, Yakovyna V, Kryvinska N (2020) An improved software defect prediction algorithm using self-organizing maps combined with hierarchical clustering and data preprocessing. Lect Notes Comput Sci 12391:414–424. https://doi.org/10.1007/978-3-030-59003-1_27

Impact of Energy-Saving Technological Changes on the Level of Competitiveness of Enterprises

Olexandr Yemelyanov, Tetyana Petrushka, Orest Koleshchuk, Kateryna Petrushka, and Viacheslav Dzhedzhula

Abstract Today, many enterprises are facing the need to increase their competitiveness. One of the possible ways of such an increase is the implementation of energy-saving technological changes (ESTC). Therefore, the aim of this study is to develop and apply an instrumentation that would allow assessing the impact of ESTC on the level of competitiveness of enterprises. The article builds a model of such an impact based on the use of relevant indicators of competitiveness of companies. In particular, one of such indicators is the relative level of profitability of the enterprise's products, i.e. the ratio of this profitability of a particular company to the average profitability of products of all competing enterprises. A method of decomposition of the indicator of product profitability growth has been developed, the use of which allows determining the degree of influence of ESTC at enterprises on this indicator. A method for assessing the impact of ESTC on the level of competitiveness of firms has been proposed. This method has two modifications (direct and two-stage assessment) and provides for assessing the statistical significance of the relationship between the scale of ESTC at enterprises and the ratio of the growth rate of a particular enterprise's revenues from sales of its products and the average growth rate of revenues of all competing enterprises. Both modifications of the developed method have been tested on samples of Ukrainian enterprises belonging to three industries and consuming natural gas. The study has shown that, according to both methodological approaches, with an increase in the level of ESTC, the value of the ratio between the growth rate of enterprise revenues from sales of products and the average growth

O. Yemelyanov · T. Petrushka · O. Koleshchuk · K. Petrushka
Lviv Polytechnic National University, Lviv, Ukraine
e-mail: Oleksandr.Y.Yemelianov@lpnu.ua

T. Petrushka
e-mail: tetiana.o.petrushka@lpnu.ua

K. Petrushka
e-mail: kateryna.i.petrushka@lpnu.ua

V. Dzhedzhula (✉)
Vinnytsia National Technical University, Vinnytsia, Ukraine
e-mail: djedjula@vntu.edu.ua

A. Semenov et al. (eds.), *Data-Centric Business and Applications*, Lecture Notes on Data Engineering and Communications Technologies 272,
https://doi.org/10.1007/978-3-032-06608-4_14

rate of revenues of all competitors' increases. It was found that the impact of ESTC on the competitiveness of the studied enterprises is statistically significant. It was also found that in all three industries the increase in product profitability was due to the implementation of ESTC. In the end, the research has shown that the implementation of ESTC at the studied enterprises not only increased the competitiveness of most of them, but also contributed to their energy-saving economic development.

Keywords Enterprise · Competitiveness · Modelling · Profitability of product · Energy-saving technological change · Energy consumption of products

1 Introduction

One of the main factors that influence the economic efficiency and volume of business activities is the competitiveness of enterprises. At the same time, maintaining a high level of competitiveness can be a challenging task that requires companies to develop and implement appropriate strategies. Such strategies should provide for the implementation of certain measures at enterprises to ensure and maintain their competitive advantages [1].

Currently, there is a significant arsenal of tools for managing the competitiveness of enterprises. Many of these tools involve significant costs for their implementation [2]. Therefore, it is important to choose those tools of competitiveness management that are currently appropriate for a particular firm to implement. In particular, this also applies to such a way of increasing competitive advantages as the implementation of progressive technological changes by enterprises.

One of the most important varieties of technological changes occurring at enterprises is the introduction of energy-saving technologies [3]. At the same time, this introduction often requires large investment [4]. In this regard, when making decisions regarding the proceedings of energy-saving technological changes (ESTC) at the enterprises, careful and comprehensive assessment of the consequences of such proceedings should be made [5]. In particular, this also applies to the assessment of the expected impact of the implementation of these projects on the competitiveness of those enterprises that provide for these projects to implement.

2 Literature Review and Setting Research Objectives

Currently, there are a number of scientific works on evaluating the impact of technological changes on certain characteristics and indicators of companies, in particular on their ecological efficiency [6], stability [7], market value [8], financial condition [9], development [10], etc. Of course, all these characteristics directly or indirectly influence the level of competitive advantages of companies. However, the importance of these advantages in terms of ensuring the effective activity of enterprises necessitates a direct assessment of the impact of technological changes on the level of competitiveness of firms.

In general, the overwhelming majority of scientific papers on such an assessment establish the existence of significant positive effects of technological upgrading on financial and economic performance and ensuring the proper level of other important characteristics of enterprises [11]. In particular, this also applies to such a characteristic of companies as the level of their competitive advantages. Thus, in [12] notes that technological innovations serve not only as a tool for increasing competitiveness, but also contribute to the movement towards full automation of production processes. In [13], the conclusion about the positive impact of technological changes on the competitiveness of enterprises is obtained for the case of green innovations. At the same time, [14] proves the importance of the introduction of such new advanced technologies as the Internet of Things and digitalisation for the competitiveness of firms. A significant positive impact of digital transformation on the competitive advantages of companies is also noted by the authors of [15].

With regard to the impact on the competitiveness of companies of such a variety of technological changes as ESTC, a number of researchers have revealed a generally positive nature of such influence. Thus, the work [16] found that energy efficient technologies significantly increase the competitiveness of companies. In particular, as noted in [16], the introduction of energy-saving technologies at enterprises leads to a decrease in costs, improving the efficiency of production processes and strengthening corporate reputation among environmentally conscious consumers.

In [17], the impact of ESTC on firms' competitiveness is considered in a broader context, namely through the prism of the study of the consequences of the transition to a low-carbon energy efficient economy. According to the results of the study, the author of this work concluded that this transition enables production enterprises to increase their stability and competitiveness. At the same time, in order to promote the decarbonisation of the economy, products with a reduced energy and energy management services are required [17].

At the same time, some authors note that a significant positive impact of technological innovations, in particular innovative energy-saving technologies, on the competitiveness of companies can occur only under certain conditions. In particular, the authors of [18] refer to such conditions as a sufficiently high prices of energy resources, which are expected to be saved. This statement seems reasonable, since only at a sufficiently high level of energy prices can the implementation of ESTC at enterprises be economically justified. A similar conclusion was reached, in particular, by the author of [19], who assessed the competitiveness of technologies that use renewable energy resources.

An important aspect of the implementation of ESTC at enterprises is the environmental impact of such implementation. In this regard, it is of interest to read [20], which considers, in particular, the impact of pro-environmental activities of small and medium-sized firms on the process of forming their competitive advantages. Among other things, the authors of this work prove the importance of proper management of energy saving at enterprises, which will facilitate their pro-environmental activities and, as a result, will increase the level of competitiveness of these enterprises.

Despite the presence in the scientific literature of a number of fruitful studies on the influence of ESTC on the competitiveness of enterprises, there is a need to further study this issue. In particular, it is necessary to increase the degree of formalization of the procedure of evaluation of such influence. In addition, there is a need for an in depth study of its mechanism. In the end, the issue of assessing the possibilities of simultaneously increasing the competitiveness of enterprises and ensuring their energy-saving economic development as a result of the implementation of ESTC (that is, ensuring economic growth, which would be accompanied by a reduction in the consumption of certain types of energy) remains insufficient [21, 22].

Considering the above, the purpose of this study is to develop and use tools that would help to evaluate the impact of energy -saving technological changes on the level of competitiveness of enterprises. To achieve this goal, the following main tasks were solved when performing this study:

(1) build a model of influence of ESTC on the competitiveness of enterprises;
(2) develop a method of evaluating such influence;
(3) to test the method of evaluation of the impact of ESTC on the level of competitiveness of enterprises by sampling of industrial companies of Ukraine.

3 Modeling the Impact of Energy-Saving Technological Changes on the Competitiveness of Enterprises

Establishing the extent of the impact of ESTC on the competitiveness of companies requires a preliminary selection of the indicator (or group of indicators) that will be used to assess the level of competitive advantages of firms. In doing so, it is necessary to take into account the existence of a number of types of enterprise competitiveness. In particular, it may relate to a certain type of product or the entire set of products manufactured by the company. Thus, to assess the existing competitiveness of a company in a particular product market for a certain type of product that this company produces, it is advisable to use the following two indicators:

(1) the share of this market occupied by the company's products;
(2) the relative economic efficiency of production and marketing of these products by this company in the relevant market. An indicator of such efficiency may be the ratio of the profitability of the products of the enterprise to its average market profitability. In this case, the profitability of the product will be assessed by the share of profit in its selling price, from which indirect taxes are deducted.

The same indicators of assessing the competitiveness of the company can be distributed to the whole set of products of a particular enterprise. Then these indicators can be calculated using the following equations:

$$I_1 = \frac{R_i}{R_s} \tag{1}$$

$$I_2 = \frac{P_i}{P_s} = \left(\frac{R_i - C_i}{R_i}\right) / \left(\frac{R_s - C_s}{R_s}\right) \tag{2}$$

where I_1 is the share of the income of the enterprise from the sale of its products in the total amount of income received by the given enterprise and its competitors in the reporting period, the share of one;

R_i is the amount of income from the sale of products received by the company in the reporting period;

R_S is the total amount of income from the sale of products received by the company and its competitors in the reporting period;

I_2 is the relative profitability of products of this enterprise;

P_i is the average profitability of the total product of a given enterprise in the reporting period;

P_S is the average profitability of the entire population produced and sold by this enterprise and its competitors, in the reporting period;

C_i is the cost of production and sale of products incurred by this enterprise in the reporting period;

C_S is the total cost of production and sale of products incurred by this enterprise and its competitors in the reporting period.

It should be noted that in the context of this study, it is possible to identify the following factors that determine the change in the average profitability of the enterprise's products in the reporting period compared to the baseline period:

(1) Change (decrease) in the cost of purchasing some type(s) of energy resources, the efficiency of which increased in the reporting period as a result of ESTC at the enterprise;
(2) Change (increase) in some other expenses of the enterprise caused by the implementation of ESTC at the enterprise. In particular, such expenses may include depreciation charges for the renovation of purchased energy-saving equipment, expenses for the purchase of other types of energy resources (if ESTC involve the replacement of one type of such resources with another), etc.;
(3) Changes in all other expenses of the enterprise that are not related to the implementation of ESTC at the enterprise;
(4) Changes in the enterprise's income from sales.

In order to eliminate the impact of changes in income from the sale of products on the change in the average profitability of products of the enterprise, it is advisable to determine other factors per unit of income. Then it is possible to decomposition the value of the average profitability of the total product of the enterprise in the reporting period compared to the base period. For this purpose, the following equations should be used:

$$\Delta P_{i1} = (c_{i10} - c_{i11}) \cdot 100/\Delta P_i \tag{3}$$

$$\Delta P_{i2} = (c_{i20} - c_{i21}) \cdot 100/\Delta P_i \tag{4}$$

$$\Delta P_{i3} = (c_{i30} - c_{i31}) \cdot 100/\Delta P_i \tag{5}$$

where $\Delta P_{i1}, \Delta P_{i2}, \Delta P_{i3}$ is the increments in the average profitability of the entire set of products of a given enterprise in the reporting period compared to the base period due to changes in the costs of purchasing a certain type (types) of energy resources, some other costs associated with the implementation of ESTG at the enterprise, and all other costs of the enterprise not related to ESTG, calculated per unit of revenue from product sales;

c_{i10}, c_{i11} is the cost of acquiring some type (types) of energy resources, the increase of efficiency of use of which occurred in the reporting period as a result of proceedings at the enterprise of ESTG, per unit of income from the sale of products of the enterprise in the base and reporting periods, %;

ΔP_i is the total increase in the average profitability of the total product of the enterprise in the reporting period compared to the base period;

c_{i20}, c_{i21} is the value of some other costs of the enterprise related to the proceeding of ESTC on it, per unit of income from the sale of products of this enterprise, respectively in the basic and reporting periods, %;

c_{i30}, c_{i31} is the value of all other costs of the enterprise that are not related to the proceedings at the enterprise ESTC, per unit of income from the sale of products of the given enterprise, respectively in the basic and reporting periods, %.

Taking into account the above, it is possible to build a model of the impact of ESTC on the competitiveness of enterprises (Fig. 1). This model is based on the assumption that there should be a sufficiently close relationship between the market share occupied by the products of a given enterprise and the relative economic efficiency of production and sales of these products. In this case, if the implementation of ESTC at the enterprise leads to an increase in the relative profitability of the enterprise's products, the growth rate of the enterprise's revenues should be higher than the average growth rate of the total revenues of the enterprise and its competitors. At the same time, under certain conditions, an increase in the relative profitability of a given enterprise's products may not lead to a significant increase in the enterprise's revenues from the sale of its products. This may be due to the lack of sufficient production, sales and/or financial potential of the enterprise. However, if the enterprise has sufficient amounts of the above-mentioned types of potential, as well as appropriate management competences, the growth of relative profitability of products, in general, should have a statistically significant positive impact on the relative (compared to competitors) increase in the enterprise's revenues.

The use of in Fig. 1 model in the practice of enterprise management will better understand the mechanism of influence of ESTG on the competitiveness of these enterprises, which is necessary when making decisions on the implementation of energy-saving technological processes.

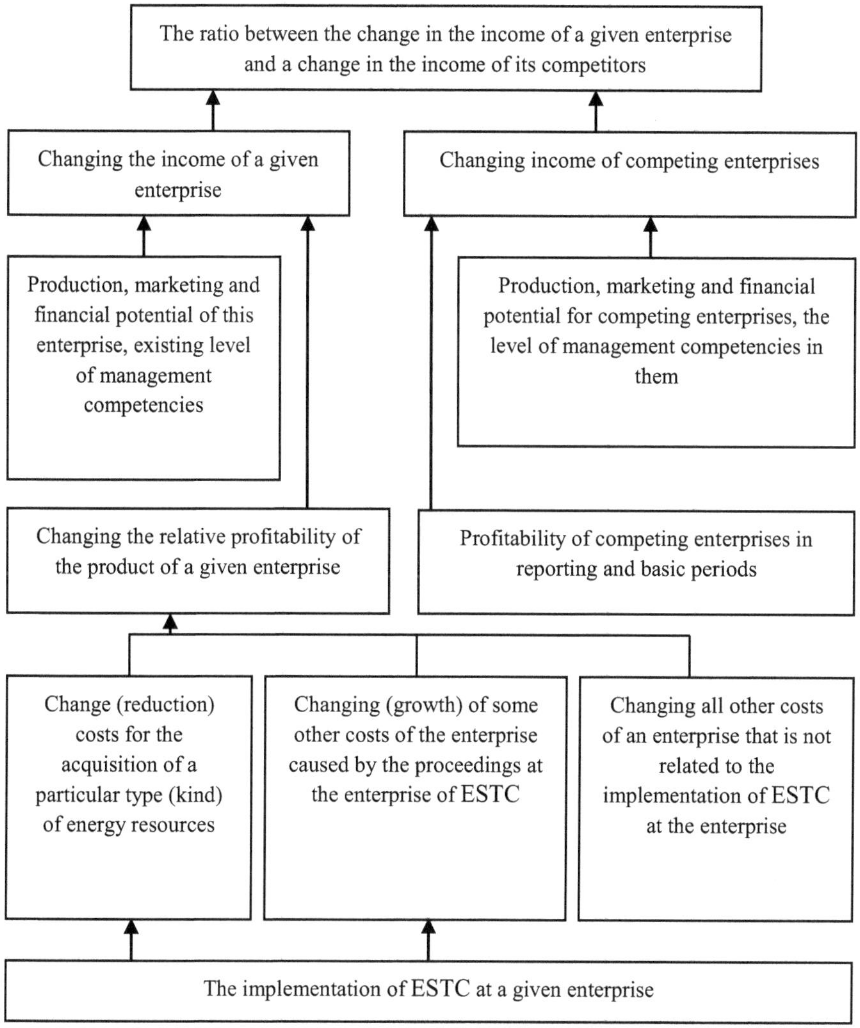

Fig. 1 Model of influence of energy-saving technological changes (ESTC) on the competitiveness of an enterprise. *Source* by authors

4 Developing a Method of Evaluating the Impact of Energy-Saving Technological Changes on the Competitiveness of Enterprises and Establishing the Conditions of Their Energy-Saving Development

When developing a method for assessing the impact of ESTC on the competitiveness of enterprises, it should be borne in mind that the market share indicator and the product profitability indicator characterise different aspects of the competitive advantages of companies. Indeed, the market share indicator characterises the competitive advantages already realised by the enterprise. At the same time, a company's high level of profitability of its products usually indicates a high potential for competitiveness of this company, but this potential may remain unrealised or not fully realised. Therefore, in the future, the main criterion for increasing the competitiveness of a particular enterprise will be the ratio between the growth rate of its income and the average growth rate of income of all competing enterprises:

$$I_c = T_i/T_s, \tag{6}$$

where I_c is the ratio between the growth rate of a particular enterprise's revenue from sales of its products and the average growth rate of revenue of all competitors;

T_i is the growth rate of a particular enterprise's revenue from sales of products in the reporting period compared to the base period;

T_s is the average growth rate of revenues of all competitors from sales of products in the reporting period compared to the base period.

Given the above, it is possible to offer two methodological approaches to assessing the impact of ESTC on the competitiveness of enterprises, namely:

1. A direct evaluation of such impact. The procedure of such evaluation involves the preliminary choice of indicator of evaluation of the level of ESTC in the enterprise. Subsequently, such an indicator will be the percentage of reducing the energy consumption of the company's products in the reporting period compared to the base period by conducting energy-saving technological changes. Then, having formed a sample of enterprises of a certain type of economic activity, it is possible to perform the division of these enterprises by the amount of the indicated indicator. For example, it is possible to distinguish companies with high, medium and low levels of ESTC that occurred during the study period. At the same time, for each such group of enterprises it is necessary to make a average value of the indicator of increasing competitiveness (6). Then, using the method of single-factor dispersion analysis, it is possible to establish the presence of statistically significant differences between these averages. If such differences exist, it means that there is a clear impact of energy-saving technological changes on the competitiveness of enterprises (Fig. 2).
2. Two-stage assessment. According to this methodological approach to assessing the impact of ESTC on the competitiveness of firms, the first step is to decompose

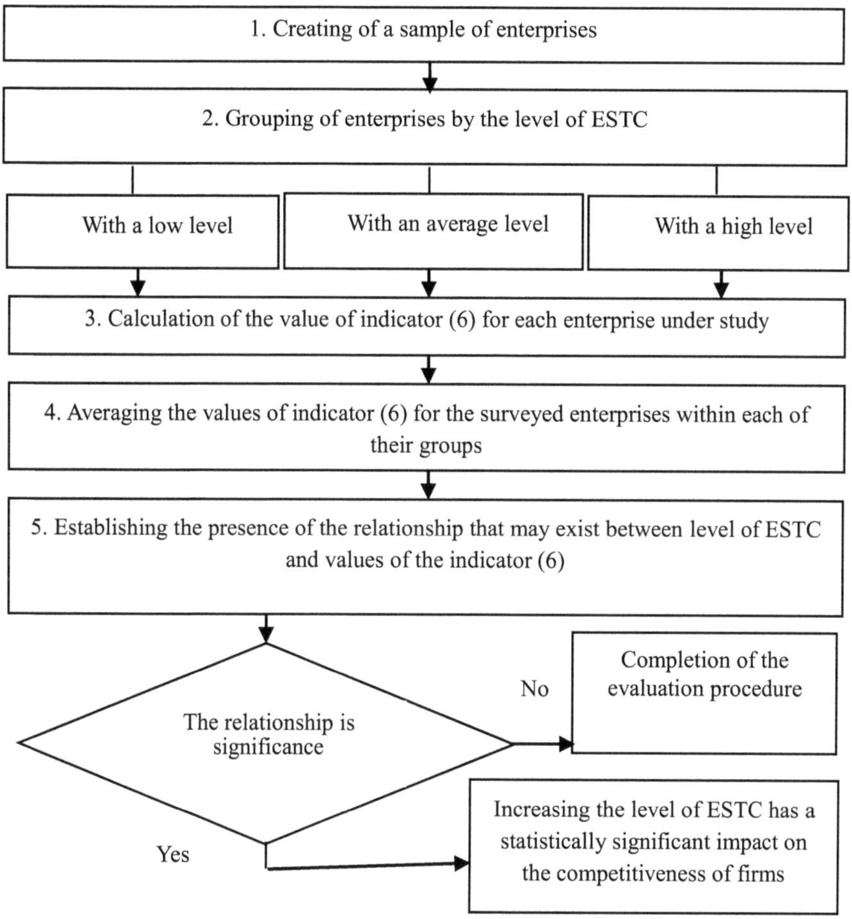

Fig. 2 Sequence of assessing the impact of the level of energy-saving technological changes (ESTC) at enterprises on their competitiveness when applying the method of direct assessment of such impact. *Source:* by authors

the increase in the average profitability of competing enterprises. For this purpose, Eqs. (3–5) should be used. This makes it possible to determine the extent to which ESTC affect changes in the profitability of companies' products. The corresponding indicator can be calculated using the following expression:

$$I_f = \frac{\Delta P_{1i} + \Delta P_{2i}}{\Delta P_i} \tag{7}$$

where I_f is an indicator of the degree of influence of ESTC on the change in profitability of a particular company's products, fractions of a unit.

If the indicator value (7) is high enough, in particular exceeding 0.5, it can be concluded that energy-saving technological changes have had a significant positive impact on the change in profitability of a particular enterprise. Then in the second stage of the calculations, you need to perform a sequence of actions similar to the Fig. 2, only instead of dividing enterprises into groups by the values of the level of ESTC, it is necessary to divide enterprises by the absolute increase of the average profitability of their products ΔP_i. If it is established that increasing the profitability of products has a statistically significant effect on the indicator (6), and ESTC, in turn, significantly affect the profitability of products, it will also indicate the significant impact of these changes on the competitiveness of at least part of the studied companies.

It should be noted that the introduction of ESTC may not always lead to savings in the respective types of energy resources in absolute terms, i.e., the rebound effect in energy consumption, known in science, is possible. This happens if the positive impact of the introduction of energy-saving technologies at enterprises on their competitiveness is so great that the growing production volumes will require more energy consumption than before the introduction of these technologies, although the specific energy consumption decreases. At the same time, this situation will not arise if the specified impact is relatively moderate. Thus, there are certain limitations to the increase in the level of competitiveness of enterprises as a result of the introduction of energy-saving technologies, if the additional goal is to ensure energy-saving economic development (in which the growth of enterprises' income is accompanied by a decrease in their consumption of a certain type(s) of energy resources).

To establish these restrictions, it is necessary that the elasticity of the enterprise's income in terms of reducing the energy intensity of its products due to the implementation of ESTC is not too high. The appropriate statement can be formalized in the form of such inequality:

$$T_i = 1 + \gamma \cdot T_e < 1/T_e \tag{8}$$

where γ is the elasticity of the enterprise's income with respect to the rate of reduction of the energy intensity of its products as a result of ESTC at this enterprise, fraction of a unit;

T_e is the rate of decrease in the energy intensity of the enterprise's products in the reporting period compared to the baseline period as a result of ESTC at this enterprise.

Then, taking into account expression (6), inequality (8) can be transformed into the following:

$$I_c \cdot T_s < 1/T_e \tag{9}$$

Thus, inequality (9) formalises the restriction on the level of growth of competitiveness of a certain enterprise I_c, so that such growth is accompanied by energy-saving economic development of this enterprise.

5 Practical Applications Developed Method of Evaluating the Impact of Energy-Saving Technological Changes on the Competitiveness of Enterprises

In order to evaluate the impact of ESTC on the competitiveness of enterprises, three samples of industrial companies of Ukraine were formed. Each of these samples contained only companies belonging to the same industry. At the same time, all these industries are natural gas consumers. Therefore, this empirical study was performed in relation to this type of energy resources. Accordingly, in the future, ESTC will be understood as those that lead to a reduction in the specific costs of natural gas.

General data on the samples of the studied enterprises are presented in Table 1. As follows from the information in this table, in 2018–2023, more than 50% of enterprises in each industry sector implemented ESTC. At the same time, more than 47% of the surveyed companies in all industries were characterised by energy-saving economic development in 2018–2023.

At the subsequent stage of the study, the impact of ESTC on the competitiveness of the studied enterprises was established by directly evaluating such influence. To this end, the enterprise was divided into groups by a relative decrease in the energy consumption of products in 2023 compared to 2018 due to the implementation of ESTC. At the same time, only consumption of natural gas enterprises was taken into account when calculating energy intensity.

The results of the calculations performed according to the sequence of actions described above in Fig. 2, shown in Table 2. As the data Table 2 follows, as the level ESTC increases the value of the indicator (6) increases by groups of companies. At the same time, the high estimated values of the F test indicate that this dependence is statistically significant. Therefore, the direct assessment of the impact of ESTC on the competitiveness of the studied enterprises showed that this influence exists and it is quite significant and positive.

Using expressions (3–5), we decomposed the increase in the average profitability of the products of the studied enterprises that implemented ESTC in 2018–2023. As follows from the data in Table 3, in all three industries, on average, more than 150% of the increase in product profitability was due to ESTC. At the same time, as can be seen from the data summarised in Table 4, for the companies studied, an increase in product profitability has a statistically significant positive impact on indicator (6). Thus, according to the two-stage assessment of the impact of ESTC on the competitiveness of the researched enterprises, this impact for the researched companies is quite strong and positive.

Of some interest is the question of whether all enterprises that experienced an increase in the values of indicator (6) during 2018–2023, that is, those that experienced an increase in the level of competitiveness, were characterized by energy-saving economic development during this period. As it follows from the data given in Table 5, the combination of increased level of competitiveness with energy-saving economic development during 2018–2023 was peculiar to all studied companies, which in this period of time improved competitiveness.

Table 1 General data on the sample of surveyed enterprises

Names of indicators	The value of indicators by industry		
	Production of metal products	Production of glass and glass products	Production of bricks and other building materials made of clay
1. Total number of surveyed enterprises	35	31	42
2. Number of surveyed enterprises that implemented ESTC in 2018–2023	20	16	22
3. Share of enterprises that implemented ESTC in 2018–2023 in the total number of surveyed enterprises, %	57.14	51.61	52.38
4. Number of surveyed enterprises that experienced economic growth in 2018–2023	27	26	37
5. Share of enterprises with economic growth in 2018–2023 in the total number of surveyed enterprises, %	77.14	83.87	88.10
6. Number of surveyed enterprises with energy-saving economic development based on natural gas in 2018–2023	18	16	20
7. The share of enterprises with energy-saving economic development in 2018–2023 in the total number of surveyed enterprises, %	51.43	51.61	47.62

Source:by authors

Table 2 Indicators of direct assessment of the impact of ESTC on the competitiveness of the studied enterprises

Names of indicators	The value of indicators by industry		
	Production of metal products	Production of glass and glass products	Production of bricks and other building materials made of clay
1. Number of enterprises belonging to groups of companies by levels of ESTC			
1.1. Enterprises with a low level of ESTC (less than 0.1%)	15	15	20
1.2. Enterprises with an average level of ESTC (from 0.1% to 0.3%)	12	9	13
1.3. Enterprises with a high level of ESTC (more than 0.3%)	8	7	9
2. Average values of indicator (6) by groups of companies by levels of ESTC			
2.1. Enterprises with a low level of ESTC	0.91	0.88	0.93
2.2. Enterprises with an average level of ESTC	1.03	1.01	1.04
2.3. Enterprises with a high level of ESTC	1.08	1.07	1.09
3. Estimated values of the *F*-test	5.93	6.26	5.78

Source:by authors

Thus, the implementation of ESTC at the studied enterprises not only increased the competitiveness of most of them, but also contributed to their energy-saving economic development.

6 Conclusions

Establishing a measure of the impact of ESTC on the competitiveness of companies requires a preliminary selection of the indicator (or group of indicators), which assesses the level of competitive advantages of firms. In particular, such indicators include the share of the income of the enterprise from the sale of its products in the

Table 3 Indicators of the decomposition of the increase in the average profitability of the products of the studied enterprises that implemented ESTC in 2018–2023

Factors of influence	Averaged by enterprises the importance of the impact of relevant factors by industry, %		
	Production of metal products	Production of glass and glass products	Production of bricks and other building materials made of clay
1. Changing the cost of purchasing natural gas per unit of income from sales of products	164.43	170.17	159.72
2. Changing some other costs related to the implementation of ESTC at the enterprise per unit of income from the sale of products	– 11.21	– 12.95	– 8.76
3. Changing all other costs of an enterprise not related to ESTC, per unit of income from sales of products	– 53.22	– 57.22	– 50.96
4. Together the first two factors	153.22	157.22	150.96

Source:by authors

total amount of income received by the company and its competitors in the reporting period. Another such indicator may be the relative profitability of the product of the enterprise, that is, the ratio of this profitability in this company to the average profitability of products in all competing enterprises.

In this case, the factors that cause the change in profitability of the enterprise in the reporting period compared to the base period, should be divided into the following groups: change (reduction) of costs for the acquisition of a certain type(s) of energy resources, increasing the efficiency of use of which occurred in the reporting period as a result of proceedings at the enterprise of ESTC; change (growth) of some other costs of the enterprise, due to the proceedings at the enterprise of ESTC; changing all other costs of an enterprise that is not related to the implementation of ESTC on it; changing the income of the enterprise from marketing products.

In order to assess the impact of ESTC on the competitiveness of enterprises, two methodological approaches developed in the process of this study should be used, namely, direct assessment of such impact and two-stage assessment. Despite some differences, both of these approaches involve assessing the statistical significance of

Table 4 Indicators of two-stage assessment of the impact of energy-saving technological changes on the competitiveness of the studied enterprises

Names of indicators	The value of indicators by industry		
	Production of metal products	Production of glass and glass products	Production of bricks and other building materials made of clay
1. Number of enterprises belonging to groups of companies by absolute increase in the average profitability of their products			
1.1. Enterprises with a small increase in the average profitability of their products during 2018–2023 (less than 0.01%)	16	17	22
1.2. Enterprises with an average increase in the average profitability of their products during 2018–2023 (from 0.01% to 0.03%)	11	7	12
1.3. Enterprises with a large increase in the average profitability of their products during 2018–2023 (more than 0.03%)	8	7	8
2. Average values of indicator (6) by groups of companies by absolute increase in the average profitability of their products			
2.1. Enterprises with a small increase in the average profitability of their products during 2018–2023	0.93	0.90	0.94
2.2. Enterprises with an average increase in the average profitability of their products during 2018–2023	1.02	1.01	1.03

(continued)

Table 4 (continued)

Names of indicators	The value of indicators by industry		
	Production of metal products	Production of glass and glass products	Production of bricks and other building materials made of clay
2.3. Enterprises with a large increase in the average profitability of their products during 2018–2023	1.09	1.08	1.11
3. Estimated values of the *F*-test	7.08	6.55	6.16

Source:by authors

Table 5 Indicators of the relationship between increasing the level of competitiveness of the studied enterprises and the availability of energy-saving economic development

Names of indicators	The value of indicators by industry		
	Production of metal products	Production of glass and glass products	Production of bricks and other building materials made of clay
1. Number of surveyed enterprises with a value of (6) greater than one	18	16	20
2. Including the number of enterprises that were characterised by energy-saving economic development in 2018–2023	18	16	20
3. The share of enterprises that were characterised by energy-saving economic development in 2018–2023 in the total number of enterprises with a value (6) greater than one, %	100.00	100.00	100.00

Source:by authors

the relationship between the scale of ESTC and the ratio of the growth rate of a particular enterprise's revenue from sales and the average growth rate of all competitors' revenue.

Both modifications of the developed method for assessing the impact of ESTC on the competitiveness of enterprises were tested on samples of Ukrainian enterprises belonging to three industries and consuming natural gas. The study has shown that, according to both methodological approaches, with an increase in the level of ESTC, the ratio between the growth rate of the enterprise's revenues from sales of its products and the average growth rate of revenues of all competitors increases. It has been found that the impact of ESTC on the competitiveness of the researched enterprises is quite significant and statistically significant.

It was also found that in all three industries the increase in product profitability was due to the implementation of ESTC. Finally, the research showed that the implementation of ESTC at the studied enterprises not only increased the competitiveness of most of them, but also contributed to their energy-saving economic development.

References

1. Hurzhyi N, Mishustina T, Kulinich T, Dashko I, Harmider L, Taranenko I (2021) The impact of innovative development on the competitiveness of enterprises. Postmod Open 12(4):141–152. https://doi.org/10.18662/po/12.4/365
2. Nagdiev O (2023) Evaluation of the innovation-investment factor for augmenting the competitiveness of construction enterprises. Econ Transp Complex 42:143–154. https://dspace.khadi.kharkov.ua/handle/123456789/17706
3. Lesinskyi V, Yemelyanov O, Zarytska O, Symak A, Petrushka T (2021) Devising a toolset for assessing the potential of loan financing of projects aimed at implementing energy-saving technologies. East-Eur J Enterp Technol 4(13):15–33. https://doi.org/10.15587/1729-4061.2021.238795
4. Lesinskyi V, Yemelyanov O, Zarytska O, Petrushka T, Myroshchenko N (2022) Designing a toolset for assessing the organizational and technological inertia of energy consumption processes at enterprises. East-Eur J Enterp Technol 6(13):29–40. https://doi.org/10.15587/1729-4061.2022.267231
5. Jalo N, Johansson I, Kanchiralla FM, Thollander P (2021) Do energy efficiency networks help reduce barriers to energy efficiency? A case study of a regional Swedish policy program for industrial SMEs. Renew Sustain Energy Rev 151:111579. https://doi.org/10.1016/j.rser.2021.111579
6. Chen Y, Jiang Z (2024) The impact of enterprise technological innovation on environmental performance—an industry perspective. Sustainability 16:6457. https://doi.org/10.3390/su16156457
7. Wu Y, Tham J (2023) The impact of environmental regulation, environment, social and government performance, and technological innovation on enterprise resilience under a green recovery. Heliyon 9(10). https://doi.org/10.1016/j.heliyon.2023.e20278
8. Yemelyanov O, Petrushka T, Symak A, Lesyk L, Musiiovska O (2022) Modelling the impact of energy-saving technological changes on the market capitalization of companies. In: Systems, decision and control in energy III. Springer, pp 89–106. https://link.springer.com/chapter/https://doi.org/10.1007/978-3-030-87675-3_5
9. Yemelyanov O, Dziurakh Y, Danylovych T, Bondarchuk N, Demianchuk O, Kharchuk Y (2024) Investing in resource-saving measures as a tool of anti-crisis financial management at enterprises. Int J Qual Res 18(1). https://doi.org/10.24874/IJQR18.01-20
10. Andrew AA, Adebayo TS, Lasisi TT, Muoneke OB (2024) Moderating roles of technological innovation and economic complexity in financial development-environmental quality nexus of the BRICS economies. Technol Soc 78:102581. https://doi.org/10.1016/j.techsoc.2024.102581
11. Ahmad N, Nguyen DK, Tian XL (2024) Assessing the impact of the sharing economy and technological innovation on sustainable development: an empirical investigation of the United Kingdom. Technol Forecast Soc Chang 209:123743. https://doi.org/10.1016/j.techfore.2024.123743
12. Kyrylenko VI, Tkachenko OV, Kotenok AG, Sarkisian HO, Sitkovska AO (2022) The role of innovative technologies in ensuring the competitiveness of the enterprise, vol 5. Naukovyi

Visnyk Natsionalnoho Hirnychoho Universytetu, pp 185–190. https://doi.org/10.33271/nvngu/2022-5/185

13. Padilla-Lozano CP, Collazzo P (2022) Corporate social responsibility, green innovation, and competitiveness—causality in manufacturing. Compet Rev 32(7):21–39. https://doi.org/10.1108/CR12-2020-0160
14. Alvarez-Aros EL, Bernal-Torres CA (2021) Technological competitiveness and emerging technologies in Industry 4.0 and Industry 5.0. Anais Da Academia Brasileira de Ciencias 93(1). https://doi.org/10.1590/0001-3765202120191290
15. Leão P, Silva MM (2021) Impacts of digital transformation on firms' competitive advantages: a systematic literature review. Strat Change 30(5):421–441. https://doi.org/10.1002/jsc.2459
16. Gavkalova N, Smereka S, Kadyrus I, Kyrylenko S, Kasatkina M (2024) Using innovation and energy efficiency technologies to drive business competitiveness. Grassroots J Nat Resour 7(3):s212–s229. https://doi.org/10.33002/nr2581.6853.0703ukr11
17. Di Foggia G (2021) Energy-efficient products and competitiveness in the manufacturing sector. J Open Innov Technol Mark Complex 7:33. https://doi.org/10.3390/joitmc7010033
18. Wei YG, Zhu RQ, Tan LY (2022) Emission trading scheme, technological innovation, and competitiveness: evidence from China's thermal power enterprises. J Environ Manag 320:115874. https://doi.org/10.1016/j.jenvman.2022.115874
19. Timilsina GR (2021) Are renewable energy technologies cost competitive for electricity generation? Renew Energy 180:658–672. https://doi.org/10.1016/j.renene.2021.08.088
20. Klemke-Pitek M, Majchrzak M (2022) Pro-ecological activities and shaping the competitive advantage of small and medium-sized enterprises in the aspect of sustainable energy management. Energies 15(6):2192. https://doi.org/10.3390/en15062192
21. Dzhedzhula V, Yepifanova I, Brych V (2024) Economic and mathematical simulation model of production in the crisis environment condition. In: 2024 14th International conference on advanced computer information technologies (ACIT), vol 11, pp 439–442. https://doi.org/10.1109/acit62333.2024.10712473
22. Yepifanova I, Dzhedzhula V (2025) Mathematical model of intellectual decision-making support for the selection of the enterprise's optimal production plan under conditions of energy restrictions. Lect Notes Data Eng Commun Technol 240:367–377. https://doi.org/10.1007/978-3-031-81557-7_21

Conceptual Strategies for Managing Innovation Processes in the Life Cycle of the EU Tourism Industry

Daria Mamotenko, Olga Korniienko, Tetiana But, Olena Gubar, and Anatolii Pavlenchyk

Abstract Due to its developed tourism potential, Europe continues to be one of the most attractive regions for international tourists. Therefore, the study aims to develop conceptual strategies for managing innovation processes in the tourism industry of the EU countries, taking into account the life cycle stages, which will ensure the industry's long-term development. The research methodology included statistical data, graphical, tabular, abstract, and logical methods, visualization of theoretical and practical provisions, and systematic and comprehensive analysis. The study identifies the stages of the EU tourism industry life cycle (ILC) by such indicators as the growth rate of tourism revenue, the growth rate of the TTDI tourism development index, the growth rate of the eco-innovation index, the growth rate of the human resources index, and the growth rate of the digitalization index. The analysis of the ILC was carried out based on the classical model of Porter M. and the TALC hypothesis, according to Butler RW. The result of the study is an analysis of the EU ILCs and the development of conceptual strategies for managing innovation processes according to the stages studied in the modern EU ILCs. The recommendations provided in this study will be helpful for global public institutions to update and develop strategies for managing innovation processes in line with market conditions.

D. Mamotenko (✉) · O. Korniienko
Department of Tourism, Hotel and Restaurant Business, Faculty of Humanities, National University Zaporizhzhia Polytechnic, Zaporizhzhia, Ukraine
e-mail: mamotenko@zp.edu.ua

T. But
Department of Tourism, Faculty of Entrepreneurship and Law, Pan-European University, Prague, Czech Republic
e-mail: tatyanabut1979@ukr.net

O. Gubar
Department of Economics and Customs, Faculty of Business Technologies and Economics, National University Zaporizhzhia Polytechnic, Zaporizhzhia, Ukraine

A. Pavlenchyk
Department of Informatics, Kinesiology and E-Sports, Faculty of Tourism, Ivan Boberskyj Lviv State University of Physical Culture, Lviv, Ukraine

A. Semenov et al. (eds.), *Data-Centric Business and Applications*, Lecture Notes on Data Engineering and Communications Technologies 272,
https://doi.org/10.1007/978-3-032-06608-4_15

Keywords Tourism industry · Innovation · Life cycle · EU · Strategies

1 Introduction

The study aims to develop conceptual strategies for managing innovation processes in the tourism industry of the EU countries, taking into account the life cycle stages, which will ensure the industry's long-term development.

The tourism industry in the EU plays a key role in shaping the region's economic development, making a significant contribution to GDP, employment, and the international image of the countries.

European tourism remained strong at the end of 2024, with the number of foreign tourists up 6.3% compared to 2019 and 6.7% compared to 2023. Overnight stays were up 5.9% compared to 2019 and 4.8% year-on-year. Despite economic difficulties and extreme weather conditions, demand remained strong, although inflation pushed tourists towards destinations with good value for money. Tourism spending increased 7.8% to €705 billion, with Western Europe accounting for almost 75% of the total [1].

However, the development of the industry is not uniform: different countries are at various stages of the industry life cycle (ILC), which necessitates differentiated management strategies. The study's conceptual framework requires further development through deeper analysis and integration of several key aspects that contribute to effectively managing innovations in the tourism industry. In particular, it is essential to pay attention to all stages of the life cycle of tourism services, from the initial to the mature [2]. This requires adapting innovative strategies to changing demand conditions and seasonality and changes in global trends such as digitalization, climate change, and social trends.

It is essential to incorporate the principles of environmental sustainability [3, 4] and social responsibility at all stages of the tourism services life cycle. This requires the development of innovations [5] that reduce the negative impact on the environment while increasing the efficiency and accessibility of tourism services [6].

Due to the development of digital technologies [7], it is necessary to focus on integrating the latest digital tools in the tourism industry, such as big data, artificial intelligence, and process automation. Conceptual strategies should focus on creating conditions for the effective use of these technologies, in particular, to optimize tourism management processes, personalize services, and improve interaction with end users. For effective management of innovation in the life cycle of the European tourism industry, it is vital to develop concepts that can respond quickly to global challenges [8] while taking into account local specifics and the needs of different markets. It should consider the collaboration between governments, businesses, academic institutions, and NGOs, which will allow mechanisms for coordinating efforts at all levels of implementing innovative strategies for developing the tourism industry [9]. Considering cultural and social characteristics to create sustainable and adaptive innovation strategies will affect European social and cultural integration.

Thus, conceptual strategies for managing innovation processes in the life cycle of the European tourism industry should combine technological, environmental, economic, and social aspects, focusing on sustainable development and global competitiveness. Integrating these strategies is a critical step toward creating an innovative and sustainable tourism industry that can effectively adapt to future challenges.

2 Literature Review

For a deep and comprehensive approach to the study of innovation processes, a range of factors should be considered to ensure a thorough analysis and high scientific validity.

The innovation approach determines the impact of innovations on the economic development of the tourism industry while identifying innovative strategies for tourism enterprises to adapt more effectively to changing market conditions and global competition [5, 7]. Another approach is based on the role of digital technologies (e.g., artificial intelligence, big data, blockchain) in transforming the tourism sector, as studied by [6]. It is essential to consider how these changes will affect the sustainability and development of the industry in the future.

The economic approach is based on the competitiveness of tourism enterprises in the international market [10]. He points out how competition stimulates tourism businesses to improve service quality, which helps attract customers. This approach has potential but requires more detailed analysis and consideration of various aspects to ensure its effectiveness and reliability.

An investment approach based on the dynamics of investments in the industry and their effectiveness identifies successful cases and strategies and can attract new investors to the sector [11]. This should consider the risks and challenges that investors may face when investing in the tourism industry.

The environmental approach affects seasonality, the popularity of tourist destinations, and natural ecosystems. In particular, tourism enterprises are forced to adapt to new conditions by introducing energy-efficient technologies and environmental standards, which can increase costs [3] and determine the impact of climate change on the tourism industry [4].

The role of «green tourism» in the industry life cycle will allow engaging the local population in tourism activities, creating jobs, and maintaining cultural heritage [12]. Thus, environmental approaches not only help to adapt the tourism industry to climate change but also create new opportunities for its development. It is essential to consider the economic consequences so that the study will use statistical data for quantitative analysis.

The management approach is based on innovative solutions in management and marketing (online platforms, virtual tourism) [6]. However, it is necessary to consider the regional peculiarities and cultural differences of countries, as well as the human factors: motivation, satisfaction, and productivity of employees, as well as the way of interacting with customers.

The consumer approach is based on changes in consumer preferences in the field of tourism, provides an understanding of the needs of tourists, and contributes to the creation of personalized offers, which increases customer satisfaction and loyalty [13]. By analyzing changes in tourist preferences, this approach allows for creating new products and services that meet modern expectations, such as ecotourism, gastronomic tours, or digital tours. At the same time, differences should be considered when developing personalized offers.

The globalization approach is based on globalization and localization, i.e., it shows the balance between international and local tourism, as studied in [8]. Localization helps to avoid the over-saturation of popular tourist destinations, reducing environmental and social pressure. At the same time, the balance between international and local tourism allows for an even distribution of revenues, supporting both major tourist centers and less popular regions. However, this approach can be a challenge for less developed areas that require significant investment to improve transport and hotel infrastructure.

Considering the principles of sustainable development in innovation processes [14] contributes to conserving natural resources and creates new opportunities for developing the tourism industry. At the same time, it is crucial to consider that the principles of sustainable development are long-term, and changes in legislation may affect the tourism industry.

Studies of the European tourism industry show effective practices that contribute to implementing high service standards and innovative solutions [15]. At the same time, EU regulatory mechanisms should be considered, as they influence innovation processes in the tourism industry and play an important role [16].

The study of the duration of the different stages of the tourism industry life cycle [2, 17] allows the industry to adapt to the new realities after the COVID-19 pandemic by taking advantage of various factors. The net benefit of visiting a destination, which characterizes the exploration, engagement, and development phases, becomes clear. This explains how the social effects associated with destination popularity characterize the consolidation and stagnation phases.

The above scientific approaches emphasize the critical role of innovation processes in ensuring sustainable development in Europe. Innovations focused on sustainable development not only reduce the negative impact on the environment but also actively promote the integration of environmentally friendly technologies and renewable energy sources into various areas of economic activity. However, it is essential to note that the studies presented often do not pay due attention to a comprehensive analysis of the relationship between the product life cycle (ILC) and innovation processes and the principles of sustainable development. In particular, the issues of adapting the ICL to digital transformation, the impact of green investments, and EU regulatory initiatives remain insufficiently studied. Moreover, there is an apparent lack of research analyzing the changes in the stages of the ILC under the influence of global challenges, such as climate change, economic and social crises (e.g., the COVID-19 pandemic), and the rapid evolution of consumer preferences, particularly in the context of tourism.

Given the importance of managing innovation processes that directly affect the life cycle of the European tourism industry, there is an urgent need to develop a new approach to creating a theoretical and practical framework for effective innovation management in this sector. Such an approach should consider not only the specifics of the life cycle of tourism services but also the dynamics of their development and the crucial challenges associated with globalization and sustainable development. This approach must integrate modern technological innovation trends and changes in consumer preferences and meet the requirements for environmental sustainability and socio-economic balance in the context of global transformations. In addition, it is necessary to consider the diversity of European markets and regional specifics, which require a flexible and adaptive innovation management strategy in the tourism industry context.

3 Methods and Technologies

In the course of the study, the following methods of economic research were used: statistical (selection of variables for analyzing the life cycle of the European tourism industry); graphical and tabular methods (for visual display of statistical research materials); abstract and logical (for theoretical generalizations and formulation of conclusions); visualization of theoretical and practical provisions; systematic and comprehensive analysis (to identify trends in the development of the European tourism market).

When studying the life cycle of an industry, the models are based on a specific indicator or group of indicators, the dynamics of which determine the sector's transition from one stage to another. Taking into account both quantitative and qualitative aspects of development, an essential stage of the analysis is the identification of key indicators that signal changes in the life cycle of the tourism industry. The key indicators include the growth rate of tourism revenue, the growth rate of the TTDI tourism development index, the growth rate of the eco-innovation index, the growth rate of the human resources index, and the growth rate of the digitalization index.

The research was based on Eurostat data, Travel & Tourism Development Index, World Tourism Organization, Trading Economics, World Economic Forum, World Travel & Tourism Council, Organisation of United Nations, IMD World Competitiveness, World Investment Report, European Tourism: Trends & Prospects. World Tourism Organization, International Tourism Highlights.

The study was conducted in the following stages:

- theoretical aspects of the life cycle of the tourism industry are studied;
- Identify the life cycle stages of the EU countries' tourism industry using key indicators. The life cycle of the tourism industry in the EU countries is analyzed, identifying the stages of development and features of each of them;

- conceptual strategies for managing innovation processes in the life cycle of the tourism industry are developed.

4 Results

4.1 Theoretical Aspects of the Industry Life Cycle

Many researchers and scholars have addressed the life cycle issue at the industry level. M. Porter approached the industry life cycle analysis from the perspective of competitive dynamics. Michael Porter defined the ILC as a process that goes through four main stages: introduction, growth, maturity, and decline [18] (Fig. 1). The industry faces different challenges and opportunities at each stage, affecting companies' strategies and competitiveness. This concept helps to analyze how industries develop and adapt over time. In terms of the strategy organizations use at different stages of the industry life cycle, Porter identifies four possible strategies in crises: exit and rapid capital withdrawal, harvesting strategy, leadership capture and retention, and niche occupation. Any other methods can be used at different stages of the industry's life cycle [18].

In his research, Simons K. considers the classical industry life cycle model, which includes the following main phases: emergence, growth, shakeout, maturity, and decline. He has researched the dynamics of industrial organization and technological change, particularly the causes and consequences of «shakeouts» in the life cycle of industries. In his work, he analyzed how technological innovation and other factors influence the evolution of industries and the survival of companies [19].

In contrast to M. Porter and K. Simons, J. Moore considers the different stages of the industry life cycle from the consumers' perspective. The model he proposed includes three stages of industry development [20]:

1. The functionality phase is characterized by consumers for whom the product's functionality is crucial, the so-called «early adopters».
2. Consumers' focus on product reliability characterizes the reliability phase. J. Moore called such consumers «the early majority.»
3. The convenience phase is characterized by a shift in innovation and competition towards convenience. At this stage, the 'late majority' of consumers enter the market. In general, this model is based on the fact that the improvement of technology can reach a level where the needs of the market as one of the indicators will be satisfied.

James F. Moore also proposed the concept of a «business ecosystem,» which views industries as dynamic systems that evolve through four stages: pioneering, expansion, leadership, and renewal or death. This approach emphasizes the interdependence of companies and the importance of cooperation for joint development and survival in a competitive environment [20, 21].

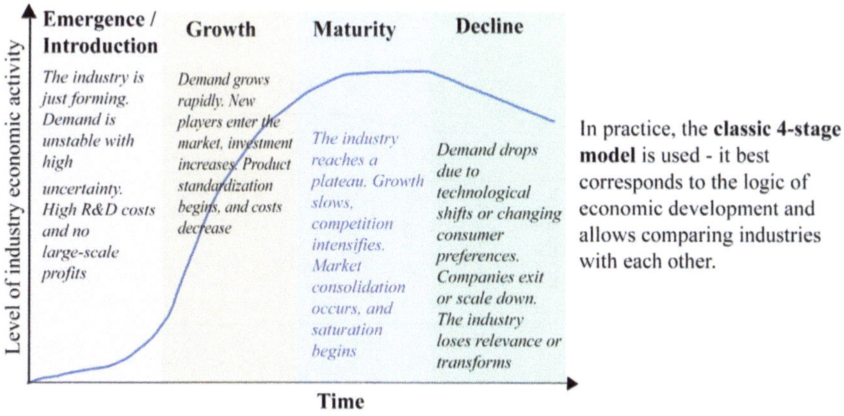

Fig. 1 The classical model of the industry life cycle is proposed by [18]

Clayton Christens discussed the «disruptive innovation» model, which describes how new technologies can change the market structure and the industry's life cycle [22].

The Tourism Area Life Cycle (TALC) hypothesis proposed by Butler [23] is probably the most essential theory discussed in the literature to describe and predict the evolution of the number of tourists in a particular tourist destination over time. Several extensions of the basic theory have been developed over the years. Still, in its simplest form, TALC states that destinations will go through different economic, social, and tourism development phases, which significantly affects the number of visitors they can receive.

In particular, each tourist destination will strive to pass through five stages, during which the number of tourists will grow more or less rapidly. At the early stage, the number of tourists will be limited and mainly represented by researchers (research). Then, as information about the destination spreads and the number of tourists increases, local people will get involved by providing tourist facilities (attraction). In the next stage, when information is widely disseminated through advertising, campaigns and tourist facilities are extensive in a destination where tourist flows will continue to grow regularly (development). Further, as promotional campaigns become less effective and the supply of tourist facilities cannot be expanded further, the increase in tourist flows will gradually decrease (consolidation) until the peak of tourism. The number of tourists is stable (stagnation).

Thus, according to the TALC hypothesis, the evolution of the number of tourists over time will follow a logistic pattern, as shown in Fig. 2.

Modern authors studying the life cycle of the industry have based their models on a specific indicator or group of indicators, the quantitative and qualitative changes that contribute to the sector's transition from one stage to another:

- Marsiglio and Tolotti analyzed the life cycle of the tourism industry through the influence of social and economic factors [2]

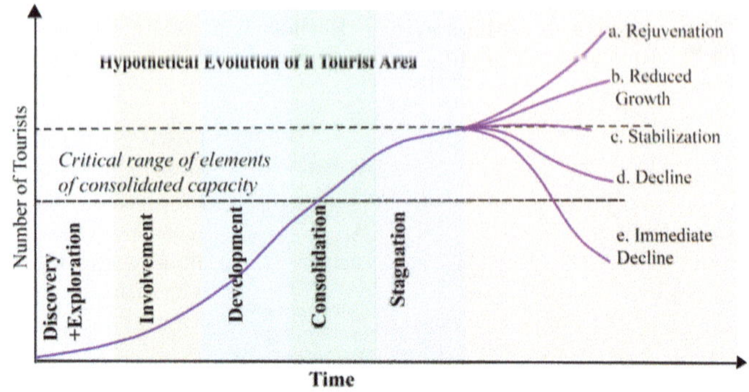

Fig. 2 The life cycle model of the tourism industry, according to Butler [23]

- The authors studied the impact of digital technologies (artificial intelligence, big data, blockchain) on the industry's transformation [6].
- The authors studied the relationship between the principles of sustainable development and the industry life cycle. The authors emphasize the commitment between environmental, economic, and social variables that is optimal in a given space and time frame to achieve sustainable tourism [24].
- The authors define the role of digital platforms in accelerating the stages of the life cycle of tourism services [25]

The economic literature offers several methods to determine an industry's life cycle stage based on specific indicators (Table 1).

The methods described in the table are based on analyzing the dynamics of market development, competitive environment, investment activity, and technological innovations as the basic factors signaling a change in the stage of the industry's life cycle. However, it should be emphasized that these indicators are not an unambiguous criterion that explains the behavior and dynamics of the industry's movement along the ILC curve, as external macroeconomic factors, the regulatory environment, and socio-demographic changes also influence its development.

The tourism industry is a complex economic system with interconnected subsystems, each impacting changes within the industry's life cycle. The interaction of these subsystems and their adaptation to modern challenges determine the strategic development of tourism. Given the current trends in digitalization, environmental sustainability, and innovative development, we will identify the main factors that influence the life cycle of the tourism industry:

1. **Economy and investment in tourism**—the development of the tourism industry depends on such indicators as tourism revenues, the share of tourism in GDP, the level of foreign direct investment, and the profitability of the hotel business and travel companies. An important factor is support from the government and international organizations in the form of programs to promote sustainable tourism.

Table 1 Methods for determining the life cycle stage of an industry

Method	Essence
Analysis of market growth rates	Determines the stage of an ILC based on the dynamics of sales, investment, and market share. High growth rates are characteristic of the growth stage, while slowdowns are typical of maturity and decline
Analysis of the competitive structure	Assesses the level of competition and the number of players in the market. In the early stages, a market has many firms, while in the mature stage, consolidation occurs
Analysis of profitability and margins	It defines the stage of the ILC by the level of profitability. High profitability is typical for the initial stages, while a decline in profitability signals maturity or decline
Technological analysis	Assesses the impact of technological innovations on the development of the industry. The active development of new technologies is observed at the stages of formation and growth, and their slowdown is observed at the stage of maturity
Analysis of consumer behavior	It studies the dynamics of changes in consumer preferences. Early adoption of innovations indicates the initial phase and mass consumption suggests the maturity stage
Analysis of investment activity	Determines the level of investment in the industry. High investment volumes are typical for growth and development stages, while a decline in investment may indicate maturity or decline
Analysis of variance	Identifies the factors that affect the life cycle of an industry and helps determine its current stage of development
Least squares method	It identifies trends in key industry indicators using regression analysis to predict the transition between the stages of the housing and utilities sector
Product lifecycle	Analyses the life cycle of key tourism products and services that shape the industry's overall development
Scenario forecasting	It assesses possible industry development trajectories based on macro and microeconomic factors
Analysis of development indices	It considers the level of digitalization, innovation potential, and environmental sustainability as key factors in the industry's development (TTDI, Global Innovation Index, Smart City Index)
Cohort analysis of consumers	It analyses changes in the behavior of different consumer groups over time, which allows us to assess the stage of the industry's life cycle
Cluster analysis	It is used to determine market heterogeneity and identify the industry's stage of development by identifying companies' strategic groups

Source: developed by the authors

2. **Digital infrastructure and technological innovation**—modern tourism increasingly depends on digital solutions. Booking systems, mobile applications, artificial intelligence, automated services, and big data analysis form the digital backbone of the industry. Smart City strategies and the «digital traveler» concept, which provides a personalized travel experience through technology, also play an essential role.

3. **Human capital and intellectual resources**—The level of training in the tourism sector is key to maintaining competitiveness. Educational programs, professional development, and the adaptation of employees to the use of digital technologies and new management approaches are critical to the effective functioning of the industry.
4. **Sustainable development and environmental initiatives**—modern tourism strategies are focused on ecological sustainability. Environmental certificates, green hotels, renewable energy sources in the tourism sector, control over carbon dioxide emissions, and compliance with ESG (environmental, social, and corporate governance) standards are increasingly important in shaping competitive advantages.
5. **Globalization and changes in consumer behavior**—modern tourists are becoming more demanding and informed. There is a growing demand for personalized travel products and new travel formats (eco-tourism, gastronomic tourism, digital nomads), and the influence of social media and bloggers on demand generation is growing.

Thus, the industry life cycle concept is a necessary analytical framework for understanding market development and strategic planning stages. Various scholars have widely studied it, and today, there are different approaches to defining its stages. In its classical form, the ILC model includes the following stages: introduction, growth, maturity, and decline. At the same time, several authors propose detailed approaches with additional stages, such as restructuring or transformation, which is especially relevant in digitalization and sustainable development. Each life cycle stage—inception, growth, maturity, decline, or transformation—requires a unique approach to investment, regulatory policy, marketing strategies, and infrastructure development. This is especially important for the tourism industry, a complex, dynamic system with a high dependence on external factors such as macroeconomic fluctuations, technological progress, changes in consumer behavior, and environmental challenges. Tourism is also closely linked to innovation, which includes digital transformation, sustainable mobility, environmental certification, and new travel behavior. Therefore, adapting strategies to the current stage of ILC in tourism is a prerequisite for achieving long-term sustainability and competitiveness in the industry.

4.2 Identification of the Life Cycle Stages of the EU Tourism Industry by Key Indicators

Europe, the world's largest destination region, recorded 747 million international arrivals in 2024 (up 1% from 2019 and 5% from 2023), driven by strong intra-regional demand. All European sub-regions have exceeded pre-pandemic levels, except Central and Eastern Europe, where many destinations are still suffering from the lingering effects of Russia's aggression against Ukraine [26]. It is essential to record the recovery of tourist flows and analyze structural changes in the industry, which are manifested through its dynamics in different countries. The analysis of key indicators allows us to assess the industry's current state and identify the stages of its life cycle. Differentiation of countries by the level of development of the tourism sector helps to identify the main trends, challenges, and prospects for growth, as well as to determine the strategic directions of state policy in the field of tourism.

To study the life cycle of the EU tourism industry, it is advisable to analyze the key factors that determine the dynamics of its development, including the economy and investment in tourism, digital infrastructure, and technological innovation, human capital and intellectual resources, sustainable development, and environmental initiatives, as well as globalization and changes in consumer behavior. Each of these factors will be represented by a respective analytical indicator, which quantifies its impact on the industry. In particular, the analysis includes such indicators as the growth rate of tourism revenue, the growth rate of the TTDI tourism development index, the growth rate of the eco-innovation index, the growth rate of the tourism human resources index, and the growth rate of the digitalization index. This approach provides a comprehensive vision and allows us to determine the stage of the industry's life cycle based on objective data.

The growth rate of tourism revenue is a dynamic indicator that promptly reflects changes in demand for tourism services, the country's ability to attract tourists, the level of economic benefits from tourism, and the effectiveness of tourism policy implementation. Unlike static absolute values of revenues, which can be high even in stagnant economies due to historically established infrastructure or brand, the growth rate demonstrates the direction the tourism industry is developing. Figure 3 shows the distribution of EU countries by the life cycle stages of the country's tourism industry (revenue growth rate in 2023/2022).

The European Union countries demonstrate different stages of development of the tourism industry, which the growth rate of tourism revenues can identify. The

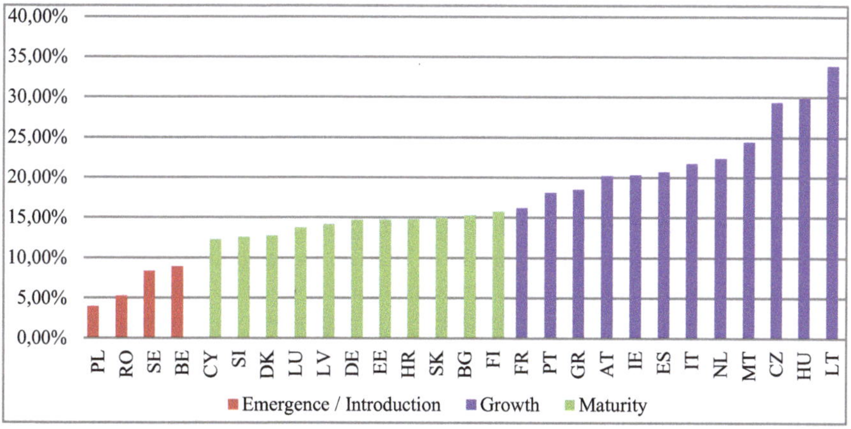

Fig. 3 Life cycle stages of the EU tourism industry by revenue growth rate in 2023/2022. *Source* developed by the authors according to [27]

Emergence/Introduction stage includes countries with low but favorable revenue growth rates (up to 10%), which indicates the initial or slow phase of recovery and expansion of the tourism sector. Among them are Poland (4%), Romania (5%), Sweden (5%) and Belgium (9%). The tourism industry in these countries is in an early growth phase after the crisis, with low demand dynamics. To accelerate development, it is advisable to intensify marketing campaigns, support the development of tourism infrastructure, expand international cooperation, and implement digital solutions.

The Growth stage covers countries with growth rates of approximately 16–35%, which indicates an active growth of interest in tourism services and an increase in the economic return from the sector. This group includes France, Portugal, Greece, Austria, Ireland, Spain, Italy, the Netherlands, Malta, the Czech Republic, and Hungary.

In these countries, tourism is actively developing, with positive income dynamics, expanding tourism product offerings, and modernization of the industry. Investing in sustainable tourism development, effective management of tourist flows, implementation of «smart tourism» concepts, and support for small businesses in the sector is recommended.

Finally, the Maturity stage includes countries that demonstrate an average level of growth (about 10–15%) and are approaching a stable functioning of the tourism industry. This category contains Cyprus, Slovenia, Denmark, Luxembourg, Latvia, Germany, Estonia, Croatia, Slovakia, Bulgaria, and Finland. In these countries, the tourism sector is already established and well-developed, with a high level of involvement, but the growth rate is gradually slowing down. At this stage, it is advisable to focus on improving the quality of tourism services, digital transformation, developing niche products, and ensuring a regional balance in tourism development.

It should be noted that the highest level of tourism revenue in the world is observed in four countries that showed significant growth in 2024 compared to 2019: the United Kingdom (+40%), Spain (+36%), France (+27%), and Italy (+23%) [28].

According to the European Commission for Tourism, tourists have increased interest in Southern European and Mediterranean destinations. This is because tourists still prefer the warm weather and value for money that can still be found in some parts of Southern Europe [29].

Among the significant destinations, Portugal reported 18% more arrivals than in 2019, Denmark 18% more than in 2019, and Denmark 17% more than in 2019, while Greece and Turkey recorded 14% more and Spain 10% more. In total, 21 European destinations exceeded pre-pandemic levels during this period [30].

The growth rate of the TTDI (Travel & Tourism Development Index) should be used to build the life cycle of the tourism industry, as this indicator reflects the dynamics of changes in the competitiveness of the country's tourism sector. The TTDI index is a comprehensive tool that considers a wide range of factors, from transport and digital infrastructure to natural and cultural resources, the level of sustainable development, the regulatory environment, safety, health, and the quality of labor. Thus, TTDI is not limited to the assessment of tourist flows or revenues but shows the real state of development of the institutional and infrastructural framework on which the tourism industry is based.

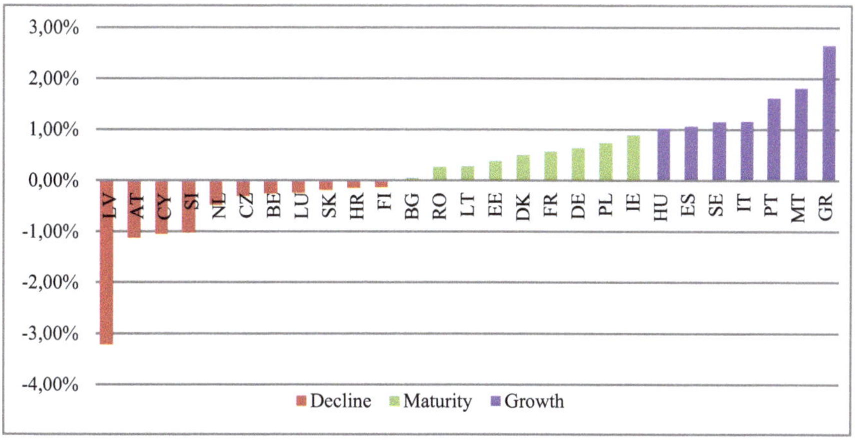

Fig. 4 Stages of the life cycle of the tourism industry in the EU countries by the growth rate of the TTDI tourism development index (2024/2021). *Source* developed by the authors according to [30]

The growth rate of the TTDI tourism development index is particularly valuable, as it allows not only to record a country's position at a particular stage but also to assess its progress over time—whether it is growing, stabilizing, or, conversely, declining. This makes it possible to place countries at different life cycle stages of the tourism industry: from growth to maturity and stagnation (Fig. 4).

Countries with a negative TTDI index are in recession or stagnation. This group includes Latvia, Austria, Cyprus, Slovenia, the Netherlands, the Czech Republic, Belgium, Luxembourg, Slovakia, Croatia, and Finland. They are characterized by a decline in tourism competitiveness, a possible reduction in investment and tourist flows, and signs of infrastructure fatigue and ineffective current tourism policies.

The maturity stage includes countries with moderate but positive TTDI growth. These include Bulgaria, Romania, Lithuania, Estonia, Denmark, France, Germany, Poland, and Ireland. The tourism sector in these countries is functioning stably but without significant breakthroughs. At the same time, there is still a high dependence on traditional markets, and there is a need to update business models and adapt to new challenges, particularly in digitalization and sustainable development.

The growth stage includes countries with high growth rates of the TTDI index. These include Hungary, Spain, Sweden, Italy, Portugal, Malta, and Greece. These countries are actively integrating digital and environmental solutions, increasing their international attractiveness, implementing successful strategic investments, and increasing the dynamics of the tourism industry. This development is evidence of effective resource management, readiness for innovation, and active participation in the transformation of the tourism sector.

The analysis of the eco-innovation index shows that the growth rate accurately indicates the stage of the life cycle of the tourism industry in each country. By dividing the EU countries according to their growth rate, we can draw a deeper conclusion

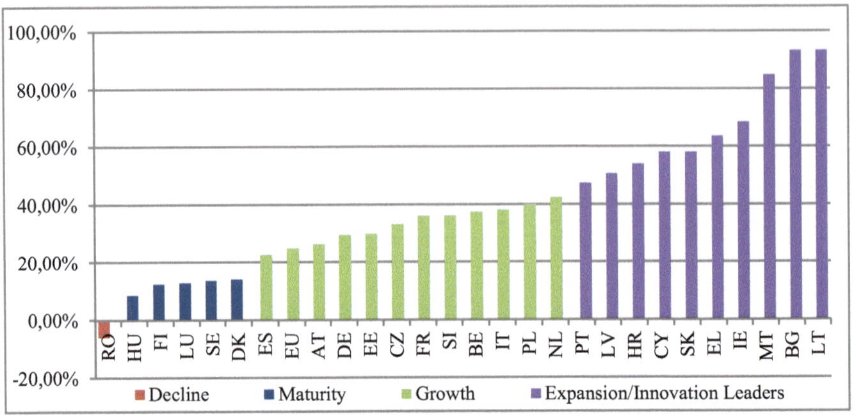

Fig. 5 Stages of the life cycle of the tourism industry in the EU countries by the growth rate of the eco-innovation index in 2024/2017. *Source* developed by the authors based on the data of [31]

about how the eco-innovation level correlates with the viability of the tourism sector (Fig. 5).

Countries with negative or minimal growth are in the Decline stage, which indicates stagnation of eco-innovation, loss of competitiveness, and a possible decline in tourist flow. Romania is the only country in this category, with a growth rate of – 5.88%. This signals serious problems in supporting the modern requirements of sustainable tourism.

The Maturity stage includes EU countries with a low growth rate (8–20%), which indicates a loss of development momentum, a decrease in innovation activity, and a possible risk of deviating from tourism trends. These are: Hungary (8.62%), Finland (12.50%), Luxembourg (12.90%), Sweden (13.79%), Denmark (14.19%). Despite the overall stability of these economies, the pace of change in the field of eco-innovation is too slow to maintain the interest of tourists, especially those focused on sustainable development.

At the Growth stage are EU countries with an active growth rate (20–45%), which have already reached the peak but still retain development inertia: Spain (22.77%), the EU average (25.00%), Austria (26.43%), Germany (29.63%), Estonia (30.00%), Czech Republic (33.33%), France (36.19%), Slovenia (36.36%), Belgium (37.50%), Italy (38.18%), Poland (40.00%), and the Netherlands (42.55%). Innovations are taking place in these cases, but they are not breakthroughs. These are countries where the tourism industry is still important, but growth requires a revision of approaches, particularly in sustainability.

The most dynamic countries with more than 45% growth are leaders in eco-innovation, likely to invest in new technologies, green tourism, digitalization, and climate change adaptation. Their tourism industry is moving into the Expansion/Innovation Leaders phase: Portugal (47.44%), Latvia (50.65%), Croatia (53.97%), Cyprus (58.06%), Slovakia (58.06%), Greece (63.64%), Ireland (68.57%), Malta (84.91%), Bulgaria (93.33%), Lithuania (93.33%). These countries are actively

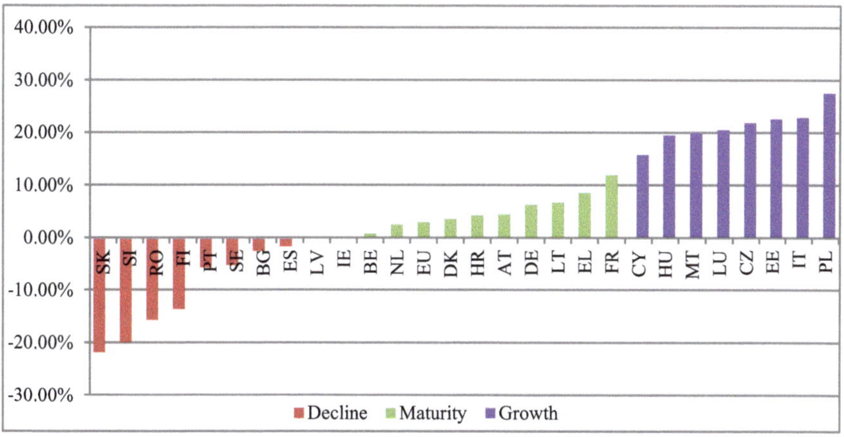

Fig. 6 Stages of the life cycle of the tourism industry in the EU countries by the growth rate of the human resources index in 2024/2017. *Source* developed by the authors according to [31]

innovating, likely turning tourism into one of the flagship areas of their economic recovery.

Thus, the policy of sustainable tourism and ecological modernization is not only an environmental but also a strategic economic necessity to extend the life cycle of the tourism industry.

Analysis of the growth rate of the human resources index of the European Union countries in 2024/2017 allows us to compare the rate of change in human resources with the stages of the life cycle of the tourism industry (Fig. 6).

Slovakia, Slovenia, Romania, and Finland show a significant decline in the Human Resources Index, indicating that the industry is in the Decline stage when it is undergoing crisis transformations, requiring new strategies and reformatting the resource base. With a slight decline or stability, Portugal, Sweden, Bulgaria, and Spain reflect the Decline stage, where the intensity of the industry's development is decreasing.

Latvia and Ireland, with stable zero growth, are on the verge of Maturity, characterized by maturity and the need for supportive change. Belgium, the Netherlands, the EU, Denmark, Croatia, Austria, Germany, Lithuania, Greece, and France show slight but positive growth, corresponding to the Maturity stage, a stabilization stage with the potential for qualitative improvements.

With a significant increase in the indicator, Cyprus, Hungary, Malta, Luxembourg, the Czech Republic, Estonia, Italy, and Poland demonstrate the Growth stage, where human capital is actively developing to support the expansion of the tourism sector. This distribution shows that investment in human potential remains a key factor in the sustainability and competitiveness of the tourism industry in the face of modern challenges.

The analysis of the growth rate of the digitalization index of the European Union countries in 2024/2017 shows a significant multi-level change in digital potential, which allows us to distribute countries by the stages of the tourism industry life cycle (Fig. 7).

Countries with negative index growth are at the stage of digital stagnation or decline. For example, Latvia shows a significant decrease in digitalization with a score of – 21.84%, and Belgium shows a slight decline at – 1.52%. This indicates a loss of momentum in implementing digital solutions or the curtailment of some digital initiatives.

The Maturity stage includes countries with low but positive growth not exceeding 10%. Such indicators were recorded, in particular, in Sweden (0.59%), Greece (3.33%), Luxembourg (4.35%), Austria (4.76%), France (5.38%), Denmark (6.40%), and the Netherlands (6.67%). This indicates the end of the active phase of digital development and the transition to supporting the existing digital infrastructure.

The stage of active digital growth (Growth) covers countries where the development of digitalization is approximately 10–40%. Here, the adoption of digital solutions is already more dynamic. For example, Lithuania has an indicator of 10.91%, Portugal—17.65%, Spain—20.81%, Ireland—22.31%, Malta—24.09%, Slovenia—24.68%, Poland and the EU average—25% each, Slovakia (29.69%), Italy (32.86%), Croatia (33.33%), Germany (33.75%), Finland (34.67%), Estonia (36.78%), and Bulgaria (37.21%) These countries are actively modernizing their tourism sector by integrating new technologies and digital services.

The stage of digital breakthrough (Expansion/Innovation Leaders) indicates a very high growth rate of digitalization—more than 45%. This shows a large-scale technological upgrade of the industry. This cluster includes Romania (48.72%), Hungary (54.29%), the Czech Republic (71.01%), and Cyprus, which shows the highest growth rate of 118.75%. These countries are an example of the rapid transition to digital ecosystems in tourism, with the active implementation of innovations

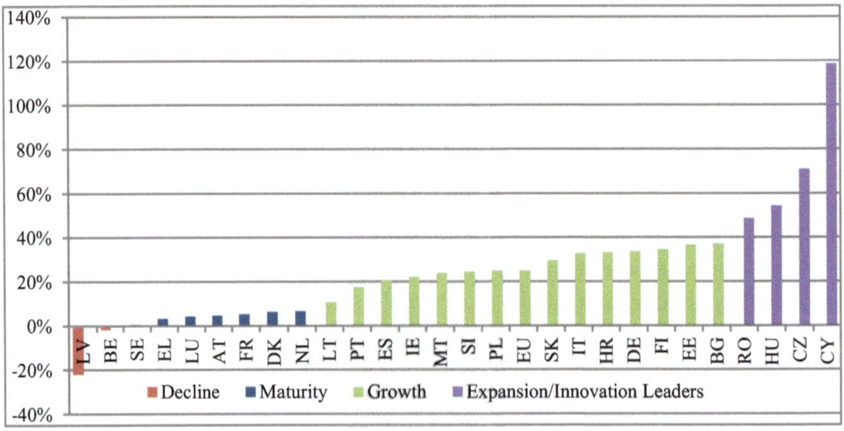

Fig. 7 Stages of the life cycle of the EU tourism industry by the growth rate of the digitalization index in 2024/2017. *Source* developed by the authors according to [31]

such as artificial intelligence, big data, mobile platforms, and digital solutions focused on sustainable development and customer personalization.

Thus, the analysis made it possible to determine the life cycles of the tourism industry in the European Union based on the dynamics of quantitative indicators.

The analysis of tourism revenue growth rates showed that the Emergence/ Introduction stage, characterized by low or initial revenue growth (within 4–10%), includes Poland (4%), Romania (5%), Sweden (5%), and Belgium (9%). These countries are showing early signs of tourism sector development, with the potential for further activation.

The Growth stage includes France, Portugal, Greece, Austria, Ireland, Spain, Italy, the Netherlands, Malta, the Czech Republic, and Hungary, countries with steady revenue growth, indicating an expanding tourism market, strengthening infrastructure, and increasing attractiveness.

Cyprus, Slovenia, Denmark, Luxembourg, Latvia, Germany, Estonia, Croatia, Slovakia, Bulgaria, and Finland represent the Maturity stage, which is traditionally associated with a slowdown in revenue growth. These countries are experiencing market saturation with stable demand.

According to the growth of the TTDI tourism development index, the EU countries are distributed as follows: Latvia, Austria, Cyprus, Slovenia, the Netherlands, the Czech Republic, Belgium, Luxembourg, Slovakia, Croatia, and Finland are in decline. These countries demonstrate negative or zero dynamics in global tourism competitiveness, which indicates a loss of positions in the worldwide tourism market system.

The Maturity stage includes Bulgaria, Romania, Lithuania, Estonia, Denmark, France, Germany, Poland, and Ireland, which maintain a stable level of competitiveness in a saturated tourist environment.

Finally, the Growth stage is headed by Hungary, Spain, Sweden, Italy, Portugal, Malta, and Greece, demonstrating high improvement rates in international tourism rankings.

The results of the growth of the eco-innovation index, as one of the key drivers of sustainable tourism development, indicate the following:

Romania is in the Decline stage (5.88%), which indicates a significant lag in eco-oriented innovations.

The Maturity stage (8–20% growth) is characteristic of Hungary (8.62%), Finland (12.50%), Luxembourg (12.90%), Sweden (13.79%), and Denmark (14.19%).

Spain (22.77%), the EU average (25.00%), Austria (26.43%), Germany (29.63%), Estonia (30.00%), the Czech Republic (33.33%), France (36.19%), Slovenia (36.36%), Belgium (37.50%), Italy (38.18%), Poland (40.00%) and the Netherlands (42.55%) represent the Growth stage with growth rates in the range of 20–45%.

At the same time, Portugal (47.44%), Latvia (50.65%), Croatia (53.97%), Cyprus (58.06%), Slovakia (58.06%), Greece (63.64%), Ireland (68.57%), Malta (84.91%), Bulgaria (93.33%), and Lithuania (93.33%) are in the Breakthrough/Innovation Leaders phase.

The human resources index growth results show that Slovakia, Slovenia, Romania, Finland, Portugal, Sweden, Bulgaria, and Spain, which show signs of stability or moderate decline, are in the decline stage.

Latvia and Ireland are on the verge of maturity, while Belgium, the Netherlands, Denmark, Croatia, Austria, Germany, Lithuania, Greece, France, and the EU are in a stable mature phase.

Cyprus, Hungary, Malta, Luxembourg, the Czech Republic, Estonia, Italy, and Poland are in the Growth phase.

The digitalization index growth results indicate that Latvia and Belgium are in the Decline stage. The Maturity stage is demonstrated by Sweden (0.59%), Greece (3.33%), Luxembourg (4.35%), Austria (4.76%), France (5.38%), Denmark (6.40%), and the Netherlands (6.67%).

The Growth phase includes Lithuania (10.91%), Portugal (17.65%), Spain (20.81%), Ireland (22.31%), Malta (24.09%), Slovenia (24, 68%), Poland (25%), Slovakia (29.69%), Italy (32.86%), Croatia (33.33%), Germany (33.75%), Finland (34.67%), Estonia (36.78%) and Bulgaria (37.21%).

Finally, the Expansion/Innovation Leaders include Romania (48.72%), Hungary (54.29%), Czech Republic (71.01%), and Cyprus (118.75%), which demonstrate exponential growth in the digital transformation of travel services.

The analysis results by EU countries showed that innovation processes affect the economic sustainability of the EU, including the financial benefits from developing tourism products, tourism consumption, and industrial development. This analysis will ensure the competitiveness and sustainable development of the tourism industry in the face of global change.

4.3 *Development of Conceptual Strategies for Managing Innovation Processes in the Life Cycle of the Tourism Industry*

It is imperative to develop innovation strategies, as they provide the industry with competitive advantages, sustainable growth, and adaptation to new challenges. For example, at the nascent stage, it is necessary to invest in research and development, while at the mature stage, innovations can be aimed at optimizing processes and increasing efficiency. Innovative transformations can help an industry find new opportunities and revive during a downturn.

Determining the life cycle stage of the tourism industry is a key task for effective strategic management, as it allows companies and government institutions to adapt their actions to market conditions. Each stage, inception, growth, maturity, decline, or transformation, requires a unique approach to investment, regulation, marketing, and infrastructure development (Table 2).

Table 2 Strategies for managing innovation processes by stages of the modern life cycle

Stage	Name of the strategy	Recommended conceptual strategies
Emergence/introduction	Strategy for innovation incentives and regulatory support	State support for R&D and development of startups in travel technology
	Legal environment strategy	Develop a regulatory framework for new tourism sectors (space tourism, digital tourism)
	Infrastructure investment strategy	Investing in infrastructure and digital platforms to support innovation
	Pilot innovation strategy	Pilot projects and experimental business models
Growth	Standardization and digitalization strategy	Standardization of travel services and digitalization of booking and management systems
	Smart infrastructure strategy	Expanding investments in smart tourism infrastructure (IoT, AI, big data)
	Public–private partnership strategy	Strengthening public–private partnerships to scale innovative solutions
	Sustainable development strategy	Implementation of sustainability standards and certification programs
Maturity	Strategy for optimizing operations	Optimizing operations through AI, automation, and big data analytics
	ESG and sustainable tourism strategy	Strengthening sustainable tourism practices (ESG compliance, reducing carbon footprint)
	Experiential tourism strategy	Differentiation through experiential tourism (personalization, cultural immersion)
	Market diversification strategy	Market diversification (focus on digital nomads, silver tourism, etc.)
Decline	Business model transformation strategy	Reassessing the business model and adapting to new trends
	Circular economy strategy	Transition to circular economy principles (waste reduction, regenerative tourism)

(continued)

Table 2 (continued)

Stage	Name of the strategy	Recommended conceptual strategies
	Technological efficiency strategy	Using technology to improve efficiency and reduce costs
	Strategy for finding new sources of income	Researching alternative sources of income
Expansion/innovation leaders	Disruptive technology strategy	Implementation of disruptive technologies (blockchain, AI-based personalization)
	Digital reorientation strategy	Rethinking value propositions and positioning in the digital economy
	Strategic merger strategy	Strategic mergers, acquisitions, and alliances to maintain competitiveness
	Impact-oriented development strategy	Transition to regenerative and impact-oriented business models in tourism

Source developed by the authors

The developed conceptual strategies for managing innovation processes in the context of the stages of the EU tourism industry life cycle demonstrate the importance of a dynamic, flexible, and cyclical approach to developing policy instruments. The Emergence/Introduction stage prioritizes strategies to create the basic conditions for growth: government incentives for innovation, creating a favorable legal environment, investing in infrastructure, and testing pilot solutions.

These measures provide the foundation for future scaling of innovations.

At the Growth stage, strategies that unify innovative practices, expand access to technology, and form sustainable partnerships are key. Standardization, digitalization, innovative infrastructure development, and public–private initiatives create a powerful multiplier effect, contributing to stable and high-quality growth.

The maturity stage characterizes the need for a strategic rethinking of operational efficiency, ESG components, and customer experience development. Optimization strategies, sustainable tourism, experiential tourism development, and geographic diversification dominate it. They are aimed at maintaining competitiveness in saturated markets.

The Decline stage requires transformation: businesses must change their business models, look for new sources of revenue, invest in efficiency technologies, and implement circular economy principles. These strategies provide an opportunity to restart or enter new markets.

The Expansion/Innovation Leaders stage reflects the transition to innovation leadership. Disruptive technologies, digital reorientation, strategic alliances, and impact-oriented management that combines economic, social, and environmental value are effective here. Such strategies can set trends for the entire industry.

Thus, an adaptive model of innovation management in tourism should consider the industry's development stage and actively predict future changes, creating conditions for the evolutionary transition between the life cycle phases.

Thus, the developed conceptual strategies for managing innovation processes, according to the studied stages of the modern industry life cycle, are essential because they allow adapting to changing market conditions, ensuring sustainable growth and competitiveness of the industry. Considering the specifics of each life cycle stage helps to allocate resources efficiently, introduce innovations, meet customer needs, and minimize risks.

This contributes to economic development and the social and cultural enrichment of the regions where tourism is developing. Thus, strategic management that considers the life cycle stage and is focused on innovation is the basis for the industry's long-term development.

5 Discussion

A general analysis of the life cycles of the tourism industry in the European Union, based on the dynamics of key indicators, shows the complex and multi-vector dynamics of the sector's transformation. The defining characteristics of each stage—from formation to restructuring—are manifested through the multidirectionality of indicators that act as diagnostic indicators of the strategic state of tourism systems. The Emergence/Introduction stage is characteristic of countries where the tourism industry is just recovering or is in the initial phase of modernization: digitalization, eco-innovation, and human capital are beginning to play a key role as growth drivers in such countries. On the contrary, Decline is recorded where there is a depletion of resource or innovation potential, a decline in the competitiveness index, or stagnation of digital transformation. At the same time, the Maturity and Growth stages demonstrate a balanced involvement of several components: moderate income growth, institutional stability, gradual integration of sustainable practices, high level of human resources, and digital maturity. Expansion/Innovation Leaders emphasize active market expansion and innovation, enabling travel companies to increase their presence in new markets, meet changing consumer preferences, improve service quality, and maintain competitiveness while contributing to regions' economic and social development.

It is important to emphasize that none of the analyzed indicators can be the only marker of the development stage—only their interaction allows us to determine the real position of the country within the industry life cycle. Thus, adaptability, multi-vector policy, digital modernization, human capital development, and eco-orientation are becoming key factors in the formation of a competitive, sustainable, and flexible tourism system in the context of the new economic paradigm of the European Union.

As we can see, according to the developed conceptual approach [5, 7], tourism enterprises should carefully assess the risks and potential benefits of introducing new innovative technologies. Introducing digital technologies [6] requires significant investment and changes in the management structure of enterprises. It is also important to consider data security and privacy issues [32]. The balance between competitiveness and business sustainability should be considered [10].

The EU countries must balance international and local tourism to ensure an even income distribution and reduce environmental pressure [8]. Taking advantage of various factors and developing conceptual strategies for managing innovation processes will allow the tourism industry to adapt to global changes in the long run [2–4, 6, 10, 11, 13–15, 17], to create regulatory mechanisms that will affect the implementation of innovation processes.

In general, innovation processes are focused on sustainable development and reducing the negative impact on the environment, so it is essential to take all these aspects into account for the successful development of the tourism industry in the long term.

Further research will study the impact of new technologies on the sustainability of tourism enterprises, analyze the effectiveness of various innovative strategies, and develop recommendations for their implementation. This will help ensure the competitiveness and sustainable development of the tourism industry in the face of global change.

6 Conclusion

The results showed that innovation processes affect the economic sustainability of the European Union countries, including the financial benefits of tourism product development, tourism consumption, and industrial development. In addition, innovation processes in tourism contribute to social development, cultural awareness, and tourists' participation in digital technologies and cultural heritage.

The stages of the life cycle of the tourism industry in the EU countries are analyzed by the following key indicators: the growth rate of tourism revenue, the growth rate of the TTDI tourism development index, the growth rate of the eco-innovation index, the growth rate of the human resources index, the growth rate of the digitalization index.

An analysis of the growth rates of tourism revenues showed that Poland, Romania, Sweden, and Belgium belong to the Emergence/Introduction stage, which is characterized by low or initial revenue growth. These countries are showing early signs of tourism sector development, with the potential for further intensification. The Growth stage includes France, Portugal, Greece, Austria, Ireland, Spain, Italy, the Netherlands, Malta, the Czech Republic, and Hungary—countries with steady revenue growth, which indicates an expanding tourism market, strengthening infrastructure, and increasing attractiveness. The Maturity stage includes Cyprus, Slovenia, Denmark, Luxembourg, Latvia, Germany, Estonia, Croatia, Slovakia, Bulgaria, and Finland. These countries are experiencing market saturation with stable demand.

According to the growth of the TTDI tourism development index of the EU countries, Latvia, Austria, Cyprus, Slovenia, the Netherlands, the Czech Republic, Belgium, Luxembourg, Slovakia, Croatia, and Finland are at the stage of stagnation or decline. These countries demonstrate negative or zero dynamics in global tourism competitiveness, which indicates a loss of positions in the worldwide tourism market system. The Maturity stage includes Bulgaria, Romania, Lithuania, Estonia, Denmark, France, Germany, Poland, and Ireland, which maintain a stable level of competitiveness in a saturated tourist environment. The Growth stage is headed by Hungary, Spain, Sweden, Italy, Portugal, Malta, and Greece, demonstrating high improvement rates in international tourism rankings.

The results of the growth of the eco-innovation index, as one of the key drivers of sustainable tourism development, indicated that Romania is in the Decline stage, which means a significant lag in environmentally friendly innovations. The Maturity stage (8–20% growth) is characteristic of Hungary, Finland, Luxembourg, Sweden, and Denmark. The Growth stage, with growth rates in the range of 20–45%, was represented by Spain, the EU average, Austria, Germany, Estonia, the Czech Republic, France, Slovenia, Belgium, Italy, Poland, and the Netherlands. At the same time, Portugal, Latvia, Croatia, Cyprus, Slovakia, Greece, Ireland, Malta, Bulgaria, and Lithuania entered the Expansion/Innovation Leaders phase.

The human resources index growth results showed that Slovakia, Slovenia, Romania, Finland, Portugal, Sweden, Bulgaria, and Spain, where signs of stability or moderate decline were recorded, are in the Decline. Latvia and Ireland are on the verge of maturity, while Belgium, the Netherlands, Denmark, Croatia, Austria, Germany, Lithuania, Greece, France, and the EU are in a stable, mature phase. The Growth phase is characterized by Cyprus, Hungary, Malta, Luxembourg, the Czech Republic, Estonia, Italy, and Poland. The result of the digitalization index growth showed that Latvia and Belgium are in the Decline stage. Sweden, Greece, Luxembourg, Austria, France, Denmark, and the Netherlands are at the Maturity stage. The Growth phase includes Lithuania, Portugal, Spain, Ireland, Malta, Slovenia, Poland, Slovakia, Italy, Croatia, Germany, Finland, Estonia, and Bulgaria. The Expansion/Innovation Leaders include Romania, Hungary, the Czech Republic, and Cyprus, which are demonstrating exponential growth in the digital transformation of travel services.

The result of the digitalization index growth showed that Latvia and Belgium are in the Decline stage. The Maturity stage includes Sweden, Greece, Luxembourg, Austria, France, Denmark, and the Netherlands. The Growth phase includes Lithuania, Portugal, Spain, Ireland, Malta, Slovenia, Poland, Slovakia, Italy, Croatia, Germany, Finland, Estonia, and Bulgaria. Finally, Romania, Hungary, the Czech Republic, and Cyprus belong to the Expansion/Innovation Leaders phase, which demonstrates exponential growth in the digital transformation of travel services.

The developed conceptual strategies for managing innovation processes, according to the studied stages of the modern life cycle of tourism development in the EU countries, will allow them to adapt to changing market conditions, ensuring sustainable growth and competitiveness of the industry, which involves a differentiated approach to strategic planning. At the Emergence/Introduction stage, the main strategies are those that create institutional and infrastructural conditions for growth: stimulating innovation, regulatory support, infrastructure investment, and pilot projects. The Growth phase is dominated by scaling tools, such as digitalization, standardization, innovative infrastructure, and public–private partnerships, which ensure sustainable, creative growth. At the Maturity stage, optimization of operations, ESG approaches, development of experiential tourism, and market diversification become relevant. They are aimed at maintaining efficiency in a saturated market. In the Decline phase, transformational strategies are needed: business model change, technological efficiency, circular economy, and search for new sources of income to avoid loss of viability. At the Expansion/Innovation Leaders stage, strategic emphasis is placed on disruptive technologies, digital reorientation, impact-oriented development, and strategic alliances that shape new vectors of industry evolution.

The results showed that innovation processes affect the economic sustainability of EU countries, including financial benefits from developing tourism products, tourism consumption, and industrial development. In addition, innovation processes in tourism contribute to social development, cultural awareness, and tourists' participation in digital technologies and cultural heritage.

References

1. European Travel Commission (2024) European tourism: trends & prospects. https://etc-corporate.org/reports/european-tourism-2024-trends-prospects-q4-2024/#:~:text=Summary&text=European%20tourism%20remained%20strong%20in,%25%20year%2Don%2Dyear. Accessed on 21 April 2025
2. Marsiglio T (2024) The tourism area life cycle hypothesis: a micro-foundation. Tour Econ 30(2):345–360. https://doi.org/10.1177/13548166221138645
3. Tsvilyi S, Kornienko O, Mamotenko D et al (2024) Transformation of the tourism industry in thecontex of the development of digitaltechnologies. In: Tatomyr I, Kvasnii L Green and digital economic transformation: a synthesis of the future: monograph. Oktan Print, Praha, p 266. https://doi.org/10.46489/gadetas-24-47
4. Gao J, Zhang L (2024) Climate policy uncertainty and corporate investment: evidence from the U.S. Tourism and hospitality sector. J Travel Res 63(2):517–530. https://doi.org/10.1177/00472875221149319
5. Xu J, Shi P-H, Chen X (2025) Curators or creators: role configurations of digital innovation strategy in museum tourism destination and the principles underlying their attractiveness. Tour Manage 106:105024. https://doi.org/10.1016/j.tourman.2024.105024
6. Siddik AB, Forid MS, Yong L et al (2025) Artificial intelligence as a catalyst for sustainable tourism growth and economic cycles. Technol Forecast Soc Chang 210:123875. https://doi.org/10.1016/j.techfore.2024.123875
7. But T, Mamotenko D, Lnenicka L et al (2023) A conceptual model for creating smart cities in czechia based on smart specialization. Sci Pap Univ Pardubic 31(2):1736. https://doi.org/10.46585/sp31021736

8. Mamotenko D (2009) Globalisation of international financial markets. Humltarian Bull Zaporizhzhya State Eng Acad 39:229–237. https://zgia.zp.ua/gazeta/VISNIK_39_20.pdf [in Ukrainian]
9. Tsviliy S, Mykhailyk D, Gurova D et al (2024) Strategy for the development of the investment potential of the tourism industry of Ukraine in the international economic system. Theor Pract Res Econ Fields 15(3):601–619. https://doi.org/10.14505/tpref.v15.3(31).08
10. But T (2024) Modern tourism: global and national trends: monograph. Oktan Print, Prague, p 217. https://doi.org/10.46489/MTGANT-24-28
11. Mamotenko D (2008) Evaluation of the efficiency of investment projects. Lviv Polytechnic National University Press, pp 209–216. https://vlp.com.ua/files/32_13.pdf [in Ukrainian]
12. But T, Mamotenko D, Korniienko O, Gubar O (2025) Financial assurance for tourism development and its impact on the economy of European countries. In: Semenov A, Yepifanova I, Kajanová J (eds) Monograph. Data-centric business and applications. Lecture notes on data engineering and communications technologies, vol 240. Springer, Cham, pp 63–80. https://doi.org/10.1007/978-3-031-81557-7_4
13. Wang X (2024) Construction of smart tourism system integrating tourist needs and scene characteristics. Syst Soft Comput 6:200168. https://doi.org/10.1016/j.sasc.2024.200168
14. Campos C, Gutiérrez D, Dias AC et al (2024) «Small-scale» tourism versus traditional tourism: Which will be the new key to achieve the desired sustainable tourism? Sci Total Environ 912:168964I. https://doi.org/10.1016/j.scitotenv.2023.168964
15. Garanti Z, Zvaigzne A, Berjozkina G (2023) Introduction: smart tourism: what developments and issues are important to the Baltic States? Worldw Hosp Tour Themes 15(5):463–467. https://doi.org/10.1108/WHATT-06-2023-0071
16. Erdiaw-Kwasie MO, Owusu-Ansah KK, Abunyewah M et al (2023) Circular economy, environmental quality and tourism receipts in Europe: a time series data analysis. PLoS ONE 18(11):e0288098. https://doi.org/10.1371/journal.pone.0288098
17. Matyushkenko OI (2010) Life cycle of an enterprise: essence, models, evaluation. problems of economics. 4:82–91 (2010). https://www.problecon.com/export_pdf/problems-of-economy-2010-4_0-pages-82_91.pdf [in Ukrainian]
18. Porter ME (1980) Competitive strategy: techniques for analyzing industries and competitors. Free Press, New York (Republished with a new introduction, 1998)
19. Simons K (2001) Product market characteristics and the industry life cycle. University of London, Manuscript
20. Moore GA (2002) Crossing the chasm. HarperCollins, p 256. ISBN 0060517123
21. Moore J (1996) The death of competition: leadership and strategy in the age of business ecosystems. HarperBusiness, New York, p 297. ISBN 0887308090
22. Christensen CM, Leslie D (2000) The innovator's dilemma. Highbridge Audio, p 140. ISBN 0-06-662069-4
23. Butler RW (2006) The tourism area life cycle: application and modifications. Channel View Publications, Clevedon, UK. ISBN 1-84541-025-4
24. Campos HC, Laso J, Cristóbal J, Fullana-i-Palmer P et al (2022) Tourism under a life cycle thinking approach: a review of perspectives and new challenges for the tourism sector in the last decades. Sci Total Environ 845:157261. https://doi.org/10.1016/j.scitotenv.2022.157261
25. Xie-Carson L, Magor T, Benckendorff P, Hughes K (2023) All hype or the real deal? Investigating user engagement with virtual influencers in tourism. Tour Manag 99:104779. https://doi.org/10.1016/j.tourman.2023.104779
26. UN Tourism (2025) UN tourism world tourism barometer 2025-Excerpt. https://www.unwto.org/news/international-tourism-recovers-pre-pandemic-levels-in-2024. Accessed 24 Apr 2025
27. World Tourism Organization (2025) World tourism barometer, vol 23(1). UN Tourism, Madrid. https://doi.org/10.18111/wtobarometereng
28. Trading Economics (2025) Income from tourism. Countries. List. Europe. https://ru.tradingeconomics.com/country-list/tourism-revenues?continent=europe. Accessed 24 Mar 2025
29. Finam (2025) Tourism's contribution to Europe's economy in 2024 will be a record €800bn. https://www.finam.ru/publications/item/vklad-turizma-v-ekonomiku-evropy-v-2024-godu-sostavit-rekordnye-800-mlrd-evro-20240711-1345/. Accessed on 21 Apr 2025

30. World econom Forum (2024) Travel & tourism development index 2024. WEF_Travel_and_Tourism_Development_Index_2024.pdf. Accessed on 3 Apr 2025
31. EU (2025) European Innovation Scoreboard 2024. https://op.europa.eu/en/publication-detail/-/publication/8a4a4a1f-3e68-11ef-ab8f-01aa75ed71a1/language-en. Accessed on 26 Apr 2025
32. Ahmed W, Rasool A, Javed AR, Kumar N, Gadekallu TR, Jalil Z, Kryvinska N (2021) Security in next generation mobile payment systems: a comprehensive survey. IEEE Access 9:115932–115950. https://doi.org/10.1109/access.2021.3105450

Ensuring the Competitiveness of Coastal Regions Through the Development of the Tourism Business

Anastasiia Bezkhlibna and Svitlana Zhuravlova

Abstract The research provides an in-depth exploration of the concept of regional competitiveness, highlighting its theoretical foundations, structural elements, and practical implications for coastal regions. Various scientific approaches to defining regional competitiveness are examined, including functional, resource-based, factor-oriented, balance-driven, innovation-centered, and institutional approaches. Through a semantic analysis, the study uncovers the dual nature of competitiveness-both as an internal potential and as the capacity to adapt to external conditions. The study proposes a detailed classification of regional competitiveness based on territorial, resource, specialisation, development level, source, and temporal criteria. This classification is aimed at clarifying how regions can differ in their competitive capacities and which elements may enhance their socio-economic development. Special attention is given to external factors influencing regional competitiveness. These include the effectiveness of public administration-measured through World Bank indices such as Voice and Accountability, Government Effectiveness, and Rule of Law-as well as broader institutional conditions, international economic integration, and innovation support. The structure of competitiveness in coastal regions is explored through the interaction of landscape characteristics, resource utilisation, and multi-level governance. Coastal regions are recognised for unique competitive advantages arising from their natural conditions (e.g., maritime resources, tourism potential, renewable energy) and institutional frameworks (e.g., public authorities, NGOs, local governance). Furthermore, the research emphasises the strategic integration of tourism into the system of regional competitiveness management. Tourism is not seen as a separate sector, but as a key structural driver that influences the broader development model of coastal regions. To this end, a tripartite model of regional competitiveness is developed, consisting of: provision subsystem (regulatory, investment, cultural, digital, and natural resource foundations), functioning subsystem (creation and monitoring of competitive advantages, sustainability), management subsystem (principles, structures, stakeholders, and competencies). The study concludes by advocating for a comprehensive management system that integrates environmental,

A. Bezkhlibna (✉) · S. Zhuravlova
National University Zaporizhzhia Polytechnic, Zaporizhzhia, Ukraine
e-mail: bezkhlibna@zp.edu.ua

A. Semenov et al. (eds.), *Data-Centric Business and Applications*, Lecture Notes on Data Engineering and Communications Technologies 272,
https://doi.org/10.1007/978-3-032-06608-4_16

systemic, functional, and institutional approaches. It underscores the importance of harmonising tourism development with socio-economic and ecological priorities to enhance the long-term competitiveness of coastal regions.

Keywords Tourism · Business · Development · Competitiveness · Economic system · Subsystem

1 Introduction

Maritime regions are objects-represented by economic actors-and subjects-represented by regional management of market relations, so they are subject to objective market laws, including free competition, which is one of the factors of market relations development. The etymology of the concept of competition in the coastal region is related to the term competition. The semantic analysis carried out in this paper allowed us to make certain generalisations. In particular, in a number of European languages, the original words from which this category is derived have different meanings. For example, in English, the word concur means "to agree, to converge", concurrent means "coinciding, converging", concurrence means "agreement". In French, concourir means "to facilitate", concours means "assistance, help". In Spanish, concurrir means "to meet, to connect; to facilitate". Competition is "rivalry in any field, the struggle to achieve better results; competition"; "the struggle between private entrepreneurs for more favourable conditions for the production and sale of goods in commodity production; the struggle between capitalists for the highest profits".

The first person to describe competition in the economy was its "father"-Adam Smith, who in the eighteenth century in his work "An Inquiry into the Nature and Causes of the Wealth of Nations" [1] provided theoretical statements on the division of production and trade using the concept of "absolute advantage". In the nineteenth century, his follower, D. Ricardo, developed the concept of "absolute advantage", which became the basis for his theory of comparative advantage, which became widespread in the description of international trade. As part of the development of international trade theories, E. Heckscher and B. Olin (XX century) transformed Ricardo's idea into the theory of factor endowment: countries export products of intensive use of surplus factors and import products of intensive use of scarce factors. In other words, the characteristic features of competition are the dynamism of its development as an economic process of functioning of economic entities, the possibility of its comparison between similar economic entities and acting as a balancing instrument between surplus and deficit factors of production. Further study of the content of the concept of "competitiveness" is based on the understanding of its close relationship with "competition", which emphasises, in fact, the qualitative and quantitative state of the socio-economic system. The evolution of the interpretation of the definition of "competitiveness" is associated with a long historical period and the functioning of various economic systems during which the study was conducted and the breadth of interpretation [2].

Regional competition is a fait accompli, and it arises in relation to attracting the best resources and specialists, obtaining better growth opportunities through favourable access to resources: sources of financing, state targeted programmes, international grants, involvement in infrastructure projects, etc.

Understanding the essence of competitiveness and developing effective tools for managing regional resources is impossible without applying an epistemological approach ("subject-object") and substantiating methodological principles. The compliance of the modern set of measures for managing competitiveness should be based on the standards of management of territories and social systems, with adequate consideration of the impact of globalisation and information factors on regional development.

Competitiveness is a policy goal of both public authorities at the national level, local authorities at the regional level and business owners at the enterprise level.

Competitiveness is usually seen as a process of competition between regions for resources and assets. Meanwhile, the acquisition of a synergistic effect in economic interregional cooperation creates a number of positive effects for the economies of the participating territories and allows us to characterise competitiveness in terms of gaining additional advantages and creating favourable conditions for the business environment of the territory and the life of its citizens.

The territories that have acquired the status of independent economic entities compete for economic resources (financial, business, labour, etc.) and conditions, defend the interests of larger structural enterprises in the region, seek opportunities to enter the international market on their own, and try to protect local producers through administrative measures.

At the end of the twentieth century, more and more attention in scientific and political circles began to be paid to the concept of regional competitiveness. "Regions and cities have no other choice but to become competitive in order to survive in the new globalised market of new competition", says American researcher Best [3]. Aiginger [4, pp. 159–169] emphasises that the ultimate goal of the state and the people is to maximise the public welfare function, in which income, social contributions and environmental protection are priorities for strategic development.

This leads to a fairly comprehensive definition of competitiveness with three immediate implications. First, low cost is not an economic policy goal and is not even an indicator of long-term competitiveness. Second, the external balance characteristic directly or indirectly affects the welfare of the population, but the impact is negligible relative to consumption.

Thirdly, competition is a dynamic issue whose performance changes over time, while investments in human capital, technology and information are a means to increase welfare and improve competitiveness. This approach to creating a framework for assessing competitiveness is consistent with the main goal of national economic development, i.e., improving living standards [5].

Distinguishing between the concepts of "regional competitiveness" and "competitiveness of the region, there is a need to clearly identify the difference between these concepts. The study of the theoretical postulates of competitiveness without reference to any entity lies in the field of regional competitiveness. The study of competitiveness in relation to the economy and applied management of the region becomes the subject of the study of regional competitiveness.

2 Defining the Essence of Regional Competitiveness

The study of regional competitiveness is one of the most urgent tasks due to various factors. Regional competitiveness as a scientific category, however, is poorly understood, and in some areas-such as competitiveness and competitiveness, implementation mechanisms, etc. (the definition is debatable). At the same time, the use of regional management levers in the areas of human capital, financial flows, knowledge culture, and adjustment of the region's energy policy should be adjusted with the strategic goal of increasing regional competitiveness.

The issues of competitiveness research are addressed in the works of foreign scholars. In particular, M. Peneder in his work links the intensive development of education with the increase in the competitiveness of each sector of the economy [6]. I. Grilo and J. Koopman studied the impact of microeconomic reforms on productivity and competitiveness in the European Union [7]. Rodriguez-Posey studied the relationship between social structure and regional growth in Eastern Europe [8].

Cellini and Socki [9] argue that "the concept of regional competitiveness is neither macroeconomic (national) nor microeconomic (based on the performance of firms): regions are neither a simple association of firms nor smaller versions of nations". Camagni [10] is of a similar opinion. He assumes that regions do compete, but on the basis of absolute competitive advantages, not comparative ones. A region can be considered to have an absolute competitive advantage if it possesses distinctive technological, social, infrastructural or institutional assets that benefit individual firms in such a way that no alternative factor prices will cause a geographical reallocation of economic activity. These assets tend to give firms in the region as a whole higher productivity than would otherwise be the case.

Doel, M. and Hubbard, P. focus on the fact that regional competitiveness cannot be viewed in the context of the distribution of national competitiveness across territories. Instead, the authors consider regional competitiveness as "successful competition among extremely open "flow spaces" [11], in attracting and retaining production factors in order to become or remain centres of national trade, investment and knowledge flows.

The hypothesis of regional competitiveness due to the synergistic effect of small and medium-sized businesses operating in the administrative territory is to combine the interests of local authorities and business entities in order to achieve the effect of overall productivity. In other words, "the prosperity of a region is determined, firstly, by the strength of its export base of all those activities that generate income in the region through the export of goods and services to the outside world" [12].

In this context, productivity is a key element of competitiveness [13]: "Competitiveness remains a poorly understood concept, despite the widespread recognition of its importance. To understand competitiveness, the source of a country's prosperity should be the starting point. A nation's standard of living is determined by the productivity of its economy, measured by the value of goods and services produced per unit of labour, natural and financial resources. Productivity depends both on the value of the nation's goods and services, measured at prices that can be obtained on free markets, and on the efficiency of production of these goods and services." The goal of increasing the productivity of entrepreneurial activity is to ensure that the state provides favourable conditions for doing business, using the assets of entrepreneurial structures, and developing value chains, the growth of which is a criterion for increasing steam productivity.

Over time, the diversity of approaches to the analysis of competitiveness in the context of the region has led to a significant number of studies in this area. The study of existing theories and concepts in order to identify the elements that form the competitiveness of a region is a necessary stage in the study of regional competitiveness in terms of characterising the processes taking place in the competitive environment of the region and describing the conditions for increasing competitive advantages [3]. In this regard, the study was aimed at determining the essential characteristics of the region's competitiveness, identifying different approaches to the interpretation of this economic phenomenon.

Table 1 shows the evolution of the main approaches to the interpretation of the definition of "regional competitiveness", highlighting the following scientific approaches:

- functional approach-focuses on interregional interaction;
- resource approach-emphasises the use of conditions and potential of the region,
- factor approach-focuses on self-organisation and distribution of factors between regions,
- balance approach-involves a dynamic process of gravity towards market equilibrium,
- innovation approach-emphasises innovation as the main factor of competitiveness,
- institutional approach-involves increasing regional competitiveness.

It should be noted that, analysing the evolution of approaches to the interpretation of the definition, the oldest approach is the functional one, whose ancestor is J. Schumpeter, while the most common are the factor, balance sheet and resource approaches. It is worth noting that the existence of different approaches to understanding regional competitiveness necessitates further research of regional competitiveness, taking into account their basic postulates on the determining influence of competitiveness factors, innovations, institutions, resources, and the tendency to balance and equilibrium.

Most researchers, such as Porter E., I. G. Kadyrus, A. Kaszmarek, Ja. Sobon classify competitiveness only by object-national, regional, production, goods and services competitiveness. Instead, the classification of the region's competitiveness

Table 1 Approaches to interpreting the concept of "regional competitiveness"

Approach, author	The essence of the hypothesis
Functional	
J. Schumpeter	Competitiveness is a dynamic process that strives for a new state, in which it is the mechanism of competition that drives unviable producers out of the market
L. Walras, V. Pareto, J. Mill, M. Alle, J. Timbergen	Optimization of interregional interactions is a tool for increasing the region's competitiveness
A. Marshall (theory of industrial districts)	Competitive advantages arise as a result of economies of scale both within the region and in interregional cooperation
Resource-based approach	
M. Porter	The region serves as an environment that creates or does not create conditions for stimulating the growth of firms" competitiveness
Y. Friedman, H. Rechko, Y. Pisarov	The competitiveness of a region is the ability to use the economic potential available in the region, the dynamics of growth in living standards and incomes of capital owners
I. Begg, Huggins	The competitiveness of a region is generally consistent with the conditions that allow firms to compete in their chosen markets within a competitive region
M. Storper	Regional competitiveness consists of the ability of a particular region to attract and retain firms with a stable or growing market share in its operations while maintaining or improving the living standards of its participants
Dijkstra	Regional competitiveness can be defined as the ability to offer an attractive and sustainable environment for firms and residents to live and work
Factor approach	
Б. B. Roberts, A. Murray, P	The self-organization of the system should use the cause-and-effect relationships of factors
E. Heschker, B	Competitive advantages arise from a more rational division of labor between regions
V. Picture	Regional competitiveness is the ability to use competitiveness factors to create a competitive position and maintain it among other regions
L.S. Shekhovtseva	Competitiveness of a region is the productivity of the use of regional resources, primarily labor and capital, compared to other regions, which results in the gross regional product per capita and its dynamics
P	Productivity growth is a key factor in the region's competitiveness
Organization for Economic Cooperation and Development	Regional competitiveness is the ability of companies, industries, regions, countries, or supranational regions to generate and remain open to international competition in a relatively high ratio of income and employment on a sustainable basis

(continued)

Table 1 (continued)

Approach, author	The essence of the hypothesis
A. Balkit, M. Tvaronavitsyn	Regional competitiveness means a steady increase in the living standards of a nation or region and a low level of involuntary unemployment
A. Markusen	Regional competition is an opportunity to attract investment and talented labor in order to maintain attractiveness for both labor and capital,
Balance sheet approach	
P. Schnipper	The competitiveness of a region is mainly determined by the ability of regional authorities to maintain a balance of various regional interests (socio-political, economic, inter-sectoral, cultural and domestic)
A. Lesch	Building models of spatial economic equilibrium allows us to identify market areas, the size of which depends on the degree of competitiveness of sales of certain products
L. Mizes	Market competition is an automatic dynamic process, not a specific market structure. The tendency towards market equilibrium is the result of business activity. An enterprise wins or loses in competition depending on the strength of its capabilities and the degree to which its supply meets market needs
Innovative approach	
T. Hegerstrand	Competitive advantages in the region are formed through the development of innovations, taking into account the specifics of innovation processes and the processes of innovation diffusion (diffusion of innovations)
F. Perru	The need to increase localization requires the search for sources of value growth and cooperation processes to create conditions for the development and application of new knowledge-intensive products, improvement of technologies, production methods, marketing and other types of improvement of economic management of regional clusters and regional development
D. Meija, N. Grosjean	The concentration of territorial production systems contributes to the allocation of industrial districts based on the use of local scientific and technical resources, where the internal cause of development is the innovative environment
Institutional approach	
Friedrich, M. Weber, J	In addition to economic factors, the competitiveness of a region is shaped by social institutions, such as government agencies, trade unions, financial institutions, socio-political organizations, property and organizational structures, and mental habits and norms of behavior

(continued)

Table 1 (continued)

Approach, author	The essence of the hypothesis
I. Lenguel	Regions and cities compete by creating a business environment that promotes productivity and concentrates regional institutions: specialized educational institutions, efficient special infrastructure, information services that facilitate innovation, and the development of research related to the cluster profile
Market approach	
J. Black, N. Ghashimzeid, G. Miles	Regional competitiveness is the ability to compete in the markets for goods and services based on a combination of price and quality

Source Developed by the authors

is proposed to be distinguished by the following classification criteria (Table 2): by territorial feature; by natural resource feature of the natural resource management economy; by specialisation of the region; by the degree of development of the region; by the nature of the source of competitiveness; by the duration of competitiveness.

Based on the semantic analysis, the study deepens the understanding of the dual nature of the region's competitiveness, which, on the one hand, determines the immanent property of the object, the readiness to attract internal competitive resources for the development of the region at a particular time in a particular place under specific conditions and is interpreted as competitiveness, and on the other hand, competitiveness as the ability of the object to develop under the influence of external factors, in the presence of conditions, circumstances, that is, competitiveness, which allows to deepen the essence of the region's competitiveness.

Therefore, from the point of view of targeting, the competitiveness of a region is a dynamic process of responding to external factors and challenges, which involves achieving the tactical goals of the business environment and social institutions of the region in order to improve the welfare of the local population, provided that the resources of the territory are used efficiently and innovative processes in the regional economy are ensured through the creation of a regional development programme (strategy), direct or indirect influence of competitors through competition with other territories.

Regarding the competitiveness of the region as an object of study, it is important to note that "regional competitiveness" is not only a property of the economic system, but also acts as a system of properties that allow the region to compete effectively to achieve its goals [14].

Since the concept of "competitiveness" is applied to different objects, approaches to the interpretation of its economic content differ. The competitiveness of a region is viewed from the perspective of efficient territorial labour organisation. The competitiveness of a region is a dynamic process, not a market structure, which ultimately leads to a balance of surplus and deficit factors of production.

Table 2 Classification of regional competitiveness

Classification feature	Types of competitiveness
On a territorial basis	Coastal region Mountainous region Forest region River basin region Steppe region
According to the natural resource characteristic of the nature management economy	Agricultural land Public land for public purposes Lands of the nature reserve fund and other environmental protection purposes Lands for health and recreational purposes Lands of historical and cultural designation Lands of forestry designation Water fund lands Land for industry, transport, communications, energy, defense and other purposes
By region specialization	Industrial region Agricultural region Agricultural region
By the degree of development of the region	A highly developed region Developed region An undeveloped region A depressed region
By the nature of the source of origin	Based on innovation Determined by the level of infrastructure development Based on natural, climatic and geographical conditions Technical (technological) Based on the development of information resources Determined by economic factors
By validity period	Long-term Medium-term Short-term

Source Developed by the authors

The competitive economy of a region is a state of competitiveness that manifests itself in a dynamic process of adaptive response to external factors and challenges of the regional economy, provides for the achievement of tactical goals of the business environment and social institutions of the region in order to improve the welfare of the local population, provided that the territory"s resources are used efficiently and innovation processes in the regional economy are ensured through the creation of a regional development programme (strategy), direct or indirect influence of competition participants.

The economic consequences of ensuring the competitiveness of the region are an increase in the standard of living of the population, income of capital owners, an increase in the level of introduction of innovations in economic activity, and a decrease in the unemployment rate in the region, provided that the regional authorities create a balance in the structure of the regional economy, when the interests of all institutions are ensured.

3 Factors Affecting the Region's Competitiveness

An important point in the study of the essence of regional competitiveness is to identify the factors on which it depends and its interrelationships with the competitiveness of the state, on the one hand, and the competitiveness of basic (local) level entities, on the other.

First, the competitiveness of a region is dual in nature in the sense that it is determined by endogenous (internal) and exogenous (external) factors. The key internal factors are: the territory of the region, geographical location, availability of natural and human resources, level of education and qualifications of the region's population, availability of fixed assets and capital equipment of employees in the main sectors of the region's economy, efficiency of local and regional authorities, attraction of investments, etc.

Among the external factors affecting the competitiveness of a region, first of all, it is worth noting the effectiveness of public administration, which is determined by the World Bank on the basis of six indices that reflect various aspects of public administration:

- Voice and Accountability;
- Political Stability and Lack of Violence;
- Government Effectiveness;
- Regulatory Quality;
- Rule of Law;
- Control of Corruption.

Other key external factors include the institutional provision of a favourable business environment in the country, the country's place in the international division of labour, export potential, the country's participation in international economic and financial organisations, state support for the innovative development of the high-tech manufacturing sector, etc.

There are significant differences in the understanding of regional competitiveness, which require a more detailed definition of the areas in which regions can compete and the elements and factors that can improve the indicators of socio-economic development of regions.

Economic rivalry of regions and determination of its scientific justification, identification of factors and factors of competitiveness growth have been of interest to economists for more than three centuries. Traditional views on the essence and content of the competitiveness of territories have been transformed over time under the influence of various factors of society development: informatisation, scientific and technological progress, and the impact of innovations. At the beginning of the development of economic thought on competitiveness (eighteenth century), the research of economists included the study of global competitiveness (world competitiveness). The characteristic of economic development of territories of countries is inherent in the researchers of the early twentieth century, the issues of competitiveness of territories (countries and regions) are devoted to the works of a later time-the middle of the twentieth century and the beginning of the twenty-first century [15].

Foreign authors Chitea and Dona [12] consider regional competitiveness as the sum of individual competitiveness of firms, which, however, depends on macroeconomic (national) competitiveness. In this regard, the peculiarities of its functioning and structure are important for studying the problems of increasing the competitiveness of coastal regions.

4 The Structure of Competitiveness of Coastal Regions

The essence of the regional competitiveness of the coastal region is manifested through its forms, internal structure and peculiarities of interaction of elements within the integral structural and logical composition of competitiveness. In this sense, it is important to identify the features of the coastal region that become its competitive advantages.

The peculiarities of the coastal region, revealed in terms of territorial structure, use of natural resources and management, are shown in Table 3.

The founder of the competition "rhombus" M. Porter and co-author K. Ketel note that the competitiveness of an economic entity is based on five forces that influence business development: buyers (customers), suppliers, existing competitors, new players and substitute goods [16]. This concept of studying competitiveness on its structural basis is based in the neoclassical paradigm on the market approach, or its subject (buyers, suppliers, competitors, new players) and product (goods, product, service) basis. In this case, the constituent elements of competitiveness (actors and products) automatically reach market equilibrium as market aggregates. Instead, regional policy, as well as the achievement of higher competitiveness in the vast majority of developed countries, clearly defines the regulatory nature of the development and implementation of measures and instruments to ensure it. The use of the extrapolation method makes it possible to "transfer" Porter"s idea to the object of research—the competitiveness of coastal regions, and leads to the following causal relationships and generalisations: (1) the constituent elements of subjective relations are residents and guests of the region; (2) representatives of business (domestic market) and authorities; (3) new types of business related to the use of the potential of coastal regions; (4) competitors represented by other regions and goods of a specific "coastal" purpose. That is, it is logical to assume that in order to analyse

Table 3 Features of the coastal region

Landscape features	Peculiarities of natural resources use	Management features
Influence of water temperature and atmospheric pressure on weather and climate in general (heat and humidity balance); seasonality	Water: maritime transport and shipbuilding, fleet, aquaculture, fisheries	Global level: international organizations and international cooperation agreements
Topography characteristics of the soil, erosion processes of the coastal zone approaching the water's edge	Air: favorable air temperatures contribute to the development of the tourist and recreational complex	Public administration: ministries, regional state administrations, state inspectorates and agencies, and the border guard service
Water flow, tides, surface and groundwater, the presence of estuaries and their condition	Land: offshore mining, coastal zones as places of recreation	Local government: regional, city, district and village councils; united territorial communities
Special flora and fauna of the coastal zone	Energy: the use of solar energy, tidal energy, wind energy, etc	Non-governmental organizations: associations, alliances, charitable foundations, unions

Source Developed by the authors

the competitiveness of coastal regions, it is advisable to proceed from the statement about the effectiveness (productivity) of its elements.

L.N. Chaynikova considers the competitiveness of a region as a system and includes the following components [17]:

- a set of elements that ensure the competitiveness of a region (competitive advantages and their sources, factors and criteria for assessing competitiveness, etc.);
- the relationship between the elements that make up the system of competitiveness of a region;
- integrity and unity of elements, their structure and hierarchy;
- clearly expressed management of the competitiveness of a region;

The use of the above theory is close to the understanding of regional competitiveness, but denies the fact that its formation carries the basis for the effective operation of high-tech industries and the maritime sector, in particular for coastal regions.

Understanding and applying the interpretation of A. Maslow's theory of the pyramid of needs in the process of studying the components of regional competitiveness allows us to distinguish the following elements:

- at the basic level, the region's competitiveness is manifested in the given geographical and climatic conditions, historical identity of the region, and the quality of public services;

- security requirements can be interpreted as the quality of healthcare and law enforcement services in the region, stability of economic development of the region, and absence of military armed conflicts;
- the need for social belonging is manifested in the conditions prevailing in the region in terms of the formation of social circles, connections, activities of public institutions, charitable foundations;
- the need for respect and recognition forms regional competitiveness by creating opportunities for improving skills, qualifications for the workforce, forming human capital, increasing opportunities for intellectual development of the region's residents, recognising cultural achievements, mentality, traditions.

The highest level of needs of the Maslow's pyramid is, on the one hand, confirmed by the existence and effectiveness of educational institutions, and on the other hand, by the recognition of the region's competitiveness at the national level, and the provision of advantages and benefits to such regions.

In our opinion, Maslow's model as a methodological construct of the coastal region's competitiveness is not productive in terms of implementation and administration and is more typical for market actors than for spatial development. More productive may be the idea of Granberg O.G. regarding the typology of regions by their inherent characteristics, in particular, territory, human resources, business processes and management.

Thus, summarising the above views of scientists, we can conclude that there is a certain systematisation of views on the competitiveness of the coastal region, due to different forms of relations, and distinguish the following approaches:

1. The market approach of Porter M. and Ketel K. (formation of relations "through the market") [16], which asserts the existence of competitive forces that influence the development of the region, determining its competitiveness—residents and guests of the region; representatives of business (domestic market) and government; new types of business related to the use of the potential of coastal regions; competitors represented by other regions and goods of specific "maritime" origin.
2. The systemic approach of L.N. Chaikova (formation of relations based on an understanding of the interconnectedness and interdependence of system elements) [18].
3. The approach of "corrected", "hard" and "soft" conditions by Sachs J., Cinnes K. and Eilat J. (formation of competitiveness relations based on managerial influences on the conditions of the region).
4. Psychological approach of A. Maslow (related to the pyramid of needs satisfaction): distribution of elements of the region's competitiveness according to the hierarchy of needs.
5. The structural–functional approach of Granberg O.G. (involves the formation of relations based on structuring the elements of competitiveness on the basis of space, human resources, business processes and regional management).

As a scientific approach to understanding the structure of competitiveness of the coastal region, the synthesis of the above approaches is chosen as the basis, which can be defined as a system-functional approach, which will provide for an understanding of competitiveness from the standpoint of the systemic interconnection of competitiveness elements, which, in turn, will be measured by indicators of the status or efficiency of management of these elements of the region's economy. It is advisable to analyse the interaction of the constituent elements through a set of competitiveness criteria.

The European Commission logically believes that "The idea of regional competitiveness should reflect the view that, despite the presence of strong competitive and non-competitive firms in each region, there are common features in the region that affect the competitiveness of all firms located there [19]. This approach demonstrates the importance of studying the mechanisms of ensuring the competitiveness of coastal regions, which, due to their geographical location, have competitive advantages that are reflected in the characteristics and efficiency of the processes of each element of competitiveness. That is, we will focus on the theoretical justification of understanding each element of competitiveness of the coastal region according to the chosen system-functional approach.

Regions, having the initial resources for competition (territory, minerals, weather, climate and natural conditions, available infrastructure, production), compete by ensuring social and economic conditions for the well-being of citizens.

Today, the study of the economic component of the competitiveness of Ukraine's regions is an urgent issue from the point of view of implementing the decentralisation reform and identifying "leading regions" and "outsider regions" in terms of socio-economic development. The functioning of the national economy is impossible without a balanced strategy of regional development of the country's territories, which will lead to sustainable growth of the regional economy and welfare of citizens, and ensure the investment attractiveness of the region. The state regional policy should be transformed to ensure high development goals. Undoubtedly, the priority of the new approach should be the policy of stimulating the development of territories based on the competitiveness of the regional economy and regional management rather than the policy of financial equalisation of regions.

5 Structure of the Regional Competitiveness Management System Based on Tourism Development

Moving on from the general scientific understanding of the category of competitiveness of the coastal region, it is worth emphasising the need to form a holistic, structured approach to managing this process. In today's economy, the competitiveness of territories, including coastal regions, is increasingly determined by their ability to dynamically adapt to global challenges, efficiently use resource potential, and develop attractive areas of specialisation, with tourism taking a leading position.

Coastal regions have specific prerequisites for strengthening their competitive position through the development of the tourism sector. The coastline, natural recreational facilities, historical and cultural heritage, as well as the traditions and mental characteristics of the local population, create unique opportunities for creating a sustainable model of economic growth. However, the availability of resources alone does not guarantee a high level of competitiveness. This requires an effective management system that ensures the rational use of the existing potential and stimulates sustainable development of the region [20].

The system of managing the economic competitiveness of a region has a complex multi-level structure consisting of three interrelated subsystems: provision, operation and management. Each of them performs specific functions, forming a single integrated mechanism aimed at maintaining and developing the region's competitive advantages. In particular, the support subsystem covers regulatory, scientific, methodological, informational, human resources, investment support, as well as digital resources, natural, cultural and social factors that form the basic conditions for the region's competitiveness.

The functional subsystem is responsible for the processes of forming and developing competitive advantages, assessing the effectiveness of the system, monitoring the interconnections between its elements, and stimulating the emergence of new forms of cooperation between subsystems. This part of the system is based on qualitative and quantitative indicators of process efficiency and the level of competitive results achieved.

At the same time, the management subsystem ensures the implementation of management influence based on a set of methodological, organisational and practical tools. These include management principles, relevant methods, factors, scientific approaches, structural elements of management, subjects of management decisions, and tools for influencing socio-economic processes. As a result, an integrated system is formed that is capable of ensuring the adaptability of the regional economy to changes in the external environment and promoting its sustainable development.

Tourism, as one of the most promising areas of specialisation that combines not only economic, but also socio-cultural, environmental and image potential, is of particular relevance in the structure of the competitiveness management system of the coastal region. The development of tourism provides a multiplier effect, activating related industries, stimulating employment and income growth, and contributing to the preservation of cultural heritage.

Integration of tourism into the system of regional competitiveness implies considering it not as a separate area, but as one of the key structural elements that influences the formation of a modern model of development of the coastal region. It is through the prism of tourism potential that the need for systemic modernisation of management approaches, including the introduction of digital technologies [21, 22], infrastructure development, improvement of the regulatory framework and mechanisms of interregional and international cooperation, is becoming more relevant.

The system of regional economic competitiveness management has a complex structure consisting of various subsystems: provision, operation and management (see Table 4).

Table 4 Subsystems of regional competitiveness based on tourism development

Subsystem for ensuring the competitiveness of the region through tourism development	Subsystem of regional competitiveness based on tourism development	Subsystem for managing the region's competitiveness based on tourism development
Regulatory and legal support	Processes of formation and development of competitive advantages	Management methodology (principles, factors, methods, scientific approaches)
Scientific and methodological support	Evaluation of the system's efficiency and effectiveness	Management structure
Information support	Monitoring of interconnection processes	Management processes
Financial support	Stimulating the creation of new processes of interaction between subsystems	Management entities
Staffing of the company	Competitiveness of regional economic entities	Controls
Investment security	Ecosystem sustainability of regional development	Competencies (best practices)
Digitalization and information resources (BIG DATE)	Potential of the region's economic facilities	
Natural resources		
Culture, customs, traditions		

Source Developed by the authors

The regulatory and legal support, scientific and methodological support, information, human resources and investment support, digitalisation and information resources, natural resources, culture, customs and traditions established in the coastal region form the subsystem of ensuring the region's competitiveness.

The processes of forming and developing competitive advantages, assessing the effectiveness and efficiency of the system, monitoring interconnection processes, and stimulating the creation of new processes of interaction between subsystems constitute the subsystem of the coastal region's competitiveness This subsystem is based on qualitative and quantitative indicators of interaction and effectiveness of processes within the system and competitive advantages.

The system of competitiveness of the coastal region forms a framework based on the mutual influence of socio-economic relations of competitiveness subsystems. The management methodology (principles, factors, methods, scientific approaches), structure, processes, subjects and means of management, competences make up the subsystem of competitiveness management of the coastal region.

The peculiarity of the competitiveness of the coastal region is to take into account the specifics of the indicators of socio-economic development of coastal regions, which are as follows:

- environmental approach to the protection, preservation and reproduction of the region's productive forces;

- systemic approach to the development of the maritime complex of industries, recreation and tourism, and overcoming seasonality;
- functional approach, i.e. management of human resources of the region (creation of conditions for education, development and improvement of skills, realisation of potential) in order to ensure favourable development of socio-economic processes in the region;
- institutional approach, i.e. management and solving problems of coastal regions' development at the level of the region, communities, industries, and enterprises.

6 Conclusion

The conducted study offers a comprehensive theoretical and practical exploration of regional competitiveness, with particular emphasis on coastal regions and the strategic role of tourism development. As a result of the research, several important conclusions have been drawn, grounded in decades of academic observation and critical analysis of socio-economic transformations.

First, the concept of regional competitiveness has undergone a significant evolution, reflecting the complexity and multidimensionality of modern economic systems. The analysis confirms that regional competitiveness should be regarded not merely as a static characteristic of territories, but as a dynamic and adaptive system shaped by both internal resources and external influences. The dual interpretation of competitiveness—as an inherent potential and as a response to external stimuli—enriches the understanding of its essence and broadens the methodological toolkit available to researchers and policymakers.

Second, the classification of regional competitiveness, developed within this study, provides a more nuanced framework for analysing territorial development. The criteria proposed—territorial, resource-based, sectoral, developmental, and institutional—allow for a more precise identification of competitive advantages and vulnerabilities across different regional contexts.

Third, the research underlines the critical importance of governance quality in shaping regional competitiveness. The effectiveness of public administration, legal frameworks, institutional coherence, and macroeconomic stability emerge as decisive factors influencing regional performance. These insights resonate with global studies, affirming the pivotal role of state capacity and institutional resilience in regional development.

Fourth, coastal regions, due to their distinct ecological, economic, and geopolitical features, possess a specific structure of competitiveness. The combination of maritime resources, favourable climatic conditions, and access to international transport corridors provides these regions with significant competitive advantages. However, these advantages can only be realised through systematic management and sustainable development practices that integrate environmental protection, infrastructure modernisation, and human capital enhancement.

Fifth, tourism is recognised as a central driver of competitiveness in coastal regions. When integrated into a comprehensive regional strategy, tourism contributes not only to economic growth but also to the formation of a positive territorial image, the preservation of cultural identity, and the stimulation of innovation. The management of tourism-based competitiveness requires the alignment of multiple subsystems—legal, informational, financial, institutional—and the application of advanced digital tools and interregional cooperation mechanisms.

In summary, the research confirms that the competitiveness of coastal regions is not a spontaneous phenomenon but a result of deliberate, coordinated action across all levels of governance and economic activity. A strategic, systems-based approach—rooted in scientific knowledge, supported by robust institutions, and oriented toward long-term sustainability—is essential for enhancing regional competitiveness in the twenty-first century.

Thus, managing the competitiveness of the coastal region through tourism development requires multilevel coordination of efforts by government, business, educational and research institutions, as well as broad public involvement. This approach allows for the creation of an effective model of the regional economy, where tourism is not only a tool for economic growth, but also a factor in the integration of local development into global processes. In this context, the structure of the competitiveness management system of the coastal region based on tourism development requires a separate theoretical and applied understanding, which will be the subject of further analysis.

References

1. Smith A (2025) An inquiry into the nature and causes of the wealth of nations. https://audiobookstore.com/audiobooks/the-wealth-of-nations.aspx. Accessed 26 Feb 2025
2. Deineha OV (2008) Methodological aspects of evaluating enterprise competitiveness [Metodolohichni aspekty otsiniuvannia konkurentozdatnosti pidpryiemstv]. http://vlp.com.ua/files/10_25.pdf (Access date: 26.02.2025)
3. Best M (2001) The new competitive advantage: the renewal of American industry. Oxford University Press
4. Aiginger K (1998) A framework for evaluating the dynamic competitiveness of countries. Struct Chang Econ Dyn 9:159–188
5. Paas T (2004) Evaluating the competitiveness of the countries in the Baltic Sea region. In Modelling the economies of the Baltic Sea Regions, Tartu, pp 19–54
6. Peneder M (2007) A sectoral taxonomy of educational intensity. Empirica 34(3):189–212
7. Grilo I, Koopman GJ (2006) Productivity and microeconomic reforms: strengthening EU competitiveness. J Ind Compet Trade 6(2):67–84
8. Rodríguez-Pose A (1998) Social conditions and economic performance: the bond between social structure and regional growth in Western Europe. Int J Urban Reg Res 22:443–459
9. Cellini R, Soci A (2002) Pop competitiveness. Banca Nazionale del Lavoro Quarterly Rev 55(220):71–101
10. Camagni R (2002) On the concept of territorial competitiveness: sound or misleading? Urban Stud 39:2395–2411
11. Doel M, Hubbard P (2002) Taking world cities literally: marketing the city in a global space of flows. City 6(3):351–368

12. Chitea MA (2016) Evaluation of territorial competitiveness. Case study: South-East region and Tulcea county. Sci Pap Ser Manage Econ Eng Agric Rural Dev 16(1)
13. Skokan K (2004) Konkurenceschopnost, inovace a klastry v regionálním rozvoji, Repronis
14. Bezkhlibna A, Nykonenko S (2017) Strengthening of regional tourism development by creating tourism clusters in the cities of Ukraine. Sci Pap Univ Pardubice Ser D Faculty Econ Adm 40:5–15
15. Cowell PJ, Stive MJF, Niedoroda AW, Swift DJP, De Vriend HJ et al (2003) The coastal tract. Part 2: applications of aggregated modelling of lower-order coastal change. J Coastal Res 19:828–848
16. Porter M, Ketel C (2003) UK competitiveness: moving to the next stage (DTI Economics Paper No. 3). Department of Trade and Industry
17. Bezkhlibna A, But T, Nykonenko S (2018) Assessment of tourism industry clustering potential. Scientific Papers of the University of Pardubice, Series D, Faculty of Economics and Administration, vol 43
18. Byelikova M, Bezkhlibna A, Polyezhayev Y, Zaytseva V, Pukhalska H (2024). Informational and Digital Business Security in Tourism as a Component of the Coastal Region Competitiveness. In: Semenov A, Yepifanova I, Kajanová J (eds) Data-Centric Business and Applications. Lecture Notes on Data Engineering and Communications Technologies, vol 195. Springer, Cham, pp 117–145. https://doi.org/10.1007/978-3-031-54012-7_6
19. But T (2024) Analysis of the tourism industry competitiveness in the Czech regions. J Competitiveness 16(3):1–25. https://doi.org/10.7441/joc.2024.03.01
20. But T, Mamotenko D, Korniienko O, Gubar O (2025) Financial assurance for tourism development and its impact on the economy of European countries. In: Semenov A, Yepifanova I, Kajanová J (eds) Data-centric business and applications. Lecture Notes on Data Engineering and Communications Technologies, vol 240. Springer, Cham, pp 63–80. https://doi.org/10.1007/978-3-031-81557-7_4
21. Engelhardt-Nowitzki C, Kryvinska N, Strauss C (2011) Strategic demands on information services in uncertain businesses: a layer-based framework from a value network perspective. In: 2011 International conference on emerging intelligent data and web technologies. IEEE, pp 131–136. https://doi.org/10.1109/eidwt.2011.28
22. Urikova O, Ivanochko I, Kryvinska N, Strauss C, Zinterhof P (2013) Consideration of aspects affecting the evolvement of collaborative e-business in service organisations. Int J Services Econ Manage 5(1/2):72. https://doi.org/10.1504/ijsem.2013.051863

Intelligent Audit Technology Knowledge Base

Tetiana Neskorodieva, Eugene Fedorov, Pavlo Kulakov, Volodymyr Kucheruk, Anait Karapetyan, Roman Lishchuk, Irina Miroshkina, Alla Nesterenko, and Anastasiia Neskorodieva

Abstract This paper presents a methodology for formalizing knowledge about IT accounting and audit system users as a component of the audit system in the context of designing an information technology DSS audit, which includes the formalization of the following components: characteristics of audit system users; structure of the enterprise as an object of audit and management; functional features of accounting for enterprise characteristics as variables of the subject area; relationships of accounting variables as rules of the subject area of audit, and functional features of accounting information technologies. This knowledge base is invariant with respect to the industry affiliation of the audit object, type of production, and accounting features, which allows it to be used to automate audit tasks.

T. Neskorodieva (✉) · P. Kulakov · V. Kucheruk · R. Lishchuk
Uman National University, Uman, Ukraine
e-mail: t.neskorodeva@udau.edu.ua

P. Kulakov
e-mail: p.kulakov@udau.edu.ua

V. Kucheruk
e-mail: kucheruk@udau.edu.ua

R. Lishchuk
e-mail: roman.lishchuk@udau.edu.ua

E. Fedorov · A. Karapetyan · I. Miroshkina · A. Nesterenko
Cherkasy State Technological University, Cherkasy, Ukraine
e-mail: fedorovee75@ukr.net

A. Karapetyan
e-mail: a.karapetian@chdtu.edu.ua

I. Miroshkina
e-mail: i.miroshkina@chdtu.edu.ua

A. Nesterenko
e-mail: a.nesterenko@chdtu.edu.ua

A. Neskorodieva
Vasyl' Stus Donetsk National University, Vinnitsa, Ukraine

A. Semenov et al. (eds.), *Data-Centric Business and Applications*, Lecture Notes on Data Engineering and Communications Technologies 272,
https://doi.org/10.1007/978-3-032-06608-4_17

Keywords Structural model · Functional model · Knowledge base · Audit subject area · Audit information technology

1 Introduction

Today, the development and improvement of digital technologies in the economies of developed countries of the world indicates the rapid development of this segment of the IT market and the spread of economy digital transformations [1, 2]. Therefore, in the context of auditing, which is not an integral part of the economy, the problem of audit automation based on relevant information technologies and decision support systems is relevant [2, 3]

The analysis of subject area data is performed on the basis of information in the subject area information model, which is a necessary and mandatory component of the information support system for decision-making processes, for the improvement of which there is a need to create methods for automated processing and analysis of information about the audit object [3].

The issue of using IT in auditing is regulated by standards [4, 5]. The use of information technologies for audit tasks is associated with the audit of the IT technologies of the audit object itself, which are described in works [6], specific practices are given in [7].

The basics of expert audit systems are presented in [8]. Approaches to audit automation are formulated in [9], determinants of the implementation of computer tools and audit methods by internal and external auditors are considered in [10]. The accounting information system when submitting audits of financial statements using automated audit methods is presented in [11].

In [12] presents cases where auditors' research hypotheses were analyzed using structural equation modeling using the least squares method. It is shown that the use of IT by internal auditors has a strong and positive impact on fraud detection in the business process "from purchase to payment", and that the intensity of this relationship does not depend on the type and size of the business entity. General approaches and specific cases of the use of IT for internal audit are described in [13].

The issue of developing and implementing information automated decision support systems and data analysis technologies, including those with elements of artificial intelligence, is considered in [14]. Bidiuk et al. [15] presents approaches to improving intelligent decision support systems for predicting nonlinear and non-stationary processes.

In [16], a method for automating the analysis of audit data based on neural networks is presented. In [17, 18], solutions to individual audit automation problems using artificial intelligence methods are presented. The use of artificial intelligence methods for detecting outliers is presented in [19], anomaly detection based on known distribution laws in [20]. Intelligent models for forecasting in the insurance and financial sectors are presented in [21–23]. Intelligent algorithms for detecting fraud in [24], money laundering in [25]. The use of an autoencoder for external audit is presented in [26] for internal audit in [27], for detecting falsification of financial statements in [28], individual audit automation tasks in [18, 29].

Thus, the studies presented above lack systematic approaches to solving the problems of analysis automation in the audit system, but only general methodological approaches are presented or individual methods, models are developed, and specific problems are solved. This leads to insufficient efficiency of data analysis in the audit subject area and the formation of recommendations for the decision-maker.

Thus, the problem of increasing the efficiency and validity of the audit of digital economy objects is relevant, based on the unification of the knowledge base of intelligent information technology for data analysis in the decision-making support system.

2 Research Results

2.1 Characteristics of Audit System Users

Analysis of the characteristics of audit system users includes:

1. Classification of audit system users;
2. Definition of common components of the audit system for all user groups;
3. Determining the functional features of different user groups.

The goals, objectives, and functional features of the audit process depend on the user group being audited. Therefore, to determine them, it is necessary to form a structure of the set of users.

In relation to the audit object (enterprise), the entire set of users can be represented as a union of two subsets:

$$G = G_{\mathrm{BH}} \cup G_e, \tag{1}$$

where G is the set of all users, G_{BH} is multiple internal users, G_e is a set of external users.

Structuring user groups of each subset defined in (1) will be performed based on the analysis of their functions in interaction with the enterprise. Each user group corresponds to a set of possible decisions made regarding the enterprise, which determine the goals, objectives, subject area of analysis and functional features of audit system users.

We will distinguish user groups based on the functions of interaction with the enterprise:

$G_l,\ l = \overline{1, 10}$, l is type of user group,

where G_1 is auditing firms, G_2 is consulting firms, G_3 is lenders, G_4 is suppliers, G_5 is buyers, G_6 is government agencies, G_7 is company personnel, G_8 is trade unions, G_9 is administration, top-level managers, G_{10} is owners.

The set of internal users is equal to the union of the last two sets: administration, top-level managers, and owners: $G_{\text{BH}} = G_9 \cup G_{10}$.

The set of external users is equal to the union of all other sets: $G_{\text{e}} = \cup_{i=1}^{8} G_i$. Based on a certain structure of audit system users and their characteristics, we will identify the general components of the audit.

Let's define the common components of auditing all user groups.

(O_1) The audit subject area for all user categories can be represented as a combination of data sets and rule sets:

$$\Pi = D \cup R, \tag{2}$$

where D is data set, R is set of rules.

(O_2) The set of data of the audit subject area can be represented as a combination of regulatory planning, reporting, accounting and external data:

$$D = N \cup \text{P} \cup \text{O} \cup \text{Q} \cup \text{B}, \tag{3}$$

where N is set of normative data, P is set of planned data,

O is set of reporting data, Q is multiple credentials,

B is set of external data.

(O_3) The data of the four financial reporting forms (F1–F4) is a subset of the subject area of analysis of all groups of external users.

$$\cup_{k=1}^{4} \Phi_k \subseteq D_l, \tag{4}$$

where Φ is set of reporting form data, k is reporting form number,

D is subject area of analysis, l is user group index.

(O_4) Fourth, in each set of rules, four subsets can be distinguished:

- Conducting business activities $R^{(1)}$;
- Application of IT in production and other business processes of the enterprise $R^{(2)}$;
- Accounting $R^{(3)}$;
- Application of IT in accounting $R^{(4)}$.

$$R = \cup_{i=1}^{4} R^{(i)}. \tag{5}$$

(O_5) Accounting Regulations (Standards) are a subset of the rules of the subject area of auditing of all user groups:

$$\tilde{R} \subseteq R. \tag{6}$$

Certain common components of an audit system define common components when designing an audit DSS: subsystems, tasks, subject area.

At the same time, we will highlight the problem of automating the verification of compliance with internal rules, which consists in the need to take into account the peculiarities of internal rules for conducting business activities $R_{\text{BH}}^{(1)}$, the use of IT in production and other business processes of the enterprise $R_{\text{BH}}^{(2)}$; accounting $R_{\text{BH}}^{(3)}$ (accounting policy), the use of IT in accounting $R_{\text{BH}}^{(4)}$.

To take into account the specifics of the users work of different groups when designing an audit DSS, it is necessary to highlight the functional features of their work.

The different goals and objectives of user groups $G_l,\ l = \overline{1, 10}$, as well as the different subject areas of analysis, determine the functional features of their work, which are as follows.

The subject area of the first two external user groups G_e: audit firms (F) and consulting firms (K) includes the subject areas of other external user groups.

$$(\Pi_{l_1} \supseteq \Pi_{l_2}) \wedge (l_1 \in \{1, 2\}) \wedge \left(l_2 \in \left\{\overline{3, 8}\right\}\right) \tag{7}$$

Analysis of external audit task statements allows us to conclude that solving the tasks of all groups of external users $G_l\ l = \overline{2, 8}$ starting from the second, it is based on the results of checking the rules that make up the tasks of the first group F (audit firms):

$$(3_1 \supseteq 3_l) \wedge (3_1 \Rightarrow 3_l), l = \overline{2, 8} \tag{8}$$

where 3 is multiple tasks, l is user group index.

The subject area of internal users G_{BH} includes the subject areas of external user groups.

$$(\Pi_{l_1\cdot} \supseteq \Pi_{l_2}) \wedge (l_2 \in L_e) \wedge (l_1 \in L_{\text{BH}}), \tag{9}$$

where L_{BH} is set of internal user indexes, L_e is set of external user indexes.

During internal audit, the set of verified rules can be presented as a combination of external R_{BH} and internal rules R_e for conducting business activities and accounting. In this case, the set of internal rules must be subordinate to the external ones (internal rules may coincide, complement, clarify and not contradict the external ones).

$$(R_{\text{BH}} \cup R_e) \wedge (R_{\text{BH}} \Rightarrow R_e). \tag{10}$$

Analysis of the internal audit task statements allows us to conclude that they consist in checking a set of internal rules R_{BH}. During an external audit, internal rules R_{BH} can only be checked for inconsistency with external ones R_e, and their implementation is not checked.

Since a set of internal rules R_{BH} must be subordinate to external ones R_e, when checking indicators in the external audit task, algorithmic features are combined with internal audit.

The task for the external audit is to verify the prerequisites. Internal audit tasks follow the same pattern, but internal conditions are checked, and tasks of increasing efficiency and capacity are solved by identifying bottlenecks.

The distinct functional features of different user groups determine audit concepts and functional requirements.

2.2 *Formalization of the Enterprise Structure as an Object of Audit and Management*

An enterprise as an object of management and audit is characterized by states and transitional processes [30, 31].

The state of the enterprise is characterized by indicators of the enterprise's economic resources and sources at the beginning, end of the inspection period and reporting points for the planned periods.

Transition processes are characterized by indicators of the transformation of economic means and sources at the middle and lower levels and indicators of activity at the upper level for the planning periods and the verification period.

The first stage of analyzing an enterprise as an audit object consists of analyzing the structure of economic resources, sources, activities, and the relationships between them in order to decompose the audit system into subsystems and determine the relationships between them.

Therefore, the analysis of the structure of the enterprise as an audit object includes:

- Analysis of the structure of economic resources and sources of the enterprise;
- Analysis of the structure of the enterprise's activities;
- Analysis of the relationships between economic resources, sources and activities of the enterprise.

The structure of the enterprise's economic assets and sources is reflected in the reporting form Φ_1 "Balance Sheet". Economic assets are reflected in the first part of the Balance Sheet—Assets, which consists of three parts $A_i, i = \overline{1,3}$ (Table 1). The second part of the Balance Sheet—Liabilities, which reflects the sources of economic assets, consists of five parts $\Pi_i, i = \overline{1,5}$.

From the definition of the structure of Assets and Liabilities of the Balance Sheet in the P(s)BO, the following relationships follow:

$$A = \cup_{i=1}^{3} A_i, A_j \cap A_i = \emptyset, i \neq j, \tag{11}$$

$$\Pi = \bigcup_{i=1}^{5} \Pi_i, \Pi_j \cap \Pi_i = \emptyset, i \neq j, A \cap \Pi = \emptyset, \tag{12}$$

Table 1 Formalization of economic resources sets and their sources

Balance sheet article	Marking
Assets	A
Non-current assets	A_1
Current assets	A_2
Deferred expenses	A_3
Passive	Π
Equity	Π_1
Securing future expenses and payments	Π_2
Long-term liabilities	Π_3
Current liabilities	Π_4
Deferred income	Π_5

At the average level for accounting according to the Accounting Regulations (Standards) set A_i, $i = \overline{1,3}$, $\prod_j$, $j = \overline{1,5}$ economic resources and their sources are presented as a union of disjoint subsets:

$$\mathrm{A}_i = \cup_{k_i=1}^{K_i} \mathrm{A}_{k_i}, i = \overline{1,3}, \mathrm{A}_{k_i} \cap \mathrm{A}_{l_i} = \emptyset, k_i \neq l_i, \tag{13}$$

$$\prod_j = \cup_{n_j=1}^{N_j} \prod_{n_j}, j = \overline{1,5}, \prod_{n_j} \cap \prod_{m_j} = \emptyset, n_j \neq m_j, \tag{14}$$

where A_{k_i}, $\prod_{n_j}$ are subsets of economic assets and sources consisting of economically homogeneous elements (determined in the accounting policy of the enterprise).

Let us denote the system of subsets of economic means and sources of economically homogeneous elements:

$$\vartheta = \left\{ \mathrm{A}_{k_i}, k_i = \overline{1, K_i}, i = \overline{1,3}, \prod_{n_j}, n_j = \overline{1, N_j}, j = \overline{1,5} \right\} = \left\{ E_n, n = \overline{1,N} \right\}, \tag{15}$$

where

$$E_n = \left\{ \frac{A_n, n = \overline{1, \sum_{i=1}^{3} K_i},}{\prod_n, n = \overline{\sum_{i=1}^{3} K_i + 1, \sum_{i=1}^{3} K_i + \sum_{j=1}^{5} N_j},} \right. ,$$

$$N = \sum_{i=1}^{3} K_i + \sum_{j=1}^{5} N_j$$

Table 2 Formalization of audit objects of the enterprise's activities

Set of activity operations	Marking
Basic	$O\Pi_1$
Another operating room	$O\Pi_2$
Financial	$O\Pi_3$
Investment	$O\Pi_4$
Emergency	$O\Pi_5$

Formulas (11)–(15) determine M_1 is model of the structure of economic resources and sources of the enterprise. Formulas (11), (12) are reflect the structure at the top level, (13)–(15) are at the middle level.

To answer the question of whether it is possible to decompose the verification of transformations of economic means and sources of an enterprise into the verification of disjoint subsets at the upper and middle levels and their elements at the lower levels, it is necessary to investigate the structure of the enterprise's activities and the relationship with it.

The activity of the enterprise consists in the transformation of economic means and sources. Therefore, to determine the structure of transformations, it is necessary to consider the structure of accounting for the activity of the enterprise as an object of audit.

In the accounting mode, the activities of the enterprise are presented in the form of operations, which are grouped by type of activity (Table 2).

According to the Accounting Regulations (Standards), each transaction of economic activity for the period $t(t \in \{t_{j_m}, T_m,\ j = \overline{1, J_m}, m = \overline{1, M}, T\})$ belongs to only one of the sets On_i, $i = \overline{1, 5}$, that is, the set of transactions for the period t satisfy the following relationship:

$$O\Pi_j \cap O\Pi_i = \emptyset, i \neq j, i = \overline{1, 5}, j = \overline{1, 5} \tag{16}$$

$$\cup_{i=1}^{5} O\Pi_i = O\Pi, \tag{17}$$

where $O\Pi$ is the set of business transactions of the enterprise for the period t, i is type of activity.

Since at the top level, types of expense and income transactions are distinguished in each set, the structure of the sets of transactions for each type of activity can be represented as a union of disjoint subsets of expense and income transactions:

$$O\Pi_i = O\Pi_i^{\mathrm{p}} \cup O\Pi_i^{\mathrm{d}}, O\Pi_i^{\mathrm{p}} \cap O\Pi_i^{\mathrm{d}} = \emptyset, i = \overline{1, 5}, \tag{18}$$

where $O\Pi^{\mathrm{p}}$—set of cost operations,

$O\Pi^{\mathrm{d}}$—a set of income transactions.

At the average level, for accounting of operations and calculation of the activity results for each set of operations by type of activity (Table 2), subsets are distinguished according to their relation to economic means and sources (Table 1). Therefore, at the average level, the set of operations for each type of activity (including the sets of expenses $O\Pi^{\mathrm{p}}$ and income $O\Pi^{\mathrm{d}}$) can be represented as disjoint subsets of operations that lead to the transformation of a pair of sets of economic means and sources:

$$O\Pi_i = \cup_{(E_{n_e}, E_{n_g}) \in \vartheta^2} O\Pi_i^+ \left(E_{n_e}, E_{n_g}\right) \cup O\Pi_i^- \left(E_{n_e}, E_{n_g}\right), i = \overline{1,5}, \tag{19}$$

where $O\Pi_i^+ \left(E_{n_e}, E_{n_g}\right), O\Pi_i^- \left(E_{n_e}, E_{n_g}\right)$ are the sets of operations that lead pairs of sets of economic means and sources $\left(E_{n_e}, E_{n_g}\right) \in \vartheta^2$ to transformation in the "forward" and "reverse" directions, respectively,

$$\vartheta^2 = \left\{\left(E_{n_e}, E_{n_g}\right) \in \vartheta \times \vartheta, n_e < n_g\right\}. \tag{20}$$

At the bottom level, the elements of each set of operations for a period t are the operations for a given period:

$$O\Pi = \{on(\tau), \tau \in [0, t]\}, t \in \left\{t_{j_m}, T_m, \ j = \overline{1, J_m}, m = \overline{1, M}, T\right\}, \tag{21}$$

where $O\Pi$ is operation, τ is the moment of the operation.

M_2 is model of the structure of the enterprise's activities (16)–(21). Formulas (16)–(18) correspond to the upper level, (19), (20) to the middle, (21) is to the lower level.

Also, based on the model M_2, we conclude that the interconnection of economic resources, sources, and the activities of the enterprise are subject to verification. We formalize the data of the interconnection in the form of a model M_3.

We present the model of relationships in the form of a graph of correspondence of economic resources and sources E_{n_e} and E_{n_g}, interconnected by an operation $O\Pi^+\left(E_{n_e}, E_{n_g}\right)$ or multiple $O\Pi^+\left(E_{n_e}, E_{n_g}\right)$ transformation operations:

$$G = (Z, R), Z = \left\{n, n = \overline{1, N}\right\}$$
$$, R = \left\{c_e^g = \left\{n_e, n_g\right\}, n_e \in \left\{\overline{1, N}\right\}, n_g \in \left\{\overline{1, N}\right\}\right\}, \tag{22}$$

where G is graph of relationships between economic resources and sources,

Z is the set of vertices of the graph,

R is the set of edges of the graph,

n is the vertex corresponding to the economic resource or source E_n,

N is the number of vertices, which corresponds to the number of types of economic resources and sources that are converted,

c_e^g is the edge that corresponds to the operation of transforming an economic asset or source E_{n_e} into E_{n_g}.

The correspondence graph of economic means and sources can characterize the types of operations of any set defined in (17) or (18), that is, by types of activities and all activities of the enterprise, as well as by subsets of expenses and income.

Based on the formed models, we will draw the following conclusions.

- The objects of the audit are the structural elements of the enterprise and their relationships formalized in the form of models M_1, M_2, M_3.
- The characteristics of the structural components of the enterprise determine the structural variables of the audit.
- The values of the characteristics of the structural components of the enterprise form the subject area of the audit of the values of structural variables D_1.
- The rules for displaying the values of the characteristics of the structural components of an enterprise form the subject area of the rules for implementing structural R_1 audit relationships.
- The set of models M_1, M_2, M_3 describes the functional structure of the enterprise as an audit object and is the basis for forming the structure of the audit system.

Based on the analysis conducted, we will highlight the following audit automation problems.

1. Mass -production, small-scale production, -piece production (components) or weight production, contractual, automatic, etc., according to the types of operations and processes at the enterprise).
1. The presence of interrelationships of the structural components of the enterprise at each level, presented in the form of a model, M_3 does not allow the decomposition of the audit process (according to the characteristics of the audit objects) into subsystems according to structural elements, represented by models M_1, M_2, which significantly complicates the verification algorithms.

2.3 Functional Features of Accounting for Enterprise Characteristics as Variables of the Audit Subject Area

The analysis of the characteristics of the enterprise includes the following points.

1. Identification of functional features of accounting for enterprise characteristics in the mode of rationing, planning, reporting and accounting.
2. Determination of the relationships between the characteristics of the functional structure of the enterprise and the variables of the accounting system.
3. Analysis of enterprise indicators in the accounting system as audit variables.

We formalize the accounting of enterprise characteristics in the mode of normalization of planning, reporting and accounting. Enterprise characteristics are generally reflected in regulatory N, planning P, reporting O, accounting Q and external data B, defined in (3).

The variables of the characteristics of the enterprise in the set of regulatory N, planning P and reporting O data depend on the characteristics of the enterprise and its accounting policy and are generally not standardized.

There are also features of accounting for data sets in the IT environment of the enterprise. Creation, processing and storage of sets of regulatory N, planning P, reporting O data can be carried out in various ways according to the structural elements of the enterprise defined in the models M_1, M_2, M_3: in manual mode (only paper documents), in semi-automatic, automatic. These subsystems can function locally and are not combined into a corporate system. IT can be used that are not compatible with each other (for example, for inventory accounting and in production), as a result of which the accounting of the relationships between the elements of the structural components described by the models M_1 and M_2 is not carried out in automatic mode.

The identified features of accounting for sets of regulatory N, planning P, reporting O, and accounting Q data make it difficult to verify the relationships between the characteristics of elements at the lower level:

1. Require a lot of time, which significantly reduces the efficiency of the inspection;
2. Make it difficult to comply with the correctness of verification procedures and, as a result, the reliability of conclusions is reduced;
3. The possibilities for automating verification procedures at this stage are limited.

To analyze a set of accounting data Q as a subject area of audit, it is necessary to determine the relationships between the structural elements of the enterprise and the accounting system.

Let's define the relationships between the functional structure of the enterprise and the variables of the accounting system. The structure of the accounting system is represented by the structure of accounts (Table 3). In class accounts $K_i, i = \overline{1,6}$ accounting of structural elements of models M_1 and M_2 enterprises are carried out. Calculation of expenses, income and results of activities for the period t is carried out on the effective accounts of classes $K_i, i = \overline{7,9}$ (class accounts K_8 may not be used).

Accounting for economic assets and sources that are not reflected in the enterprise's balance sheet (goods taken on commission, etc.) is carried out on class

Table 3 Compliance of structural elements of models M_1 and M_2 enterprise and accounting system

Element of the accounting system		Enterprise element
Name	Marking	
Non-current assets	K_1	$E_n \subset A_1, O\Pi_i, i \in \{\overline{1,5}\}$
Stocks	K_2	$E_n \subset A_2 \cup A_3, O\Pi_i, i \in \{\overline{1,5}\}$
Funds, accounts and other assets	K_3	
Equity and collateral	K_4	$E_n \subset \prod_1 \cup \prod_2, O\Pi_i, i \in \{\overline{1,5}\}$
Long-term liabilities	K_5	$E_n \subset \prod_3 \cup \prod_5, O\Pi_i, i \in \{\overline{1,5}\}$
Current liabilities	K_6	$E_n \subset \prod_4, O\Pi_i, i \in \{\overline{1,5}\}$
Revenue and performance	K_7	$P_i(t), m(On_i^d(t)), i \in \overline{1,5}$
Costs by element	K_8	–
Operating expenses	K_9	$m\left(On_i^{r(vid)}(t)\right), i \in \overline{1,5}$
Off-balance sheet accounts	K_0	–

accounts K_0. We denote the set of all accounts by K. From the definition of account classes, the following relations follow:

$$K = \cup_{i=0}^{9} K_i \;\; K_i \cap K_j = \emptyset, i \neq j, i = \overline{0, 9}, j = \overline{0, 9} \tag{23}$$

Each account has a number, which we will denote s_e^k, $k = \overline{0, 9}$, $e = \overline{0, e^*}$, where k is account class number, e is account number in this class. The total number of accounts is denoted by s^*. To simplify further entries, the indices e and k will be indicated only if necessary. The subsystem of each account consists of two parts: debit d_s and credit k_s, which reflect the values of the balance and the amount of transactions.

The relationships of structural elements in the accounting system M_3 are reflected in the form of correspondence of accounts defined above (Table 3), which is formalized in the form of a graph:

$$\begin{aligned} &G = (Z, R), Z = \left\{s, s = \overline{1, S}\right\}, \\ &R = \left\{c_e^g = \left\{s_e, s_g\right\}, s_e \in \left\{\overline{1, S}\right\}, s_g \in \left\{\overline{1, S}\right\}\right\} \end{aligned} \tag{24}$$

where G is transaction correspondence graph,

Z is the set of vertices of the correspondence graph of operations,

R is the set of edges of the correspondence graph of operations,

s is the vertex that corresponds to the account with number s,

S is the number of vertices, which corresponds to the number of accounts,

c_e^g is edge that corresponds to the transfer operation $op\left(l_e^g\right)$ from the loan k_{s_e} in debit d_{s_e}.

The graph of correspondence of operations (24) can characterize the types of operations and their sets defined in (17) or (18), that is, by types of activities and all activities of the enterprise, as well as by subsets of expenses and income.

The rules of correspondence (Table 3) and (24) describe the relationships between the structural elements of the enterprise and the system at all levels. These rules form a verification model, which we denote by M_4:

1. The correctness of the correspondence of accounts and the types of economic assets and sources that are taken into account in them.
1. Determining transaction correspondence.

To analyze the relationships between enterprise characteristics and accounting variables, it is necessary to determine the correspondence between them.

We will analyze the characteristics of the enterprise in the accounting system as audit and management variables. At the lower level, the elements of the enterprise are characterized by a set of characteristics of the following types:

1. Numerological (document numbers; nomenclature of raw materials, materials, finished products, suppliers, buyers, technologies, workshops, etc.);
2. Numerical (price, cost, quantity in physical units, transaction amount, etc.);
3. Logical (conditions for raw material supply, payment, etc.);

4. Semantic (names of raw materials, product materials, operations).

We will denote this set of lower-level indicators as:

$$\overline{\mu(e)}|\tau, \tag{25}$$

where $\overline{\mu(e)}|\tau$ is vector of indicators of the state of an economic asset or source at a point in time τ,

$$\overline{m(\text{O}\Pi(E_{n_e}, E_{n_g}), t)}, \tag{26}$$

where $\overline{m(O\Pi(E_{n_e}, E_{n_g}), t)}$ is vector of indicators of the operation of transformation of an economic asset or source in a period of time t.

These characteristics are reflected in the following primary accounting documents:

- POR is payment order requirement,
- CNRM is consignment note for raw materials,
- RMP is internal invoice for the release of raw materials into production,
- PFP is internal invoice for posting finished products,
- SFP is consignment note for the shipment of finished products.

Variables for accounting of primary documents depend on the characteristics of the object, operations at each enterprise. The variables also depend on the characteristics of economic activity and the accounting system of different enterprises: with different functional structures (for production—mass, small-scale, for raw materials—piece (components) or weight, for settlements with suppliers—contractual, automatic, etc., for types of operations and processes at the enterprise).

At the medium level, subsets of enterprise elements are characterized by a set of numerical, logical, semantic, and structural indicators (estimates in monetary and physical terms, name, logical discount conditions, membership in sets, etc.). We will denote this set of medium-level indicators as:

$$\overline{\mu(E_{n_e})}|\tau, \tag{27}$$

where $\overline{\mu(E_{n_e})}|\tau$ is vector of indicators of the state of a subset of economic assets or sources E_{n_e} at the time τ,

$$\overline{m(O\Pi(E_{n_e}, E_{n_g}), t)}, \tag{28}$$

where $\overline{m(O\Pi(E_{n_e}, E_{n_g}), t)}$ is vector of indicators of the set of operations for the time period t, which accompanies the transformation of economic assets or sources E_{n_e} y E_{n_g}.

At the top level, the set of enterprise elements is characterized by a set of numerical, logical, semantic, structural indicators (estimates in monetary terms, name, logical discount conditions, membership in sets of expenses or income, etc.). We will denote this set of top-level indicators:

$$\overline{\mu(E_n)}|\tau\,, \tag{29}$$

where $\overline{\mu(E_n)}|\tau$ is vector of indicators of the state of a set of economic assets or sources E_n at a point in time τ,

$$\overline{m\big(O\Pi^d, t\big)}, \tag{30}$$

where $\overline{m\big(O\Pi^d, t\big)}$ is indicators vector of the income transactions set for the time period t,

$$\overline{m(O\Pi^r, t)}, \tag{31}$$

where $\overline{m(O\Pi^r, t)}$ is vector of indicators of the set of cost operations for the time period t.

Among the sets of indicators defined at three levels in (25)–(31) we will distinguish indicators in monetary and natural units. General accounting rules are defined for these indicators. We will determine the correspondence between them and the accounting variables. We will denote:

$C_s(\tau)$ is the value of the account balance with the number at the time τ,

$l_e^g(t)$ is transaction number for the time period t with correspondence $\big(s_e, s_g\big)$,

$L_e^g(t)$ is number of operations per time t with correspondence $\big(s_e, s_g\big)$,

$op(l_e^g(t))$ is loan transfer operation k_{s_e} in debit d_{s_e} during the time period t, $s_e \neq s_g$, $s_e = \overline{1, s^*}$, $s_g = \overline{1, s^*}$.

$\delta(l_e^g(t))$ is the amount transferred from credit k_{s_e} to debit d_{s_e} during the time period t,

$S_e^g(t)$ is the amount transferred from credit k_{s_e} to debit d_{s_e} over a period of time t,

$Ob_{s_e}^k(t)$ is loan turnover k_{s_e} for the period of time t,

$Ob_{s_e}^d(t)$ for the period of time t.

At the lower and middle levels, the characteristics of the enterprise are reflected in balance sheet accounts (see Table 4).

The results of activities at the top level are reflected in the profit and loss accounts (see Table 3). Let us determine the correspondence between the characteristics of the main and financial activities of the enterprise and at the top level.

The turnover value on the debit of account 901 "Cost of sales" reflects the cost $SS'(t)$ of products sold for the period t:

$$Ob_{901}^d(t) = SS'(t). \tag{32}$$

The turnover value on the debit of account 704 "Deductions from income" reflects the costs of paying deductions from income $OP(t)$:

$$Ob_{704}^d(t) = OP(t). \tag{33}$$

Table 4 Rules for matching accounting variables and enterprise characteristics at the lower and middle levels

Accounting variable	Company characteristics	
	$E_{n_e} \subset A$	$E_{n_e} \subset \prod$
$C_s^k(\tau)$	–	$\mu(E_n)\|\tau$
$C_s^d(\tau)$	$\mu(E_n)\|\tau$	–
$\delta(l_e^g(t))$	$m(on(l^-((E_{n_e}, E_{n_g}), t)))$	$m(on(l^+((E_{n_e}, E_{n_g}), t)))$
$\delta(l_g^e(t))$		$m(on(l^-((E_{n_e}, E_{n_g}), t)))$
$S_e^g(t)$	$m(On^-((E_{n_e}, E_{n_g}), t))$	$m(On^+((E_{n_e}, E_{n_g}), t))$
$S_g^e(t)$	$m(On^+((E_{n_e}, E_{n_g}), t))$	$m(On^-((E_{n_e}, E_{n_g}), t))$
$S_{e(i)}^g(t), i \in \overline{1,5}$	$m(On_i^-((E_{n_e}, E_{n_g}), t))$	$m(On_i^+((E_{n_e}, E_{n_g}), t))$
$S_{g(i)}^e(t), i \in \overline{1,5}$	$m(On_i^+((E_{n_e}, E_{n_g}), t))$	$m(On_i^-((E_{n_e}, E_{n_g}), t))$
$Ob_{s_e}^k(t)$	$\sum_{E_{n_g} \in \vartheta} [m(On^-((E_{n_e}, E_{n_g}), t))]$	$\sum_{E_{n_g} \in \vartheta} [m(On^+((E_{n_e}, E_{n_g}), t))]$
$Ob_{s_e}^d(t)$	$\sum_{E_{ne} \in \vartheta} [m(On^+((E_{n_e}, E_{n_g}), t))]$	$\sum_{E_{ne} \in \vartheta} [m(On^-((E_{n_e}, E_{n_g}), t))]$

The turnover value on the debit of account 951 "Interest on loans" reflects the costs of using short-term loans:

$$Ob_{951}^d(t) = RK(t). \tag{34}$$

The structures of the relationship between indicators of sets of operations at the top level are schematically presented in (see Figs. 1, 2, 3 and 4).

Counts $G^{(k)}$, $k = \overline{1,4}$ form a model M_6 of structural relationships of enterprise characteristics at the top level.

At the next stage, we will analyze data transformations at the middle level. Transformations of indicators at the middle level characterize the transformations

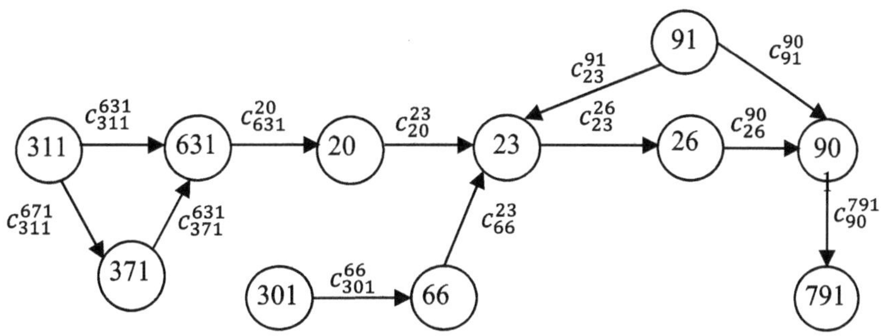

Fig. 1 Graph $G^{(1)}$ of indicators conversion when calculating the cost of goods sold SS

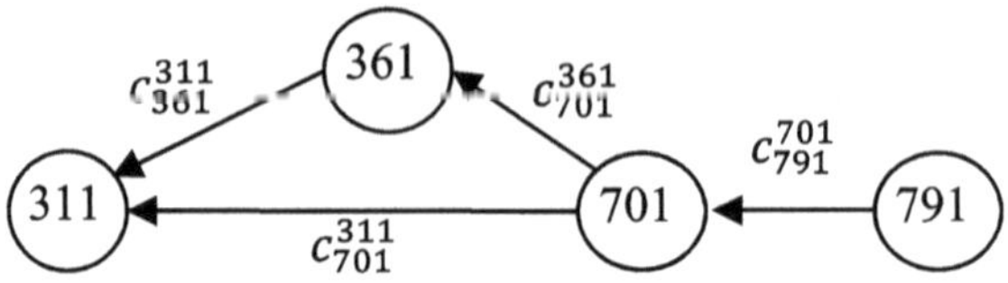

Fig. 2 Graph $G^{(2)}$ of indicators conversion when calculating income from product sales *SP*

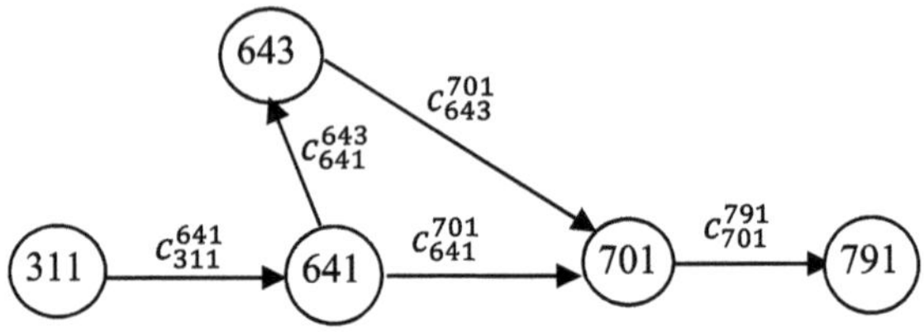

Fig. 3 Graph $G^{(3)}$ of indicators conversion when calculating deductions from income *OP*

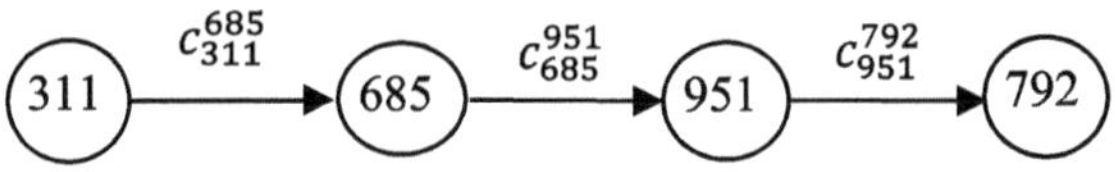

Fig. 4 Graph $G^{(4)}$ of indicators conversion when calculating costs from the use of short-term loans *RK*

of subsets of operations. Since the correspondence relations of subsets of economic means and sources are single-valued, the transformations of indicators of subsets of operations (28) have a more complex structure and are not decomposed into disjoint graphs. Since the vertices of the graph correspond to subsets of economic means and sources (subaccounts in the accounting system), the dimensionality of the graph also increases. We will consider the formalization of data transformations at the middle level in the next section.

At the next stage, we will analyze the data transformations at the lower level. Transformations at the lower level characterize the relationship between the indicators of individual operations of all subsets. The number of operations of each subset of the middle level of medium and large enterprises that are subject to mandatory audit is large. The number of indicators of each operation can reach several dozen. The dimensionality of the graph and the number of relationships characterizing each edge are increasing. Verification of these relationships is the main problem of the audit, since the accounting of these relationships is not 100% automatic.

Thus, when checking the correct implementation of relationships at each level, the following problems arise:

1. Significant growth in dimensions and complexity of the structure of relationships at lower levels;
2. An increase in the number and types of relationships between variables in the structures specified in paragraph 1;
3. The need to verify multidimensional heterogeneous interconnected data, the accounting of the relationships between which is not automated.

To analyze the audit subject area, it is also necessary to analyze the relationships of accounting variables between levels as audit elements.

We formalize the relationships between accounting variables at each level. The calculation of the characteristics of the enterprise at three levels (see Table 4, and Eqs. (32)–(34)) is carried out from the lower level to the upper, by step-by-step generalization over subsets and sets of the structural components of the enterprise.

At the lower level, the transaction amount is calculated based on the transaction data.

At the next stage, the double-entry rule is used to record the amounts of transactions in the accounting system: it is reflected on the credit of the account s_e and debits of the corresponding account s_g. After accounting for the transaction's amounts $\delta(l_e^g(t))$, the values of the amounts $S_e^g(t)$ transferred from the credit of accounts s_e to the debit of accounts s_g for the period of time are calculated t using the formula:

$$S_e^g(t) = \sum_{l_e^g(t)=1}^{L_e^g(t)} \delta(l_e^g(t)), \left(s_e, s_g\right) \in R^{(2)}. \tag{35}$$

were $\delta(l_e^g(t))$, $(s_e \in K)$ is transactions amounts.

Credit turnover of accounts is calculated s_e over time t t using the formulas:

$$Ob_{s_e}^k(t) = \sum_{s_g=1}^{s^*} S_e^g(t), s_e \in K. \tag{36}$$

Turnover on the debit of each account according to the formula:

$$Ob_{s_e}^d(t) = \sum_{s_e=1}^{s^*} S_e^g(t), s_e \in K. \tag{37}$$

and the final values of debit and credit account balances according to the formulas:

$$C_s^k(t_0 + t) = C_s^k(t_0) - Ob_s^d(t) + Ob_s^k(t), \tag{38}$$

$$C_s^d(t_0 + t) = C_s^d(t_0) + Ob_s^d(t) - Ob_s^k(t) \tag{39}$$

in accordance.

Formulas (38)–(41) form the model M_7 functional relationships of enterprise characteristics between levels.

Since the implementation of model relationships M_7 is carried out automatically, the task of verifying them can be solved by testing the accounting system.

2.4 Functional Features of Accounting Information Technologies

Let us define the features of accounting for data in the audit subject area according to the models $M_j, j = \overline{1,7}$.

Let us highlight the following features of the accounting system as an object of audit.

The stage of entering data into primary documents can be performed manually for some or all primary accounting variables (see Fig. 5).

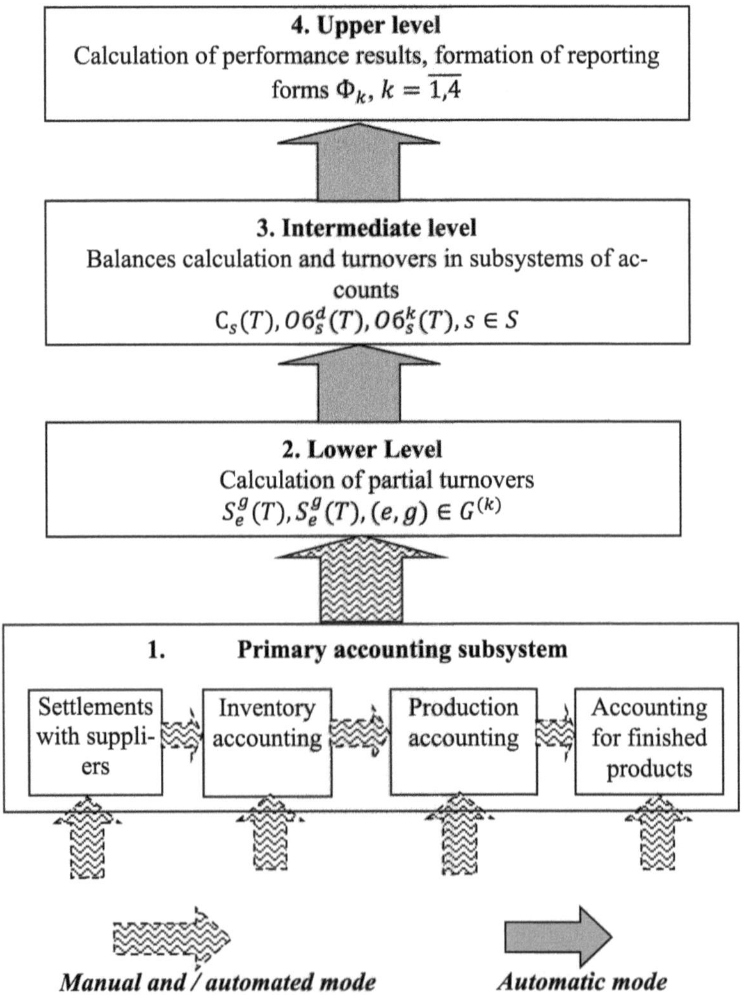

Fig. 5 Scheme of implementation of heterogeneous data relationships of the audit subject area by models $M_j, j = \overline{1,7}$ in the enterprise's IT environment

At the stage of entering data from primary documents, accounting for correspondence relationships between them is usually carried out manually or generalized across all documents for the period.

Loading primary accounting data into the first-level subsystem (assignment of account correspondence) is usually performed manually and/or summarized across all documents for the period.

The upper-level relationships are formalized in the form of graphs (see Fig. 1, 2, 3 and 4) and are formed on the basis of the middle-level relationships, which in turn are formed on the basis of the lower-level data. Accounting for these relationships at the lower levels depends on the accounting policy and the use of IT and, due to the peculiarities of the operation in which two subjects participate, is carried out in most cases in manual mode (although in a computer environment). That is, the double-entry recording of the operation is not carried out automatically, in addition, as a rule, there is a time gap between the records. The exception is the fully automated movement of raw materials, semi-finished products and finished products at the enterprise (the enterprise staff does not participate in the issuance of documents, loading, unloading, etc.).

The decomposition of the relationships of the upper-level data sets formalized in the form of graphs (see Figs. 1, 2, 3 and 4) at the middle and lower levels has a complex structure and depends on the structure of the sets of economic means and sources at the middle levels and the accounting variables of the lower level, which are determined by the internal accounting policy of the enterprise. This leads to problems of checking the compliance of such indicators of the upper level as the set of income and expense transactions, and of the middle level—indicators of the sets of transactions of payment for raw materials, release of raw materials to production, posting of finished products, that is, data sets of parallel-sequential operations.

That is, to check formulas (40) and (41) for calculating the account balance:

$$C_s^k(t_0 + t) = C_s^k(t_0) - Ob_s^d(t) + Ob_s^k(t), \tag{40}$$

$$C_s^d(t_0 + t) = C_s^d(t_0) + Ob_s^d(t) - Ob_s^k(t) \tag{41}$$

it is necessary to check the correspondence between debit $Ob_s^d(t)$ and credit turnovers $Ob_s^d(t)$ (which can be done only on the basis of checking the correspondence of the first-level data on the basis of which these turnovers were calculated). Since even in the case of checking the initial $C_s(t_0)$ and final balances $C_s(t_0 + T)$, Eqs. (40) and (41) contain two unknowns: debit $Ob_s^d(t)$ and credit turnover $Ob_s^k(t)$.

Also, for accounts s and $s + 1$, which are adjacent in Figs. 1, 2, 3 and 4, it is necessary to check the correspondence between credit turnovers $Ob_s^k(t)$ on the account $s + 1$ and $Ob_{s+1}^d(t)$ debit turnovers on the account $s + 1$ (based on checking the correspondence of the first level data).

These considerations answer the question of why even a complete arithmetic check (testing the accounting system) does not confirm the reliability of reporting indicators. The reason is that the sets of correspondences at the lower level are taken into account manually. In this case, correspondences are established only between documents (their numbers), and the verification of relationships between document variables is not carried out.

For example, a payment document for payment of raw materials and which has two consignment notes for the supply of raw materials. The consignment note will contain the payment document number (possibly in manual mode) to which the consignment notes correspond. The nomenclature and payment amount are not checked automatically for each such pair of documents. The amounts are verified for the planned period (usually a month).

In general, regarding the IT environment of the accounting system, it can be concluded that the accounting system is not automatic in accounting for all data and relationships. The reflection of the characteristics of the enterprise in the set of accounting data Q at most (but not all) stages is automated and standardized and is the last generalizing stage of accounting and calculating reporting indicators F1–F4. Since accounting data Q is stored in the accounting system database, this allows you to automate their analysis and verification, taking into account the mode (automatic or automated) of their reflection and transformation.

The above-mentioned features of the IT environment for accounting for sets of regulatory N, planned P, reported O.

- Data by type of input or calculation (manual, automated, automatic);
- Relationships by type of their implementation (manual, automated, automatic).

Solving selected IT audit tasks before conducting audit procedures will reduce the scope of the audit by automatically testing relationships, the implementation of which is carried out automatically, and highlighting data sets that are entered or calculated automatically.

Therefore, we will highlight the preliminary (recommended) stage is of IT audit of the data entry mode and the implementation of relationships.

3 Conclusions

Thus, the knowledge base of the intelligent audit system includes:

- Model and M_1, M_2, M_3 structural elements of the enterprise and model M_4 their relationships formalized at three levels;
- Model M_5 of correspondence between accounting variables and enterprise characteristics;
- A model M_6 of structural relationships of enterprise characteristics at the top level;
- Model M_7 of functional relationships of enterprise characteristics between levels.

A further direction of development of the work is the creation of models, methods and corresponding intelligent technologies for automating the analysis of relationships between multidimensional heterogeneous (four types of data are highlighted), which will allow solving analysis problems by gradually reducing the search area at each stage of analysis.

References

1. Global Center for Digital Business transformation website. https://www.imd.org/dbt/digitalbusiness-transformation.
2. Ukraine 2030E A Country with A Developed Digital Economy. https://strategy.uifuture.org/kraina-z-rovinutoyu-cifrovoyu-ekonomikoyu.html
3. Burak M, Yaremko I (2024) Automation of financial reporting audit during wartime. Econ Soc 60. https://doi.org/10.32782/2524-0072/2024-60-13
4. Professional Standards That Apply to Information Technology (Audit, Security, and Privacy Issues) (2008) Information technology control and audit, 3rd edn., pp 641–675. https://doi.org/10.1201/9781420065541.ax3
5. The Audit Committee and Information Technology (2012) Audit committee essentials. Portico, pp 200–212. https://doi.org/10.1002/9781119201472.ch13
6. Weber R (1999) Information systems control and audit. Prentice-Hall Inc., Upper Saddle River, p 1013
7. Otero AR (2018) Information technology environment and IT audit. In: Information technology control and audit, pp 3–30. https://doi.org/10.1201/9780429465000-1
8. Van Dijk JC, Williams P (1990) Expert systems in auditing. Stockton press, NY, p 192
9. Zavgorodniy VP (1998) Automation of accounting, control, analysis and audit. A.S.K, Kyiv, p 768
10. Ivakhnenkov SV (2005) Computer audit: control methods and technologies. Znannya, K, p 286
11. Azizah N, Sudarto F (2010) Sistem informasi akantansi dalam penyajian audit financial report dengan menggunakan computer assisted audit techniques (CAATs). CommIT (Commun Inf Technol) J 4(1):12. https://doi.org/10.21512/commit.v4i1.530
12. Samagaio A, Diogo TA (2022) Effect of computer assisted audit tools on corporate sustainability. Sustainability 14(2):705. https://doi.org/10.3390/su14020705
13. Coderre DG (2000) Computer Assisted Fraud Detection. Intern Audit 57(4):253
14. Zgurovsky MZ, Zaychenko YP (2017) The fundamentals of computational intelligence: system approach. In: Studies in computational intelligence. Springer International Publishing. https://doi.org/10.1007/978-3-319-35162-9
15. Bidiuk P, Prosyankina-Zharova T, Diakon V, Diakon D (2023) The improvement of the intelligent decision support system for forecasting non-linear non-stationary processes. Technol Audit Prod Reserv 4(2(72)):37–46. LOCKSS. https://doi.org/10.15587/2706-5448.2023.28651
16. Neskorodieva T, Fedorov E (2021) Automatic analysis method of audit data based on neural networks mapping. In: VII International Conference IT&I, information technology and interaction, vol 2833, 02–03 Dec 2020, Kyiv, Proceedings. CEUR-WS, p 60–70. http://ceur-ws.org/Vol-2833/Paper_6.pdf
17. Neskorodieva T, Fedorov E, Smirnov O, Rudakov K, Neskorodieva A (2021) Method detection audit data anomalies on the basis restricted Cauchy machine. In: Proceedings of the cybersecurity providing in information and telecommunication systems II. Volume I co-located with international conference on problems of in communications. Science and technology (PICST 2021), vol 3187, Oct 26, 2021 (online). CEUR Workshop Proceedings, Kyiv, Ukraine, pp 1–12. https://ceur-ws.org/Vol-3187/paper1.pdf
18. Neskorodieva T, Fedorov E (2022) Neural network models ensembles for generalized analysis of audit data transformations. Math Model Simul Syst 263–279. https://doi.org/10.1007/978-3-030-89902-8_21
19. Aggarwal ChC (2017) Outlier analysis, 2nd edn. Springer, Cham, p 465

20. Eskin E (2000) Anomaly detection over noisy data using learned probability distributions. In: Proceedings of the seventeenth international conference on machine learning. Morgan Kaufmann Publishers Inc., pp 255–262
21. Tkachenko R, Izonin I, Greguš ml M, Tkachenko P, Dronyuk I (2019) Committee of the SGTM neural-like structures with extended inputs for predictive analytics in insurance. In: Youna SM, Awan I, Benbernou S (eds) Big data innovations and applications. Innovate-data 2019. Communications in computer and information science, vol 1054. Springer, Cham, pp 121–132. https://doi.org/10.1007/978-3-030-27355-2
22. Bodyanskiy G, Boiko O, Zaychenko Y, Hamidov G, Zelikman A (2020) The hybrid GMDH-neo-fuzzy neural network in forecasting problems in financial sphere. In: Proceedings of the 2nd International conference on system analysis & intelligent computing (SAIC), Kyiv, Ukraine. IEEE, pp 1–6. https://doi.org/10.1109/SAIC51296.2020.9239152.
23. Bustos O, Pomares-Quimbaya A (2020) Stock market movement forecast: a systematic review. Expert Syst Appl 156:113464. https://doi.org/10.1016/j.eswa.2020.113464
24. Alex G, Pereira G, Pappa L (2018) A customized classification algorithm for credit card fraud detection. Eng Appl Artif Intell 72:21–29
25. Salehi A, Ghazanfari M, Fathian M (2017) Data mining techniques for Anti money laundering. Int J Appl Eng Res 12(20):10084–10094. ISSN 0973-4562
26. Schultz M, Tropmann-Frick M (2021) Autoencoder neural networks versus external auditors: detecting unusual journal entries in financial statement audits. In: Proceedings of the 53rd Hawaii international conference on system sciences (HICSS-2020), Maui, Hawaii, USA, pp 5421–5430. https://doi.org/10.24251/hicss.2020.666
27. Nonnenmacher J, Kruse F, Schumann G, Marx G (2021) Using autoencoders for data-driven analysis in internal auditing. In: Proceedings of the 54th Hawaii international conference on system sciences, Maui, Hawaii, USA, pp 5748–5757. https://doi.org/10.24251/hicss.2021.697.
28. Craja P, Kim A, Lessmann S (2020) Deep learning for detecting financial statement fraud. Decis Support Syst 139:113421. https://doi.org/10.1016/j.dss.2020.113421
29. Neskorodieva T, Fedorov E, Neskorodieva A, Sichko T, Utkina T (2023) Automated analysis of production audit with returnable waste and semi-products by deep simple recurrent network with losses. Math Model Simul Syst 143–157. https://doi.org/10.1007/978-3-031-30251-0_12
30. Kryvinska N, Auer L, Strauss C (2008) Managing an increased service heterogeneity in a converged enterprise infrastructure with SOA. Int J Web Grid Serv 4(4):440. https://doi.org/10.1504/ijwgs.2008.022546
31. Stoshikj M, Kryvinska N, Strauss C (2013) Efficient managing of complex programs with project management services. Glob J Flex Syst Manage 15(1):25–38. https://doi.org/10.1007/s40171-013-0051-8

Features of the Functioning of the National Repository of Scientific and Research Data as Part of a Multi-DB Scientific Ecosystem for the Development of Open Science in Ukraine

Anatolii Shyian, Liliia Nikiforova, Iryna Dohtieva, and Serhii Zharinov

Abstract A tuple-oriented model of information technology has been developed in the form of a multi-DB scientific ecosystem, which will contribute to the expansion of international scientific and technical cooperation and the increase in the participation of Ukrainian scientists in international grant programs. This IT will function within the framework of a specialized web portal in the Ukrainian System NAUKA. The proposed IT model can be used in two ways: general–limited; registered–extended. The second method provides extended access to the use of a library of specialized databases, which is the basis of the developed model. The tuple-oriented IT model is based on 11 main databases, among which the National Repository of Scientific and Research Data should be highlighted. To increase the integration of the National Repository of Scientific and Research Data into the international scientific space, it is proposed to deploy it based on the Dataverse technology, which is successfully used in the USA and the European Union. The proposed tuple-oriented IT model is described from the point of view of a scientist as the main participant in international scientific and technical cooperation. It is substantiated that the developed tuple-oriented model of organizing scientific activity will evolve, as in developed countries, to a decentralized distribution between universities, scientific laboratories, state, public and private structures, etc. The paper also develops an ER-model for the correlation of databases in the tuple K and provides an example of the use of the database library by Ukrainian scientists as a graphic scheme.

Keywords National repository · Scientific and research data · Scientific ecosystem · Open science · Development · Ukraine

A. Shyian · L. Nikiforova (✉)
Vinnytsia National Technical University, Vinnytsia, Ukraine
e-mail: nikiforova@vntu.edu.ua

I. Dohtieva · S. Zharinov
State Scientific and Technical Library of Ukraine, Kyiv, Ukraine

A. Semenov et al. (eds.), *Data-Centric Business and Applications*, Lecture Notes on Data Engineering and Communications Technologies 272,
https://doi.org/10.1007/978-3-032-06608-4_18

1 Introduction

A repository is a concentration of information for the users that needs to satisfy a particular need. As a result, repositories differ according to the tasks/needs that the user has. The main task of each repository is to maximize the possibilities of access to its databases for as many users as possible. Today, it is possible to create repositories that have in their storage not only files with scientific texts, but also audio and/or video files, databases, and many other materials necessary for the user in digital form.

The current requirements for scientific and research data, known as the FAIR principles, were formulated in 2016 [1]. They include: Findable (includes 4 requirements), Accessible (includes 2 requirements), Interoperable (includes 3 requirements), and Reusable (includes 3 requirements). Every repository of scientific and research data must meet all of the FAIR principles. It is through the implementation of these principles and requirements that scientific and research data gain the ability to be integrated into international science and become the foundation for the further development of society.

Cataloging of modern digital repositories is carried out by structuring the databases available in them, as well as a system of hierarchy of links. Relational databases are often used for this, which allows the use of sets of filters required by the user. The organization of links (i.e., connections between databases) satisfies the needs of the user in performing the scientific and business processes he needs.

Of particular importance today are electronic repositories focused on supporting scientific activity. It is a scientific activity that is the soil (foundation) on which innovations grow, thanks to which the development of society is carried out.

Open scientific data repositories are distinguished separately. Today, they include open scientific data with full access and open scientific data with limited access. However, these repositories require specific skills for their use. Therefore, their use for the development of scientific activity has its characteristics. One of the important features is manifested in the field of financing science, research, and development. Today, the functioning of science, the development of technologies, and the creation of databases are ensured mainly through grant funding. In particular, for the main subjects of such processes, scientific teams, individual scientists, the need for access to such funding is determined.

The formation of international temporary teams of scientists is a main trend in the development of science. Such teams jointly receive new scientific results, which are subsequently brought to technical implementation [2] by one or another international team. Thus, today, the creation of universal database libraries, which are focused on the effective inclusion of Ukrainian science in the international scientific environment, as well as on supporting grant funding, is an important element in the formation and development of Ukraine.

A significant part of the financing of scientific and innovative projects is carried out based on competitive and grant financing of fundamental, applied research, and development. Financing of the scientific sphere by one country creates a significant burden on the state budget, and therefore, to achieve results, different countries cooperate to jointly finance projects and use the results of scientific and innovative research. This is evidenced by the prevalence of the creation of international programs and competitions for financing scientific research and development, the result of which is the development of international scientific and technical cooperation.

An important component of grant financing is informing the scientific community about the scientific results obtained in the process of scientific activity. Today, for example, scientists often conduct a large number of experiments to write a single article with a presentation of experimental results. But in the article, they usually use only a very small amount of experimental data. Thus, a large amount of scientific data is lost to science, which experimental scientists receive during the often several years of daily experiments. These data can be of significant importance to both theoretical scientists (who can analyze them) and experimental scientists engaged in other scientific tasks. In addition, such "intermediate" results can play an important role in the development of technologies for their use in business.

The Ukrainian scientific community is still at the stage of integration into this field, as evidenced by the relatively low involvement of both Ukrainian scientists and Ukrainian scientific institutions in international research and development (R&D) funding programs. One of the problems why Ukrainian teams do not participate in these programs is the low level of awareness of foreign scientists and funding institutions of the scientific results obtained by Ukrainian scientists. On the other hand, Ukrainian scientists are practically not informed about the scientific results of foreign scientists (especially at the stage of their receipt) and about existing funding programs, conditions of participation in them, and requirements for participation in international teams.

The basis of the state repository of scientific data should be the presentation of the database catalog in the format of a tuple, where the attributes are the database of a certain object of the information resource. This approach should make it possible to build a database and consistently work with any information resource, depending on the needs and current tasks of the scientist.

Currently, the international community, in particular the EU, pays great attention to creating a single space for international scientific and technical cooperation (ISTC) and expanding the participation of scientists from all countries of Europe and the world in order to develop science and innovation [2–4]. For example, the repository of scientific data [5] is actively used within the framework of EU grant programs. However, Ukrainian scientists still lag behind foreign (for example, European) scientists in their participation in ISTC [6].

For more effective involvement of Ukraine in ISTC, it is necessary to apply appropriate information technologies both for creating a state repository of scientific data and for informing and training Ukrainian scientists in modern technologies of joint research and development. First of all, such technologies are focused on a scientist who actively publishes his scientific results in international scientific journals, constantly monitors foreign scientific journals and international databases of scientific preprints and repositories of scientific data, and maintains scientific correspondence with foreign scientists whose scientific results are in the field of his interests [7]. To increase the number of such scientists in Ukraine, it is necessary to create appropriate advisory information technologies and databases that would help them participate in ISTC. Such advisory tools, repositories of scientific data, and databases of scientific results can be created, for example, based on scientific libraries. This is possible because digital technologies introduce changes in the main subject, the scope of their activities, the processes of creating, storing, and distributing information [8] and can be used as one of the elements of supporting the integration of Ukrainian scientists into international scientific and technical cooperation.

In the work [9], analyzing the functioning of the world's leading universities, the head of the Higher Education Department of the World Bank Salmi Jamili emphasizes that for the successful work of a scientist, appropriate conditions must be created. First of all, the conditions are organizational, focused primarily on bringing the necessary information to the scientist (it should be noted that the book [9] itself successfully ensures the fulfillment of such conditions). An important component of such conditions should be repositories of scientific data. In Ukraine, these can be state repositories, since science in Ukraine is financed mainly from the state budget.

The rapid development of innovative startups, the commercialization of scientific research and their internationalization require intensification and targeting of their financing. In such realities, the requirements for grant financing of the activities of both individual scientists and scientific teams are increasing [10]. At the same time, there is a need to form temporary teams to obtain grant financing for the project, as a rule, on an international basis [11].

All this requires the emergence of a new, specific communicative environment. The appropriate environment will primarily include representatives of research and development (R&D) funding, researchers and structures that support scientific activity, broad areas of business and society at the regional and international levels [12]. Ensuring the presence of such an environment in the ISTC field can be achieved with the help of specially developed information technologies [13, 14].

In the process of preparing an application for grant funding, the need for consulting and education/training of scientists, a kind of interface between stakeholders who shape new R&D, financial institutions that provide grants, business and society is rapidly growing [6, 14, 15].

Social networking sites are suitable tools for sharing and showcasing scientific works in order to increase citations and specialized databases [16, 17]. The participation of researchers in social networks (ResearchGate, LinkedIn, WhatsApp) effectively increases citation rates. Thus, greater activity in professional social networks can have favorable results for improving the integration of Ukrainian scientists into international scientific and technical cooperation.

The work [18] emphasizes that against the background of increasing open access to full-text scientific research and open scientific data, all search engines on the Internet require specially created metadata and, accordingly, specialized library databases where they could be stored in open access. Open access to academic literature and the latest scientific achievements through Internet search engines calls into question the traditional role of specialized databases. The production of scholarly research and open scientific data outside of academic journals and peer-reviewed publications broadens the scope of scholarly research in a given field. The resulting volume of scholarly work raises the question of whether scholars prefer the supposed greater accessibility of web search engines or the precision of research databases in a given field [19]. The widespread use of web search engines and open access initiatives suggests that open scientific databases will be integrated into the evolving web infrastructure that supports discovery and access to scholarly literature [19, 20].

Science is becoming increasingly global, and the number of publications in international co-authorship is growing rapidly, which activates international scientific cooperation with the aim of further cooperation. The amount of scientific data is growing even more rapidly, not all of which is reflected in the publications of scientists. Using bibliographic data from the Web of Science, Scopus databases, and using social network analysis in the work [21], the structure, dynamics and determinants of international scientific cooperation networks were studied. And since international scientific cooperation (ISC) is gaining increasing importance as a new organization of scientific research and development, the article [22] considers the intellectual database, the main research trajectories and intellectual communities of the scientific field IRC, within the framework of which a qualitative review and quantitative analysis were conducted, including the analysis of the co-citation network, the analysis of the main path, and the analysis of bibliographic relationships of databases, and the correlation of their results with the development of ISC.

For the most part, Ukrainian scientists are not very familiar with the functioning of grant funding for scientific activities. The majority of Ukrainian scientists have never participated in international grant projects. Specialized information technologies [23] will allow us to reach a large number of scientists. The most effective format is the National ISTC Portal, which will concentrate official information on grant support for research and development in the field of ISTC.

The purpose of this study is to develop a methodology for forming and using a repository of open scientific and research data to increase the efficiency of Ukrainian science's entry into the international scientific environment based on FAIR principles and support domestic and international grant funding for scientific research and development in Ukraine.

2 Material and Methods

The purpose of this study is to develop a methodology for forming and using a repository of open scientific and research data to increase the effectiveness of Ukrainian science entering the international scientific environment on the principles of FAIR and supporting domestic and international grant funding of scientific research and development in Ukraine. First of all, it is necessary to form a list of key needs and tasks of a scientist that can be satisfied through the use of a repository of open scientific data and to develop scenarios for using this repository, presented in a tuple format.

This study is conducted based on structuring data in information resources on the field of open scientific data, including those obtained within the framework of international scientific and technical cooperation for individual objects. To this end, the needs of potential users and their user experience when working with these information resources are studied. Special attention is focused on taking into account the behavior of scientists when searching for grant programs, choosing a particular program, preparing and submitting applications for participation in the program. This is necessary precisely because the creation of international scientific groups for joint R&D is today, within Ukraine, a powerful mechanism for integrating Ukrainian science into the world.

From the point of view of a scientist, as the main participant in the ISTC, the components of the information content of such a portal can be represented by a tuple (digital notation is used to shorten the record and for further algorithmization):

$$K = \langle 1, 2, 3, 4, 5, 6, 7, 8, 9, 10, 11 \rangle \tag{1}$$

Here 1 is a database of names and summaries of requirements of grant programs for foreign funding of scientific R&D. This is a source of primary information for the scientist.

2 is a database of complete information on the requirements for a grant application, methods and features of financing, along with templates of necessary documents, requirements for their completion, as well as deadlines, necessary addresses for submitting applications, etc. We emphasize that it is within the framework of preparing a grant application that a Ukrainian scientist will most effectively use both international repositories and the state repository of open scientific data. He will need this information, for example, to substantiate the relevance of the proposed project and to substantiate the global uniqueness of the results obtained as a result of the project.

3 is a database of Ukrainian scientists, which provides information about each scientist regarding the scientific topics he has worked on, his grant history, major publications (for example, such information may contain links to his web pages in the ORCID, Scopus, Web of Science, etc. databases), his willingness to cooperate, etc. The presence of a state repository of open scientific data will allow a scientist to provide an example of previously obtained scientific results by him (and the research group as a whole), which will be extremely important information for obtaining reliable expertise.

4 is a database is available in Ukraine to ensure scientists' access to scientific instruments, laboratory equipment, etc. A link to the scientific results provided in the state repository by scientists (research group) allows experts to obtain reliable information about the capabilities of Ukrainian scientists and assess their contribution to the international project.

5 is database of support available to Ukrainian scientists in the world (scientific instruments, laboratory equipment, etc.). The results presented here in the state repository of scientific data will allow obtaining objective information about the level of development of modern international techniques by Ukrainian scientists for conducting experiments and processing the obtained scientific results.

6 is database of examples of filling out documents for applications for open grant programs.

7 is database of volunteer consultants (Ukrainian and foreign, including National Contact Points [15]), who can provide advice on the preparation of grant applications (for each consultant, it is necessary to indicate the field of research, main publications, time available for providing advice, including the waiting time for a response, etc.).

8 is database of ISTC grant funding trainings (dates, start time and duration of training, conditions, organizer (Ukrainian and/or foreign scientists, representatives of grant organizations, experts in reviewing submitted grants, etc.), experience of trainers, etc.

9 is database of scientists' offices, access to which can be obtained through registration. Full access to the tuple (1) is provided only to registered users from database 9. Also, in the personal account, the scientist is provided with automatic information about the start of grant programs, trainings, search for partners, offers of access to scientific equipment, etc., in the field of his scientific interests. In each scientist's account, links are provided to his scientific results in international and state repositories of scientific data that he received throughout his career.

10 is database of analytical materials on the participation of Ukrainian scientists in ISTC, both current grant programs and completed ones. Links to scientific results obtained by scientists and research groups (including international ones) will include international and state repositories of scientific data.

11 is the National Repository of Open Scientific and Research Data itself. It is used at all stages of applying for a grant, examining the application, and implementing the grant. At the stage of applying for a grant, scientists use the National Repository of Open Scientific and Research Data to argue for the possibility of implementing the proposed scientific project, providing evidence of the scientific and research results they have already obtained. Also, at this stage, the need to include specific scientists is argued by describing the contribution they will make to the project. At the stage of examining the application, experts gain access to the scientific and research results already obtained by the applicants, which critically increases the reliability of the expert results. At the stage of implementing the project, using the National Repository of Open Scientific and Research Data allows the grantor to exercise current control over the implementation of the project, as well as make operational decisions on managing project financing.

Description of the methods of using the Ukrainian tuple by Ukrainian scientists (1) in introductory mode (without registration and creation of a personal account). A Ukrainian scientist can use a tuple (1) in two ways.

2.1 First Method

This is the use of "general access" in introductory mode, without the need to register on the portal and create your personal account.

This method is very important. It is aimed not only at scientists, but also at university teachers, students and graduate students. Here, the portal itself acts as an element of the educational process, providing the user with the opportunity to get acquainted with the features of research and development in the ISTC. This is an important element of a scientist's training. At the same time, the scientist decides whether he is personally allowed to participate in the ISTC.

Of course, in this case, the user gets access only to a certain (sometimes relatively small) part of the relevant databases. This forms a tuple with which the portal user works (the index "wr" means "without registration").

$$Kwr = \langle 1_{wr}, 2_{wr}, 6_{wr}, 7_{wr}, 8_{wr}, 10_{wr}, 11_{wr} \rangle \quad (2)$$

It should be noted that such content is selected for the Kwr tuple that would motivate the Ukrainian scientist to further active participation in the ISTC (primarily by applying for a grant). To do this, he needs to go to registration on the portal and create a personal account.

A typical trajectory of a Ukrainian scientist in introductory mode looks like this.

$$\text{Scientist} \rightarrow 1\text{wr} \rightarrow 10\text{wr} \rightarrow 2\text{wr} \rightarrow 6\text{wr} \rightarrow 8\text{wr} \rightarrow 7\text{wr} \rightarrow \text{Personal account} \quad (3)$$

In the process of gaining knowledge about participation in grants within the framework of the ISTC, Ukrainian scientist can assess the level of his readiness to submit such an application, which has a real chance of success. He can complete his acquaintance at any stage of the process (3).

An important element of the Ukrainian scientist's trajectory (3) is that the scientist reveals the knowledge that he or she needs to acquire to participate in the ISTC. These may be, for example, the need for a wider acquaintance with the results of foreign scientists, the acquisition of the ability to formulate one's thoughts briefly and highlighting their essence, acquaintance with new methods and tools for conducting research, etc.

It is also important that the involvement of students who are planning an academic career in the future allows them to master the international rules and norms of academic behavior at the very beginning of their scientific career. For the current state of functioning of science in Ukraine, this is an important element. Therefore, it is promising to include process (3) in the relevant curricula, for example, within the discipline "Fundamentals of Research Work" (it is an element of almost every bachelor's or master's program in Ukraine).

2.2 Second Method

To use the tuple (1) is to register the scientist and create his personal account. In this office, the scientist provides information about himself, about his scientific results and about the methods and tools of scientific activity he uses. This information is important for a scientist to be able to carry out scientific communication with other scientists in Ukraine and the world. This information will be important for the creation of temporary research teams, the purpose of which is to achieve the result planned in the grant application. In fact, it is this information that is decisive for the participation of a scientist in the ISTC: other scientists of the world decide to start cooperation on the basis of exactly the information that is provided in the personal account.

3 ER-Model for the Correlation of Databases

Let us consider the model for the functioning of the tuple (1). The roles for tuple *K* can be distinguished as follows.

«Scientists». Register on the portal and creates a personal account and gets access to the databases of the entire tuple. It is in this role that the National Repository of Open Scientific and Research Data comes into play.

«Administrator». Has access to all the databases that manages the portal. The administrator also fills the databases within the framework of separate protocols. For example, it imports and maintains up-to-date data in the databases of tuple *K*. Manages registration and information protection.

«Owners_research_nfrastructure». These can be organizations of various forms of ownership that have equipment that can be used for scientific activities. They provide, in particular, information on the parameters and characteristics of scientific equipment and the conditions of access to it.

«Experts». These are scientists who already have experience in participating in the implementation of grants. They can provide examples of filling out documents for applications for relevant grant programs, offer consulting assistance to other scientists in filling out documents for a grant application, offer training programs.

«National contact points» (NCP). Provide information about their work, consulting and training activities, examples of filling out grant application documents, etc.

«Coaches». Provide information about upcoming trainings and the conditions for their implementation, materials of previous trainings, etc.

«Analysts». Receive statistical information from tuple *K* and analyze it.

The ER model for the correlation of databases in tuple *K* is shown in Fig. 1.

The diagram shown in Fig. 1 corresponds to the "entity-relation" scheme. This model has the features of a conceptual one, without detailing its attributes. Independent entities are defined by Administrator, Analysts, Scientists, NCP, Grants, Research_infrastructure.

Separately, it is worth highlighting the combined entity of the Research_ infrastructure, since the theoretical material of the article provides for the division into separate databases: a database on the provision available in the world for Ukrainian scientists and a database on the provision existing in Ukraine.

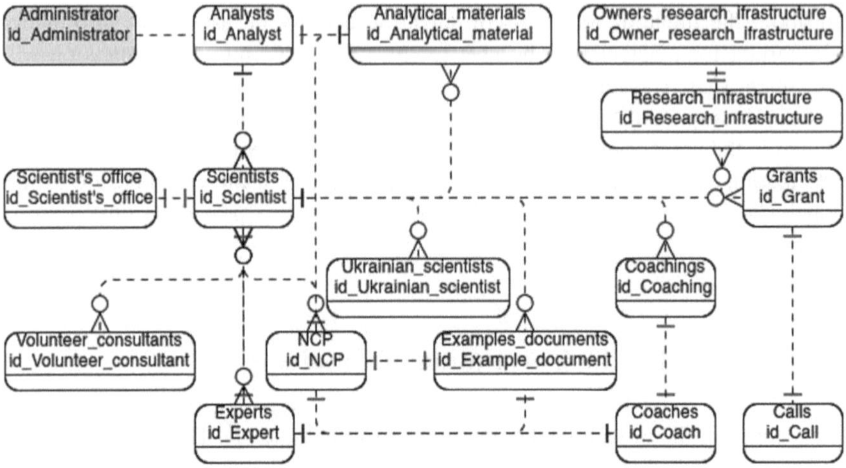

Fig. 1 ER-model for the correlation of databases in tuple K

Dependent entities include: Scientist's_offices, as a personal account for each registered scientist; Experts from Scientists, since this category is a sample, are scientists who already have experience in participating in the implementation of grants; Coaches based on NCP and Experts entities; Calls linked to Grants, a database of complete information on grant application requirements; Ukrainian_scientists that are an expanded database of Ukrainian Scientists; Owners_research_nfrastructure that are part of the information of the database of scientists' access to scientific instruments, laboratory equipment, etc.

Dependent entities also include the database of trainings on grant funding of Coaching, which is a derivative of the creation of the dependent entity Coaches, and the database of analytical materials on the participation of Ukrainian scientists in the ISTC Analytical_materials, which are prepared by Analysts and NCPs, as well as partially this applies to Examples_documents, which are offered by representatives of NCP and Experts.

In part, dependent entities also include Volunteer_consultants that are formed on the basis of Ukrainian and foreign, including National Contact Points.

Primary keys are assigned to all entities, although they are usually not specified for dependent entities, foreign (secondary) keys determine how the tables are joined.

The one-to-one (1:1) type of communication is used to capture the dependencies described above.

One-to-many (1:N) relationships are used overwhelmingly for an entity's access to databases, for example, in the case of Scientists to: Grants, Research_infrastructure, Coaching, Examples_documents, Analytical_materials.

It is worth noting that such a list, according to the theoretical description, is somewhat wider and is provided by other transitions through links, in particular for Scientists who have access to the databases of the entire tuple, for example: Calls, Owners_research_nfrastructure.

Relationships for Administrator and Analysts are also displayed in a similar way, there is a relationship with the Scientists entity that has the relationships mentioned above, and in addition direct and indirect relationships with other entities that display roles for the tuple.

The many-to-many (N:M) type of communication is used to provide counseling, in particular for pairs of Scientists and NCPs, Scientists and Experts, Scientists and Volunteer_consultants.

4 An Example of Using the Database Library by Ukrainian Scientists

Let's describe an example of using the database library (1) to apply for a grant under Horizon Europe (Fig. 2).

Stage 1. A Ukrainian scientist creates a personal account in DB 9 and DB 11. To do this, he turns to DB 9, fills out the necessary forms and receives advice (if necessary).

Stage 2. From the personal account of the Ukrainian, the scientist first turns to DB 1 for brief information, choosing the grant funding he needs. After selecting a grant, he goes to DB 2, which provides complete information about the selected grant and provides templates to fill out.

Stage 3. Since the Horizon Europe grant funding program requires a collaboration of scientists from different countries, Ukrainian scientist is forced to use such databases to prepare an application. In DB 3 and DB 11, he receives information about Ukrainian scientists working on related scientific topics. He also uses DB 10 to obtain information about grants already made in Ukraine on related scientific topics. He can contact them if necessary and invite them to the team. In DB 4, DB 5 and DB 11 Ukrainian the scientist/team finds information about the existing in Ukraine and available in the world provision of scientists' access to scientific instruments, laboratory equipment, etc. If necessary, he/the team will conduct preliminary

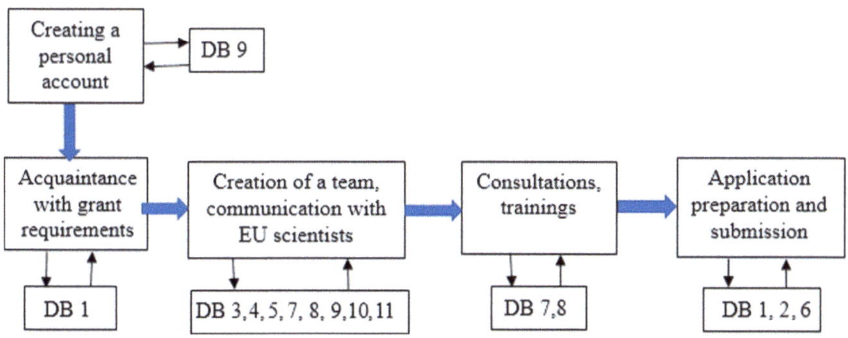

Fig. 2 An example of using the database library (1) by Ukrainian scientists

negotiations. If necessary, the team turns to consultants on specific issues related to the preparation of a grant application and undergoes the necessary trainings. Using DB 9, the scientist/team collects the information necessary for the grant application.

Stage 4. After gathering the necessary information, Ukrainian scientist/team in DB 7 and DB 11 looks for the necessary consultants and in DB 8 searches for and undergoes the necessary training.

Stage 4. Using DB 1, DB 2 and DB 6 with information about the requirements of the required grant program and templates of the required documents, as well as examples of filling out documents for current grant programs, Ukrainian the scientist/team prepares and submits an application for grant funding.

In this example, it is taken into account that within the framework of the Horizon Europe Ukrainian program, a scientist can either participate as an involved individual in a team of scientists from Europe, or participate as part of an involved team of Ukrainian scientists. These two possibilities require different documentation in the application.

5 Discussion and Outlook

The vast majority of scientists in Ukraine do not have the knowledge, skills, and experience of participating in both the implementation of grants and the submission of their scientific data to the state repository of scientific data and the use of foreign repositories of scientific data. National grant systems are very bureaucratic and are focused mainly on administrative staff as managers of the main applicants for grants. Foreign and international repositories of scientific data began their activities relatively recently, and most Ukrainian scientists are simply not informed about them. As a result, the main task of information technology to support Ukrainian scientists in ISTC is to provide education and consultations, and training on the preparation and submission of grant applications. It is also important to support scientists during the implementation of the grant and the preparation of the final report. Thus, today a powerful National Repository of Scientific and Research Data should be created in Ukraine.

This repository should be integrated with both existing international repositories of scientific and research data, and with all information technologies of the country that support scientific activity in Ukraine. Therefore, as a platform for creating a National Repository of Scientific and Research Data, it is proposed to choose one of the existing foreign platforms, adapting it to the conditions of functioning of science and scientific activity and the existing regulatory framework in Ukraine. For example, online storage, sharing, and registration of research data, during the research period and after its completion. That is [24], which uses the Dataverse software [25], developed by Harvard University and used worldwide.

Integration of Ukrainian science into the international scientific space can be carried out using information technologies that are oriented to the tuple (1). At the same time, part of the tuple (1) databases in them can be dispersed and fragmented between individual universities and scientific laboratories, state, public, and private structures, etc. This is because in developed countries, for many decades, scientific and technical activity has been international, oriented towards implementation in socio-economic life [26]. In addition, scientific activity in developed countries is sufficiently funded, which allows individual structures supporting science to create specialized information systems. In Ukraine, unfortunately, this method of organizing scientific and technical activity is only beginning to be implemented. Therefore, at the initial stage, there should be some centralization of the library and its information technologies for accessing the tuple databases (1) so that Ukrainian scientists receive both the necessary information and access to examples of filling out documents, consultations, and training. This is necessary to familiarize Ukrainian scientists with international scientific and technical cooperation standards and support their integration into the international scientific and research space.

In the future, in Ukraine, the tuple-oriented model of organizing scientific activity (1) will evolve, as in developed countries, to a decentralized distribution between universities, scientific laboratories, state, public and private structures, etc. To provide information support to Ukrainian scientists in obtaining grant funding, it is proposed to create a multi-DB scientific ecosystem that will function within the framework of a specialized web portal in the System NAUKA [27]. The basis of such a multi-DB scientific ecosystem will be the proposed tuple-oriented model of organizing scientific activity. One of the first steps towards creating such a multi-DB scientific ecosystem is the development of the National Repository of Scientific and Research Data, which is one of the important elements of the tuple-oriented system of organizing scientific activity. The article describes an example of the methodology for using such a multi-DB scientific ecosystem in various components of the web portal in the System NAUKA, formed to increase the efficiency of participation of Ukrainian scientists in international scientific activity.

6 Conclusion

A tuple-oriented model of information technology has been developed in the form of a multi-DB scientific ecosystem, which will contribute to the expansion of international scientific and technical cooperation and the increase in the participation of Ukrainian scientists in international grant programs. This IT will function within the framework of a specialized web portal in the Ukrainian System NAUKA.

The proposed IT model can be used in two ways: general–limited; registered–extended. The second method provides extended access to the use of a library of specialized databases, which is the basis of the developed model. The tuple-oriented IT model is based on 11 main databases, among which the National Repository of Scientific and Re-search Data should be highlighted. To increase the integration of the National Repository of Scientific and Research Data into the international scientific space, it is proposed to deploy it based on the Dataverse technology, which is successfully used in the USA and the European Union.

The proposed tuple-oriented IT model is described from the point of view of a scientist as the main participant in international scientific and technical cooperation. It is substantiated that the developed tuple-oriented model of organizing scientific activity will evolve, as in developed countries, to a decentralized distribution between universities, scientific laboratories, state, public and private structures, etc.

The paper also develops an ER-model for the correlation of data-bases in the tuple K and provides an example of the use of the database library by Ukrainian scientists as a graphic scheme.

References

1. FAIR Principles. https://www.go-fair.org/fair-principles/.
2. Bray DA et al (2021) Report of the Commission on the Geopolitical Impacts of New Technologies and Data. Atlantic Council GeoTech Center. Commission on the Geopolitical Impacts of New Technologies and Data
3. A new ERA for Research and Innovation {SWD (2020) 214 final} (2020) Communication from the Commission to the European Parliament, The Council, The European Economic and Social Committee and the Committee of the Regions. Com 628 final, Brussels: 21
4. Reding DF, Eaton J (2020)Science & technology trends 2020–2040. Exploring the S&T Edge. NATO Science & Technology Organization
5. ZENODO. https://zenodo.org/
6. Nikiforova L, Azarova A, Shyian A, Zhitkevich O (2022) Using international grant programs Horizon 2020 and Horizon Europe in the digital economy. Innov Sustain 3:176–185 (in Ukrainian). https://doi.org/10.31649/ins.2022.3.176.185
7. Dachs B, Zahradnik G (2022) From few to many: main trends in the internationalization of business R&D. Transnational Corporations 29(1):107–134
8. Zharinov S (2020) The role of the library in the digital economy. Inf Technol Libr 39(4):17. https://doi.org/10.6017/ital.v39i4.12457
9. Davidson CN (2022) The new education: how to revolutionize the university to prepare students for a world in flux. Basic Books
10. Brears RC (2024) Financing the blue economy: investment strategies and partnerships for sustainable ocean development. Independently published
11. Fitzpatrick K (2024) Leading generously: tools for transformation. Johns Hopkins University Press
12. Wingreen SC, Samandari A (2024) Information Technology security and risk management. CRC Press
13. Nikiforova L (2022) The use of web-oriented information systems for the development of international scientific and technical cooperation. Innovations and promising ways of information technology development. ChDTU, Cherkasy, pp 78–80 (in Ukrainian)
14. Rawlings CM (2023) Network analysis (Structural analysis in the social sciences). Cambridge University Press
15. Shyian A, Nikiforova L, Dohtieva I, Zharinov S (2022) Digital technologies of intensification of international scientific and technical cooperation. Marketing Digital Technol 6(4) 81–98 (in Ukrainian). https://doi.org/10.15276/mdt.6.3.2022.8
16. Valizadeh-Haghi S, Nasibi-Sis H, Shekofteh M, Rahmatizadeh S (2022) ResearchGate metrics' behavior and its correlation with RG score and scopus indicators: a combination of bibliometric and altmetric analysis of scholars in medical sciences. In: Information technology and libraries, vol 12. https://doi.org/10.6017/ital.v41i1.14033

17. Guo Y, Lam AHC, Chiu DKW, Ho KKW (2022) Perceived quality of reference service with whatsapp: a quantitative study from user perspectives. Information Technol Libraries 41(3). https://doi.org/10.6017/ital.v41i3.14325
18. Prasanth A, Sharma V, Dhanaraj RK, Raj P (2025) Blockchain-based digital twins: research trends and challenges. Apple Academic Press
19. Borgatti SP, Everett MG, Johnson JC, Agneessens F (2024) Analyzing social networks. SAGE Publications Ltd
20. Aidi A, Rosli M (2019) Bibliometric analysis of global scientific literature on web accessibility. Int J Recent Technol Eng (IJRTE) 7(6S2):250–258
21. Qinchang G, Chengliang L, Debin D (2019) Globalization of science and international scientific collaboration: a network perspective. Geoforum 105(10). https://doi.org/10.1016/j.geoforum.2019.06.017
22. Chen K, Zhang Y, Fu X (2019) International research collaboration: an emerging domain of innovation studies? Research Policy 48(1):149–168. https://doi.org/10.1016/j.respol.2018.08.005
23. Shyian A (2022) Information technologies to support Ukrainian scientists in international scientific and technical cooperation. Modern computer systems and networks in management: a collection of scientific works. Vyshemyrskyi V. S. Publishing House, Kherson, pp172–173 (in Ukrainian)
24. DataverseNL https://dataverse.nl/
25. The Dataverse Project. https://dataverse.org/
26. Rauer JN, Kroiss M, Kryvinska N, Engelhardt-Nowitzki C, Aburaia M (2021) Cross-university virtual teamwork as a means of internationalization at home. Int J Manage Educ 19(3):100512. https://doi.org/10.1016/j.ijme.2021.100512
27. System NAUKA. https://nauka.gov.ua/

The Optical Dynamic Random Access Memory for Neurocomputers

Andrii Semenov, Serhii Tsyrulnyk, Olena Semenova, Natalia Kryvinska, Serhii Baraban, Andrii Rudyk, and Oleksandr Stalchenko

Abstract In the rapidly evolving landscape of image processing, conventional techniques prove insufficient and require the exploration of new approaches. This change is particularly crucial given the growing demand for optical memory devices and their integration into emerging computer and information technology infrastructures. However, existing methods in image processing cannot meet the evolving requirements of modern applications. This poses a significant challenge in optimising the technical capabilities of memory devices and extending their usefulness in various fields. Moreover, the transition to nanotechnology-based multiprocessor systems, especially in geoinformation energy control systems, is a further obstacle in terms

A. Semenov (✉) · O. Semenova · O. Stalchenko
Faculty of Information Electronic Systems, Vinnytsia National Technical University, Vinnytsia, Ukraine
e-mail: semenov.a.o@vntu.edu.ua

O. Semenova
e-mail: semenova.o.o@vntu.edu.ua

O. Stalchenko
e-mail: stalchenko.o.v@vntu.edu.ua

S. Tsyrulnyk
Department of Computer Sciences and Digital Economy, Vinnytsia National Agrarian University, Vinnytsia, Ukraine

N. Kryvinska
Department of Information Management and Business Systems, Comenius University in Bratislava, Bratislava, Slovakia
e-mail: natalia.kryvinska@uniba.sk

S. Baraban
Faculty of Computing and Telecommunications, Poznan University of Technology, Poznan, Poland
e-mail: serhii.baraban@put.poznan.pl

A. Rudyk
Department of Automation, Electrical Engineering and Computer-Integrated Technologies, National University of Water and Environmental Engineering, Rivne, Ukraine
e-mail: a.v.rudyk@nuwm.edu.ua

A. Semenov et al. (eds.), *Data-Centric Business and Applications*, Lecture Notes on Data Engineering and Communications Technologies 272,
https://doi.org/10.1007/978-3-032-06608-4_19

of efficiency and scalability. To address these challenges, the research focuses on the utilisation of dynamic optical electronic memory deployed in fibre optic lines within advanced optical electronics systems. Using the latest optoelectronic technologies, it aims to improve the technical performance of memory devices while expanding their scope of application. Furthermore, a signal model has been developed to clarify the functioning of dynamic memory on fibre optic lines, allowing for a complete analysis of its light energy efficiency. Through these efforts, he tried to pave the way for the seamless integration of optical memory devices into the fabric of modern computers and information technology by strengthening cybersecurity defence walls in edge computer environments.

Keywords Optical dynamic random access memory · Fiber-optic line · Waveguide splitter · Optical neurocomputers · Neural network

1 Introduction

For the neural network hardware implementation, a new information medium is required, which can be a photon [1–5]. If data carriers in a neurocomputer are photons, this enables to increase sharply the efficiency by two or three orders due to the following factors [6–8]:

- Applying light provides physical organization of a processed information in the form of two-dimensional frames of large dimension, in contradistinction electricity enabling the information organization in the form of one-dimensional time pulse series;
- Light rays intersect in three-dimensional space without interacting with each other, this allows performing complex operations on matrices;
- An optical element base (LCLV) enables arithmetic operations of addition, subtraction and nonlinear transformation to be performed in parallel, therefore, a two-dimensional output signal can be calculated fully optically in parallel;
- An optical element base (LCLV, LCTV, FLC SLM, MO SLM, photorefractive crystals, matrixes of nonlinear elements, etc.) enables to operate not only with binary two-dimensional signals, but also with analog two-dimensional signals;
- Optical neurocomputers can input directly and optically a two-dimensional signal;
- A limited number of independent and parallel communication channels can be organize;
- Information as a two-dimensional optical signal is propagated inside a neurocomputer at a speed close to the light one;
- Geometric dimensions of active and passive elements of the optical element base for a neurocomputer are determined only by the length of the wave carrying information (1…100 microns). Since in optics information processing is performed in three-dimensional space, it is obvious that the volume is much more efficient than

in electronics. As a result, optical neurocomputers can be much more compact than their electronic analogs;

- Passive elements of optics, such as two-dimensional and three-dimensional holograms and diffraction lattice, require no energy. That is why the optical neurocomputer consumes less energy than the electronic one to process the same (large) amount of information;
- The optical element base is less sensitive to defects of materials and to radiation defects;
- The cost of the optical element base is comparable to the cost of an electronic element base, at the same time, the performance of optical neurocomputers in solving special problems exceeds the performance of electronic neurocomputers at least by 5–10 orders of magnitude [9, 10].

Optical neurocomputers can be used as specialized high-speed computing devices in such ares as [4, 9]:

- Pattern recognition,
- Adaptive learning and self-learning neural networks for solving tasks of pattern recognition and adaptive control systems in a dynamically changing environment;
- signal processing: adaptive control of phased array antennas, multi-purpose support, parallel tracking of multiple-purpose trajectories in real-time mode, operational spectral analysis of optical or infrared signals, image compression, path selection, filtration of background and interference;
- An adaptive intellectual associative memory for organization of dynamic databases and expert systems;
- Specialized optoelectronic processors for fast solving large-scale standard problems: matrix rotation, solving systems of linear equations, solving systems of linear inequalities, optimization problems, and solving differential equations.

Moreover, neurocomputers are applied to improve the operability of telecommunication systems [11–13]. According to [10], a neural network architecture contains a memory that can be divided into short-term and long-term ones. The short-term memory saves little information, loads fast, saving time is up to 30 s. The long-term memory is a read-only memory with long-term storage of a data base. The short-term memory can be considered as an analogue of a RAM [14]. One of the easiest ways to build the short-term memory is to implement time delays on an optical element that is a fiber optic line (FOL) [15–17].

Therefore, this study aims to investigate an optical FOL-based memory device capable to operate in optical information systems, optical neurocomputers, and geographic-information energy systems.

If $M + 1 = 2^N$ copies of a input signal with a delay period τ_d must be made, then the delay time of the first delay line should be τ_d, that of the second delay line is $2\tau_d$, and that of the j-th one is $2^{j-1}\tau_d$. That is the delay time of each subsequent line increases by 2 times (Fig. 1).

A zero copy of the input signal corresponds to a direct transmission from DWC_1 to DWC_{N+1} bypassing all delay lines. The first copy is formed by branching to the

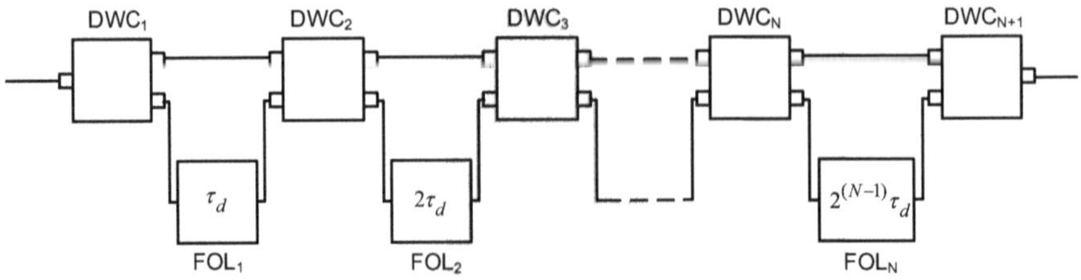

Fig. 1 Dynamic storage device on fiber optic structures

DWC_1 part of the delayed signal τ_d. The signal goes from the output of FOL1 to the DWC_2 and further without delay. A third copy of the signal is generated due to the delay on the FOL_1 and FOL_2. The last copy of the signal has a total delay time $M\tau_d = (2N - 1)\cdot\tau_d$ and passes through all delay lines. Thus, 2^N copies of the input signal are formed at the output.

Modern fiber-optic technologies, high technical and technological characteristics of the FOL contribute to the of FOL-based optical RAM construction for information and computer systems of future generations [18, 19].

2 The Block Diagram of the Base Element of the Optical Dynamic Random Access Memory (ODRAM)

The rise of low-loss optical fibers with a large product of length and bandwidth (about 100 GHz × km), as well as advances in the optical element technology open new prospects of optical signal processing. Such operations as convolution, computation of correlation functions, operations on matrices, filtering, generation of pulses and code combinations, compression of pulses, consistent filtering provide the development of fiber-optic devices [20, 21]. Furthermore, fiber-optics is a supporting infrastructure for 5G networks, which are now being widely deploied, researched and improved [22–25].

However, dynamic storage of light pulses for optical computing machines and optical information transmission systems is a crucial task [26, 27]. In fact, a light beam may have a spectrum width comparable to its carrier frequency, it means that t_u may be about 10^{-13}–10^{-14} s. The question is whether such pulses can be used to the information transmission and storage.

A block diagram of a base element of the optical dynamic random access memory (ODRAM) on the FOL shown in Fig. 2 enables an optical signal to be input either through U_{in1} or through U_{in2}. The information can be read either at U_{out1} or U_{out2}. There are several technologies for creating optical switches. According to them, there are eight types of such switches [20, 28]: mechanical optical, electro-optical, thermo-optical, SOA-based optoelectronic, integrated active-wave, on photonic crystals, on multilayer light-guide liquid crystal matrices, on integrated circuits with optoelectronic valve arrays coupled by optical beam. Construction, operation and utilization of optical switches are considered in [18, 20, 28].

Criteriafor choosing an optical switch are the maximum speed and possibility of integral execution. The criteria are met by the $LiNbO_3$ electro-optical switch, which uses directional splitters to focus a light flux at one of the outputs by controlling the coupling coefficient (Table 1). The electro-optical switch schematic for using in the ODRAM is shown in Fig. 3 [18, 20, 28].

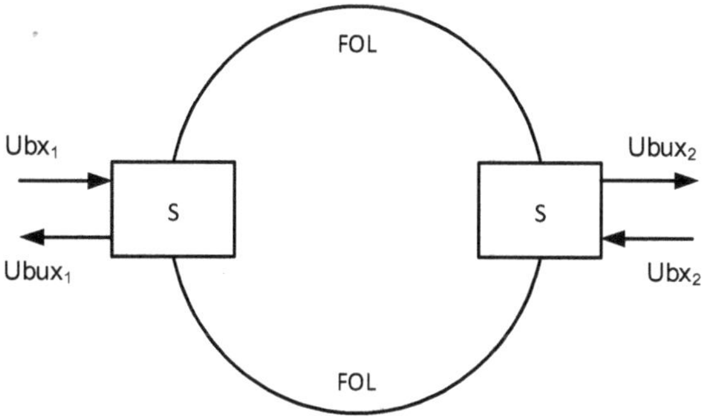

Fig. 2 Block diagram of the base element of the optical dynamic random access memory

Table 1 Comparative characteristics of the basic optical switches

Switch type	Insertion loss, dB	Transient attenuation, dB	Polarization loss, dB	Switching time
Mechanical	3	55	0,2	10 ms
Electro-optical $LiNbO_3$	4	40	0,5	10 ps
Optoelectronic with SOA	0	40	small	1 ns
Active-wave IP	0	30	–	1 ns
Thermo-optical: quartz	10	15	small	2 ms
Polymeric	10	30	small	2 ms

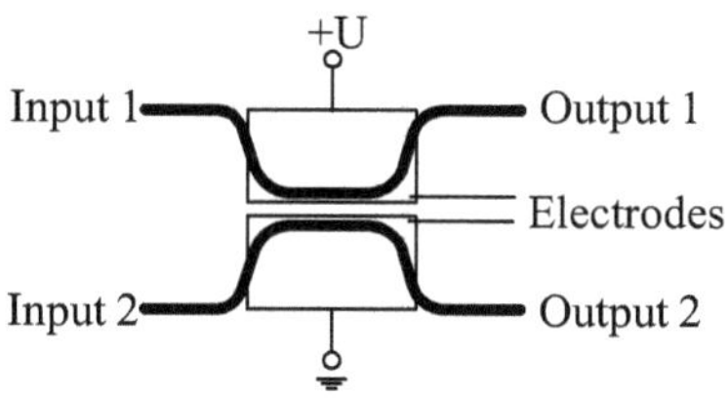

Fig. 3 The 2 × 2 electro-optical switch

3 The Model of the Memory Structure on Delay Lines

A great deal of telecommunication networks transmit data between computers, while performance and efficiency of modern personal computers is related to the elementary base and their architecture [29–31]. The characteristics of modern computers are limited by characteristic features due to types of information exchange between the processor and the storage device. The most vulnerable place is the "processor—RAM" link, since information flowing along the path requires a large time-consuming operation of the RAM. This complicates the real-time implementation and indicates that the information processing time and the memory are the main resources of computer systems. The speed of the processor depends significantly on the speed at which the memory provides it with commands and data. The performance can be improve by using parallel data processing methods [32, 33].

The main purpose of the memory for parallel structures is to convert static information to dynamic, which causes the output signal to depend on time [34, 35].

A RAM can be implemented in either continuous or discrete time (a continuous time is denoted by the symbol t, a discrete time is denoted by the symbol n). A useful tool for working with discrete-time systems is the Z-transformation [8, 18]. Suppose $\{x(n)\}$ is some sequence in discrete time that can be extended into an infinitely distant past. Z is a transformation of such sequence and defined as follows:

$$X(z) = \sum_{n=-\infty}^{\infty} x(n)z^{-1}, \tag{1}$$

where z^{-1} is the unitary delay operator. Applying it to $x(n)$, we generate a copy of the delayed signal $x(n–1)$. Suppose that $x(n)$ is used in a discrete-time system with a pulse response $h(n)$. Then the output of the system is determined by the convolution sum:

$$y(n) = \sum_{k=-\infty}^{\infty} h(k)x(n-k). \tag{2}$$

If $x(n)$ is equal to a unitary impulse, then $y(n)$ is reduced to the impulse response $h(n)$ of the system. If we denote the Z-transformation of sequences $\{h(n)\}$ and $\{y(n)\}$ as $H(z)$ and $Y(z)$, respectively, the application of the Z-transformation to formula (2) leads to a relation:

$$Y(z) = H(z)X(z), \tag{3}$$

or equally:

$$H(z) = \frac{Y(z)}{X(z)}. \tag{4}$$

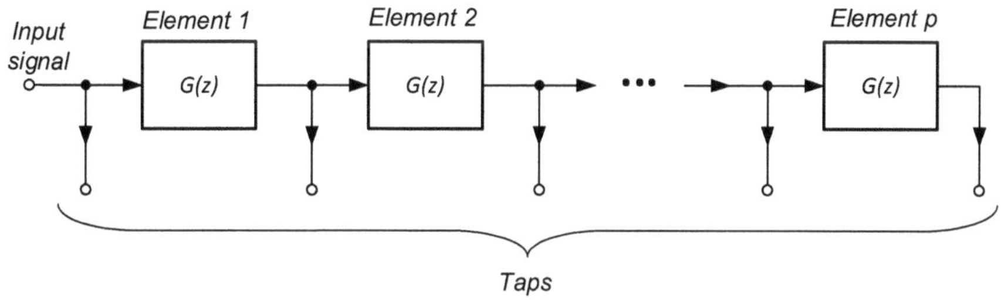

Fig. 4 Generalized memory of p-order

The function $H(z)$ is called the transfer function. A block diagram of a discrete-time memory consisting of p identical blocks with cascaded connection is shown in Fig. 4.

If we consider each of the delay blocks as an operator, it can be described by the transfer function $G(z)$.

From Fig. 4 the RAM can be formally defines as a linear, invariant in time, a single-input and multiple-output system, its core satisfying the requirements above.

Let us denote the total memory impulse response as $g_P(n)$, which is defined as p successive convolutions $g(n)$.

A diagram of the simplest form of memory based on delay line with taps is shown in Fig. 5. It consists of p unitary delay operators, each characterized by the function $G(z) = z^{-1}$.

The total memory impulse is:

$$g_p(n) = \delta(n - p). \tag{5}$$

Figure 6 demonstrates the signal transmission graph G(z)for the main functional block used in a structure called the gamma memory [5]. Each block of such memory consists of a loop with a unitary delay and a parameter μ that can be tuned. The transfer function of each such block is

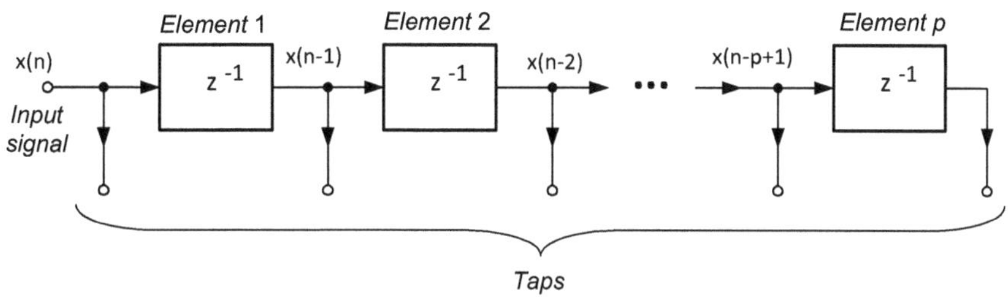

Fig. 5 Memory based on the delay line with taps of p-order

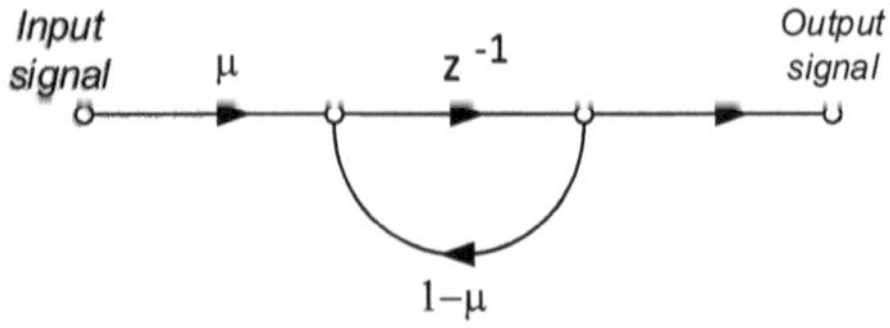

Fig. 6 Graph of signal transmission in the gamma block

$$G(z) = \frac{\mu z^{-1}}{1 - (1 - \mu)z^{-1}} = \frac{\mu}{z - (1 - \mu)}. \tag{6}$$

To provide stability, the only pole of *G (z)* at $z = 1\text{-}\mu$ must be inside of the unitary circle on the z-plane. This, in turn, requires $0 < \mu < 1$.

The total impulse response of the gamma memory is the reverse z–transformation of the total transfer function:

$$Gp(z) = \left(\frac{\mu}{z - (1 - \mu)}\right)^p, \tag{7}$$

that is

$$g_p(n) = \binom{n-1}{p-1}\mu^p(1 - \mu)^{n-p}, \qquad n \geq p, \tag{8}$$

where (:) is the binomial coefficient determined for integers *n* and *p* by the expression:

$$\binom{n}{p} = \frac{n(n-1)\ldots(n-p+1)}{p!}. \tag{9}$$

The total impulse response $g_P(n)$ for variable *p* is a discrete version of a sub-integral expression of a gamma function. At $\mu = 1$, the memory parameters are drawn to corresponding memory values on the delay line with taps.

4 Mathematical Modeling of Processes in Optical Dynamic Random Access Memory Devices

The FOL ring where with pulses circulating and two splitters for inputing and outputong optical information (Fig. 2) is employed as a dynamic memory element.

The output of the delay element equals the input signal delayed in time [17, 18]. If the input signal *f*(*k*) is a discrete, the output signal is *f*(*k-1*), where *k* is a time index. The block diagram and the impulse response (intensity response to the impulse input signal) are shown in Fig. 7a. The refractive index in the fiber is assumed to be about 1.5, and we obtain a propagation delay about 5 ns/m. Different devices can be employed: directional bumpers, light-emitting diode, etc. for diversity of light

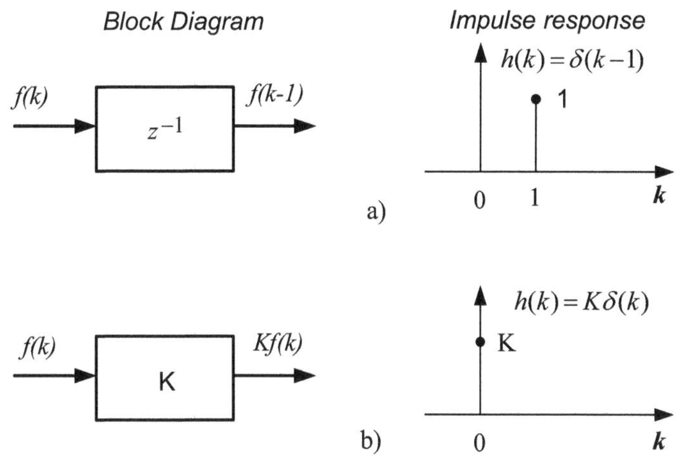

Fig. 7 Block diagram and impulse response: **a** delay element, **b** splitter

signals [36, 37]. The output at the splitter differs mathematically from the input in K times, where K is a constant (Fig. 7b).

The ODRAM, a dynamic optical memory element (DEMO) can be implemented on directional splitters and a single-mode optical fiber (Fig. 8a). The DEMO, a basic element of the ODRAM, consists of the single-mode fiber-optic circuit (with the delay T), which is self-looped through the directional splitters, and a part of the light from the second splitter through the fiber circuit is fed again to the first splitter. As a result, optical signals received at the input X_1 will circulate repeatedly in the circuit and thus will partially come to the outputs Y_1, Y_2. The impulse response of such system consists of a series of descending maxima, which are from each other at the same distance in time equaling the delay in the circuit T (Fig. 8b).

A relation between the output and the input of the directional splitter can be described by a complex 2×2 transfer matrix (Fig. 9):

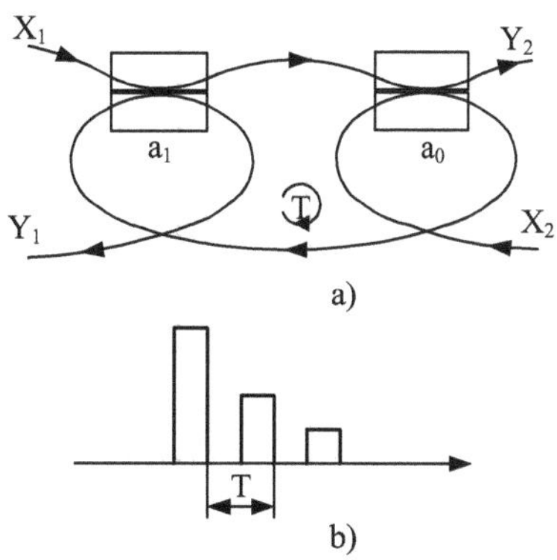

Fig. 8 **a** Two-splitter recirculation delay line, **b** impulse response of the two-splitter recirculation delay line

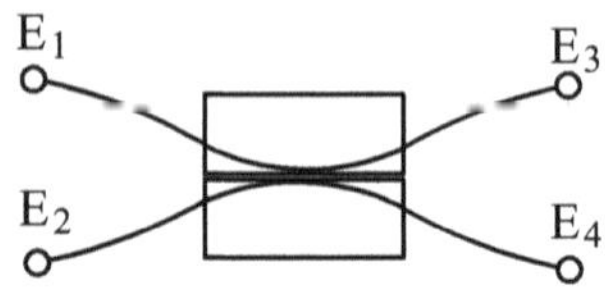

Fig. 9 Fiber optic directional splitter

$$\begin{pmatrix} E_3 \\ E_4 \end{pmatrix} = (\gamma)^{1/2} \begin{pmatrix} A & B \\ C & D \end{pmatrix} \cdot \begin{pmatrix} E_1 \\ E_2 \end{pmatrix}, \tag{10}$$

where (E_1, E_2) and (E_3, E_4) are the complex amplitudes of the voltage vectors at input and output of the splitter, and for simplicity it is assumed that the field polarization in all cases is the same.

The splitter characteristics depend slightly on the input field polarization [38]. Since the backwards splitter can be neglected, the matrix can be roughly represented [38] as the product of a general amplitude transmittance $(\gamma)^{1/2}$, ($0.9 < \gamma < 1$) and the unitary matrix. Following relations hold for the unitary matrix:

$$|A|^2 + |B|^2 = |C|^2 + |D|^2 = 1, \tag{11}$$

$$CA* + DB* = 0,$$

where the asterisk means a complex conjugate value. In the case of a symmetric splitter A = D, B = C, relation (10) can be rewritten as

$$\begin{pmatrix} E_3 \\ E_4 \end{pmatrix} = (\gamma)^{1/2} \begin{pmatrix} \sqrt{1-K} & j\sqrt{K} \\ j\sqrt{K} & \sqrt{1-K} \end{pmatrix} \cdot \begin{pmatrix} E_1 \\ E_2 \end{pmatrix}, \tag{12}$$

where $\sqrt{K}$ is the coefficient of coupling by amplitude; $j = \sqrt{-1}$ signifies the 90^0 signal shift.

When a non-coherent source is appplied, the relation between input and output signal intensity can be described as:

$$\begin{pmatrix} I_3 \\ I_4 \end{pmatrix} = (\gamma) \begin{pmatrix} 1-K & K \\ K & 1-K \end{pmatrix} \cdot \begin{pmatrix} I_1 \\ I_2 \end{pmatrix}, \tag{13}$$

where (I_1, I_2) is the signal intensity at the splitter input; (I_3, I_4) is the signal intensity at the splitter output; γ is the total intensity transmittance; K is the intensity ratio.

In order to investigate processes in the ODRAM, we apply the Z-transformation method [28, 38]. Its application is justified by the fact that the considered system is, firstly, linear and invariant in time, and secondly, discrete in time (Fig. 10). The latter contributes to introducing a basic delay in time T, in relation to which other delays can be expressed by multiple numbers. In other words, the impulse response of the system consists of a series of maxima at the same distance. This simplifies

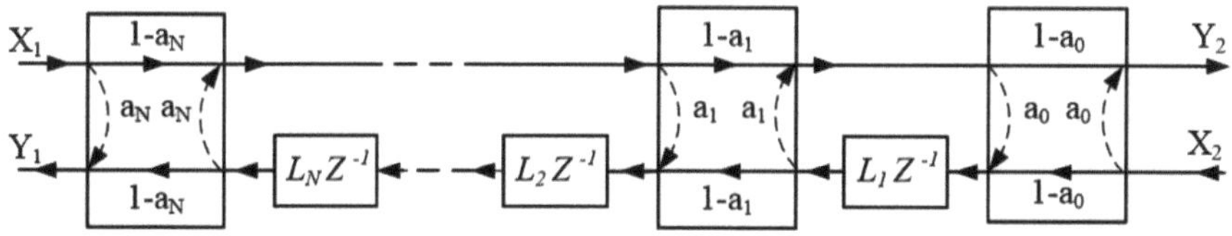

Fig. 10 Equivalent circuit of the N-order fiber-optic structure with (N + 1) splitter and N delay elements (L is the intensity transmittance in the corresponding link)

the mathematical analysis by assuming that the signals in the system exist only at discrete time moments [39]. Therefore, in this case, the input and output signals are described by discrete instantaneous values, these systems are called discrete or pulsed.

Mathematical analysis of impulse systems is basically similar to analysis of continuous systems, but at the same time it is simpler and physically clearer [40, 41].

For the signal $f(k)$, the Z-image of $F(z)$ is defined as the result of the Z-transformation:

$$F(z) \cdot \sum_{k=0}^{\infty} f(k) z^{-k}, \tag{14}$$

where k is the time index equal to a multiple of the basic delay time $T(k = nT)$; z is theconversion variable that is a unit of time advance (z^{-1} corresponds to a unit of delay).

The connection between the input and output signals is described by the transfer function of the system, which equals to the ratio of the Z-image of the output signal to the Z-image of the input signal. The frequency response of the system can be calculated by substituting $z = e^{-jwT}$ into the transfer function, which determines the circle of unitary radius centered at the origin on the Z-plane (Fig. 11). To maintain the system's stability, all poles of its transfer function must be in the middle of the unitary radius circle. The FOL memory is stable if it contains no amplifying elements.

Applying the definition of Z-transform (14), we find the expression for the transfer function $H(z)$ of the recirculation delay line with two splitters (Fig. 12).

(a) For a system, where the input is X_1 and the output is Y_1:

Fig. 11 Z-plane with unitary radius circle

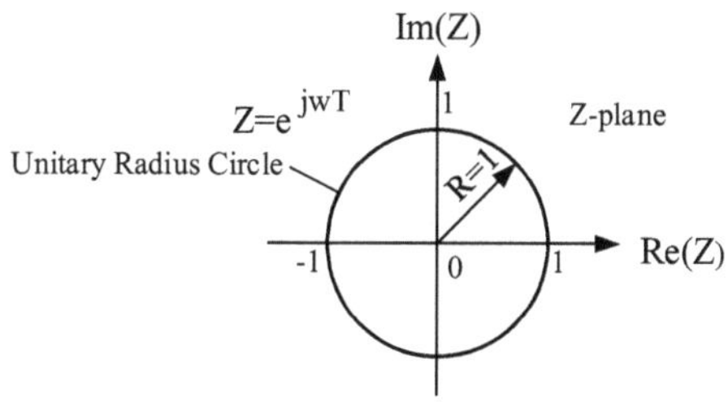

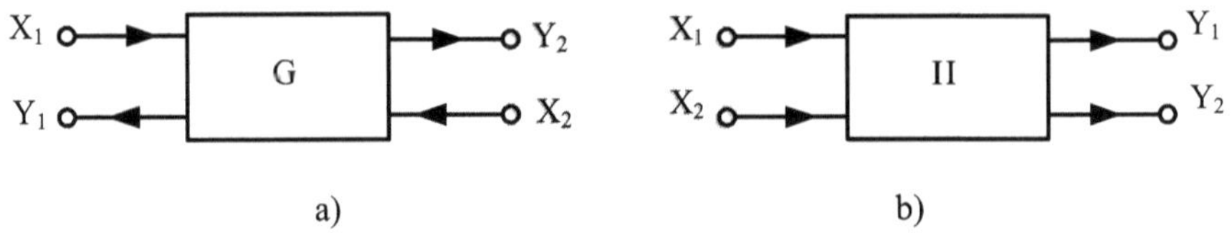

Fig. 12 **a** Two-pair system: feedback scheme, **b** direct scheme

$$H_{11}(z) = \frac{a_1 + (1 - 2a_1)a_0 l_1 z^{-1}}{1 - a_1 a_0 l_1 z^{-1}}, \tag{15}$$

where a_1 and a_0 are the coefficients of coupling intensity; l_1 is the intensity transmittance of the system circuit.

As can be seen from (15), the system has a single zero located in

$$z = \frac{(2a_1 - 1)}{a_1} a_0 l_1,$$

as well as the only pole at the point $z = a_1 a_0 l_1$.

(b) For a system, where the input is X_1 and the output Y_2 is

$$H_{21}(z) = \frac{(1 - a_1)(1 - a_0) l_{11}}{1 - a_1 a_0 l_1 z^{-1}}, \tag{16}$$

where l_{11} is the intensity transmittance of the lightguide in the direct link.

The system contains one zero at the origin and one pole at $z = a_1 a_0 l_1$.

Consider a linear invariant system with two ports, each containing two inputs. The first and second ports have signals (X_1, Y_1) and (X_2, Y_2) as shown on Fig. 12. We will call this structure a two-pair feedback system. If there are (X_1, X_2) and (Y_1, Y_2) signals on the first and second pairs, then this system will be called a two-pair direct system.

The input (X) and output (Y) variables are interrelated with a transfer matrix

$$\begin{pmatrix} Y_1(z) \\ Y_2(z) \end{pmatrix} = \begin{pmatrix} H_{11}(z) \; H_{12}(z) \\ H_{21}(z) \; H_{22}(z) \end{pmatrix} \cdot \begin{pmatrix} X_1(z) \\ X_2(z) \end{pmatrix}, \tag{17}$$

where Z is the variable of the Z-transformation.

The function $H_{\mathrm{mn}}(z)$ (12) is a transfer function associateing the input variable X_{n} with the output variable Y_{m}.

To analyze the two-pair feedback systems, we can aplly a chain matrix linking the pairs of variables (X_1, Y_1) and (X_2, Y_2):

$$\begin{pmatrix} X_1(z) \\ Y_1(z) \end{pmatrix} = \begin{pmatrix} G_{11}(z) \; G_{12}(z) \\ G_{21}(z) \; G_{22}(z) \end{pmatrix} \cdot \begin{pmatrix} Y_2(z) \\ X_2(z) \end{pmatrix}. \tag{18}$$

Both the transfer and chain matrices have dimensions of 2×2 and are uniquely determined to specify the input and output variables. The elements of the transfer matrix for the two-pair feedback system are related to the elements of the chain matrix for the same system by the following relations:

$$H_{11} = \frac{G_{21}}{G_{11}}, H_{12} = \frac{G_{11}G_{12} - G_{12}G_{21}}{G_{11}}, H_{21} = \frac{1}{G_{11}}, H_{22} = \frac{G_{12}}{G_{11}}. \tag{19}$$

If condition $G_{11}G_{22} - G_{12}G_{21} = 1$ is satisfied, the system is called reciprocal. For the cascade two-pair feedback system, the chain matrix of the component system is equal to the product of the individual chain matrices:

$$G_{total} = G_1G_2G_3...G_N. \tag{20}$$

A similar rule for multiplication of matrices taken in reverse order holds for the transfer matrix of the two-pair direct system:

$$H_{total} = H_N...H_3H_2H_1. \tag{21}$$

For the directional splliter with coupling coefficient, the transmission matrix a and the chain matrix Hc are specified by the following expressions:

$$Hc = \begin{pmatrix} 1-a & a \\ a & 1-a \end{pmatrix}; \tag{22}$$

$$Gc = \frac{1}{1-a}\begin{pmatrix} 1 & -a \\ a & 1-2a \end{pmatrix}. \tag{23}$$

Similarly, for matrices of the simplest system of two delay lines with the single delay time of z^{-1} and the intensity transmittance of l_{11}, l_{12} (Fig. 13):

$$Hd = \begin{pmatrix} l_{11} & 0 \\ 0 & l_{12}z^{-1} \end{pmatrix}; \tag{24}$$

$$Gd = \frac{1}{l_{11}}\begin{pmatrix} 1 & 0 \\ 0 & l_{11}l_{12}z^{-1} \end{pmatrix}. \tag{25}$$

A two-splitter delay line (TSDL) is two cascaded splitters and a line with two delay elements. Therefore the circuit matrix of the recirculation delay line and the transfer matrix of the two-splitter delay line on the basis of expressions (20)–(25):

$$Gb = Gc_1 \cdot Gd \cdot Gc_0 = \frac{1}{l_{11}(1-a_1)(1-a_0)} \cdot \begin{pmatrix} G'_{11} & G'_{21} \\ G'_{12} & G'_{22} \end{pmatrix}, \tag{26}$$

where

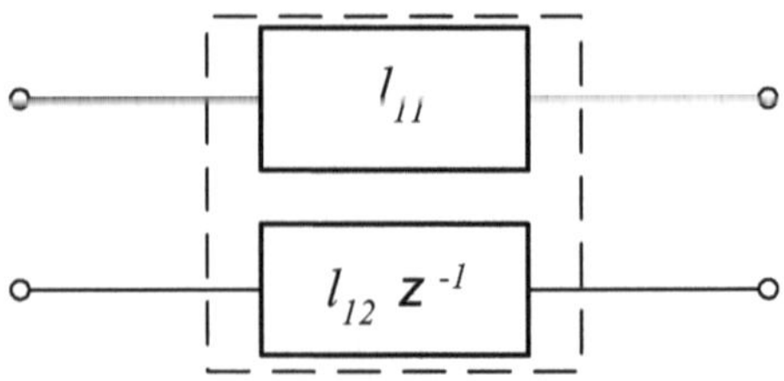

Fig. 13 Simplified two-lie delay system (l_{11} and l_{12} are intensity transmittances for the first and second fiber lines; z^{-1} is the unitary delay time)

$$\begin{cases} G'_{12} = 1 - a_1 a_0 l_1 z^{-1}; \\ G'_{12} = -a_0 - a_1(1 - 2a_0) l_1 z^{-1}; \\ G'_{21} = a_1 + (1 - 2a_1) a_0 l_1 z^{-1}; \\ G'_{22} = -a_1 a_0 + (1 - 2a_1)(1 - 2a_0) l_1 z^{-1}, \end{cases} \quad (27)$$

where $l_1 = l_{11}\ l_{12}$ is the intensity transmittance of the circuit; Gc_1 and Gc_0 is are chain matricess of the splitter (expression (23)) with the coupling coefficient a_1 and a_0; Gd is the chain matrix of the two-element delay line specified with expression (25). In addition, the elements corresponding to the transfer matrix can be found from expression (28):

$$Hb = Hc_1 \cdot Hd \cdot Hc_0 = \begin{pmatrix} H_{11} & H_{12} \\ H_{21} & H_{22} \end{pmatrix}, \quad (28)$$

$$\begin{cases} H_{11} = (1 - b_1)(1 - b_0) l_{11} + b_1 b_0 l_{12} z^{-1}; \\ H_{12} = b_1(1 - b_0) l_{11} + (1 - b_1) b_0 l_{12} z^{-1}; \\ H_{21} = (1 - b_1) b_0 l_{11} + b_1(1 - b_0) l_{12} z^{-1}; \\ H_{22} = b_1 b_0 l_{11} + (1 - b_1)(1 - b_0) l_{12} z^{-1}. \end{cases} \quad (29)$$

We use the shown above approach in combination with the Z-transformation method to obtain general expressions for the transfer functions of the studied structures. The generalized linear system of n order with discrete time in modern control theory is described in state space by the state transition equation and the initial equation of the form:

$$\begin{cases} x(k+1) = A(k)x(k) + B(k)u(k); \\ y(k) = C(k)x(k) + D(k)u(k). \end{cases} \quad (30)$$

The uppercase letters are matrices, the lowercase ones are vectors. In expression (30), the vector $x(k)$ is a n-dimensional state vector; $u(k)$ is a m-dimensional input vector; $y(k)$ is a p-dimensional output vector, and A, B, C, D are matrices of the corresponding size $(n \times n)$, $(n \times m)$, $(p \times n)$, $(p \times m)$. For systems invariant to time shift, the indicated matrices do not depend on time, but depend on system parameters such as the coupling coefficient or the loss coeficient.

In the mathematical model described by expression (30), it is generally assumed that the system has m-inputs and p-outputs. If the system has a single input and a single output, then expression (30) will look like:

$$\begin{cases} x(k+1) = A(k)x(k) + b(k)u(k); \\ y(k) = C(k)x(k) + d(k)u(k), \end{cases} \tag{31}$$

where b and C are vectors and d is a scalar value.

Taking the initial values of the delay elements as state parameters, we obtain the following expressions for the matrices A, B, C, D describing structures of N-order with feedback:

$$A = \begin{pmatrix} a_1 a_0 l_1 & a_2(1-a_1)a_0 l_1 & \cdots & a_N \prod_{k=1}^{N-1} (1-a_k) a_0 l_1 \\ (1-a_1) l_2 & a_2 a_1 l_2 & \cdots & a_N \prod_{k=2}^{N-1} (1-a_k) a_0 l_1 \\ & \ddots & & a_N (1-a_{N-1}) l_{N-1} \\ 0 & & & (1-a_{N-1}) \ln a_N a_{N-1} l_N \end{pmatrix};$$

$$B = \begin{pmatrix} \prod_{k=1}^{N-1} (1-a_k) a_0 l_1 & (1-a_0) l_1 \\ \prod_{k=2}^{N-1} (1-a_k) a_0 l_1 & 0 \\ \vdots & \vdots \\ (1-a_N) a_{N-1} l_N & 0 \end{pmatrix}; \tag{32}$$

$$C = \begin{pmatrix} 0 & 0 & \cdots & 1-a_N \\ a_1(1-a_0) & a_2(1-a_1)(1-a_0) & \cdots & a_N \prod_{k=0}^{N-1} (1-a_k) \end{pmatrix}$$

$$D = \begin{pmatrix} a_N & 0 \\ \prod_{k=0}^{N} (1-a_K) & a_0 \end{pmatrix}$$

The input and output vectors, as well as the state vector, correspond to

$$u = \begin{pmatrix} x_1 \\ x_2 \end{pmatrix} \quad Y = \begin{pmatrix} Y_1 \\ Y_2 \end{pmatrix} \quad x = \begin{pmatrix} x_1 \\ x_2 \\ \vdots \\ x_N \end{pmatrix} \tag{33}$$

Rise of the indices corresponds to the shift from right to left in the network. It is easy to notice that all matrix elements and components of vectors are positive values, which means that the system is a positive optical system [38]. The theory of positive systems is closely related to the theory of positive matrices, which is developed enough to allow us to use its results for studying properties of positive systems. Codsidering the state space model for linear systems, a linear positive system is defined as a linear system whose state parameters (as well as output signals) are positive.

5 Analysis of Light Energy Efficiency

Loss of energy in directional splitters and optical fibers, as well as loss due to multiple outputs, cause that insertion loss that does not equal zero. Thus, there is a practical interest in amount of light energy reachig the output.

To calculate the light flux at output at a given flux at input, we apply the transfer function of the system $H(z)$ calculated by expressions (19, 21). It is sufficient to consider that the light flux is proportional to the zero component of a spectrum ($f = 0$) in the modulation frequency band. In the transfer function, this component corresponds to $z = 1$ ($z = e^{j2\pi fT}$), we obtaine the following relation between values of the light flux at input and output of the system:

$$Po = [H(1)]Pi, \tag{34}$$

where *Po*, *Pi* is the light flux at output and input of the system, respectively.

Thus, the ratio of the output luminous flux to the input is equal to the transfer function value at $z = 1$:

$$\eta \triangleq \frac{Po}{Pi} = H(1). \tag{35}$$

For the memory element built on a recirculation line with two splitters, we have:

$$\eta_1 = \frac{a_1 + (1 - 2a_1)a_0 l_1}{1 - a_1 a_0 l_1}, \tag{36}$$

$$\eta_2 = \frac{(1 - a_1)(1 - a_0)l_{11}}{1 - a_1 a_0 l_1}, \tag{37}$$

where η_1 is the efficiency when X_1 is the input and Y_1 is the output; η_2 is the efficiency when X_1 is the input and Y_2 is the output; a_1, a_0 are the intensity coupling coefficients of splitters; l_1 is the intensity transmittance of the optical fiber ring; l_{11} is the intensity fiber transmittance in the direct line.

According to formulas (36) and (37), graphs of $\eta_1 = F(a_1)$ and $\eta_2 = F(a_2)$ were constructed at different values of a_0 (Figs. 14 and 15).

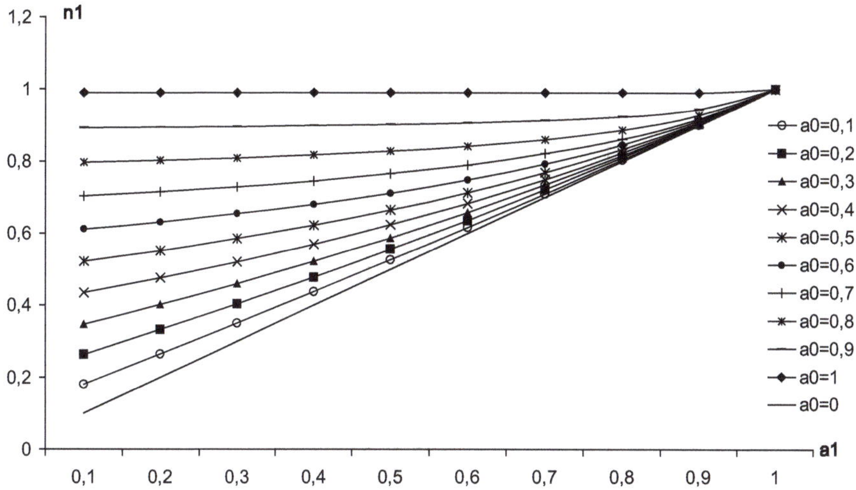

Fig. 14 Dependance of the efficiency η_1 on the coupling coefficient a_1 at different values of the coupling coefficient a_0 ($l_1 = 0.99$)

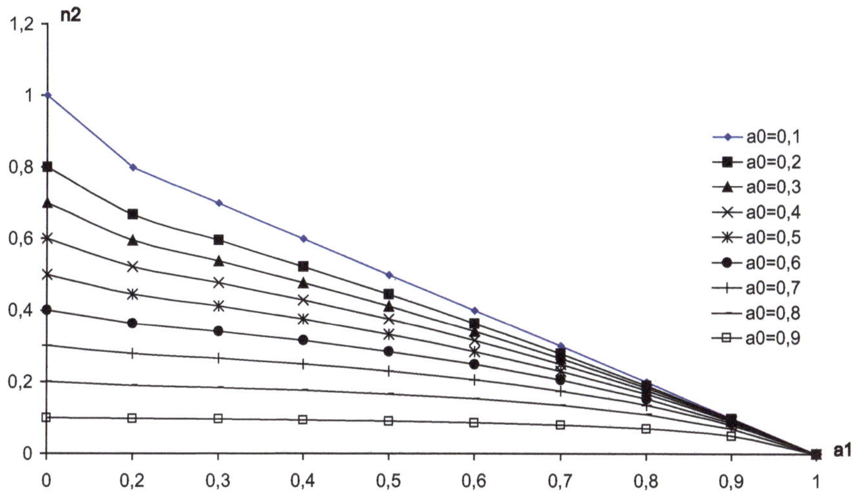

Fig. 15 Dependance of the efficiency η_2 of the coupling coefficient a_1 at different values of the coupling coefficient a_0 ($l_1 = 0.99$)

Thus, it follows from expressions (36), (37) and Figs. 14 and 15 that when a_0 and a_1 increase the luminous flux at the output Y_2 decreases and those at the output Y_1 increases. If the first directional splitter is employed for entering the information sequence (input X_1) and the second directional splitter is employed for drawing the stored optical information sequence (output Y_2) according to Fig. 15, the intensity coupling coefficients of the splitters a_0 and a_1 must be as small as possible. If the first directional splitter is employed for entering the information sequence (input X_1) and for drawing the stored optical information sequence (output Y_1) according to Fig. 15, the intensity coupling coefficient of the first splitter a_1 must be as large as possible and the intensity coupling coefficient of the second splitter a_0 must be as small as possible.

6 Conclusion

The study examines the optical dynamic random access memory on fiber-optic lines. The proposed mathematical model of the base element of the optical dynamic random access memory enables to obtain its transfer functions and to calculate light energy efficiency. The mathematical model was applied to analyze the light energy efficiency of the base element of the optical dynamic random access memory. These characteristics allow designing optical memory systems on fiber-optic lines for high-speed optical information systems, optical neurocomputers, geographic-information energy networks.

References

1. Huang TJ (2017) Imitating the brain with neurocomputer a "new" way towards artificial general intelligence. Int J Autom Comput 14:520–531. https://doi.org/10.1007/s11633-017-1082-y
2. Alexoudi T, Terzenidis N, Pitris S et al (2018) Optics in computing: from photonic network-on-chip to chip-to-chip interconnects and disintegrated architectures. J Lightwave Technol 37(2):363–379. https://doi.org/10.1109/JLT.2018.2875995
3. Mourgias-Alexandris G, Totović A, Tsakyridis A et al (2020) Neuromorphic photonics with coherent linear neurons using dual-IQ modulation cells. J Lightwave Technol 38(4):811–819. https://doi.org/10.1109/JLT.2019.2949133
4. Zhang Y (1996) Optical/digital neuroprocessor: an approach to neurocomputer. In Proceedings SPIE 2778, 17th Congress of the international commission for optics: optics for science and new technology, p 27786V (1 Sept 1996). https://doi.org/10.1117/12.2299130
5. Bayro-Corrochano E. (2019) Applications in neurocomputing. In: Geometric algebra applications, vol I. Springer, Cham. https://doi.org/10.1007/978-3-319-74830-6_18
6. Tsakyridis A, Alexoudi T, Miliou A et al (2019) 10 Gb/s optical random access memory (RAM) cell. Opt Lett 44(7):1821–1824. https://doi.org/10.1364/OL.44.001821
7. Mourgias-Alexandris G, Vagionas C, Tsakyridis A, Maniotis P, Pleros N (2018) All-optical 10Gb/s ternary-CAM cell for routing look-up table applications. Opt Express 26(6):7555–7562. https://doi.org/10.1364/OE.26.007555
8. Goswami D (2003) Optical computing. Reson 8:56–71. https://doi.org/10.1007/BF02837869
9. Stroke GW (1972) Optical computing. IEEE Spectr 9(12):24–41. https://doi.org/10.1109/MSPEC.1972.5218639
10. Boy GA (2001) Organisational memory systems. IFAC Proc Vol 34(16):393–404. https://doi.org/10.1016/S1474-6670(17)41556-4
11. Semenova O, Semenov A, Kulakov P (2017) Access neuro-fuzzy controller for W-CDMA networks. In: 2017 4th International scientific-practical conference problems of infocommunications science and technology, (PIC S&T), Kharkov, 10–13 Oct 2017. https://doi.org/10.1109/INFOCOMMST.2017.8246344

12. Semenov AA et al (2016) Routing in telecommunication networks using fuzzy logic. In: 2016 17th International conference of young specialists on micro/nanotechnologies and electron devices (EDM), Erlagol, 30 June—4 July 2016. https://doi.org/10.1109/EDM.2016.7538719
13. Semenova O et al (2013) Access fuzzy controller for CDMA networks. In: 2013 International Siberian conference on control and communications (SIBCON), Krasnoyarsk, 12–13 Sept 2013. https://doi.org/10.1109/SIBCON.2013.6693644
14. Evtikhiev NN, Starikov RS, Scherbakov IB et al (1995) Hybrid optoelectronic neurocomputer: variants of realization. In: Proceedings SPIE 2492, applications and science of artificial neural networks, 6 April 1995. https://doi.org/10.1117/12.205102
15. Polyakov AV (2012) Information parameters of DWDM fiber optic dynamic loop memory. In: Proceedings SPIE 8498, optics and photonics for information processing VI, p 849813, 15 Oct 2012. https://doi.org/10.1117/12.929492
16. Malevich IA, Polyakov AV, Chubarov SI (2011) Multichannel fiber optic recirculating memory. In: 2011 11th International conference on laser and fiber-optical networks modeling (LFNM), 5–9 Sept 2011, Kharkov, Ukraine. https://doi.org/10.1109/LFNM.2011.6145024
17. Sarrazin DB, Jordan HF, Heuring VP (1990) Digital fiber optic delay line memory. In: Proceeding SPIE 1215, Digital Optical Computing II, 1 July 1990. https://doi.org/10.1117/12.18084
18. Zhang J, Guzzon RS, Coldren LA, Yao J (2018) Optical dynamic memory based on an integrated active ring resonator. Opt Lett 43(19):4687–4690. https://doi.org/10.1364/OL.43.004687
19. Vagionas C, Tsakyridis A, Mourgias-Alexandris G, Pitris S, Maniotis P, Pleros N, Alexoudi T (2019) Optical memory architectures for fast routing address look-up (AL) table operation. J Phys Photonics 1(4):1–14. https://doi.org/10.1088/2515-7647/ab4368
20. Kitayama K, Maruta A, Yoshida Y (2014) Digital coherent technology for optical fiber and radio-over-fiber transmission systems. J Lightwave Technol 32(20):3411–3420. https://doi.org/10.1109/jlt.2014.2310461
21. Ibrahimov BG, Hasanov MH (2018) Analysis of methods increasing the quality of the operation optical signal transfer systems in telecommunication systems. In: 2018 Systems of signal synchronization, generating and processing in telecommunications (SYNCHROINFO), 4–5 July 2018, Minsk, Belarus. https://doi.org/10.1109/SYNCHROINFO.2018.8457007
22. Al-Maitah M et al (2018) A hybrid approach to call admission control in 5G networks. Adv Fuzzy Syst 2018:1–7. https://doi.org/10.1155/2018/2535127
23. Semenova O, Semenov A, Voitsekhovska O (2019) Neuro-fuzzy controller for handover operation in 5G heterogeneous network. In: 2019 3rd International conference on advanced information and communications technologies (AICT), Lviv, 2–6 July 2019. https://doi.org/10.1109/AIACT.2019.8847898
24. Bondarenko O, Ageyev D, Mohammed O (2019) Optimization model for 5G network planning. In: 2019 IEEE 15th International conference on the experience of designing and application of CAD systems (CADSM), Polyana, 26 Feb—2 March 2019. https://doi.org/10.1109/CADSM.2019.8779298
25. Mariia S, Oleksandr R (2018) The structure of a mobile provider network with network functions virtualization. In: 2018 14th International conference on advanced trends in radioelecrtronics, telecommunications and computer engineering (TCSET), Slavske, 20–24 Feb 2018. https://doi.org/10.1109/TCSET.2018.8336370
26. Berrettini G, Potì L, Bogoni A (2011) Optical dynamic random access memory (ODRAM). In: Optical fiber communication conference/National fiber optic engineers conference 2011, OSA Technical Digest (CD) (Optical Society of America, 2011), paper JWA036. https://doi.org/10.1364/NFOEC.2011.JWA036
27. Srivastava R, Singh RK, Singh YN (2007) Regenerator based fiber optic loop memory. In: TENCON 2007—2007 IEEE Region 10 conference, 30 Oct—2 Nov 2007, Taipei, Taiwan. https://doi.org/10.1109/TENCON.2007.4428878
28. Li X, Shao Z, Zhu M, Yang J (2018) Forward the next generation supercomputer. in: fundamentals of optical computing technology. Springer, Singapore. https://doi.org/10.1007/978-981-10-3849-5

29. Skulysh MA et al (2019) Managing the process of servicing hybrid telecommunications services. Quality control and interaction procedure of service subsystems. In: Pejaś J, El Fray I, Hyla T, Kacprzyk J (eds) Advances in soft and hard computing. ACS 2018. Advances in intelligent systems and computing, vol 889. Springer, Cham, pp 244–256. https://doi.org/10.1007/978-3-030-03314-9_22
30. Ageyev D et al (2019) Infocommunication networks design with self-similar traffic. In: 2019 IEEE 15th International conference on the experience of designing and application of CAD systems (CADSM), Polyana 26 Feb—2 March 2019. https://doi.org/10.1109/CADSM.2019.8779314
31. Radivilova T et al (2020) The methods to improve quality of service by accounting secure parameters. In: Hu Z, Petoukhov S, Dychka I, He M (eds) Advances in computer science for engineering and education II. ICCSEEA 2019. Advances in intelligent systems and computing, vol 938. Springer, Cham, pp 346–355. https://doi.org/10.1007/978-3-030-16621-2_32
32. Mourgias-Alexandris G et al (2018) A 10Gb/s all-optical match-line for optical content addressable memory (CAM) rows. In: 2018 Optical fiber communications conference and exposition (OFC), pp 11–15 March 2018, San Diego, CA, USA
33. Pilipovich VA, Esman AK, Goncharenko IA et al (1995) Circulatory fibre-optic memory loop with a built-in service channel. Quantum Electron 25(10):984. https://doi.org/10.1070/QE1995v025n10ABEH000515
34. Ponzini F et al (2005) An optical memory cell based on erbium-doped fiber. In: Proceedings of 2005 IEEE/LEOS workshop on fibres and optical passive components, 2005, pp 22–24 June, Palermo, Italy. https://doi.org/10.1109/WFOPC.2005.1462115
35. Belkin ME et al (2017) Continuously accessible long-term fiber optic memory of microwave signal. In: 2017 International topical meeting on microwave photonics (MWP), pp 23–26 Oct 2017, Beijing, China. https://doi.org/10.1109/MWP.2017.8168695
36. Fukushima T, Tanaka T, Sakamoto T (2000) Stability of optical dynamic memory operation under period-two state of chaotic system. Jpn J Appl Phys 39(1R):114. https://doi.org/10.1143/jjap.39.114
37. Semenov AO, Baraban SV, Osadchuk OV, Semenova OO, Koval KO, Yu Savytskyi A (2019) Microelectronic pyroelectric measuring transducers. IFMBE Proc 77:393–397. https://doi.org/10.1007/978-3-030-31866-6_72
38. Semenov A, Voznyak O, Osadchuk O, Baraban S, Semenova O, Rudyk A, Klimek J, Orazalieva S (2019) Development of a non-standard system of microwave quadripoles parameters. In: Proceedings SPIE 11176, photonics applications in astronomy, communications, industry, and high-energy physics experiments 2019, p 111765N. https://doi.org/10.1117/12.2536704
39. Haykin S (1998) Neural networks: a comprehensive foundation. Prentice Hall, Upper Saddle River
40. Semenov A, Semenova O, Osadchuk O (2015) The UHF oscillators based on a HEMT structure with negative conductivity. Int Sib Conf Control Commun (SIBCON) 2015:1–4. https://doi.org/10.1109/SIBCON.2015.7147215
41. Semenov A (2016) Reviewing the mathemetical models and electrical circuits of deterministic chaos transistor oscillators. Int Sib Conf Control Commun (SIBCON) 2016:1–6. https://doi.org/10.1109/SIBCON.2016.7491758

Analysis and Systematization of Gas Concentration Measurement Methods

Igor Dudatiev, Andrii Semenov, Kostyantyn Ovchynnykov, Maksym Prytula, and Serhii Maltsev

Abstract The article provides a comprehensive analysis and classification of the main methods for measuring gas concentrations used in ecology, industry, medicine, and safety systems. Five main groups of technologies are considered: physico-chemical, optical, electrochemical, acoustic, and thermochemical methods. Each group is described in terms of its operating principles, advantages, disadvantages, fields of application, and relevant mathematical models that characterize sensor performance. Particular attention is given to a comparative analysis of the methods based on sensitivity, selectivity, response time, equipment cost, suitability for operation under various conditions, and maintenance complexity. Based on the analysis, it is concluded that electrochemical methods offer the most balanced option for practical use. This article may be useful for specialists involved in the development of gas analyzers, as well as for students of technical disciplines and researchers in the field of chemical sensor analysis. Controlling gas concentrations in the environment, industrial facilities, medical institutions, and household settings is a critical challenge in the modern world. Air pollution, leaks of toxic or explosive gases, and the need for precise gas dosing in technological processes all require reliable and highly sensitive methods for gas environment analysis. There exists a wide range of physical, chemical, and combined methods for measuring gas concentrations, each with its own advantages and limitations. Depending on the application area and requirements for accuracy, selectivity, response time, cost, and operating conditions, the appropriate approach is chosen. The aim of this work is to systematize and classify modern methods of gas concentration measurement according to physico-chemical principles, as well as to conduct their comparative analysis in terms of key metrological and operational characteristics. Additionally, mathematical models

---

I. Dudatiev · A. Semenov (✉) · K. Ovchynnykov · M. Prytula · S. Maltsev
Vinnytsia National Technical University, Vinnytsia, Ukraine
e-mail: semenov.a.o@vntu.edu.ua

K. Ovchynnykov
e-mail: ovkos1980@gmail.com

M. Prytula
e-mail: pritulamo@ukr.net

A. Semenov et al. (eds.), *Data-Centric Business and Applications*, Lecture Notes on Data Engineering and Communications Technologies 272,
https://doi.org/10.1007/978-3-032-06608-4_20

describing the working principles of sensors in each category are considered. The proposed approach allows for a justified selection of the optimal gas control method for specific tasks and promotes the further development of hybrid sensor platforms and multi-analytical systems.

Keywords Gas measurement · Gas analysis · Chemosensors · Spectroscopy · Electrochemical sensors · Acoustic methods · Thermochemical sensors · Gas concentration · Gas chromatography · Mathematical modeling · Method classification · Environmental monitoring · Analytical chemistry

1 Introduction

In the current context of rapid technological progress and escalating environmental challenges, the accurate and timely monitoring of gas concentrations in various media has become not merely a scientific and technical objective but a critical societal imperative. As the global population expands, industrial activity intensifies, and the consequences of climate change grow increasingly severe, the need for precise and real-time gas detection systems has reached unprecedented urgency. Air quality degradation, greenhouse gas accumulation, toxic emissions, and gas-related safety incidents pose serious threats to environmental sustainability, human health, and economic stability. In this context, gas analysis technologies are not only essential for routine monitoring but also for early warning systems, regulatory compliance, and scientific research [1–5].

The challenges related to atmospheric pollution, industrial emissions, leaks of toxic or explosive gases, and the control of indoor and workplace air quality are driving the demand for robust, reliable, and sensitive gas detection methods [5]. These applications span a wide range of environments—from open-air urban monitoring stations and industrial chimneys to closed laboratory settings, vehicles, and wearable safety devices. Each scenario imposes specific technical requirements on measurement accuracy, detection limits, response time, environmental resilience, and cost-efficiency. As a result, a multitude of gas detection technologies have been developed over the past few decades, each with distinct operational principles and performance characteristics [6–9].

Existing methods for measuring gas concentrations vary significantly in their foundational principles [10]. These include physical, chemical, optical, electrical, thermal, and acoustic mechanisms, with many modern systems incorporating hybrid or multi-modal approaches [11–15]. Furthermore, the rapid advancement of materials science—particularly the development of nanomaterials, metal–organic frameworks (MOFs), and novel polymers—has opened new pathways for sensor enhancement. Simultaneously, improvements in microelectronics, data acquisition systems, and signal processing algorithms have made it possible to build compact, energy-efficient, and highly sensitive gas detectors suitable for deployment in complex and dynamic environments [16–19].

Despite this progress, the diversity of methods and the increasing specialization of their applications underscore the need for a systematic, comprehensive, and scientifically grounded comparison. The selection of an appropriate gas measurement method is often influenced by numerous factors, including the specific gas species to be detected, the required measurement range, operational conditions (e.g., temperature, humidity, pressure), interference tolerance, and the desired level of automation [20–22]. Without a structured understanding of the capabilities and limitations of each approach, the risk of suboptimal sensor deployment—and consequently, inaccurate measurements or system failures—becomes significant.

Theoretical exploration and practical implementation of gas detection technologies must therefore go hand in hand. From a research perspective, the development of new gas-sensing materials, transduction mechanisms, and data interpretation models remains an active and interdisciplinary area of investigation. From an engineering standpoint, the design of sensor systems that are reliable, cost-effective, and adaptable to different contexts requires ongoing optimization. Hence, a comprehensive understanding of the existing gas analysis methodologies and a clear framework for comparing them is crucial for researchers, developers, regulatory agencies, and end users alike [23–26].

Over the past several decades, the scientific literature has documented a steady and dynamic evolution in gas measurement technologies. There is a discernible trend toward increased sensitivity—the ability to detect lower concentrations of gases—and improved selectivity, or the capacity to distinguish between chemically similar species. In parallel, efforts to enhance sensor stability, repeatability, and resistance to environmental interference continue to progress. Among the most rapidly advancing areas are optical and electrochemical methods, both of which have benefited greatly from the integration of nanostructured materials, advances in photonics, and the miniaturization of sensing elements [26–29]. The use of laser-based techniques, for instance, has expanded significantly in applications requiring real-time, non-invasive gas detection [25, 26, 29].

However, while novel technologies are becoming more prevalent, certain classical methods—such as titrimetric, colorimetric, or gravimetric techniques—are experiencing a decline in practical use. Their limitations, including relatively low sensitivity, reliance on manual operation, and susceptibility to subjective interpretation, restrict their applicability in modern industrial and environmental settings. Nevertheless, these methods still retain value in educational laboratories and in field conditions where simplicity and portability are prioritized over precision [30].

At the same time, it is important to recognize that no single gas detection method is universally superior [31]. Rather, the relevance and effectiveness of a given method are strongly context-dependent. In some scenarios, high precision and selectivity are paramount—such as in atmospheric research or trace gas analysis in medical diagnostics. In other cases, robustness and cost-effectiveness may take precedence, as in industrial leak detection or household gas alarms. Therefore, the versatility and adaptability of sensing technologies are key factors in their widespread adoption.

It is also noteworthy that the current trajectory of progress in gas analysis is closely linked to advancements in adjacent disciplines, particularly electronics, software engineering, and data science. The integration of artificial intelligence and machine learning into gas sensing platforms is enabling smarter, more autonomous systems capable of self-calibration, drift correction, and predictive analytics. These developments not only improve measurement accuracy but also extend the operational lifetime and reliability of sensor devices.

In conclusion, the study and systematization of gas concentration measurement methods remain highly relevant in light of global environmental priorities and technological developments. There is a clear need for a structured overview of current techniques, highlighting their respective advantages, drawbacks, and application domains. Such an analysis provides the foundation for informed decision-making in the selection, design, and deployment of gas detection systems. It also helps guide future research efforts toward the development of more effective, sustainable, and universally applicable solutions for gas analysis in both scientific and practical contexts.

2 Materials and Methods

Measurement of gas concentrations is a critically important task in many scientific and industrial fields. The need to detect and quantify gases arises in a broad range of applications, including but not limited to environmental protection, industrial process automation, medical diagnostics, agriculture, chemical engineering, and public safety. In each of these areas, accurate and timely information about the presence and concentration of specific gases is essential for ensuring human health, optimizing industrial performance, monitoring environmental quality, and preventing potentially dangerous situations. As such, gas analysis has become a central topic of research and technological development. To accurately determine the amount of gases in air or enclosed environments, various methods are used, which can be classified based on physical principles, accuracy, response speed, equipment cost, and other parameters [1–31].

The process of gas concentration measurement involves determining the amount of a particular gas component in a mixture, most commonly in air or controlled environments. This task can be extremely challenging due to factors such as the complexity of gas mixtures, the presence of interfering substances, the low concentration levels of target gases (often at parts-per-million or parts-per-billion), and the need for fast response times. Therefore, numerous measurement methods have been developed, refined, and adapted for specific applications.

Gas concentration measurement methods are conventionally classified based on the physical principles they utilize, their level of accuracy, response speed, operational reliability, selectivity toward specific gases, and economic factors such as cost of equipment and maintenance. These classifications serve as a useful framework for understanding the underlying mechanisms of gas detection and for choosing the appropriate method depending on the goals of a given measurement task.

One fundamental class of gas detection technologies involves physicochemical methods. These methods are based on the interaction of gas molecules with a surface, sensor material, or chemical substance that undergoes a measurable change when exposed to certain gases. These changes may involve variations in mass, color, electrical conductivity, or other physical parameters. The core advantage of such techniques lies in their ability to combine high sensitivity with chemical specificity. For example, gas chromatography is one of the most precise methods available, enabling the separation, identification, and quantification of individual gas species within a complex mixture. It relies on the differential migration of gas components through a chromatographic column and often incorporates detectors such as flame ionization or thermal conductivity sensors to analyze the eluted compounds. Despite its accuracy, gas chromatography is generally suited for laboratory settings due to its complexity and the need for sample preparation.

Chemosensors, commonly referred to as gas sensors, also belong to the category of physicochemical methods. These devices typically operate based on changes in electrical properties, such as resistance or capacitance, when a specific gas interacts with a reactive surface. Metal oxide semiconductors, such as tin dioxide (SnO_2), are widely used materials in these sensors. More recently, advanced materials like graphene and carbon nanotubes have been explored for enhancing sensitivity and selectivity. Chemosensors are widely used in commercial gas detectors, air quality monitoring stations, and portable safety devices.

Another major group of gas detection techniques is formed by optical methods. These methods are based on the interaction between gas molecules and electromagnetic radiation. Depending on the application and the physical phenomena involved, optical methods may exploit light absorption, fluorescence, scattering, or refractive index changes. Among the most prevalent optical techniques is infrared (IR) absorption spectroscopy, which relies on the fact that gas molecules absorb light at specific wavelengths corresponding to their vibrational and rotational transitions. By directing IR light through a gas sample and analyzing the absorption spectrum, it is possible to determine both the identity and concentration of gas species with high accuracy.

A particularly sophisticated implementation of optical sensing is Tunable Diode Laser Absorption Spectroscopy (TDLAS). This method uses a laser that can be finely tuned across the absorption line of a specific gas. It offers high selectivity, low detection limits, and rapid response times, making it ideal for real-time gas monitoring in industrial processes, combustion diagnostics, and environmental studies. Optical techniques are generally immune to electromagnetic interference, which enhances their reliability in complex operational environments.

Electrochemical methods form a separate and widely used category of gas measurement techniques. These methods are based on the generation or measurement of electrical signals resulting from chemical reactions between the target gas and an electrolyte or electrode. Electrochemical sensors are commonly found in applications such as indoor air quality monitoring, occupational safety, and medical diagnostics. A typical example is the galvanic sensor, in which the interaction between the gas and a reactive electrode produces an electric current proportional to gas concentration. Amperometric and potentiometric sensors, which measure current or voltage changes due to electrochemical reactions, are also widely used for gas detection, particularly for gases such as oxygen, carbon monoxide, and nitrogen dioxide.

In addition to physicochemical, optical, and electrochemical methods, there are acoustic methods that utilize the interaction between sound waves and gas molecules. One of the most well-known techniques in this category is photoacoustic spectroscopy, which relies on the generation of pressure waves when gas molecules absorb modulated light. These pressure waves can be detected using sensitive microphones or acoustic sensors. The amplitude of the acoustic signal is directly related to the concentration of the absorbing gas. Other acoustic methods, such as those based on changes in the speed of sound or sound attenuation in a gas medium, can be used to infer gas composition and density.

Thermochemical methods represent another distinct group of gas detection techniques. These methods are based on the principle that gas interactions can alter the thermal properties of the surrounding medium or trigger exothermic or endothermic reactions. For instance, changes in thermal conductivity due to the presence of specific gases can be used to measure their concentrations. Catalytic sensors, which detect heat generated from combustion reactions between the target gas and a catalyst, are widely employed for detecting flammable gases like methane or hydrogen. While these methods may lack the selectivity of optical or electrochemical sensors, they are highly robust and suitable for use in harsh industrial environments.

In summary, the field of gas concentration measurement encompasses a wide range of physical, chemical, and hybrid techniques. Each method offers distinct advantages and limitations, and the choice of technique is often dictated by the specific requirements of the application, including sensitivity, selectivity, cost, and environmental conditions. The rapid development of materials science, optics, and electronics continues to drive innovation in this area, leading to more compact, accurate, and versatile gas sensors. The classification and systematization of these methods not only aid in understanding their operational principles but also in guiding the design of integrated systems for multi-gas detection and real-time monitoring.

A systematization of methods and submethods for gas concentration measurement is presented in Fig. 1.

The gas concentration measurement methods visualized in Fig. 1, along with their advantages and disadvantages, are presented in the form of a table (Table 1).

The physical and chemical processes underlying each method [1–31] are presented in Table 2.

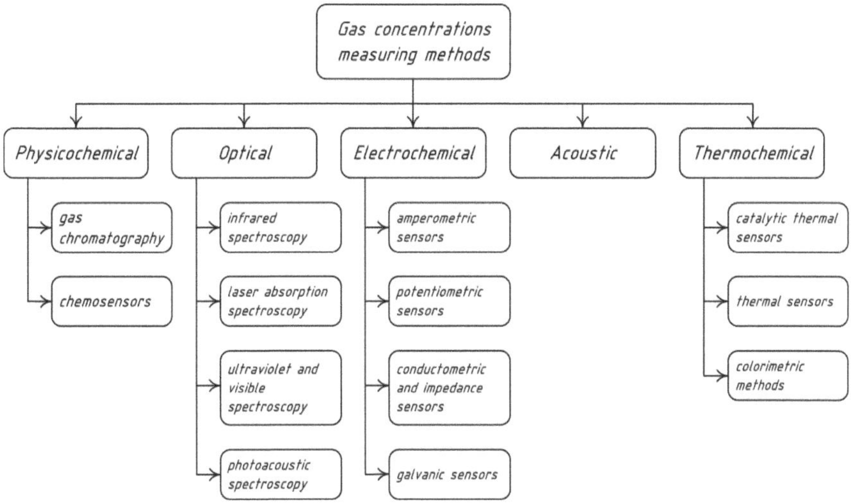

Fig. 1 Systematization of gas concentration measurement methods

3 Results and Discussion

A comparative analysis of the considered gas concentration measurement methods according to the following parameters: sensitivity, selectivity, response speed, resistance to interference, equipment cost, suitability for field conditions, and maintenance complexity is presented in Table 3.

Thus, the comparative analysis shows that no single method is universal [32, 33]. However, electrochemical methods are the most balanced in terms of accuracy, cost, and simplicity. Optical methods are the most accurate and selective, especially laser-based techniques, but they are expensive, require a clean environment, and have high maintenance demands [34, 35]. Physicochemical methods are optimal for laboratory conditions or multisensor systems but are less suitable for autonomous field use [36, 37]. Acoustic methods are promising for specialized conditions but require good noise isolation. Thermochemical methods are the simplest and cheapest, suitable for coarse detection but inadequate for tasks demanding high selectivity.

Table 1 Systematization of gas concentration measurement methods

Method	Principle of operation	Advantages	Disadvantages	Applications
Gas chromatography	Separation of gas mixture components based on their different interactions with the stationary phase in a chromatographic column	High accuracy and selectivity. Ability to measure multiple components simultaneously	High equipment cost. Sample preparation required. Long analysis time	Environmental monitoring, industrial analytics, forensics, air quality control
Chemosensors (Solid-State Gas Sensors)	Sensor responds to gas by changing electrical properties (conductivity, capacitance) upon interaction of gas molecules with a sensitive layer (often metal oxides like SnO_2, ZnO)	Simple design. Low cost. Fast response	Limited selectivity (sensitivity to several gases simultaneously). Requires periodic calibration. Sensitive to temperature and humidity	Household gas detectors, industrial leak sensors
Colorimetric and titrimetric methods	Color change of reagent upon interaction with target gas (e.g., NO_2, NH_3, CO). In titrimetry, gas is absorbed in solution and titrated with a reagent	Simple implementation. Possible manual analysis	Low accuracy. Subjective color perception (if without instruments)	Laboratory analysis, indicator tubes

(continued)

Table 1 (continued)

Method	Principle of operation	Advantages	Disadvantages	Applications
Infrared spectroscopy (IR)	Gas molecules absorb infrared radiation at specific wavelengths corresponding to vibrational or rotational transitions. Absorption magnitude is proportional to gas concentration (Beer–Lambert law)	High sensitivity and accuracy. Non-contact measurement. Selectivity to specific molecules	High cost. Requires calibration. Sensitive to optical path contamination (dust, condensation)	Industrial control, exhaust analysis, greenhouse gas monitoring
Tunable diode laser absorption spectroscopy (TDLAS)	Uses narrowband laser radiation passing through gas flow. Absorption at specific wavelength allows precise gas concentration measurement	Extremely high accuracy (down to ppb level). Fast response. Real-time measurement. Suitable for harsh environments	High equipment cost. Sensitive to mechanical shifts	Energy sector, pollutant emission control, gas turbines, nuclear energy
Ultraviolet and visible spectroscopy	Certain gases absorb light in UV or visible ranges (e.g., NO_2 has characteristic UV absorption lines). Measuring absorption intensity determines gas concentration	Applicable to a wide range of gases. Moderate cost. Use of inexpensive LEDs possible	Lower accuracy compared to IR. Limited selectivity	Air monitoring, automotive diagnostics, atmospheric studies
Photoacoustic spectroscopy	Gas absorbs modulated light and converts it to heat, causing an acoustic wave. A microphone detects pressure changes proportional to gas concentration	High sensitivity. Compact device size. High selectivity	Noise interference. Requires acoustic isolation	Explosive gas detectors, portable devices, occupational safety

(continued)

Table 1 (continued)

Method	Principle of operation	Advantages	Disadvantages	Applications
Amperometric sensors	Gas diffuses through a membrane to the electrode where it is oxidized or reduced, generating a current proportional to gas concentration	High sensitivity and signal linearity. Relatively simple calibration	Limited electrolyte lifetime. Temperature dependence	Sensors for oxygen (O_2), carbon monoxide (CO), ozone (O_3)
Potentiometric sensors	Measures potential difference between two electrodes, one sensitive to a specific gas. Gas concentration affects chemical potential, altering electrical voltage	Low power consumption. Long-term stability (under controlled conditions)	Lower sensitivity compared to amperometric. Accuracy affected by pH or humidity	Ion-selective electrodes for ammonia (NH_3) or hydrogen (H_2)
Conductometric and impedance sensors	Change in conductivity (conductometry) or electrical resistance (impedance) of sensitive material in the presence of target gas	Simple design. Suitable for miniaturization	Low selectivity. Influenced by environmental factors	Measurement of H_2S, NH_3, NO_x concentrations
Galvanic sensors	Sensors generate electric current without external power. Gas passes through membrane causing redox reaction on electrodes, producing current	Autonomous operation. High selectivity (especially for O_2)	Limited lifespan (up to 2 years). Irreversible degradation	Toxic/ explosive gas leak control (CO, H_2S, NO_2), metallurgical and chemical industries, oxygen monitoring in processes

(continued)

Table 1 (continued)

Method	Principle of operation	Advantages	Disadvantages	Applications
Photoacoustic spectroscopy (PAS)	Gas absorbs modulated (often IR or laser) light, heats up causing microscopic pressure changes (sound waves), detected by sensitive microphones. Acoustic signal intensity proportional to gas concentration	High sensitivity (down to ppb). Compact sensors. High selectivity with correct wavelength choice	Requires stable conditions (affected by background noise). Sensitive to vibrations and acoustic disturbances	CO, CH_4, NH_3 detectors; biomedical devices; industrial monitoring
Ultrasonic methods	Ultrasound passes through or reflects off gas interfaces. Velocity or attenuation depends on gas physical properties (density, viscosity, temperature) which vary with target gas concentration	Non-contact measurement. Works in aggressive/dusty environments. Fast response	Low selectivity to specific gases (suitable for mixtures). Requires temperature compensation	Air composition control in ventilation, agriculture, smoke and exhaust analysis
Resonance methods	Gas volume changes the resonant frequency of an acoustic cavity. Measuring resonance shifts allows gas concentration estimation	High sensitivity to gas composition changes. Suitable for long-term monitoring	Calibration complexity. High system stability requirements	Air quality monitoring in industrial and atmospheric applications

(continued)

Table 1 (continued)

Method	Principle of operation	Advantages	Disadvantages	Applications
Catalytic thermal sensors (Combustion Catalysts)	Gas reacts with oxygen on catalyst surface (usually at elevated temperature). The exothermic reaction produces heat; temperature change is measured by thermal sensors (thermistors, thermocouples)	Simple design. Suitable for explosive gases. High stability under harsh conditions	Requires constant heating. High power consumption. Low selectivity—responds to all combustible gases	Measurement of CH_4, H_2, CO, propane concentrations
Thermal conductivity sensors (Thermal Conductometers)	Different gases have different thermal conductivities. A heated element (e.g., platinum wire) loses heat at rates depending on gas mixture composition; this effect is electrically measured	Non-reactive method. Good for binary gas mixtures	Low selectivity. Strong dependence on temperature and pressure	Laboratory analyzers, gas purity control in production
Calorimetric methods	Reaction between gas and reagent in a chamber releases or absorbs heat, measured as a temperature change in the medium	High sensitivity to concentration changes. Adaptable to specific media	Slow response. Limited repeatability of reactions	Control of CO_2, NH_3 content in sealed systems

4 Conclusion

As a result of the conducted study, a classification and comparative analysis of the main methods for measuring gas concentrations were carried out: physicochemical, optical, electrochemical, acoustic, and thermochemical. For each group, the principle of operation, typical devices, advantages and disadvantages were considered, as well as relevant mathematical models describing the operation of sensor systems.

Table 2 Physicochemical and mathematical analysis of gas concentration measurement methods

Method	Physical basis	Transformation equation	Components of the transformation equation
Gas chromatography	Mathematical model based on the equation for component distribution in the column:	$C(t) = \frac{1}{\sqrt{4\pi \cdot D \cdot t}} \cdot \exp(-\frac{(x - v \cdot t)^2}{4 \cdot D \cdot t})$	$C(t)$—concentration at output, D—diffusion coefficient, v—gas flow velocity, x—column length
Chemosensors (Semiconductor)	Change in electrical properties of semiconductor material upon interaction with gas molecules	$R = R_0 \cdot \exp(\frac{aC}{T})$	R—sensor resistance, R_0—resistance in clean air, C—gas concentration, T—temperature, α—empirical coefficient
IR Spectroscopy/ TDLAS/UV methods	Beer–Lambert–Bouguer law	$I = I_0 \cdot \exp(-\varepsilon \cdot C \cdot l)$	I_0—source intensity, I—transmitted intensity, ε—molar absorption coefficient, C—gas concentration, l—optical path length
Photoacoustic spectroscopy	Based on detection of acoustic waves generated by modulated light absorption by gas molecules	$P_a \infty I_0 \cdot \alpha(\lambda) \cdot C$	P_a—acoustic pressure, $\alpha(\lambda)$—absorption coefficient at wavelength λ, C—gas concentration
Amperometric Sensors	Electrochemical cell principle: target gas reacts at electrode surface producing electric current	$I = n \cdot F \cdot A \cdot D \cdot \frac{C}{\delta}$	I—current, n—number of electrons, F—Faraday constant, A—electrode area, D—diffusion coefficient, δ—diffusion layer thickness, C—gas concentration
Potentiometric sensors	Nernst equation	$E = E^0 + \frac{R \cdot T}{n \cdot F} \cdot \ln(C)$	E—potential, E^0—standard electrode potential, R—gas constant, T—temperature, F—Faraday constant, n—number of electrons, C—gas concentration

(continued)

Table 2 (continued)

Method	Physical basis	Transformation equation	Components of the transformation equation
Acoustic methods	Photoacoustic effect:	$S = k \cdot C \cdot P_{\text{mod}}$	S—signal amplitude, k—system sensitivity, C—concentration, P_{mod}—source modulation power
Ultrasonic methods	Ultrasonic velocity depends on gas properties	$\upsilon = \sqrt{\frac{\gamma \cdot R \cdot T}{M}}$	υ—sound velocity, γ—adiabatic index, R—universal gas constant, T—temperature, M—molar mass of mixture (depends on concentration)
Thermochemical methods	Catalytic combustion	$\Delta T = \frac{Q_{reaction} \cdot C}{C_p \cdot m}$	ΔT—temperature change, $Q_{reaction}$—combustion heat, C—gas concentration, C_p—heat capacity, m—catalyst mass

Table 3 Comparison of gas concentration measurement methods

Method	Sensitivity	Selectivity	response speed	Cost
Infrared spectroscopy	High	High	Medium	High
Chemosensors	Medium	Low	High	Low
Gas chromatography	Very high	High	Low	Very high
Electrochemical	Medium	Medium	High	Medium
Acoustic methods	High	High	Medium	Medium

The current state of development in this field was analyzed, noting trends toward miniaturization, automation, and integration with digital technologies. It was established that electrochemical methods represent the most balanced option for practical use due to their combination of high sensitivity, relatively low cost, and adaptability to various conditions.

The results of this study can serve as a theoretical foundation for the development of new generations of gas analyzers, as well as for the optimal selection of methods in practical applications across various industries.

References

1. Fuśnik Ł, Szafraniak B, Paleczek A, Grochala D, Rydosz A (2022) A review of gas measurement set ups. Sensors 22(7):25–57
2. Yunusa Z, Hamidon MN, Kaiser A, Awang Z (2014) Gas sensors: a review. Sensors and Transducers 168(4):61–75
3. Qiao S, He Y, Sun H, Patimisco P, Sampaolo A, Spagnolo V, Ma Y (2024) Ultra highly sensitive dual gases detection based on photoacoustic spectroscopy by exploiting a long wave solid state laser. Light: Sci Appl 13:100
4. Hu Q, Ai Y, Sima C, Sun Y, Feng Z, Li T, Tong C, Cao X, Wang W, Fan R (2024) High precision low cost mid infrared photoacoustic gas sensor using aspherical beam shaping for rapidly measuring greenhouse gases. Photonics 11(7):590
5. Cassidy DT, Reid J (1982) Atmospheric pressure monitoring of trace gases using tunable diode lasers. Appl Opt 21(8):1276–1282
6. Werle P, Slemr F, Maurer K, Kormann R, Mücke R (2019) Near and mid infrared laser optical sensors for gas analysis. Opt Lasers Eng 120:97–107
7. Brinkman U (1975) Air pollution studies using laser. In: Waidelich W (ed) Laser 75: opto-electronics conference proceedings, Munich, pp 268–270
8. Hall RI, Eckbreth AC (1984) Coherent Antistokes Raman spectroscopy: application to combustion diagnostics. In: Ready J (ed) Laser applications, vol 5. pp 114–118
9. Birnbaum M (1976) Laser excited fluorescence techniques in air pollution monitoring. In: Vehry EL (ed) Modern fluorescence spectroscopy, vol 1. pp 121–157
10. Okamoto T, Kawata S, Minami SA (1986) Photodiode array fourier transform spectrometer based on a birefringent interferometer. Appl Spectrosc 40(5):691–695
11. Wiens RH, Zwick HH (1977) Trace gas detection by correlation spectroscopy. In: Mattson JS, Mark HB, Mac Donald HG (eds) In: Infrared correlation and fourier transform spectroscopy. Marcel Dekker, New York, pp 119–190
12. Kay RB (1967) Absorption spectra apparatus using optical correlation for the detection of trace amounts of SO_2. Appl Opt 6(4):776–778
13. Perrner SS, Sulzmann KGP, Chen HK (1973) Tunable-laser derivative spectroscopy on spectral lines with combined Doppler and collision broadening. J Quant Spectrosc Radiat Transfer 13:705–707
14. Edner H, Sunesson A, Svanberg S, Uneus L, Wallin S (1986) Differential optical absorption spectroscopy system used for atmospheric mercury monitoring. Appl Opt 25(3):403–409
15. Newcomb GS, Millan MM (1970) Theory, applications and results of the long-line correlation spectrometer. IEEE Trans Geosci Electron GE-8(3):148–157
16. Platt U, Perner D, Patz HW (1979) Simultaneous measurement of atmospheric CH_2O, O_3 and NO_2 by differential optical absorption. J Geophys Res 84(C10):6329–6335
17. Platt U, Perner D (1983) Measurement of atmospheric trace gases by long path differential UV/visible absorption spectroscopy. In: Optical and laser remote sensing, springer series in optical sciences, Berlin, pp 97–105
18. Kreuzer LB (1971) Ultralow gas concentration infrared absorption spectroscopy. J Appl Phys 42(9):2934–2943
19. Melugailis J, Moradian A (1975) Tunable semiconductor diode lasers and applications. In: Jacobs S, Sargent III M, Scott JF, Scully MO (eds) Laser applications to optics and spectroscopy, Addison-Wesley, pp 1–51
20. Peterson NC, Kurylo MJ, Braun N, Bass AM, Keller RA (1971) Enhancement of absorption spectra by dye-laser quenching. J Opt Soc Am 61(6):746–750
21. Fung KN, Lin HB (1986) Trace gas detection by laser intracavity photothermal spectroscopy. Appl Opt 25(5):749–752
22. Lintz J, Simonett DS (Eds) (1976) In: Remote sensing of environment. Addison-Wesley, pp 694
23. Inaba H, Kobayasi T (1972) Laser-raman radar. Optoelectronics 4(4):101–123

24. Ward TH, Zwick HH (1975) Gas cell correlation spectrometer Gaspec. Appl Opt 14(12):2896–2904
25. Uchino O, Maeda M, Hirono M (1979) Applications of excimer laser to laser-radar observations of the upper atmosphere. IEEE J Quantum Electron QE-15(10):1094–1107
26. Fredriksson K, Galle B, Nyston K, Svanberg S (1981) Mobile lidar system for environmental probing. Appl Opt 20:4181–4189
27. Murray ER, Van der Laan JE (1978) Remote measurement of ethylene using a CO_2 differential-absorption lidar. Appl Opt 17(5):814–817
28. Grant WB (1986) He-Ne and cw CO_2 laser long-path systems for gas detection. Appl Opt 25(5):709–749
29. Abrean VJ (1980) Lidar from orbit. Opt Eng 19:489–493
30. Huang AA, Ellis EC, Games LV et al (1983) Remote plume measurements by use of correlation spectrometer and a differential absorption lidar. Sci Total Environ 29:87–99
31. Black FM, High LE, Lang JM (1980) Composition of automobile evaporative and tailpipe hydrocarbon emissions. J Air Pollut Control Assoc 30(11):1217–1221
32. Semenov AO, Savytskyi AY, Bisikalo OV, Kulakov PI (2018) Mathematical modeling of the two-stage chaotic colpitis oscillator. In: 2018 14th International conference on advanced trends in radioelecrtronics, telecommunications and computer engineering (TCSET), IEEE, pp 835–839. https://doi.org/10.1109/tcset.2018.8336327
33. Semenov A, Voznyak O, Osadchuk O, Baraban S, Semenova O, Rudyk A, Klimek J, Orazalieva S (2019) Development of a non-standard system of microwave quadripoles parameters. In: Romaniuk RS, Linczuk M (eds) Photonics applications in astronomy, communications, industry, and high-energy physics experiments 2019. SPIE, pp 80. https://doi.org/10.1117/12.2536704
34. Kucheruk V, Kulakov P, Storozhuk N (2016) Measurement of the number servings of milk and control of water content in milk on stall milking machines. In: Advances in intelligent systems and computing. Springer International Publishing, pp 435–447. https://doi.org/10.1007/978-3-319-48923-0_46
35. Semenov AO, Baraban SV, Osadchuk OV, Semenova OO, Koval KO, Savytskyi AY (2019) Microelectronic pyroelectric measuring transducers. In: IFMBE proceedings. Springer International Publishing, pp 393–397. https://doi.org/10.1007/978-3-030-31866-6_72
36. Dzhedzhula VV, Yu Yepifanova I (2024) Increase in the energy efficiency of the biogas production process through vibration activation of heat and mass exchange processes in the bioreactor. In: IOP conference series: earth and environmental science, vol 1415(1). pp 012017. https://doi.org/10.1088/1755-1315/1415/1/012017
37. Dzhedzhula VV, Yepifanova IY (2023) Neural network model of investment process of biogas production. In: IOP conference series: earth and environmental science, vol 1254(1). pp 012103. https://doi.org/10.1088/1755-1315/1254/1/012103

Business Environment of Ukraine in War Conditions: Risks, Stabilization Factors, and Management Methods

Olga Gonchar and **Serhii Moroz**

Abstract The results of a study of the business environment of Ukraine in the current conditions of war and global market transformations are presented. It was found that one of the few industries that demonstrated faster adaptation to new realities and working conditions is retail and services. Since 2023, there has been a partial restoration of the main indicators of enterprise activity, some enterprises have been relocated from regions of active hostilities to safer areas and have continued to work and provide services, including developing e-commerce. In addition, since 2023, entrepreneurs' expectations regarding positive opportunities for business development have been increasing. Factors that form threats to business activity have been identified and their impact has been ranked. The most significant factors for business development have been established as: export development, increased investment (in particular, in innovation), promoting growth in population demand for goods and services, optimizing the credit rate, promoting employment of the population. The dependence of the volume of activity on economic and financial factors is substantiated and a multifactor econometric model is built. The econometric model is proposed to be used to predict the impact of factors and choose risk management methods. Based on the correlation model, a forecast of the volume of sales of products and services by enterprises of Ukraine for 2025 is made, with a justification of the economic situation and changes in factor indicators.

Keywords Entrepreneurial activity · Business environment · Management methods · Forecasting · Risks · Threats · Market · Influencing factors · Diagnostics · Stabilization · Forecasting · Financial results · Efficiency

O. Gonchar (✉) · S. Moroz
Khmelnytskyi National University, Khmelnytskyi, Ukraine
e-mail: gonchar.o.i@khmnu.edu.ua

S. Moroz
e-mail: frostsv2015@ukr.net

A. Semenov et al. (eds.), *Data-Centric Business and Applications*, Lecture Notes on Data Engineering and Communications Technologies 272,
https://doi.org/10.1007/978-3-032-06608-4_21

1 Introduction

The activities of business structures and the state of the business environment in Ukraine are undergoing turbulent changes today. Russia's military aggression on the territory of Ukraine, the processes of globalization, internationalization and European integration directions of the country's economic development, total digitalization and the use of Internet resources require the use of new approaches to diagnostics and methods of managing business activities. At the same time, as a result of the war in Ukraine, there is a tendency to worsen the financial results of enterprises in most sectors of the national economy, which in the future may lead to an increase in insolvency and bankruptcy of domestic enterprises. Therefore, the current problem today is: establishing interdependencies between influencing factors and the results of enterprise activity, the efficiency of entrepreneurial activity; assessing the negative impact of factors on the main performance indicators of enterprises and the transformation of the business environment; forecasting risks and threats and identifying opportunities for improvement; substantiating scenarios for the development of events for the purpose of optimization.

Various aspects of the study of entrepreneurial activity are covered by the works of such scientists as: I. Ansoff, E. Beltyukov, Z. Varnalii, S. Dovbnya, P. Drucker, S. Ilyashenko, G. Kozachenko, O. Kuzmin, O. Lyashenko, P. Mykytyuk, E. Mnykh, V. Nyzhnyk, A. Cherep, O. Sheremet, Y. Shumpetkr, I. Chervyakov and others [1–5]. At the same time, it is relevant to deepen the study of conceptual foundations and develop applied tools to identify factors of transformation of the business environment in modern dynamic conditions; their impact on the effectiveness of entrepreneurial activity; diagnostics of risks and threats; justification of development prospects and promotion of competitiveness through the formation of an effective management system.

The purpose of the study is to investigate the main negative factors influencing the results of the functioning of enterprises in Ukraine in modern conditions. To establish a connection between them and the results of the functioning of enterprises. To assess the impact of factors on the performance of enterprises and to substantiate the tools for forecasting this process for the purpose of optimization.

2 Diagnosis of War Risks for Ukrainian Business and the Beginning of Stabilization of Business Conditions

Currently, there is no industry in Ukraine that has not suffered losses from Russian aggression, both direct and indirect, due to disruption of logistical connections, destruction of supporting infrastructure, loss of suppliers. Most enterprises in the East and South of Ukraine have suffered large-scale destruction of production infrastructure, therefore, the volume of sales of products, goods and services by enterprises of Ukraine in 2022 decreased by 2,583,774.9 million UAH, which amounted to almost

19% in relative terms. In 2023, a partial recovery was observed, the volume of industrial products sold increased by 23.04% compared to 2022, reaching 13,574,895 million UAH [6, 7].

According to the State Statistics Service, at the end of 2022, almost all aggregated types of industrial activity showed a decrease compared to 2021. In particular, the extractive industry decreased by 27.1%, the processing industry by 40.3% (food production – "minus" 19.4%, textile production – "minus" 31.5%, production of coke, oil refining products – "minus" 62.1%, production of chemicals and chemical products – "minus" 62%, metallurgical production – "minus" 62.1%, mechanical engineering – "minus" 42.6%), the supply of electricity, gas, steam and conditioned air decreased by 27.9% [6].

Military operations, mass migration of the population, a decline in production and economic activity were the main factors that negatively affected the activities of the transport sector. Thus, in 2022, cargo turnover decreased by 39.4%, and passenger turnover by 51.4% compared to 2021. In 2023, cargo turnover amounted to about 163.36 billion ton-kilometers, which is 2% less than in 2022 and 43.6% less than in 2021. Passenger turnover increased by 26.5% compared to 2022, reaching 29.5 billion passenger-kilometers [6]. Like most other industries, construction also suffered significant losses, which in 2022 decreased by 65.3% compared to 2021. In particular, the construction of buildings decreased by 62.5%, and the construction of engineering structures by 67.6%. In 2023, the construction industry demonstrated significant growth, with the volume of construction work performed increasing by 41.5% compared to 2022, reaching UAH 162.7 billion [8] According to the Ministry of Agrarian Policy, agro-industrial enterprises, which before the war provided about 40 percent of foreign exchange earnings from exports of goods, harvested more than 60 million tons of grain and oilseed crops in 2022, which is almost half the amount harvested in 2021 (108 million tons). The agricultural sector also showed positive dynamics in 2023, with Ukrainian farmers harvesting 78.7 million tons of grain and oilseed crops, which is 11% more than in 2022 [9].

One of the few industries that has demonstrated a faster adaptation to new realities and working conditions is retail trade, whose turnover in 2022 decreased by only 20.1 percent compared to 2021, which is the lowest rate of decline compared to other aggregated types of economic activity in the real sector. Retail trade also recovered faster than other industries in 2023, retail turnover amounted to over UAH 1.82 trillion, which is 30.5% more than in 2022 and 26.2% more than in 2021 [6].

Thus, in 2023, the Ukrainian economy showed signs of recovery in most sectors, although the level of activity remains lower than the pre-war level (Table 1).

The assessment of the financial performance of Ukrainian enterprises for the period 2018–2023 demonstrates significant fluctuations in key macroeconomic indicators, especially under the influence of the full-scale Russian invasion in 2022.

In 2022, a significant reduction in the volume of products, goods and services sold by enterprises was recorded—from UAH 13.6 trillion in 2021 to UAH 11.03 trillion, which is a decrease of 18.98% or UAH 2.58 trillion. This indicates deep disruptions in production, logistics and trade chains due to hostilities, loss of territories, destruction of infrastructure and reduced demand.

Table 1 Assessment of the financial results of Ukrainian enterprises by type of economic activity, 2018–2023 and their variability under the influence of the war (compiled by [6, 8, 11])

Indicators	2018	2019	2020	2021	2022	2023	Deviation 2023–2022
Sales volume of products, goods and services by enter-prises, UAH million	9,206,049.5	9,639,731	10,049,871	13,616,793	11,033,018	13,574,895	3,935,164
Profitability of operating activities of enterprises, %	8.1	10.2	6.2	12.6	3.3	8	– 2.2
Profitability of all activities, %	4.5	7.6	0.9	10.1	– 3.2	4.5	– 3.1
Financial result before tax, UAH million, incl	369,212.3	613,044	134,734.3	1,034,013	– 216,595	563,114.7	– 49,929.3
Agriculture, forestry and fisheries	70,728.2	93,501.3	81,997.9	238,464.1	85,702.9	61,637.8	– 31,863.5
Industry	154,138	166,753	-22,289.6	449,515.2	-228,844	105,344	-61,409.5
Construction	6,433.9	15,974.8	-2,047.8	8,370.6	– 17,584.5	3,983.1	– 11,991.7
Wholesale and retail trade; transport repair	86,290.5	129,113	45,272.0	145,861.9	50,349.4	177,839	48,725.8
Transport, warehousing, postal and courier activities	– 22,661.6	14,414.8	17,013.3	31,246.8	– 23,035.5	47,108.7	32,693.9
Temporary accommodation and catering	2,886.5	4,648.1	– 4,440.9	5,347.2	– 8,594.7	3,682.5	-965.5
Information and telecommunications	18,181.0	24,838.8	16,961.1	34,002.5	17,396.7	52,625.6	27,786.8
Financial and insurance activities	24,750.0	17,520.9	28,428.0	42,821.6	13,975.5	56,453.0	38,932.1
Real estate operations	16.3	47,317.5	– 36,027.1	16,729.3	– 60,191.7	3,211.7	– 44,105.8
Professional, scientific and technical activities	26,147.7	84,025.1	– 9,583.4	44,370.1	– 48,959.7	25,354.6	– 58,670.6
Administrative and support service activities	482.5	11,446.8	875.3	9,477.1	– 5,459.1	6,072.8	– 5,373.9

(continued)

Table 1 (continued)

Indicators	2018	2019	2020	2021	2022	2023	Deviation 2023–2022
Education	178.6	377.7	143.8	126.7	− 208.9	− 74.0	− 451.8
Healthcare	1,249.3	1,990.8	19,616.0	8,466.7	10,964.8	3,824.1	1,833.3
Arts, sports, entertainment and recreation	− 496.2	120.3	− 1,540	− 2,608	− 3,807.7	12,008	11,887.3
Other service activities	137.0	460.6	123.2	302.8	146.5	329.2	− 131.4

The financial result (balance) of enterprises before tax also suffered a significant drop—from a profit level of UAH 1,034,013 million in 2021 to a loss of (− 216,595) million in 2022. The level of profitability of operating activities of enterprises decreased from 12.6% in 2021 to 3.3% in 2022, and the profitability of all activities from 10.1% to (− 3.2%), i.e. the activities of most enterprises became unprofitable.

In 2023, there was a partial recovery of the main indicators of the activities of enterprises. The volume of sales of products, goods and services by enterprises increased to 13.6 trillion UAH, approaching the pre-war level of 2021. The financial result before tax became positive again and amounted to 563,114.7 million UAH, which indicates the restoration of the profitability of enterprises. At the same time, the level of profitability of operating activities increased to 8%, and of all activities to 4.5%, i.e. almost to the levels of 2018. An analysis of the financial results of enterprises indicates a deep recession in 2022 under the influence of the war and a gradual economic recovery in 2023.

However, the profitability structure remains uneven, a significant part of industries has not yet reached pre-war levels, and business risks remain quite high. A significant decline in the main financial indicators of enterprises has led to an increase in the number of loss-making enterprises in Ukraine (Table 2).

Analysis of the dynamics of the number of enterprises in Ukraine, their profitability and the volume of financial losses during 2015–2023 (Table 2) allows us to trace key changes in the structure and financial condition of domestic business, in particular in the context of a full-scale war.

In the pre-war period, the total number of enterprises demonstrated moderate growth: from 343,440 in 2015 to a maximum of 380,597 in 2019. In 2022, there was a sharp reduction in the number of enterprises—to 261,924, which is 108,910 units (or − 29.37%) less compared to 2021. In 2023, partial stabilization allowed

Table 2 Number of loss-making enterprises in Ukraine and the amount of economic losses for 2015–2022 (compiled by [6])

Years	Total number of enterprises	Number of loss-making enterprises	Enterprises that suffered losses	
			In % to the total number of enterprises	Financial result, UAH million
2015	343,440	90,325	26.3	815,447.8
2016	306,369	81,494	26.6	545,960.9
2017	338,256	92,006	27.2	399,554.9
2018	355,877	91,460	25.7	317,401.2
2019	380,597	98,955	26.0	268,310.2
2020	373,822	106,913	28.6	545,403.4
2021	370,834	99,013	26.7	232,826.5
2022	261,924	88,792	33.9	963,952.2
2023	307,852	88,969	28.9	490,528,829

the number of registered enterprises to be restored to 307,852, but this indicator still did not reach the pre-war level.

The number of loss-making enterprises in absolute terms in 2022 was 88,792, which is 1221 enterprises less than in 2021. At the same time, in relative terms, this indicator increased from 26.7% in 2021 to 33.9% in 2022. Thus, almost every third enterprise in Ukraine operated at a loss in 2022. The amount of losses in 2021 amounted to UAH 232,826.5 million, while in 2022 this indicator increased to UAH 963,952.2 million, their increase amounted to UAH 731,125.7 million, or 314.02%. In 2023, there was a decrease in the volume of losses to UAH 490,528.8 million, which indicates the gradual adaptation of the economy to new realities.

Unfortunately, businesses do not see any prospects for improving the economic situation and business environment in the near future. What the survey results show [11–14].

The UBI (Ukrainian Business Index) business activity index as of the beginning of 2025 is 38.6 (out of 100 possible). It has increased slightly compared to December 2024 (36.8), and the local minimum of September 2022 (33.9) (Fig. 1).

UBI shows the activity of business, its ability to increase turnover, create jobs. If the indicator is below 50, then this indicates negative business expectations for further developments. Therefore, in our opinion, the above-mentioned increase in the index should be assessed as "*business fatigue*" from uncertainty and a desire to intensify, rather than an improvement in the economic situation: the number of orders from customers remains low, but production volumes are growing and demand for labor resources is slightly recovering (Table 3).

Thus, at the beginning (February) of 2025, the business activity of enterprises in various sectors of the economy demonstrated heterogeneous dynamics. Industry recorded growth due to increased consumer demand and expanded production, which

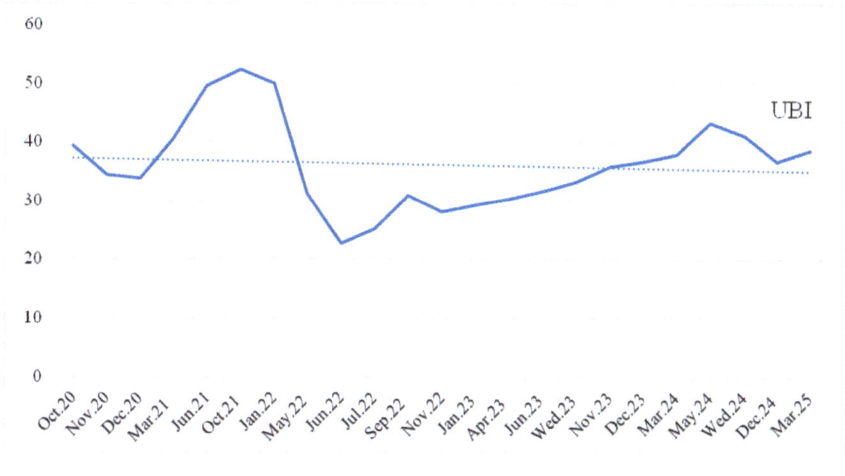

Fig. 1 Dynamics of the Ukrainian business activity index for the period from 2020 to the beginning (March) of 2025 (compiled by [6, 13, 15])

Table 3 Diffusion index of business activity - UBI (compiled by [6, 13, 15])

Sector	Weight structure for calculating IODA, %	Diffusion index (sectoral)		
		January 2025	February 2025	Deviations
Industry	35.2	42.0	50.2	8.2
Construction	3.2	37.2	44.7	7.5
Trade	26.2	40.0	49.2	9.2
Services	35.4	41.1	42.2	1.1
General	100	IODA (countrywide)		
		41.0	46.9	5.9

affected the increase in the volume of new orders, in particular export ones. At the same time, the growth in raw material prices slowed down, which contributed to a decrease in the rate of increase in the price of products.

The construction industry, although it demonstrated improved estimates, remained restrained due to seasonal factors and labor shortages. Despite expectations of increased purchases of materials and services, difficulties with the availability of contractors and the high cost of their services persisted.

Trade significantly improved its estimates due to the revival of domestic demand, which contributed to the growth of trade turnover and purchases. However, companies expected a further decrease in trade margins.

The services sector remained the most cautious in its forecasts due to rising business costs and staff shortages, which negatively affected employment expectations. Despite some improvement in overall expectations, businesses expressed more concern about the future financial and economic condition of their enterprises. Economic instability caused by external factors remains a significant challenge to development.

3 Identification of Threats and Factors Influencing Business Activities

Summarizing the conducted research, it is obvious that the most significant deterrent factor continues to be military actions and their consequences, which significantly limit the opportunities of enterprises to expand production and increase investments. Due to difficult conditions, many companies are forced to review their business strategies, optimize costs and adapt to changing circumstances.

Despite these difficulties, enterprises continue to look for ways to maintain competitiveness, implement anti-crisis measures and work to stabilize their activities, hoping for an improvement in the economic situation in the future.

Based on the results of the study of the state of the business environment and a survey of entrepreneurs [14–16], we will form a table with the main factors that prevent business structures from recovering and functioning successfully (Table 4).

In 2024, there was an increase in the share of companies that reported a significant increase in production volumes - such enterprises amounted to 5.1% compared to 2.8% in May. However, the overall picture remains ambiguous: although the weighted average indicator of work volumes to the level of the same period in 2023 reached 91.8%, almost half of the companies (48.5%) experienced a decrease in production volumes. Business prospects for 2024 remain subdued. Only 29% of companies expect growth, while 37.1% of respondents predict a deterioration in financial results. At the same time, the weighted average indicator of work volumes, compared to last year, improved to 98% compared to 94.1% in May. Despite this, concerns about the future are growing: 5.5% of companies believe that they will be forced to cease operations by the end of the year, which exceeds the current level of 2.8%. The overall forecasts of companies remain virtually unchanged since the previous survey.

Businesses are trying to retain staff: almost half of companies (48.5%) maintain a stable number of employees. However, the overall dynamics of staff turnover remains negative—the weighted average reduction is – 4.4%. However, there are positive

Table 4 Classification of the impact of war-related factors on business activities in Ukraine (February 2025) (compiled by [14–16])

Factors of influence	Percentage of respondents who indicated the influence of this factor, %
Uncertainty of the development of the situation in the country	59.1
Lack of a sufficient number of solvent clients in the country	56.7
Unforeseen actions of the state that may worsen the state of the business	56.0
Lack of a sufficient number of qualified employees	47.6
Lack of sufficient capital	42.5
High taxes and fees	36.9
Dumping of competitors, reduced profitability in the market	28.6
Fatigue of the owner or manager	26.6
Obstacles from regulatory and fiscal authorities	24.2
Unfair competition	18.7
Unavailability of credit funds, including programs 5,7, 9, etc	18.7
Delay of funds for settlements	16.3
Unavailability of grant programs announced by the Government of Ukraine	11.5

developments: in August, fewer companies were forced to reduce salaries (8.3% versus 12.9% in May) or send staff on forced leave (6.6% versus 7.5%). At the same time, the shortage of personnel remains an acute problem–29.1% of companies report a shortage of labor.

The topic of mobilization continues to cause concern in business. The vast majority of companies (80.6%) do not have a single booked specialist, and only 10.8% have used the possibility of electronic booking. Among them, 48.3% are satisfied with this mechanism, while 39.7% believe that the system needs improvement.

The task of our study is to assess the business environment and confirm the impact of negative factors cited by business representatives on the financial results of enterprises in Ukraine. The main generalizing indicator of enterprise activity is the volume of products, goods and services sold, on the value of which all performance indicators depend.

Let's examine the impact of effective demand. For which, in our opinion, it is necessary to analyze the amount of income and expenses of the population of Ukraine (Table 5).

Analysis of the data in Table 4 allows us to conclude that nominal incomes of the population increased, the amount of which increased even during the war in 2022 by

Table 5 Income and expenses of the population of Ukraine for 2018–2022, million UAH (compiled by [6])

Indicators	2018	2019	2020	2021	2022	Deviation 2022–2021
Income—total	3,248,730	3,744,060	4,045,191	4,863,519	5,257,066	393,547
including:						
Wages	1,529,367	1,758,588	1,832,294	2,211,720	2,750,808	539,088
profits and mixed income	572,065	678,217	740,334	964,851	692,862	– 271,989
Property income received	91,164	117,003	104,986	107,241	58,925	– 48,316
Social assistance	1,056,134	1,190,252	1,367,577	1,579,707	1,754,471	174,764
Expenditures and savings—total including:	3,248,730	3,744,060	4,045,191	4,863,519	5,257,066	393,547
Purchases of goods and services	2,884,971	3,356,993	3,520,546	4,305,425	3,922,442	-382,983
Property income paid	29,975	39,299	45,788	46,976	39,499	-7,477
Current income taxes	302,237	360,225	414,937	503,468	731,380	227,912
Disposable income	2,470,325	905,821	3,117,891	3,725,498	3,847,601	122,103
Real disposable income, as a percentage of the corresponding period of the previous year	110.9	109	104.5	109.2	85.9	-23.3

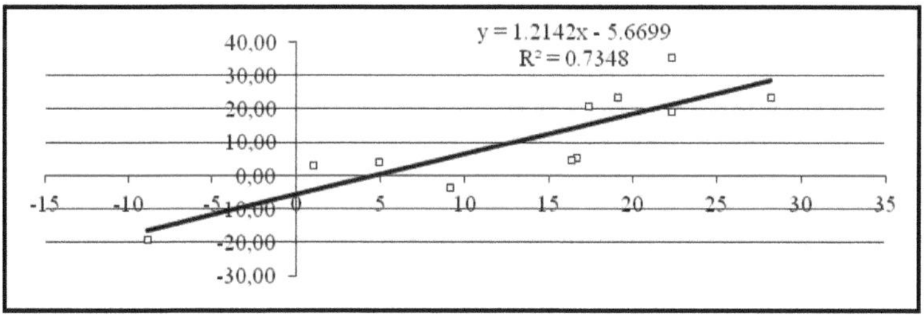

Fig. 2 Correlation between the volume of products, goods and services sold and the amount of costs for the purchase of goods and services

393,547 million UAH and amounted to 5,257,066 million UAH. Wages and social benefits and other current transfers received remain the main source of income for most of the population of Ukraine during 2018–2022.

Despite the increase in the total amount of expenses in 2022 by 393,547 million UAH, the amount of expenses for the purchase of goods and services decreased during this period by (– 382,983) million UAH, which was caused by an increase in the amount of current taxes on income, property and other paid current transfers by 227,912 million UAH. As a result, the amount of disposable income, taking into account the social transfers received in 2022, is less than the nominal amount, as evidenced by the dynamics of disposable real income, which decreased in 2022 by -23.3% and amounted to 85.9% compared to 2021.

Let us analyze the correlation between the volume of products sold and the amount of expenses for the purchase of goods and services (Fig. 2).

The equation of the linear dependence of the volume of products, goods and services sold and the amount of costs for the purchase of goods and services has the following form:

$$y = 1.2411 \quad x - 29.759 \tag{1}$$

The resulting model can be interpreted as follows: with an increase in spending on the purchase of goods and services by 1%, the volume of sold products, goods and services will increase by 1.24%. The closeness of the relationship between these indicators is 0.7605, which indicates a significant impact of the indicator of spending on the purchase of goods and services.

Another negative factor in doing business during the war, which entrepreneurs complained about, is problems with exporting products. Business concerns about the reduction in exports are objective. With the beginning of the war in Ukraine, air flights were canceled, the airspace over Ukraine was closed to civil aviation flights. This makes it impossible for both Ukrainian and foreign airlines to operate, in particular for cargo transportation.

The difficult situation with sea transportation is caused by the loss of port infrastructure and ports, Berdyansk and Mariupol, and the ports of Mykolaiv and Kherson have practically ceased their activities. The naval blockade caused by the war also negatively affected the volume of cargo transportation. Only in 2023 did Ukraine restore the sea corridor to the Black Sea ports of the Odessa region, which allowed to increase sea exports of goods.

Exports during 2022–2023 literally went on wheels, the main means of delivery were rail and road transportation through the western borders of Ukraine. But here too, problems arose due to the blockade of the borders by first carriers and then farmers from Poland, Hungary and Slovakia.

Let us analyze the dynamics of export–import, for which we will compile analytical Table 6.

Table 6 Foreign trade balance of Ukraine from 2005 to 2023 (million UAH) (compiled by [17])

Years	Nominal gross domestic product (GDP), million UAH	Export of goods and services		Import of goods and services		Balance (export—import)	
		Million UAH	Percentage to GDP, %	Million UAH	Percentage to GDP, %	million UAH	share in GDP, %
2011	1,316,600	707,953	53.8	– 779,028	– 59.2	– 71,075	– 5.4
2012	1,408,889	717,347	50.9	– 835,394	– 59.3	– 118,047	– 8.4
2013	1,454,931	681,899	46.9	– 805,662	– 55.4	– 123,763	– 8.5
2014	1,566,728	770,121	49.2	– 834,133	– 53.2	– 64,012	– 4.1
2015	1,979,458	1,044,541	52.8	– 1,084,016	– 54.8	– 39,475	– 2.0
2016	2,383,182	1,174,625	49.3	– 1,323,127	– 55.5	– 148,502	– 6.2
2017	2,982,920	1,430,230	47.9	– 1,618,749	– 54.3	– 188,519	– 6.3
2018	3,558,706	1,608,890	45.2	– 1,914,893	– 53.8	– 306,003	– 8.6
2019	3,974,564	1,636,416	41.2	– 1,947,599	– 49.0	– 311,183	– 7.8
2020	4,194,102	1,637,399	39.0	– 1,681,526	– 40.1	– 44,127	– 1.1
2021	5,459,574	2,224,704	40.7	– 2,286,067	– 41.9	– 61,363	– 1.1
2022	5,191,028	1,840,563	35.5	– 2,712,325	– 52.3	– 871,762	– 16.8
2023	6,537,825	1,868,904	28.6	– 3,237,014	– 49.5	– 1,368,110	– 20.9

Analysis of the data in Table 6 allows us to conclude that exports of goods and services from Ukraine increased during 2011–2021 and decreased in 2022. In 2022, there was a sharp decline in export volumes due to the full-scale invasion of the Russian Federation, the occupation of territories, destroyed logistics chains, the blockade of ports, and a general decline in economic activity. The volume of exports decreased by UAH 38,441 million or 17.27% compared to 2021 and amounted to UAH 1,840,563 million, which was only 35.5% of nominal GDP—the lowest figure in recent years. In 2023, despite a slight increase in exports to UAH 1,868,904 million, its share in GDP decreased to 28.6%. The key categories of Ukraine's commodity exports are food (mostly grain) and metals, which are highly dependent on ease of logistics.

A somewhat similar trend was observed with imports of goods and services into the territory of Ukraine, but growth was recorded throughout the entire study period from 2011–2023. Imports of goods and services in 2022 continued to grow—from UAH 2,286,067 million in 2021 to UAH 2,712,325 million in 2022, which amounted to 52.3% of GDP, in 2023 the volume of imports increased to UAH 3,237,014 million. This increase was due to the need to purchase weapons, fuel, generators, humanitarian aid and recovery goods.

The most indicative indicator of external economic imbalance is the foreign trade balance. It remained negative throughout the period under review, which indicates the predominance of imports over exports. In 2022, this indicator deteriorated to a record value—minus 871,762 million UAH, which is 14.2 times higher than the value in 2021 (– 61,363 million UAH). The share of the balance in GDP was – 16.8%, which is an unprecedented value for recent years. In 2023, the negative balance increased to -1,368,110 million UAH, which was 20.9% of GDP. This level of foreign trade deficit indicates a deep disproportion between domestic production and the needs of the economy, in particular during wartime, when export opportunities are limited, and imports perform a critically important function of supporting the viability of the state.

Let us analyze the correlation between the volume of products sold and the volume of exports of goods and services (Fig. 3).

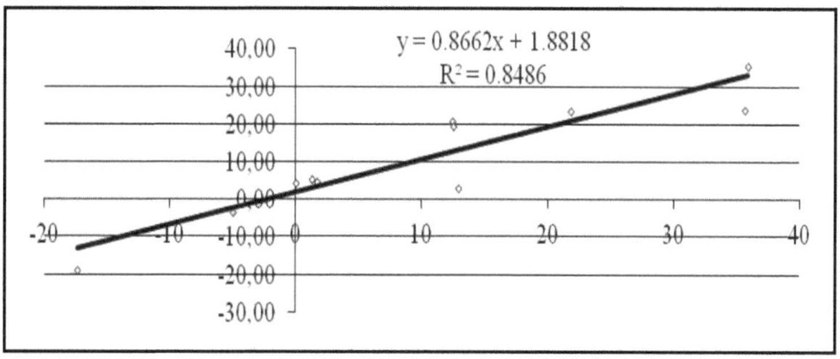

Fig. 3 Correlation between the volume of products sold and the value of exports of goods and services

The equation of the linear dependence of the volume of products sold and the value of exports of goods and services has the following form:

$$y = 0.8662 \quad x + 1.8818 \tag{2}$$

The resulting model can be interpreted as follows: with an increase in the value of exports of goods and services by 1%, the volume of sold products, goods and services will increase by 0.86%. The closeness of the relationship between these indicators is 0.8486, which indicates a high degree of influence of exports on the value of the volume of sales of goods and services.

The loss and damage caused by the war to production assets and infrastructure, disruption of supply chains have significantly weakened the potential of the labor market, which has led to the loss of employment opportunities. Since the beginning of the full-scale invasion, the State Statistics Service has not published data on the unemployment rate in Ukraine.

The demand for labor as a result of the war decreased due to: the inability to work in combat zones, a decrease in demand for goods and services, problems with logistics, and uncertainty about the future. Relocation and forced migration have led to a change in the structure of the labor market, as well as the structure of offers and saturation with specialists by region. The imbalance between labor supply and demand has also become significant (Table 7).

Analysis of the data in Table 7 allows us to substantiate the conclusion that significant transformations have occurred in the labor market of Ukraine during 2018–2023. In 2018–2019, the market was characterized by a labor shortage: demand exceeded supply by 49.8 thousand people in 2018 and by 128.9 thousand people in 2019. This indicates economic growth and a high level of demand for workers. However, starting in 2020, the situation changed dramatically, labor supply exceeded demand by 417.5 thousand people. This dynamics was caused by the economic slowdown, the COVID-19 pandemic, and a decline in business activity. In 2021, the excess supply reached its

Table 7 Indicators of labor market dynamics in Ukraine for 2018–2023 (million UAH) (compiled by [6, 17])

Years	Labor demand, thousand people	Labor supply, thousand people	Labor market situation	Number of employed population, thousand people	Employed at enterprises, thousand people	Output per employed person
2018	1,114.0	1,064.2	– 49.8	16,360.9	5,959.5	1,544.8
2019	1,153.3	1,024.4	– 128.9	16,578.3	6,407.5	1,504.4
2020	829.7	1,247.2	417.5	15,915.3	6,366.1	1,578.7
2021	705.8	1,191.0	485.2	15,610.0	6,391.7	2,130.4
2022	396.3	867.6	471.3	–	5,382.4	2,049.8
2023	329.4	483.2	153.8	–	5,052.3	2,686.9

maximum for the period—485.2 thousand people: labor demand amounted to 705.8 thousand people, while the supply reached 1,191.0 thousand people.

In 2022, the demand for labor decreased by 309.5 thousand people, or 43.85% - from 705.8 thousand people in 2021 to 396.3 thousand people. The supply also decreased - from 1191.0 thousand people to 867.6 thousand people, that is, by 323.4 thousand people or 27.15%.

In 2023, the demand in the labor market amounted to 329.4 thousand people and it continued to decline, the decrease was 66.9 thousand people 16.88%. The market stabilized after the initial shock of a full-scale war and is gradually adapting to structural changes in the economy. The labor supply in 2003 amounted to 483.2 thousand people, and during the year it decreased by 384.4 thousand. people, which amounted to 44.31%. There are fewer and fewer free workers in Ukraine: a significant part of Ukrainians are abroad, and about a million men and women are defending the country in the ranks of the armed forces of Ukraine. The number of employed workers at Ukrainian enterprises had a steady upward trend during 2018–2021, reaching its maximum in 2021 at 6391.7 thousand people. In 2022, the number of employed workers at enterprises decreased by (– 1,009.3) thousand people, (– 15.79)% to 5,382.4 thousand people. In 2023, the reduction in employed workers amounted to 330.1 thousand people or 6.13%. A significant part of qualified workers left the country and the rest participate in the defense of the state. The reduction in the number of qualified workers in the industrial sectors of Ukraine implies a shift in the composition of the total employed workforce in favor of workers with medium and low qualifications, as well as unskilled workers. In the structure of vacancies submitted through the employment service, the demand for unskilled workers prevails, while a significant part of the unemployed has higher education. This creates a situation of "professional imbalance" and complicates employment. The shortage of workers in key sectors, such as medicine, education, IT, where special training is required, is especially noticeable. This situation on the labor market is becoming a serious problem that affects the development of industries and the economy of Ukraine.

Let us analyze the correlation between the volume of products sold and the number of employees at enterprises (Fig. 4).

The equation of the linear dependence of the volume of products sold on the number of employees at enterprises has the following form:

$$y = 1.0914 \quad x + 14.053 \tag{3}$$

The resulting model can be interpreted as follows: with an increase in the number of employees at enterprises by 1%, the volume of sold products, goods and services will increase by 1.09%. The closeness of the relationship between these indicators is 0.2461, which indicates a significant impact on the volume of sales of goods and services of other factors, and this is obvious since the volume of sales is growing and the number of employees has changed insignificantly. Therefore, in our opinion, it is necessary to analyze the impact of changes in the amount of capital employed on the volume of sold products (Table 8).

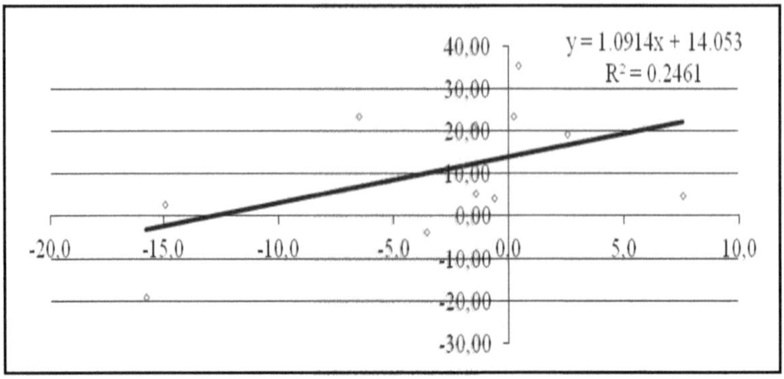

Fig. 4 Correlation between the volume of products sold and the number of employees at enterprises

Table 8 Working and non-working capital of enterprises, UAH million Working and non-working capital of enterprises, (million UAH) (compiled by [6, 17])

Years	Working capital of enterprises	Fixed capital	Total capital of enterprises
2011	2,506,632.2	2,439,813.2	4,946,445.4
2012	2,921,135.8	2,539,813.2	5,460,949.0
2013	3,069,041.6	2,639,813.2	5,708,854.8
2014	3,271,954.0	2,717,620.3	5,989,574.3
2015	4,108,602.7	3,960,148.9	8,068,751.6
2016	5,772,816.5	4,212,813.1	9,985,629.6
2017	5,650,817.4	4,303,282.8	9,954,100.2
2018	6,285,066.6	4,584,315.6	10,869,382.2
2019	6,640,282.4	4,944,512.9	11,584,795.3
2020	7,360,593.6	5,312,301.4	12,672,895.0
2021	8,411,902.5	5,650,750.3	14,062,652.8
2022	9,024,338.0	5,622,236.9	14,646,574.9
2023	10,193,400.9	6,241,270.3	16,442,568.9

Analyzing the data in Table 8, we conclude that there is a steady trend of growth in the total capital of Ukrainian enterprises during 2011–2023. During this period, its volume increased from UAH 4,946,445.4 million in 2011 to UAH 16,442,568.9 million in 2023, more than three times.

It should be noted that in 2022, despite the hostilities, the total capital of enterprises increased by UAH 583,922.1 million, which amounted to 104.15%, compared to 2021 and reached UAH 14,646,574.9 million. In 2023, growth continued even more intensively, its increase amounted to UAH 1,795,984.0 million, or 112.3%, which indicates the adaptation of the economy to new conditions and the preservation of business activity.

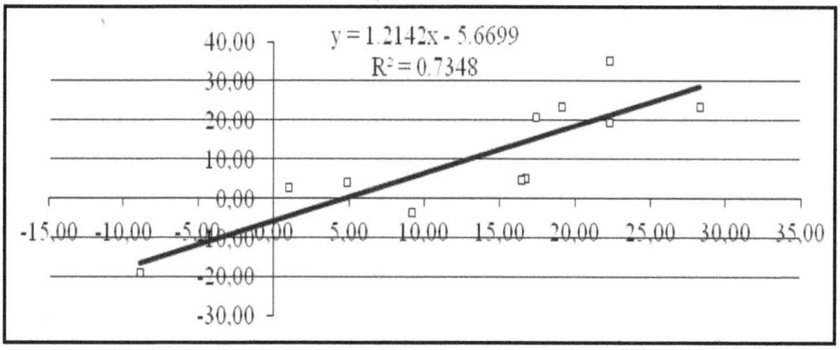

Fig. 5 Correlation between the volume of products sold and the amount of fixed capital at enterprises

Similar dynamics were observed with the volume of working capital of Ukrainian enterprises, growth was recorded throughout the entire research period from 2011–2023 and its growth in 2022 by 107.28%, which amounted to 612,435.5 million UAH, in 2023 it amounted to 10,193,400.9 million UAH and its relative growth was 112.26%.

Along with the general trend of capital growth, in 2022 there is a negative trend of reducing the value of non-current capital by 28,513.4 million UAH, which is explained by the destruction of fixed assets, loss of assets in temporarily occupied territories and a decrease in long-term investments. At the same time, in 2023, non-current capital increased to 6,241,270.3 million UAH, which in relative terms amounted to 110.99%, this indicates a partial restoration of business investment activity.

In general, the positive dynamics of aggregate capital indicates the flexibility of the Ukrainian economy, the ability of business to adapt to crisis conditions, and the gradual restoration of production potential.

Let's analyze the correlation between the volume of products sold and the amount of fixed capital (Fig. 5).

The equation of the linear dependence of the volume of products sold and fixed capital at enterprises has the following form:

$$y = 0.1792 \quad x + 8.6177 \tag{4}$$

The resulting model can be interpreted as follows: with an increase in the value of non-current assets by 1%, the volume of sold products, goods and services will increase by 0.1792%. The closeness of the relationship between these indicators is 0.02461, which indicates a very significant impact on the volume of sales of goods and services of other factors. Therefore, it is necessary to analyze the impact of changes in the value of working capital on the volume of sold products (Fig. 6).

The equation of the linear dependence of the volume of products sold and working capital at enterprises has the following form:

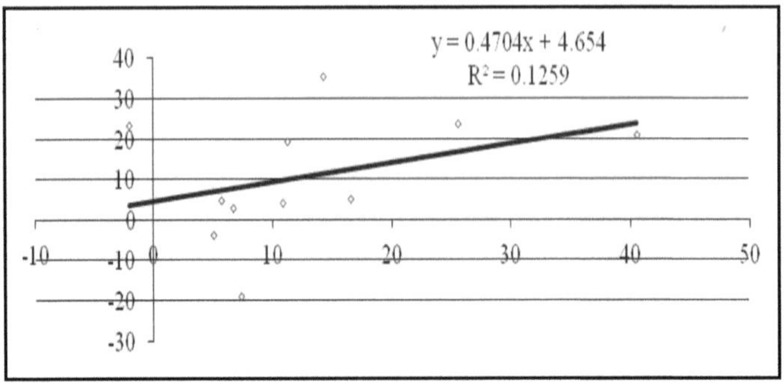

Fig. 6 Correlation between the volume of products sold and the amount of working capital at enterprises

$$y = 0.4704 \quad x + 4.654 \tag{5}$$

The resulting model can be interpreted as follows: with an increase in the value of current assets by 1%, the volume of sold products, goods and services will increase by 0.4704%. The closeness of the relationship between these indicators is 0.12459, which indicates a moderate impact on the volume of sales of goods and services of other factors.

According to Table 9, we conclude that there is a steady trend of filling the Ukrainian economy with credit resources.

Table 9 Dynamics of the lending rate and the amount of loans granted by commercial banks of Ukraine for 2011–2023 (million UAH) (compiled by [6, 17, 19])

Years	Loans granted	Loans to business entities	Loans to individuals	Average lending rate in national currency, %
2011	825,320	580,907	174,650	15.9
2012	815,327	609,202	161,775	18.4
2013	911,402	698,777	167,773	16.6
2014	1,006,358	802,582	179,040	17.6
2015	1,009,768	830,632	175,711	21.8
2016	1,005,923	847,092	157,385	18.5
2017	1,042,798	870,302	170,938	16.4
2018	1,118,618	919,054	196,634	19.1
2019	1,033,539	822,020	206,761	19.8
2020	963,664	752,503	199,556	14.1
2021	1,065,347	795,513	242,633	14.3
2022	1,036,129	801,327	209,944	22.0
2023	1,153,080	861,944	277,789	19.6

The total volume of loans granted increased from UAH 825,320 million in 2011 to UAH 1,153,080 million in 2023, their absolute growth amounted to UAH 397,760 million, in relative terms by 148.2%. It should be noted that the relationship between lending volumes and the average lending rate is not always observed, for example, in 2015 the rate reached 21.8%, but the volume of loans remained high—over UAH 1 trillion. In 2022, as a result of the war, the volume of loans granted decreased by UAH 29,218 million, which amounted to (–2.74%) compared to 2021. This decrease was due to a decrease in lending to —by UAH 32,689 million, which was caused by mass migration of the population abroad, loss of solvency and high risks. At the same time, loans to business entities increased by UAH 5814 million, which indicates increased business demand for financial resources to support operational activities, recovery and adaptation to new conditions. In 2022, there was also a significant increase in the average lending rate, it increased to 22.0% (the highest figure since 2015), however, in 2023 the rate decreased to 19.6%, which was accompanied by a significant revival of lending: the growth in the total volume of loans granted amounted to UAH 116,951 million, and their relative growth was 111.3%, lending to businesses increased by UAH 60,517 million, and to the population—by UAH 67,845 million. Such dynamics are the basis for concluding about the gradual stabilization of the banking system, growing confidence in the financial sector, and a moderate recovery of economic activity after the most acute phase of the war.

In the next step, we will analyze the correlation between the volume of products sold and the average lending rate (Fig. 7).

Then, the equation of the linear dependence of the volume of products sold and the average lending rate has the following form:

$$y = -0.2462x + 11.946 \tag{6}$$

The resulting model can be interpreted as follows: with an increase in the average credit rate by 1%, the volume of sold products, goods and services will decrease by 0.2462%. The closeness of the relationship between these indicators is 0.1279, which indicates an intermediate impact on the volume of sales of goods and services

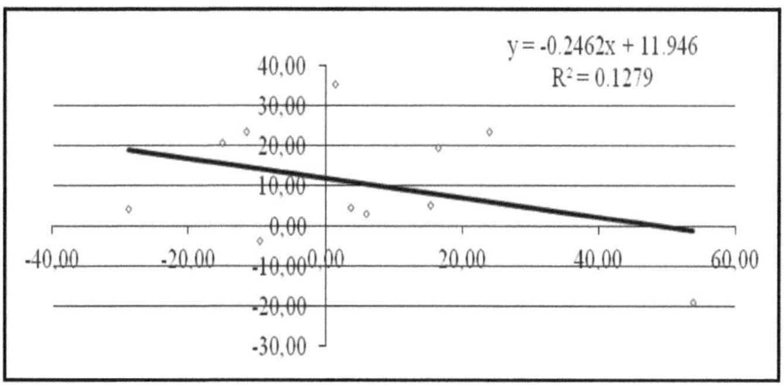

Fig. 7 Correlation between the volume of products sold and the average lending rate

of other factors. In order to assess the cumulative impact of factors on the volume of sold products by enterprises, we will build a multifactor econometric model, the initial data for which are given in Table 10.

Using the indicators in Table 10, it is possible to derive a regression equation for the dependence of the volume of sales of goods and services by enterprises in Ukraine on economic and financial factors. A multifactor econometric model was constructed:

$$Op = 12.11543 + 0.27929E + 0.37811Kn + 0.79726Bmn - 1.03247C\kappa - 0.20874\text{Ч}зn \quad (7)$$

where:

Op growth rate of sales of products and services by enterprises;
E growth rate of exports of goods and services;
Kn growth rate of capital of enterprises;
Bmn growth rate of expenditures on the purchase of goods and services by the population;
Cк rate of change of the average credit rate.
Чзn growth rate of the number of employees employed at enterprises.

With the multiple correlation coefficient R = 0.902603 and the coefficient of determination R^2 = 0.814691. The normalized value of the multiple correlation

Table 10 Dynamics of growth in the volume of sales of goods and services by enterprises and the main factors influencing it

Years	Growth rate of, %				Average lending rate in national currency, %	Growth rate of the number of employees at enterprises, %
	Sales of goods and services by enterprises	Export of goods and services	Capital of enterprises	Expenditures on the purchase of goods and services by the population		
2011	18.57	28.87	5.06	22.19	15.96	– 2.08
2012	5.31	1.33	10.40	16.65	18.40	– 1.46
2013	– 3.64	– 4.94	4.54	9.14	16.63	– 3.56
2014	2.97	12.94	4.92	0.98	17.62	– 14.96
2015	23.70	35.63	34.71	19.09	21.83	– 6.49
2016	20.90	12.45	23.76	17.35	18.54	– 1.50
2017	23.57	21.76	– 0.32	28.24	16.40	0.20
2018	19.44	12.49	9.20	22.25	19.10	2.52
2019	4.71	1.71	6.58	16.36	19.80	7.52
2020	4.25	0.06	9.39	4.87	14.10	– 0.65
2021	35.49	35.87	10.97	22.29	14.30	0.40
2022	– 18.97	– 17.27	4.15	– 8.90	22.00	– 15.79
2023	23.03	1.54	12.26	–	19.60	– 6.13

coefficient $R^2 = 0.682328$, i.e. 68.23% of the variation in the growth rate of the volume of sales of products and services by enterprises is due to the variation of *E*, *Кп*, *Втп*, *Ск*, *Чзп*, and other factors account for 31.77% of the variation. Since the calculated value is greater than F_{α, k_1, k_2}, then with a probability of $p = 0.95$ it can be stated that the econometric model is adequate to the statistical data.

Thus, interpreting this equation, we can say:

- with an increase in the growth rate of exports of goods and services by 1%, the growth rate of sales of products and services by enterprises will increase by 0.27929%,
- with an increase in the level of capital of enterprises in Ukraine by 1%, the growth rate of sales of products and services by enterprises will increase by 0.37811%;
- with an increase in the growth rate of expenditures on the purchase of goods and services by the population of Ukraine per year by 1%, the growth rate of sales of products and services by enterprises will increase by 0.79726%,
- with an increase in the average interest rate on loans by 1%, the growth rate of sales of products and services by enterprises will decrease by 1.03247%.
- with an increase in the growth rate of the number of employees at enterprises by 1%, the growth rate of the volume of sales of products and services by enterprises will increase by 0.2087%.

In general, if we simulate that the growth of working capital, the number of employees at enterprises, and the costs of purchasing goods and services by the population in Ukraine is 0%, while the change in the value of exports is 0%, and the interest rate on loans will not change, then the growth rate of the volume of sales of products and services by enterprises can be 12.11543%.

4 Substantiation of the FORECAST of the Transformation of the Business Environment of Ukraine and the Effectiveness of Management Methods

Summarizing the results of research on trends in socio-economic changes, it can be noted that in the coming years, the pace of economic growth in Ukraine will remain moderate and will significantly depend on security conditions and external factors. Different institutions give somewhat different forecasts, which reflects the level of uncertainty and complexity of the economic and political situation in the country.

According to the latest report of the European Bank for Reconstruction and Development (EBRD) [20], in 2025, the growth of Ukraine's gross domestic product will slow down to 3.3%. Previously, this figure was forecasted higher—at 4.7%. At the same time, the financial institution sets a certain optimistic scenario for 2026: under conditions of cessation of hostilities, the country's economy will grow by 5%.

The World Bank provides somewhat different calculations: according to its forecasts, if active hostilities continue, GDP growth in 2025 will be only 2% (for comparison, in 2024 this figure was 3.2%). At the same time, global economic and production growth in the coming years, according to the forecasts of this institution, will remain stable at the level of 2.7% [20]. The National Bank of Ukraine also worsened its GDP growth forecasts for 2025—from 4.3% to 3.6%. This reflects the moderate pessimism of the main realist economists regarding the near-term prospects of the country's economy and reinforces the conclusion about the significant impact of security risks and external conditions on the trajectory of Ukraine's economic and production recovery.

The European Commission has revised its forecasts for Ukrainian exports of goods and services in the coming years, noting a deterioration in this outlook. According to European experts, the near-term dynamics of exports will remain rather weak [21, 22]. This is due, in particular, to the reduction in industrial production, especially in the most energy-intensive sectors, which have suffered from increased energy prices. It is also predicted that agricultural exports will decline due to a decrease in stocks after a poor harvest in 2024, caused by adverse natural conditions. According to the forecast, in 2025, export volumes will grow by only 1.8%, and in 2026 this figure will improve to 11.4%. At the same time, imports will increase by 5.5% in 2025 and by 6.0% in 2026, which will worsen the foreign trade balance and somewhat slow down the overall pace of economic recovery.

According to the representative of the Ministry of Economy of Ukraine Taras Kachka, the government maintains a balanced forecast: foreign exchange earnings from exports in 2025 will increase by approximately 7% compared to the previous year—despite lower expected yields of late crops. Thus, realistic forecasts for export earnings in 2025 have varied from a slight increase of 1.8%, as noted by the EC, to an ambitious increase of 7%, as predicted by the Ministry of Economy. However, external challenges, on the eve of EU policies, global prices and domestic agricultural conditions, may damage earnings or exacerbate the deficit of the foreign trade balance.

According to the results of the Info Sapiens study, a noticeable improvement in consumer sentiment among the Ukrainian population was recorded. The consumer sentiment index at the beginning of 2025 reached 84.3 points, which is 8 points higher than the previous month. This positive trend is observed in all components of this index, reflecting the improvement in the perception of the economic and financial situation by households and businesses [22, 23].

The current situation index increased to 60.5 points, its increase compared to February 2025 was 7.2 points. In particular, the indicator of the current personal financial situation increased by 8 points—to the level of 54.6 points, and the index of the feasibility of large purchases increased by 6.4 points and reached 66.4 points. This gives grounds to conclude that consumer opportunities and households' willingness to spend have improved.

The economic expectations index showed a pronounced positive movement, its growth amounted to 8.6 points and reached 100.2 points. In terms of its structural components, the most noticeable growth was recorded in expectations for the country's economy in the next five years—this indicator increased by 13 points and reached the level of 117.4 points. At the same time, the index of expected economic development in 2025 improved by 11.1 points and is 87.9 points, and the indicator of expected improvement in personal financial situation increased by 1.7 points–to 95.3 points.

Thus, the analysis of consumer sentiment indicates an improvement in the economic expectations of Ukrainians. Respondents expressed moderate optimism about the near-term prospects of the economy and the standard of living, which gives grounds to consider these changes as a manifestation of gradual stabilization and improvement of socio-economic conditions in the country.

In the National Bank of Ukraine, the main instrument of influence on the level of inflation remains the discount rate. Its size determines the cost of credit resources and the level of profitability of deposits, which, in turn, is directly reflected in economic processes in the country. The formation of this rate is one of the most effective mechanisms for ensuring price stabilization and supporting the economy in conditions of increased risks.

From March 7, 2025, the National Bank increased the discount rate to 15.5%. This was done in order to reverse the upward inflationary trend and ensure a gradual slowdown in price growth this year. At the same time, in May the inflation rate exceeded the discount rate, which reflects the constant pressure of temporary and structural factors—in particular, insufficient harvest and increased production costs.

According to the National Bank's forecasts, the coming months will be the peak of inflation, after which a gradual decrease in the rate of price growth will begin. By the end of 2025, the inflation rate should slow down to 8.4%, and by 2026—to 5%. These improvements will be possible thanks to a set of monetary measures, an improvement in the situation in the agricultural sector, a strengthening of the energy system, a reduction in the budget deficit, and more moderate external price pressures.

Labor market research suggests a significant imbalance between the economy's need for labor and the level of labor supply in the coming years. According to forecast calculations, in the period from 2025 to 2030, labor demand will grow at a faster pace—up to 10%, while the supply of labor will expand more slowly—up to 7%. This will lead to increased competition for personnel and create tension in the labor market.

This situation will be most pronounced in the agricultural and industrial sectors of the economy, where the gap between the need and the ability to satisfy this need will reach 30–40%. These imbalances will become a significant obstacle to ensuring economic growth and achieving the goal of increasing gross domestic product by 4% annually.

A study by leading scientists, presented at the Ministry of Economy of Ukraine, shows that the main factor that most significantly hinders the activities of enterprises today is the lack of qualified personnel. This phenomenon encourages businesses to improve remuneration conditions and compete for employees. As a result, in 2024, the level of real salaries increased by more than 15% [according to 6, 12, 22, 23].

This growth is still ongoing, reflecting the severity of the problem and the willingness of employers to improve material conditions in order to attract the most valuable employees.

The situation with personnel selection is complicated by migration processes and mobilization, which significantly narrow the possibilities of hiring. At the same time, there is an improvement due to the expansion of sources of replenishment of labor resources: enterprises are paying more attention to students, people with disabilities, veterans and pensioners—these groups were previously under-involved in the labor market. This makes it possible to mitigate the personnel shortage and ensure the stable operation of many sectors of the economy.

The migration factor was also taken into account in the calculations: the forecast is based on the assumption of the return of at least a third of compatriots who left the country after February 24, 2022.

At the same time, these migration flows are insufficient to cover the need for labor resources. In the conditions of such an imbalance, the task arises of developing and implementing systemic measures to increase labor productivity in the most critical industries and expand the participation of economically inactive segments of the population in labor activity.

Considering the above scenarios for the development of the Ukrainian economy, we will try to make a forecast of sales volumes by Ukrainian enterprises for 2025. For this, we will form a table with the values of the model indicators and their justification (Table 11).

Taking into account the forecast values and the derived regression equation of the dependence of the volume of sales of goods and services by enterprises in Ukraine on economic and financial factors and the constructed multifactor econometric model (Eq. 7), we will calculate the forecast:

$$\begin{aligned} Op = 12.11543 + 0.27929 \times 6.5 + 0.37811 \times 5.5 + 0.79726 \\ \times 6.5 - 1.03247 \times (-1) - 0.20874 \times (-2.5) = 22.74697 \end{aligned} \quad (8)$$

Thus, the forecast growth rate of sales of products and services by enterprises of Ukraine during the calendar year for the given value of factor data will be 22.74697%, relative to 2023. In absolute terms, the increase will be 3,087,877.29 million UAH in 2024, and 3,790,275.81 million UAH in 2025, the forecast value of the sales of products and services by enterprises in 2025 will be 20,453,048.1 million UAH.

5 Conclusion

The results of the study confirmed the business assumptions about the main negative factors that formed during the war and affect business activities. The constructed correlation-regression model showed the most significant impact of the growth rate of expenditures on the purchase of goods and services by the population and the

Table 11 Initial data of indicators for the model of forecasting the volume of sales of goods and services by enterprises in Ukraine

Indicator	Justification for indicator change	Predicted value of indicator change, %
E—growth rate of exports of goods and services;	Exports of goods will decrease according to the above forecasts and the trade balance will worsen	6.5
Kn—growth rate of capital of enterprises	According to the above forecasts, Ukrainian enterprises will be characterized by a capital deficit	5.5
Bmn—growth rate of expenditures on purchase of goods and services by the population;	The level of wages and the amount of social payments will increase, but not significantly	6.5
Ск—rate of change of average credit rate	The NBU discount rate will remain at a relatively stable level, changes in terms of reduction are insignificant, therefore the reduction in the loan rate will also be insignificant	-1
Чзп—growth rate of number of employed workers at enterprises	The population will decrease, but the forecast unemployment rate will also decrease, therefore the number will change insignificantly	-2.5

growth rate of employment of employees on the expansion of sales volumes of products and services by Ukrainian enterprises. This is explained by the fact that the stronger the positions of households and enterprises, the less they rely on external financing and credit.

Somewhat weaker, but still emphatically positive, affect the results of business activities and the pace of export development and the pace of change in the capitalization of the enterprise. At the same time, there is a feedback loop between the size of the average credit rate and the pace of expansion of sales volumes—with increasing interest rates, this indicator decreases. This worsens the business's ability to provide the necessary level of working capital and expand production.

Using the above model makes it possible to develop effective scenarios for improving the economic and production situation of enterprises and business in general. On its basis, a forecast of sales volumes of products and services by Ukrainian enterprises for 2025 was formed, taking into account the current state of the economy and changes in factor indicators.

This study concluded that justified structural transformations are necessary to ensure economic recovery and strengthen the industry, in particular:

- Development of a specialized financing mechanism and preferential lending for projects to expand production and improve production technology at economically justified interest rates, for example, expansion of the "5–7–9" program;
- Provision of training and advanced training programs for personnel to increase productivity and expand opportunities for hiring new employees;
- Co-financing projects to expand sales and improve logistics, in particular, the development of new transport corridors and unblocking the busiest checkpoints;
- Development of support programs for startups and innovative enterprises to ensure a constant expansion of the range and improvement of product quality;
- Application of targeted tax benefits to those enterprises that create new high-paying jobs and improve the employment level of the population;
- Development of a platform for cooperation between business, educational institutions and research to improve the skills of specialists and reveal the best practices.
- Restoration and activation of exports by improving logistics and expanding transport corridors.

Thus, expanding the volume of sales of products and services is currently a national task, which can only be solved by coordinated efforts of business and the state. This task is systemic in nature and affects the economy of every Ukrainian, and therefore requires substantiated research and development of targeted programs for the restoration and strengthening of the Ukrainian economy.

References

1. Brych V, Manzhula V, Borysiak O, Liakhovych G, Halysh N, Tolubyak V (2020) Communication model of energy service market participants in the context of cyclic management city infrastructure. In: 2020 10th International conference on advanced computer information technologies (ACIT), pp 678–681. https://doi.org/10.1109/ACIT49673.2020.9208902
2. Buiak L, Matiukh S, Gonchar O, Yemchuk L, Dzhulii L, Bilorusets L (2023) Methods and models in management decision-making. In: 2023 13th International conference on advanced computer information technologies (ACIT), Wrocław, Poland, pp 284–288. https://doi.org/10.1109/ACIT58437.2023.10275466. https://ieeexplore.ieee.org/document/10275466
3. Sharko M, Petrushenko N, Gonchar O, Vorobyova K, Zakryzhevska I (2022) Information support of intelligent decision support systems for managing complex organizational and technical objects based on markov chains. In: CEUR Workshop proceedings this link is disabled, pp 986–998
4. Oleynikova L, Cherep O, Gonchar O, Cherep A, Iryna D, Anoshina S (2024) Experience of developed countries of the world regarding the formation of investment and innovation policy of ukraine taking into account the experience of leading countries. In: Mansour N, Bujosa Vadell LM (eds) Green finance and energy transition. Contributions to Finance and Accounting. Springer, Cham. https://doi.org/10.1007/978-3-031-75960-4_32
5. Voynarenko M, Gonchar O (2018) Transformation of socio-economic systems under conditions of integration transformations. J Europ Econ 2, том 17,—Режим доступу http://jeej.tneu.edu.ua/index.php/enjee/article/view/1336
6. State Statistics Service of Ukraine (2024). https://www.ukrstat.gov.ua

7. The Human Capital Index 2020 Update: Human Capital in the Time of COVID-19/World Bank (2020). https://openknowledge.worldbank.org/handle/10986/34432
8. Ukrainian Council of Shopping Centers (2024). https://www.ucsc.org.ua/obsyagy-budivnyczt va-v-ukrayini-zrosly-na-235-u-2024-roczi/, https://www.ucsc.org.ua/obsyagy-budivnycztva-v-ukrayini-zrosly-na-235-u-2024-roczi/
9. Russian forces shell center of city: interfax (2024). https://en.interfax.com.ua/ https://en.int erfax.com.ua/
10. Sharko M, Gonchar O, Tkach M, Polishchuk A, Vasylenko N, Mosin M, Petrushenko N (2022) Intellectual information technologies of the resources management in conditions of unstable external environment. In: Lecture notes on data engineering and communications technologies Том 77, pp 519–533. https://doi.org/10.1007/978-3-030-82014-5_35, https://link.springer.com/chapter/10.1007/978-3-030-82014-5_35
11. Business r Activity Expectations Index (BAEI) in May 2024. [Електронний ресурс].— Режим доступу: https://bank.gov.ua/en/news/all/biznes-zberigaye-pozitivni-otsinki-schodo-rezultativ-svoyeyi-diyalnosti--rezultati-opituvannya
12. Sustainable Development report 2024/Includes the SDG Index and Dashboards/The Decade of Action for the Sustainable Development Goals. https://ucap.io/wp-content/uploads/-report.pdf
13. Results of the study of the state of business in Ukraine: https://business.diia.gov.ua/analytics/research/rezultaty-doslidzhennia-stanu-biznesu-v-ukraini
14. Olga G, Brych V, Moroz S, Bohatchyk L (2024) Factors that shape business efficiency in the conditions of the war. In: 2024 14th International Conference on advanced computer information technologies (ACIT), Ceske Budejovice, Czech Republic, pp 365–369. https://doi.org/10.1109/ACIT62333.2024.10712593 https://www.researchgate.net/public ation/384992400_Factors_that_Shape_Business_Efficiency_in_the_Conditions_of_the_War
15. Action: Business: https://business.diia.gov.ua/analytics/research/rezultaty-doslidzhennia-stanu-biznesu-v-ukraini-v-hrudni-2024-roku
16. MVF pohirshyv prohnoz zrostannia ekonomiky Ukrainy na 2025 rik. https://www.unian.ua/economics/other/mvf-pogirshiv-prognoz-zrostannya-ekonomiki-ukrajini-na-2025-rik-129 60597.html
17. Ministry of Finance. https://index.minfin.com.ua/ua/economy/gdp/eximp/
18. Olga G, Brych V, Moroz S, Bohatchyk L (2024) Factors that shape business efficiency in the conditions of the war. In: 2024 14th International conference on advanced computer information technologies (ACIT). https://doi.org/10.1109/ACIT62333.2024.10712593
19. National Bank of Ukraine: Financial sector statistics https://bank.gov.ua/ua/statistic/sector-fin ancial#1ms
20. Forbes Ukraine: EBRD downgrades Ukraine's GDP growth forecast for 2025. https://forbes.ua/news/ebrr-pogirshiv-prognoz-zrostannya-vvp-ukraini-na-2025-rik-z-35-do-33-13052025-29682
21. European Commission: Directorate-General for Economic and Financial Affairs. https://eco nomy-finance.ec.europa.eu/document/download/e9de23c8-b161-40d0-9ad7-e04a25500023_en?filename=ip318_en.pdf#page=178
22. Stan sfery posluh v Ukraini pid chas viiny. [Elektronnyi resurs]. https://hub.kyivstar.ua/articles/chy-ye-perspektyvy-dlya-rozvytku-sfery-poslug-v-ukrayini-pid-chas-vijny
23. Info Sapiens: https://www.sapiens.com.ua/ua/publication-single-page?id=346

Transforming Financial Crime Prevention through Artificial Intelligence

Oleksii Ilchenko

Abstract In an era where monetary transactions can be executed instantaneously, the potential for financial exploitation continues to escalate every year. Technological advancements have made it easier to manage banking activities remotely, whether from home, via smartphones, or abroad. Concurrently, there has been a significant rise in financial crimes, often linked to terrorism financing, drug trafficking, illicit migration, corruption, and tax evasion. For financial institutions, it has become increasingly challenging to process the vast amounts of data they receive daily. This is where Artificial Intelligence (AI) and advanced algorithms can play a crucial role in curbing such illicit activities. However, the focus should not be solely on AI; Know Your Customer (KYC) procedures remain equally vital to maintaining high banking standards and efficiently detecting suspicious transactions. This study provides a literature analysis intended to serve as a guideline for further research into the use of AI in preventing financial crimes.

Keywords Artificial intelligence (AI) · Financial crime · Money laundering (ML) · Know your customer (KYC) · Machine learning · Fraud prevention

1 Introduction

Financial crimes have persisted for centuries; the only aspect that has evolved is the methodology employed. More than two thousand years ago, Chinese merchants were already known to launder money by sending funds abroad or purchasing movable assets to conceal their wealth from the government [27]. Pirates operating between the sixteenth and eighteenth centuries are similarly regarded as early pioneers of money laundering, particularly in the trade of gold [27].

The ongoing battle against financial crime has continued for over fifty years [23]. For example, Al Capone, a notorious gangster, became the first high-profile criminal

O. Ilchenko (✉)
Department of Information Technology, Comenius University in Bratislava, Bratislava, Slovakia
e-mail: alexeyilchenko83@gmail.com

A. Semenov et al. (eds.), *Data-Centric Business and Applications*, Lecture Notes on Data Engineering and Communications Technologies 272,
https://doi.org/10.1007/978-3-032-06608-4_22

to be convicted of tax evasion in 1931, as the authorities could only prove his failure to pay taxes on illegal earnings [23]. Criminals have consistently adapted their tactics more rapidly than regulatory bodies and industries can respond [23]. As Markus Bender noted in an interview with Springer Professional, the structure of traditional banks often impedes their ability to respond swiftly to evolving criminal activities. Banks typically lack the agility necessary to keep pace with increasingly sophisticated fraud schemes [16]. Moreover, simply expanding compliance departments does not guarantee greater effectiveness, nor is it a sustainable long-term solution [16].

The global financial crisis demonstrated that fraud tends to increase during economic downturns [12]. According to a 2014 report by the Association of Certified Fraud Examiners (ACFE), organizations lose approximately 5% of their annual revenue to fraud, amounting to an estimated USD 3.7 trillion globally [12]. The United Nations Office on Drugs and Crime (UNODC) further estimates that 2–5% of global GDP is laundered each year, corresponding to EUR 715 billion to EUR 1.87 trillion [16]. Financial crimes have far-reaching impacts on economies and societies, diverting resources that could otherwise fund education, healthcare, or tax relief. In 2017, for instance, the German Commerzbank identified transactions exceeding EUR 100 million with the support of AI-based monitoring tools [46].

Money laundering is a global phenomenon, though its prevalence varies by country [29]. Notably, some jurisdictions benefit by hosting laundered funds originating from crimes committed elsewhere [29]. Professional intermediaries such as accountants, lawyers, and real estate agents often facilitate money laundering operations in complex ways [4]. Understanding how illicit funds are laundered requires analyzing the most common methods used worldwide. Many countries harbor criminal networks, such as the Cosa Nostra in the United States, the Chinese Triads and Japanese Yakuzas in Asia, and the Sicilian Mafia in Europe [4].

Criminals launder money primarily to eliminate incriminating evidence and prevent the confiscation of their assets during investigations [15]. The laundering process typically operates in cyclical stages [61]. Illicit proceeds are first distanced from their criminal origin, then channeled through multiple transactions involving banks, casinos, businesses, or shipping routes, both domestically and internationally [61]. Ultimately, the funds re-enter the economy appearing legitimate and untraceable [61]. Each stage involves intricate networks that make detection and tracing highly challenging [31]. Criminals continually adapt to emerging technologies, rarely relying on a single method for long, which contributes to class overlap and concept drift in fraud detection systems [58].

Despite technological advances and industry efforts, incidents of fraud and money laundering remain widespread [52]. Financial institutions continually seek to strengthen customer trust and ensure secure transactions [11]. However, vulnerabilities persist, providing loopholes that criminals exploit [11].

The attacks of September 11, 2001, highlighted the critical importance of anti-money laundering (AML) and counter-terrorist financing measures across jurisdictions [26]. Similarly, the 2015 Paris attacks underscored the need for robust Know Your Customer (KYC) and AML systems [30]. In that case, a European branch of a North African bank, which had recently implemented advanced AML technology, successfully blocked a suspicious transaction by an individual later linked to the attacks [30]. However, the bank's headquarters, lacking comparable systems, failed to investigate when the customer protested the blocked transfer, ultimately enabling the transaction to proceed [30]. This oversight tragically contributed to an attack that resulted in more than 120 deaths and hundreds of injuries [30]. Such cases illustrate the critical role that comprehensive KYC and AML frameworks play in supporting risk and compliance teams [30]. Unfortunately, as of 2016, only a few banks globally had adopted advanced analytics and AI to enhance these functions [30].

Traditional methods of managing the volume and complexity of financial transactions are labor-intensive, costly, and prone to errors. This is where artificial intelligence becomes indispensable [62, 63]. During a panel discussion at the University of Vienna, Ewald Nowotny, Governor of the National Bank of Austria, emphasized that effective KYC is fundamental, asserting that combating financial crime without KYC is impossible. He also noted that new technologies, such as blockchain and cryptocurrencies, are transforming the financial crime landscape by increasing anonymity and reducing traceability.

AI and machine learning leverage algorithms to detect behavioural patterns and flag anomalies or suspicious transactions [45]. These systems adapt rapidly as patterns evolve [45]. The principal advantage of AI is its capacity to process vast amounts of data far more quickly than human analysts ever could [8]. However, AI systems cannot operate without human oversight; they lack the intuition and contextual understanding necessary for nuanced decision-making [8]. As Mark Wiesman observed, "AI needs to be combined with human intelligence" [14]. Because algorithms struggle with ambiguous or borderline cases, the integration of human expertise remains vital [14].

A robust AML framework must include an effective monitoring system to identify suspicious transactions concealed within legitimate activity [29]. To combat fraud more effectively, institutions can employ a combination of free and commercial software that utilizes statistical analysis, data visualisation, data mining, and filtering tools—an approach that has proven valuable across multiple industries [12].

2 Methodology

In this section, the step-by-step procedure for searching, filtering, and analyzing relevant publications is described [1]. This paper presents a literature review of scientific research within a field that has experienced significant growth in recent years [1]. An overview of the procedure is provided in Table 1 [1].

Table 1 Overview of the stepwise procedure for the systematic literature review

Process #	Process step	Described in	Activities	Results of the activities
1st	Online query for relevant publications;	Section 2.1	Search in databases with multiple search terms	593,929 hits
2nd	Online query for relevant publications	Section 2.1	Further specification of keyword combination	3542 hits
3rd	Data cleansing and preselection of publications for the sample	Section 2.2	Elimination of duplicates and evaluation of title and abstract; rough check for relevance	319 preselected publications
4th	Review of preselected publications in full text	Section 2.3	Critically analysed the abstract and checked for relevance	49 publications in final sample + 10 other sources

2.1 *Online Search for Relevant Publications*

First, an initial exploration was conducted to develop a clearer understanding of the chosen topic, using various keyword combinations and a preliminary search on Google Scholar [1]. However, the literature reviewed in this paper primarily draws on five major databases: Taylor & Francis, IEEE, ScienceDirect, SpringerLink, and Wiley Online Library [1].

Given the broad scope of the research field, the initial keyword combinations—including terms such as *financial crime*, *AI*, *AML*, and *financial sector*—produced an unmanageable volume of results, as indicated in Table 1 Step 1 with 593,929 hits [1]. To refine the scope and ensure greater relevance, additional, more targeted keyword combinations were developed [1]. Selected articles were reviewed iteratively to improve topic familiarity and inform the refinement process [1]. With the revised combinations, the number of relevant results was reduced to 3542 (see Table 1 Step 2) [1]. This more focused set of publications then formed the basis for the subsequent data cleansing stage.

2.2 *Data Cleansing and Preselection of Publication*

The data cleansing process was applied to the initial set of 3,542 publications [1]. This step aimed to identify duplicate entries across databases and to assess the relevance of each publication [1]. Keyword matches were verified, paper titles were screened, and abstracts were reviewed and evaluated for their suitability [1]. As a result of this procedure, 319 publications were preselected for inclusion in the final analysis sample [1]. Publications were excluded from further consideration if they appeared multiple times, lacked relevance, or did not align with the objectives of this study [1].

2.3 *Review of Preselected Publications in Full Text*

The remaining papers were critically reviewed based on their abstracts and, where necessary, full texts [1]. Four main inclusion criteria were applied:

(i) relevance of the human factor in connection with AI;
(ii) the role of AI in supporting anti-money laundering (AML);
(iii) the impact of AI on the financial sector; and
(iv) insights into how money laundering (ML) processes operate [1].

Using these criteria, a further 270 papers were excluded due to the broad scope and varied contexts of the research field [1]. Additionally, ten relevant publications not captured in the initial database searches were identified and included to supplement the final sample in Sect. 2.1. In total, 59 publications formed the basis for the final literature review; these are summarized in Table 9.

A detailed breakdown is provided in Tables 2, 3, 4, 5, 6, 7 and 8, which present the specific keyword combinations and any applied filters for each database. These tables illustrate the number of publications identified and how they contributed to the final sample of 59 papers.

At the Taylor & Francis database, as presented in Table 2, highly specific keyword combinations were required. More general keywords produced overly broad results, making the search process challenging; therefore, a more targeted approach was necessary. By refining the keyword terms, no additional filters were needed, which resulted in the identification of seven relevant publications.

Table 2 The literature search within the Taylor & Francis database

Taylor & Francis	
Money laundering and machine learning and financial crime and suspicious transactions and anomaly detection	1
Financial crime and money laundering and financial fraud and neural networks and information technology	2
Money laundering and abuse and internet and fraud	1
Financial havens and banking secrecy and money-laundering	1
Financial havens and banking secrecy and money-laundering	1
Political power and information and anti-money laundering and counter-terrorist financing	1
Artificial intelligence and financial regulation and augmented intelligence and financial sector	1
Total	7

Table 3 The literature search within the Wiley database

Wiley	
Credit cards and cybercrime and economic crime and financial crime and fraud	1
Fraud prevention and information confidentiality integrity and availability (CIA) and fraud risk assessment enterprise risk management	1
Artificial intelligence and anti-money laundering and financial crime and fraud detection and algorithm	1
Total	3

Table 4 The literature search within the science direct database

Science direct	
Artificial intelligence and financial crime and money laundering and KYC Filter: Research article	2
Money laundering and machine learning and financial crime and suspicious transactions and detect fraud and neural networks and fabricated data	1
Money laundering and machine learning and data mining and fraud detection and financial fraud and techniques and performance Filter: Year 2019, Research article	1
Financial fraud detection and data mining and anomaly detection money laundering and machine learning and classification and computational intelligence Filter: Review article + computer & security	1
Financial fraud detection and data mining and anomaly detection money laundering and machine learning and classification and business intelligence Filter: Research article + Year 2019	1
Anti-money laundering and financial security and underground economy and data mining and survey	1
Detecting and preventing fraud and data analytics and data analytics and preventing fraud and IT systems	1
Total	8

In contrast, Table 3 summarises the findings from the Wiley Online Library. Due to the large volume of results generated by general keyword searches in this database, the keywords were refined and tailored to specific research areas. This approach produced a manageable set of relevant publications.

The search in ScienceDirect resulted in eight relevant publications, as shown in Table 4. To refine the search, additional filters were applied to the initial keyword combinations. Given the large number of publications in recent years, various filters were necessary to capture different perspectives and ensure the inclusion of the most relevant studies.

In the SpringerLink database, additional filters were also applied, as even with relatively specific keyword combinations, a broader overview was achievable when filters were used. The applied filters and keyword search terms are detailed in Table 5.

Table 5 The literature search within the Springer Link database

Springer Link	
Artificial intelligence and financial crime and money laundering and KYC Filter: Economic + Article Filter: Computer Science + Article Filter: Banking	3
AML-systems and machine learning and financial crime and antifraud algorithm	1
Anti-money laundering and suspicious transactions and data mining and classification customer clustering and client profiling and multi-agent based	1
Money laundering and crime and cash-smuggling and terrorist financing and high-denomination banknotes Filter: Criminology and criminal justice	1
Money laundering and regulatory minefield and suspicious transactions and reporting institution Filter: Article	1
AML directive and non-financial sector and money laundering and German sector and analysis of anti-money laundering Filter: Article and business management	1
AI and machine learning and risk management and market risk and operational risk and RegTech and credit risk Filter: Business and management + Chapter	1
AI and law and AI-crime and dual-use and ethics and machine learning	1
AML and risk of money laundering and implications for practitioners and research challenges Filter: Article + Criminology and criminal justice	1
Money laundering and banking and regulation and criminal liability and civil liability and crime and financial institutions and collapse and threat Filter: Article and finance banking	1
Internet of things and challenges for business and management and artificial intelligence and new paradigm of learning and machine learning	1
Economic and societal and ethical and legal and digital revolution and deep learning and big data and manipulative technologies Filter: Popular Science + Chapter	1
War against money laundering and terrorism financing and financial action task force and suspicious transaction and wire transfer Filter: Article + finance + banking	1
Follow the money method and crime control policy and organize crime and critique and money laundering and underground economy and legal economy and wash-out Filter: Article + Energy + Criminology	1
Total	16

Table 6 The literature search within the IEEE database

IEEE	
Artificial intelligence and financial crime and money laundering	2
Anti-money laundering and machine learning and decision tree and banking	1
Machine learning and algorithm and anti-money laundering and fraud detection and suspicious transactions	1
Machine learning and artificial neural networks and classification algorithm and fraud detection and banking and data analysis	1
Money laundering and banking and artificial intelligence and banking and data security and intelligence systems and regulators	1
Machine learning and algorithm and fraud detection and banking and user patterns	1
Total	6

Table 7 The literature search in google scolar database

In google scholar through first play with keywords	
Sage Anti-money laundering and timeline and regulatory and effectiveness and relentless	1
AAAI Financial crimes enforcement and AI network system and identifying potential money laundering and large cash transactions	1
ResearchGate Multiagent based approach and money laundering and detection and prevention	1
ResearchGate Detecting money laundering and terrorist financing and design of a monitor	1
ResearchGate State-of-the-art and big data and application techniques and financial crime and survey	1
Emerald Insight International and anti-money-laundering and financial action task force	1
The Royal Society Publishing Governing artificial intelligence: ethical, legal and technical opportunities and challenges	1
Allied Business Academies Journal and legal and ethical and regulatory issues	1
CiteSeerX International money laundering in the banking sector	1
Total	9

Overall, SpringerLink produced the highest number of relevant publications, with a total of sixteen identified studies.

In contrast, the IEEE database required a much narrower and more precise keyword strategy to locate relevant papers. General keyword combinations proved insufficient; therefore, more specific search terms were necessary to achieve meaningful results. The final keywords used are presented in Table 6, which lists the six relevant publications identified through this approach.

Table 7 lists the publications identified through Google Scholar. These papers were located during the initial stages of the research process. Following a critical analysis, they were deemed suitable for inclusion in the final sample. Some publications originate from less common or alternative databases. Further details are provided in Table 7.

Table 8 The literature search in other databases

Other sources		
Springer professional	"Finanzkriminalität ist ein Katz-und-Maus-Spiel "	
Financial times article	Banks use AI to catch rogue traders before the act	
Forbes magazine	How AI could protect your business from financial fraud	
Forbes magazine	Is technology helping or hindering our fight against financial crime?	
AI business	Fighting financial crime with AI	
The international banker	How AI is disrupting the banking industry	
Fintech News	How artificial intelligence is fighting financial crime	
Handelsblatt	Kampf gegen Kriminalität: Im Kampf gegen Betrüger setzen Banken auf künstliche Intelligenz	
ZDNet	Fight against financial crime requires both artificial and human intelligence	
Total		10

Table 8 presents the articles identified from non-scholarly sources. These publications provided a broader contextual overview and contributed to a deeper understanding of the topic, which informed the filtering process and the criteria applied during the selection of the final sample.

3 Literature Analysis

Table 9 presents the full list of 59 publications included in the final sample [1]. All 49 peer-reviewed papers and 10 additional articles were systematically analysed using four key analytical categories: Title, Challenge/Problem, Methodology, and Field/Discipline.

3.1 Research Lifecyle of AI Preventing Financial Crimes (1994–2019) Based on Publications

The research focused on the past 25 years to examine trends emerging in the early twenty-first century and to track changes over time. As illustrated in Fig. 1, the majority of publications have been released within the past three years [30]. This increase may be attributed to the growing prevalence of financial crime, driven by rapid technological advancements and heightened public awareness following events such as the leak of the *Panama Papers* [30].

Table 9 Analysis of every publication and article

#	Title	Challenge/problem	Methodology	Field/discipline
1	Financial havens, Banking secrecy and money-laundering	Need for more understanding how AI can support the fight against ML	Conceptual/ analytical research	Criminal science, Economic, Political
2	Governing artificial intelligence: ethical, legal and technical opportunities and challenges	Examines how the financial sector can benefit from AI	Conceptual/ analytical research	Engineering science, Mathematical science
3	Artificial intelligence crime: an interdisciplinary analysis of foreseeable threats and solutions	Highlights the need for a framework that addresses financial crime and KYC	Conceptual/ analytical research	Science, Engineering, Ethics
4	Intelligent financial fraud detection: a comprehensive review	Integrates algorithms with KYC processes to prevent financial crime	Literature analysis; Conceptual and Analytical research	Computer science
5	State-of-the-art in big data application techniques to financial crime	Analyses detection approaches for financial fraud in combination with human factors	Survey research; Analytical research	Engineering, Computer science, Financial science
6	The application of data mining techniques in financial fraud detection: a classification framework and an academic review of literature	Actions to be taken by financial and political institutions against money laundering	Literature analysis	Computer science, Finance, Economics
7	Integrating client profiling in an anti-money laundering multi-agent-based system	How KYC and human factors can support the learning component	Conceptual and analytical research	Computer science
8	The international anti-money-laundering regime: the role of the financial action task force	Analysis of how criminals may adapt more quickly than large corporations	Conceptual/ analytical research	Finance, Political science

(continued)

Table 9 (continued)

#	Title	Challenge/problem	Methodology	Field/discipline
9	Financial crimes enforcement network AI system (FAIS): identifying potential money laundering from reports of large cash transactions	Discussion on future technological changes and implications	Literature analysis; Conceptual and analytical research	Technology, Economics
10	Journal of legal, ethical and regulatory issues	Overview of global frameworks addressing money laundering	Literature analysis	Law, Ethics
11	Tracking dirty proceeds: exploring data mining technologies as tools to investigate money laundering	Evaluation showing that machine learning alone is insufficient without KYC and human oversight	Analytical/ conceptual research	Criminal science, Finance, Computer science, Law
12	Financial security against money laundering: a survey (Chapter 36)	Outlines certain ethical challenges faced in anti-money laundering practices	Analytical review; Data mining techniques	Finance, Computer science, Technology
13	An anti-money laundering timeline and the relentless regulatory response	Highlights challenges and difficulties in implementation	Literature analysis	Finance, Law
14	Analysis of banking data using machine learning	Need for a stronger understanding of anti-money laundering (AML) in practice	Literature review	Computer science and engineering
15	Combining Benford's law and machine learning to detect money laundering: an actual spanish court case	Provides insights into potential disadvantages that could arise from using AI	Case study (Court Case)	Finance industry, Technology
16	Design of a monitor for detecting money laundering and terrorist financing	Overview of practical problems and consequences for affected countries	Case study; Conceptual and analytical research	Technology, Engineering
17	International money laundering in the banking sector	Integrates AI and KYC approaches to prevent money laundering	Conceptua/ analytical research	Financial science, Economics

(continued)

Table 9 (continued)

#	Title	Challenge/problem	Methodology	Field/discipline
18	A money laundering risk evaluation method based on decision tree	Highlights the need to understand why AML intelligence systems developed in certain countries may not be directly applicable elsewhere	Literature analysis; Conceptual and analytical research	Computer science and engineering, Mathematical science
19	Cash, crime and anti-money laundering	Need to further elaborate on how AI can be leveraged effectively	Conceptual/ analytical research	Crime, Finance
20	An analysis of anti-money laundering in the german non-financial sector	Supports understanding of shifts in regulatory focus	Grounded theory; Survey research	Management, Finance, Government
21	AI fights money laundering	Calls for more detailed insights into the current situation and effectiveness	Conceptual/ analytical research	Adaptive operational analytics
22	How artificial intelligence is fighting financial crime	Highlights lack of coverage on practical AML implementation	Brief review	Information technology, Computer science
23	Fighting financial crime with AI	Emphasises the need for deeper understanding of how financial crimes occur	Brief review	Financial industry
24	Fight against financial crime requires both artificial and human intelligence	Examines how alternative approaches might improve outcomes	Survey research; Analytical research	Business, Technology
25	Kampf gegen Kriminalität: Im Kampf gegen Betrüger setzen Banken auf künstliche Intelligenz	Explores the challenges that algorithms may encounter	Brief review	Economics, Computer science
26	How AI Is disrupting the banking industry	Provides an analytical overview and identifies the need for deeper understanding of economic impacts	Brief review	Banking, Financial science
27	Banks use AI to catch rogue traders before the act	Discusses ethical challenges that may arise as AI improves	Brief review	Banking, Finance, Economics

(continued)

Table 9 (continued)

#	Title	Challenge/problem	Methodology	Field/discipline
28	Is technology helping or hindering our fight against financial crime?	Proposes frameworks for feasible alternatives alongside AML measures	Brief review	Computer science
29	How AI could protect your business from financial fraud	Offers deeper insight into how financial crimes are perpetrated in business contexts	Brief review	Law enforcement, Economics
30	Conceptual modelling and development of an intelligent agent-assisted decision support system for anti-money laundering	Addresses a research gap by providing a framework for AI challenges in AML	Literature review	Law enforcement, Finance
31	Machine learning techniques for anti-money laundering (AML) solutions in suspicious transaction detection: a review	Insight into how ML is applied in different systems	Conceptual/ analytical research	Mathematical science, Computer science, Economics
32	Money laundering, Regulatory strategy and Internal corporate controls	Ideas for implementing recommendations	Conceptual/ analytical research	Law, Compliance
33	Statistical methods for fighting financial crimes	Framework combining human expertise with ML	Conceptual/ analytical research	Technical science, Statistics
34	Use and abuse of the internet in fraud and money laundering	Analyses detection approaches for financial fraud in combination with human factors	Conceptual/ analytical research	Law, Computer science, Technology
35	AI and augmented intelligence collaboration: regaining trust and confidence in the financial sector	How ML and KYC strengthen systems	Case study research/literature analysis	Technology, Law
36	The power of UDA to improve fraud detection and prevention	Framework linking fraud and financial crime	Conceptual/ analytical research	Law, Social science, Ethics, Criminal science
37	Fraud prevention, detection, and response	How the financial sector manages evolving tech	Conceptual/ analytical research	Criminal science, Economics, Technology
38	Concepts and dynamics of fraud	Challenges for AI and alternative solutions	Conceptual/ analytical research	Economics, Crime, Finance

(continued)

Table 9 (continued)

#	Title	Challenge/problem	Methodology	Field/discipline
39	Everyday rules and embodied information: AML/CTF practices and RFID as security politics	How ML impacts the economy	Analytical research	Computer science
40	Five trends shaping the fight against financial crime	Framework on ethical KYC challenges	Conceptual/ analytical research	Computer science, Economics
41	Performance of ML techniques in financial fraud detection	Human factor in AML	Literature review	Computer science, Financial science
42	Detecting and preventing fraud with data analytics	Lack of research on future trends	Conceptual/ analytical research	Engineering, Finance
43	BMF—Money laundering and terrorist financing	How AI helps prevent ML	Brief review	Government, Finance
44	CoDetect: financial fraud detection with anomaly feature detection	Human factor combined with AI	Experimental research	Computer networks, Big data, Finance, Trading science
45	User pattern based online fraud detection and prevention using big data analytics and self-organizing maps	Ethical challenges in big data and cloud	Literature analysis / conceptual/ analytical research	Financial science, Computer science
46	ML and sampling scheme: empirical study of ML Detection	Market change using different approaches	Empirical analysis	Computer science, Economics, Mathematical science
47	Application of machine analysis algorithms for combating criminal money laundering	Ways laundered money flows	Conceptual/ analytical research	Finance, Digital transformation
48	Historical overview	AI and KYC impact on financial crime	Analytical research	Finance, Economics
49	On the frontline against money-laundering: regulatory minefield	Historical, need for future-focused view	Literature analysis	Law, Social science, Criminal science
50	Wash-out: critique of follow-the-money methods	Comparison across countries	Conceptual/ analytical research	Law enforcement
51	Assessing the risk of money laundering	Consequences for banking sector	Methodological approach	Research, Criminal policy, Finance

(continued)

Table 9 (continued)

#	Title	Challenge/problem	Methodology	Field/discipline
52	Financial institutions and money laundering	Framework on human contribution to AML	Conceptual/ analytical research	Banking, Economics
53	Proposals to improve the war against ML and terrorism financing	State-level problems and lack of strict regs	Conceptual/ analytical research	Finance, Political science
54	ML and AI for risk management	Framework on alternative strategies	Conceptual/ analytical research	Digital business, Technology
55	Societal, Economic, Ethical and Legal challenges of the digital revolution	How societal change impacts financial crime	Conceptual/ analytical research	Social Science, Computer Science
56	A scan statistics based suspicious transactions detection model for AML	How ML helps AML	Research/empirical experiment	Multimedia communications
57	"Finanzkriminalität ist ein Katz-und-Maus-Spiel"	How institutions can be proactive	Interview	Financial science, economics
58	Integrating client profiling in an AML multi-agent system	Effective monitoring system design	Literature analysis	Technological science, Finance
59	Internet of things: new paradigm of learning. challenges for business	How IoT can influence financial crime	Conceptual/ analytical research	Computer science, Economics

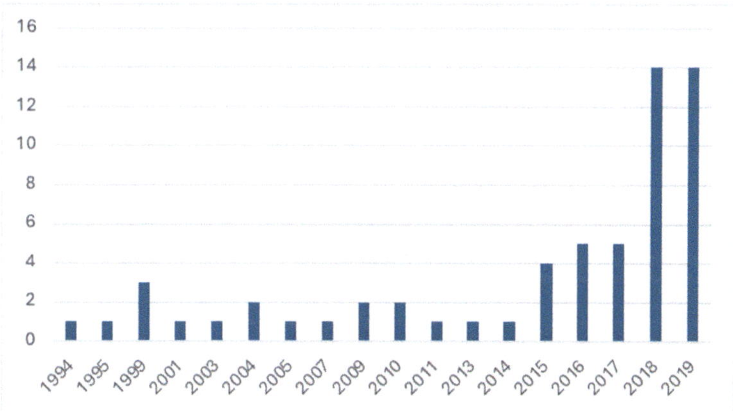

Fig. 1 Development of AI and financial crime related publications over time

Moreover, Fig. 1 suggests that the number of publications tends to rise periodically, potentially correlating with financial crises or significant economic shifts occurring before or after such events [30]. Increased awareness of financial crimes among organisations and stakeholders has likely also stimulated further research and publication in this field [30].

In conclusion, although this research area is not entirely new, it remains relatively underexplored. Should current trends persist, a substantial increase in the number of related publications can be anticipated in the coming years.

3.2 Research Lifecycle of AI and Financial Crime (1994–2019) Based on Publications Frequency

Some years are not represented in Fig. 1, as no relevant publications were identified for these periods within the scope of the keyword search. Consequently, the years 1996, 1997, 1998, 2000, 2002, 2006, 2008, and 2012 are not included in the Fig. 1.

3.3 Analysis of Aggregated Findings

Figure 2 shows that more than two-thirds of the publications are based on non-empirical findings. Additionally, three publications employed both empirical and non-empirical research methods and have therefore been counted in both categories in Fig. 2. The limited volume of empirical research in this field may be attributed to the inherent challenges associated with conducting such studies, particularly given the stringent regulatory and privacy requirements that often restrict access to organisational case studies.

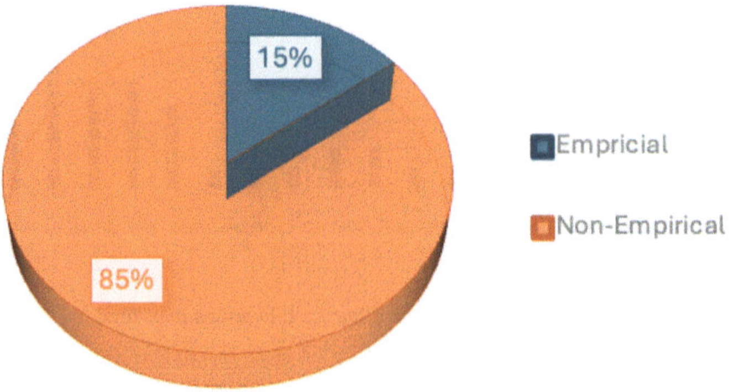

Fig. 2 Overview of empirical versus non-empirical studies within the analysis

Figure 3 indicates that the majority of empirical research has been conducted within the past two years. Notably, in 2018, half of the publications were empirical in nature. This trend suggests a gradual shift towards more empirical investigations, which may be attributed to the increasingly advanced possibilities for conducting such research.

Figure 4 illustrates that case studies were the most commonly employed method in empirical research, followed by survey research and experimental research, with two published papers each. The pronounced tendency towards conceptual and analytical research may correlate with the relative scarcity of empirical studies in this field [n]. As noted above, conducting empirical research in this domain remains challenging due to practical and regulatory constraints.

The reviewed publications encompass a diverse range of disciplinary backgrounds in connection with AI and financial crime. Only two authors appear multiple times: Ms Gelemerova [23, 27] and Mr Alexandre [6, 7], both of whom have published papers independently as well as in collaboration with other researchers. Consequently, an overview of authors has not been included, as all other papers were authored by different researchers. While each author adopts a distinct approach, there is considerable thematic overlap across the topic areas. Notably, several authors indicate an intention to pursue further research, which may lead to the formation of more cohesive research groups in the future.

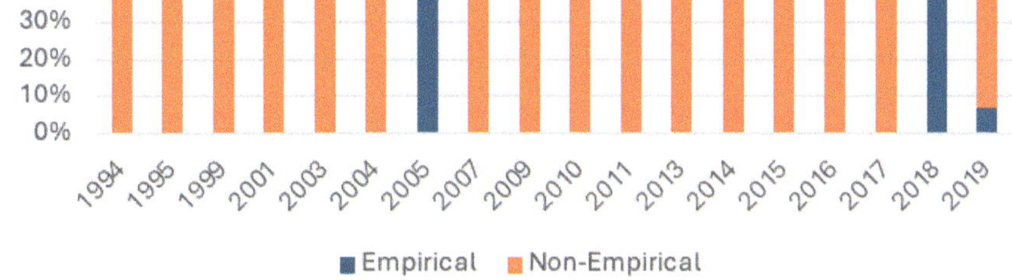

Fig. 3 Development of empirical versus non-empirical methods from 1994 to 2019

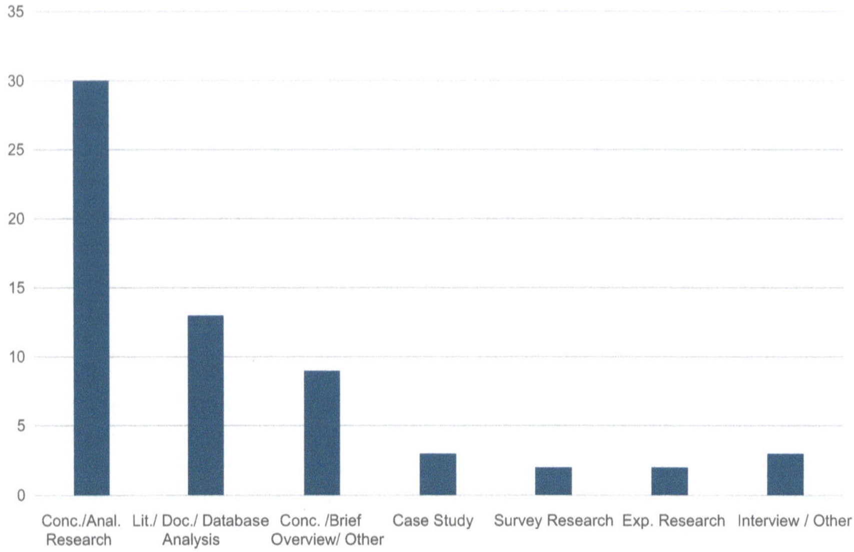

Fig. 4 Overview of the frequency with which the various research methodologies were applied

4 Conclusion

The aim of this paper was to analyse the existing research on AI and financial crime. As observed, there has been a clear increase in scholarly work within this domain, underscoring its growing relevance. With the rise of cryptocurrencies, financial crimes are likely to escalate due to the anonymity these technologies afford, thereby facilitating money laundering—a concern highlighted by Mr Nowotny during the panel discussion [3]. Consequently, further research examining the intersection of financial crime and cryptocurrencies remains crucial, as this topic was not addressed in the present study.

Despite ongoing advancements in AI, human expertise will continue to play a critical role in the prevention of financial crimes, particularly as the banking sector evolves towards a more robust Know Your Customer (KYC) framework, as also affirmed by Mr Nowotny [3]. While AI can support the detection and management of financial crime, it cannot replicate the human judgement required for nuanced decision-making [38]. A world entirely free from fraud is unattainable; therefore, organisations must implement effective fraud detection systems to mitigate criminal activity [19].

Another notable trend is the deployment of chatbots, which currently represent one of the most visible applications of AI across various industries [39]. Chatbots enhance customer convenience by enabling issues to be addressed online at any time and from any location via smartphones or laptops [39]. For the banking industry in particular, chatbots offer significant advantages by reducing costs and improving operational efficiency and responsiveness [39]. As a result, firms are increasingly investing in AI-driven solutions to gain a competitive advantage [39].

Nevertheless, organisations should be cautious not to concentrate solely on distant technological possibilities [56]. Instead, it is vital to address immediate and near-term challenges, prioritising proactive fraud prevention over reactive measures [56]. The primary challenge remains to facilitate the free exchange of information while simultaneously restricting opportunities for deception [42].

As already anticipated by Toffler [54], the increasing trend towards a post-modern digital economy, where innovation and customization become the most thriving business drivers, led to a merging of producers' and consumers' boundaries. Facilitated by the wide range of advancements made in ICT, the traditional roles of consumers and producers are readjusted through various forms of new capabilities for collaboration, participation and customer engagement [55, 56]. Consumers today are more willing to provide information while at the same time being increasingly empowered to collaborate with producers in exchange for more customised products and services that better satisfy their needs, desires and preferences [35].

With the consumer's increased capability to participate in the value creation process through enhanced communication technologies, they can co-develop dynamically, together with producers, an active system of shared meanings [33]. These shared meanings are constituted by an ongoing process of reciprocal negotiation between producers and other stakeholders at a symbolic socio-cultural level [31]. As a result, value is constituted as the locus of these shared symbolic meanings [57]. This perspective resonates with *FP6* and *FP7* of *SDL* (see Table 2), which highlight the interactive nature of reciprocal value-propositions by producers and consumers.

References

1. Bauer C, Strauss C (2016) Location-based advertising on mobile devices: a literature review and analysis. Managem Rev Quarter 66:159–194
2. Kryvinska N, Olexova R, Dohmen P, Strauss C (2019) The S-D logic phenomenon-conceptualization and systematization by reviewing the literature of a decade(2004–2013). SpringerLink. Abgerufen 17 Mai 2019. https://link-springer-com.uaccess.univie.ac.at/article/10.1007/s12927-013-0002-0
3. OeNB Panel Discussion (2019) Blockchains and cryptocurrencies: fad or game changer?" with Ewald Nowotny, Governor of the Oesterreichische Nationalbank, and OeNB Visiting Professor Bruno Biais (Toulouse School of Economics) (21 May 2019). https://www.youtube.com/watch?v=ZuvtInFb48s
4. Agarwal JD (2004) International money laundering in the banking sector1. Indien
5. Alexander K (2001) The international anti-money-laundering regime: the role of the financial action task force. J Money Laundering Control 4(3):231–248
6. Alexandre C, Balsa J (2015) A multiagent based approach to money laundering detection and prevention. In: Proceedings of the international conference on agents and artificial intelligence. Lisbon, Portugal, SCITEPRESS—Science and and Technology Publications, S. 230–35
7. Alexandre C, Balsa J (2016) Integrating client profiling in an anti-money laundering multi-agent based system. In: von Rocha Á, Correia AM, Adeli H, Reis LP, Mendonça Teixeira M

(eds) New advances in information systems and technologies, Advances in intelligent systems and computing. Springer International Publishing, S. 931–41
8. Angel S (2019) Council post: how AI could protect your business from financial fraud. Abgerufen 27 Apr 2019. https://www.forbes.com/sites/forbesfinancecouncil/2019/02/14/how-ai-could-protect-your-business-from-financial-fraud/#33c390936116
9. Aziz S, Dowling M (2019) Machine learning and ai for risk management. In: von Lynn T, Mooney JG, Rosati P, Cummins M (eds) Disrupting finance: FinTech and strategy in the 21st century, Palgrave studies in digital business and enabling technologies, Cham, Springer International Publishing, S. 33–50
10. Badal-Valero E, Alvarez-Jareño JA, Pavía JM (2018) Combining Benford's law and machine learning to detect money laundering. an actual Spanish court case. Forensic Sci Int 282:24–34
11. Balasupramanian N, Ephrem BG, Al-Barwani IS (2017) User pattern based online fraud detection and prevention using big data analytics and self organizing maps. In: 2017 International conference on intelligent computing, instrumentation and control technologies (ICICICT), S. 691–94
12. Bănărescu A (2015) Detecting and preventing fraud with data analytics. Proc Econ Finan 32:1827–1836
13. Bauer C, Strauss C (2016) Location-based advertising on mobile devices: a literature review and analysis. Managem Rev Quarter 66:159–94
14. Bindi T (2017) Fight against financial crime requires both artificial and human intelligence, ZDNet. Abgerufen 26 Apr 2019. https://www.zdnet.com/article/fight-against-financial-crime-requires-both-artificial-and-human-intelligence/
15. Blum JA, Michael Levi R, Naylor T, Williams P (1999) Financial havens, banking secrecy and money-laundering. Criminal Justice Matters 36(1):22–23
16. Breinich-Schilly A (2018) Finanzkriminalität Ist Ein Katz-Und-Maus-Spiel. Springerprofessional.De. Abgerufen 4 Mai 2019. https://www.springerprofessional.de/en/compliance/risikosteuerung/--finanzkriminalitaet-ist-ein-katz-und-maus-spiel-/16287880
17. Cath C (2018) Governing artificial intelligence: ethical, legal and technical opportunities and challenges. Philosophical Trans Series A, Mathemat Phys Eng Sci 376(2133)
18. Chen Z, Khoa LDV, Teoh EN, Nazir A, Karuppiah EK, Lam KS (2018) Machine learning techniques for anti-money laundering (AML) solutions in suspicious transaction detection: a review. Knowl Inf Syst 57(2):245–285
19. Crain MA, Hopwood WS, Pacini C, Young GR (2017) Fraud prevention, detection, and response. In: Essentials of forensic accounting. Wiley & Sons, Ltd, S. 211–43
20. Sanctis D, Martin F (2013) Proposals to improve the war against money laundering and terrorism financing. In: von De Sanctis FM (ed) Money laundering through art: a criminal justice perspective. Cham, Springer International Publishing, S. 193–213
21. Dobronauteanu MC (2019) Fighting financial crime with AI. The world's number one portal for artificial intelligence in business. Abgerufen 26 Apr 2019. https://aibusiness.com/fighting-financial-crime-ai-mimiro/
22. Dorofeev D, Khrestina M, Usubaliev T, Dobrotvorskiy A, Filatov S (2018) Application of machine analysis algorithms to automate implementation of tasks of combating criminal money laundering. In: von Alexandrov DA, Boukhanovsky AV, Chugunov AV, Kabanov Y, Koltsova O (eds) Digital transformation and global society, communications in computer and information science. Springer International Publishing, S. 375–85
23. van Duyne, Petrus C, Harvey JH, Gelemerova LY (2018) Historical overview. In: van Duyne PC, Harvey JH, Gelemerova LY (eds) The critical handbook of money laundering: policy, analysis and myths. London, Palgrave Macmillan UK, S. 41–89
24. Fisse B (1994) Money laundering, regulatory strategy and internal corporate controls. Curr Issues Crim Just 6(2):177–195
25. Friedrich C, Quick R (2019) An analysis of anti-money laundering in the German non-financial sector. Springerprofessional.De. Abgerufen 1 Apr 2019. https://www.springerprofessional.de/en/an-analysis-of-anti-money-laundering-in-the-german-non-financial/16470948?searchResult=1.Cash,%20Crime%20and%20Anti-Money%20Laundering&searchBackButton=true

26. Gao S, Xu D (2009) Conceptual modeling and development of an intelligent agent-assisted decision support system for anti-money laundering. Expert Syst Appl 36(2, Part 1):1493–1504
27. Gelemerova L (2009) On the frontline against money-laundering: the regulatory minefield. Crime Law Soc Chang 52(1):33–55
28. Helbing D (2015) Societal, economic, ethical and legal challenges of the digital revolution: from big data to deep learning, artificial intelligence, and manipulative technologies. arXiv: 1504.03751 [physics]
29. Helmy TH, Zaki M, Salah T, Badran K (2005) Design of a monitor for detecting money laundering and terrorist financing. vol 12
30. Holzenthal F (2017) Five trends shaping the fight against financial crime. Comput Fraud Secur 2017(3):5–9
31. Huang D, Mu D, Yang L, Cai X (2018) CoDetect: financial fraud detection with anomaly feature detection. IEEE Access 6:19161–19174
32. Isson JP (2018) The power of UDA to improve fraud detection and prevention. In: Unstructured data analytics. Wiley & Sons, Ltd., S. 157–90
33. KA (2016) BMF—money laundering and terrorist financing. Abgerufen 4 Mai 2019. https://english.bmf.gv.at/financial-sector/Money-Laundering-and-Terrorist-Financing.html
34. King TC, Aggarwal N, Taddeo M, Floridi L (2019) Artificial intelligence crime: an interdisciplinary analysis of foreseeable threats and solutions. SpringerLink. Abgerufen 22 März 2019. https://link.springer.com/article/10.1007/s11948-018-00081-0
35. Kingdon J (2004) AI fights money laundering. IEEE Intell Syst 19(3):87–89
36. Lawlor-Forsyth E, Michelle Gallant M (2018) Financial institutions and money laundering: a threatening relationship? J Bank Regul 19(2):131–148
37. Liu X, Zhang P (2010) A scan statistics based suspicious transactions detection model for anti-money laundering (AML) in financial institutions. In: 2010 International conference on multimedia communications, S. 210–13
38. Lui A, Lamb GW (2018) Artificial intelligence and augmented intelligence collaboration: regaining trust and confidence in the financial sector. Inform Commun Technol Law 27(3):267–283
39. Manning J (2018) How AI Is disrupting the banking industry. International Banker. Abgerufen 26 Apr 2019. https://internationalbanker.com/banking/how-ai-is-disrupting-the-banking-industry/
40. Marlin-Bennett R (2016) Everyday rules and embodied information: anti-money laundering/counter-terrorist financing practices and radio frequency identification tags as security politics. Critical Stud Secur 4(2):169–186
41. Marria V (2019) Is technology helping or hindering our fight against financial crime? Forbes. Abgerufen 27 Apr 2019. https://www.forbes.com/sites/vishalmarria/2019/01/16/is-technology-helping-or-hindering-our-fight-against-financial-crime/
42. Morris-Cotterill N (1999) Use and abuse of the internet in fraud and money laundering. Int Rev Law Comput Technol 13(2):211–228
43. Naylor RT (1999) Wash-out: a critique of follow-the-money methods in crime control policy. Crime Law Soc Chang 32(1):1–58
44. Ngai EWT, Yong H, Wong YH, Chen Y, Sun X (2011) The application of data mining techniques in financial fraud detection: a classification framework and an academic review of literature. Decis Support Syst 50(3):559–569
45. Noonan L (2019) Banks use AI to catch rogue traders before the act. Financial Times. Abgerufen 26 Apr 2019. https://www.ft.com/content/be7a5584-2ee7-11e9-80d2-7b637a9e1ba1
46. OA (2018) Kampf gegen Kriminalität: Im Kampf gegen Betrüger setzen Banken auf künstliche Intelligenz. Handelsblatt. Abgerufen 26 Apr 2019. https://www.handelsblatt.com/finanzen/banken-versicherungen/kampf-gegen-kriminalitaet-im-kampf-gegen-betrueger-setzen-banken-auf-kuenstliche-intelligenz/22798886.html
47. Kryvinska N, Olexova R, Dohmen P, Strauss C (2019) The S-D logic phenomenon-conceptualization and systematization by reviewing the literature of a decade (2004–2013). SpringerLink. Abgerufen 17 Mai 2019. https://link-springer-com.uaccess.univie.ac.at/article/10.1007/s12927-013-0002-0

48. Omolara AE, Jantan A, Abiodun OI, Singh MM, Anbar M, Dada KV (2018) State-of-the-art in big data application techniques to financial crime : a survey. pp 12
49. Palshikar GK, Apte M (2014) Chapter 36—Financial security against money laundering: a survey. In: Akhgar B, Arabnia HR (eds), Emerging trends in ICT security, Boston, Morgan Kaufmann, S. 577–90
50. Patil PS, Dharwadkar NV (2017) Analysis of banking data using machine learning. In: 2017 International conference on I-SMAC (IoT in Social, Mobile, Analytics and Cloud) (I-SMAC), S. 876–81
51. Riccardi M, Levi M (2018) Cash, crime and anti-money laundering. In: King C, Walker C, Gurulé J (eds) The Palgrave handbook of criminal and terrorism financing law, Cham, Springer International Publishing, S. 135–63
52. Ross I (2015) Concepts and dynamics of fraud. In: Exposing fraud. Wiley & Sons, Ltd, S. 55–142
53. Sadgali I, Sael N, Benabbou F (2019) Performance of machine learning techniques in the detection of financial frauds. Proc Comput Sci 148:45–54
54. Savona EU, Riccardi M (2019) Assessing the risk of money laundering: research challenges and implications for practitioners. Eur J Crim Policy Res 25(1):1–4
55. Senator TE, Goldberg HG, Wooton J, Cottini MA, Umar Khan AF, Klinger CD, Llamas WM, Marrone MP, Wong RWH (1995) Financial crimes enforcement network AI system (FAIS) identifying potential money laundering from reports of large cash transactions. AI Mag 16(4):21–21
56. Smith LM (2016) J Legal Ethical Regulatory Issues 103
57. Soumik R (2019) How artificial intelligence is fighting financial crime. FintechNews. Abgerufen 26 Apr 2019. https://www.fintechnews.org/how-artificial-intelligence-is-fighting-financial-crime/
58. Sudjianto A, Nair S, Yuan M, Zhang A, Kern D, Cela-Díaz F (2010) Statistical methods for fighting financial crimes. Technometrics 52(1):5–19
59. Sułkowski Ł, Kaczorowska-Spychalska D (2019) Internet of things—new paradigm of learning. Challenges for business. In: Ayaz H, Mazur L (eds) Advances in neuroergonomics and cognitive engineering, advances in intelligent systems and computing. Springer International Publishing, S. 307–18
60. Turner S, Bainbridge J (2019) An anti-money laundering timeline and the relentless regulatory response—Sue Turner, Jonathan Bainbridge. Abgerufen 28 März 2019. https://journals.sagepub.com/doi/full/10.1177/0022018318773205
61. Wang S, Yang J (2007) A money laundering risk evaluation method based on decision tree. In: 2007 international conference on machine learning and cybernetics. Bd. 1, S. 283–86
62. Watkins RC, Michael Reynolds K, Demara R, Georgiopoulos M, Gonzalez A, Eaglin R (2003) Tracking dirty proceeds: exploring data mining technologies as tools to investigate money laundering. Police Pract Res 4(2):163–178
63. West J, Bhattacharya M (2016) Intelligent financial fraud detection: a comprehensive review. Comput Secur 57:47–66
64. Zhang Y, Trubey P (2018) Machine learning and sampling scheme: an empirical study of money laundering detection. Comput Econ

MIX
Papier aus verantwortungsvollen Quellen
Paper from responsible sources
FSC® C105338

If you have any concerns about our products,
you can contact us on
ProductSafety@springernature.com

In case Publisher is established outside the EU,
the EU authorized representative is:
Springer Nature Customer Service Center GmbH
Europaplatz 3, 69115 Heidelberg, Germany

Printed by Libri Plureos GmbH
in Hamburg, Germany